Essentials of Oceanography

Essentials of Oceanography

Fourth Edition

Harold V. Thurman

Mt. San Antonio College

Macmillan Publishing Company
New York

Maxwell Macmillan Canada
Toronto

Maxwell Macmillan International
New York Oxford Singapore Sydney

Cover photo: Coral reef, Solomon Islands. Copyright © Howard Hall/HHP
Editor: Robert A. McConnin
Production Editor: Mary Harlan
Art Coordinator: Peter A. Robison
Photo Researcher: Diane Austin
Text Designer: Debra A. Fargo
Cover Designer: Robert Vega
Production Buyer: Patricia A. Tonneman

This book was set in Garamond by York Graphic Services, Inc. and was printed and bound by R.R. Donnelley & Sons Company. The cover was printed by Phoenix Color Corp.

Macmillan Publishing Company
866 Third Avenue
New York, New York 10022

Macmillan Publishing Company is part of the Maxwell Communication Group of Companies.

Maxwell Macmillan Canada, Inc.
1200 Eglinton Avenue East, Suite 200
Don Mills, Ontario M3C 3N1

Library of Congress Cataloging-in-Publication Data

Thurman, Harold V.
    Essentials of Oceanography/Harold V. Thurman.—4th ed.
        p.   cm.
    Includes bibliographical references and index.
    ISBN 0-02-420802-7 (pbk.)
    1. Oceanography.   I. Title.
    GC11.2.T49   1993
    551.46—dc20   92-16309
                                                         CIP

Printing: 2 3 4 5 6 7 8 9   Year: 3 4 5 6 7

Preface

The fourth edition of *Essentials of Oceanography* has been revised with the same basic goals as previous editions. It is designed for a course in oceanography taught to students with no formal background in mathematics or science. An effort has been made to present the relationship of scientific principles to ocean phenomena in a way that can be clearly understood.

Changes in this edition are designed to increase the appeal of the book to students and make it easier for teachers to cover the material in a fifteen- or sixteen-week semester. For courses taught on a ten-week quarter system the teacher may need to select those chapters that cover the topics of primary relevance to the course.

A significant increase in the appeal of the book to the student should result from the many new line drawings and photographs, many of which are in full color. Improvements in the clarity of the writing will help students understand concepts that are presented.

The order of presentation of material has not been altered appreciably. A short Introduction is followed by Chapter 1, "History of Oceanography." Chapters 2 and 3 in the previous edition have been combined in this edition as Chapter 2, "The Ocean: Its Origin and Provinces." Global plate tectonics and marine sediments are covered in chapters 3 and 4, respectively. The nature and potential of exploitable marine resources, such as metals, petroleum, sand and gravel, sulfur, and phosphorite, are covered in these chapters.

The physical oceanography section begins with Chapter 5 on properties of water and is followed by Chapter 6 on "Air-Sea Interaction." Additional aspects of physical oceanography are covered in Chapters 7, 8, and 9, "Ocean Circulation, "Waves," and "Tides." The concepts presented to this point are incorporated into the discussion of their role in developing the nature of features discussed in chapters 10 and 11, "Coastal Geology" and "The Coastal Ocean." The discussion of ocean pollution is concentrated in these chapters, which focus on those parts of the marine environment most affected by this problem.

Chapter 12, "The Marine Habitat," introduces marine biology with a discussion of the changing physical conditions throughout the marine environment and some of the general adaptations required for life in the oceans. Chapter 13, "Biological Productivity—Energy Transfer," completes the introduction to marine ecology, while chapters 14 and 15 focus on the ecology of "Animals of the Pelagic Environment" and "Animals of the Benthic Environment," respectively.

This format has reduced the number of chapters from eighteen to fifteen, to focus the coverage on the essential concepts important in marine science.

As we crowd more and more of us onto the planet, there is increasing evidence that human actions are having a significant negative impact on essentially all components of Earth's ecology. It thus seems appropriate to broaden the scope of such survey courses as this to include a greater ecological focus. This edition includes discussion of the role of the oceans in the possible increased greenhouse effect of the atmosphere (Chapter 6) and the possible relationship between the El Nino–Southern Oscillation and changes in climate throughout the world (Chapter 7). Also considered are the problems of coastal development (Chapter 10) and exploiting marine resources such as hydrocarbons and other aspects of coastal pollution (Chapter 11).

In presenting environmental considerations, every effort is made to include only factual information and point out controversial aspects of any discussions. However, the overall theme is that we must take much more care in activities that can modify the environment and keep that modification to a minimum.

In the text, important terms are introduced with bold print and defined. A summary reviews the major concepts discussed in each chapter, and the key terms are listed at the end of that chapter. The "Questions and Exercises" section provides the student with an opportunity to focus on the major concepts. The "Suggested Readings" provides a guide to articles on relevant topics that have a popular presentation in *Sea Frontiers* or a some-

v

what more demanding presentation in *Scientific American*.

To assist the student further, there are a number of appendixes and a glossary that includes the definition of all of the key terms (except the names of individuals) and additional terms that students may wish to look up.

The Instructor's Manual provides not only answers to end-of-chapter questions and a selection of possible exam questions, but also special-feature articles, covering a wide range of interesting topics, that may be duplicated and distributed to students. These features could be used simply to add breadth and interest to the coverage of the material in a chapter or serve as a basis for student reports. Other teaching aids available are 50 color transparencies for use with an overhead projector and a set of 100 slides of selected illustrations from the book.

Acknowledgments

Many individuals have provided advice and assistance for previous editions of this work, including William H. Hoyt, University of Northern Colorado; Jack H. Hyde, Tacoma Community College; Paul Pinet, Colgate University; Constantine N. Raphael, Eastern Michigan University; Erwin Seibel, San Francisco State University; and Yin S. Soong, Millersville University. Their contributions are very much appreciated and have also been included in this edition.

I especially want to thank those who provided valuable suggestions while reviewing the fourth edition of *Essentials of Oceanography:* Fritz Osell, Leeward Community College; Richard D. Little, Greenfield Community College; Alan P. Trujillo, Palomar College; Robert W. Adams, State University of New York at Brockport; and Walter C. Dudley, University of Hawaii at Hilo.

Over the course of the revision, many members of the production staff contributed to the successful completion of this project. I particularly want to thank Mary Harlan, who oversaw most of the project and kept it organized so that steady progress could be made. Mary Donaldson did an excellent job of copy editing, Debra Fargo designed the text, and Pete Robison successfully coordinated the introduction of four-color art, as well as the total art program.

Contents

Chapter 8

Waves   171

Chapter 9

Tides   190

Chapter 10

Coastal Geology   209

Chapter 11

The Coastal Ocean   228

Appendixes 352

Glossary 364

Index 385

Introduction

We hope the world will be a better place for our having been here. As good world citizens, we have that responsibility. If you approach the study of the oceans with such a goal in mind, I'm sure the oceans and the world will benefit from your presence.

When I first began writing texts in the early 1970s, I wanted to help students develop an appreciation for the oceans by learning about oceanic processes, both physical and biological—that is, develop an ecological awareness. Some of you may even find a life-long interest in the ocean and continue to study it formally or informally.

This book can help you grasp the basic principles underlying marine phenomena. Such an understanding is essential to comprehending the interrelationships between the oceans and the life forms they support. With increasing population density in coastal areas, the threat to coastal ecological systems can be greatly reduced if significant numbers of this population have an informed sensitivity to the need to protect the coastal environment. With the inevitable increase in technological capability, our power to modify the marine environment will continue to expand. Only an informed populace can help decide whether this increased capability is used to increase the damage to our oceans or is used to greatly reduce the ecological threat.

INTERRELATIONSHIPS OF THE OCEAN

Since prehistoric times, people have used the ocean as a means of transportation and as a source of food, but the importance of its processes has been studied technically only since 1930. The impetus for this study began then in a search for petroleum, continued with the emphasis on ocean warfare during World War II, and more recently has been expressed in the concern for ocean ecology. Of course, fishermen have always known to go where the physical processes of the ocean offer good fishing. But how life interrelates with ocean chemistry, geology, and physics was more or less a mystery until scientists in these disciplines began to investigate the ocean and use high technology.

For example, a group of geologists investigating global plate tectonics decided in 1977 to research hot water anomalies in the deep eastern Pacific Ocean. (The process of tectonics, confirmed in 1968, involves the creation of new ocean floor along underwater mountain ranges.) When they searched the hot-water vents along a range, they discovered many large tube worms, clams, mussels, and crabs. Normally the deep-ocean floor supports relatively small forms of life because there isn't much food. But the vents provided energy for bacteria to support abundant life. This is a striking example of how the physical nature of the ocean determines the distribution of life within it.

Scientists have always been interested in how the atmosphere and the ocean affect each other. Our present technology uses satellites to view ocean circulation and life distribution. Within the next decade we will be gathering data from every ocean, every day. Such simultaneous data will show how the water reacts to wind speed and air temperature and how the air responds to oceanic changes.

POPULATION IMPACTS ON COASTAL ECOLOGY AND THE OPEN OCEAN

Population studies now show that three-fourths of all Americans live within 80 kilometers (50 miles) of the coast or the Great Lakes. This migration to the coasts will further mar a delicate balance between the ocean and the shore, resulting in more harbor and channel dredging; sea disposal of sewage, industrial waste, and dredge spoils; the use of ocean water for cooling of power plants and industries; and the filling in of marshes vital to the cleansing of runoff waters and the maintenance of coastal fisheries (figure I–1). In the open ocean, deep-ocean mining and nuclear waste disposal are planned. How do we deal with the increased demands on the environment; how do we regulate the ocean's use?

Further, the National Academy of Sciences in 1980 predicted a rise in sea level. As carbon dioxide and other gases that increase the atmosphere's ability to hold heat reach higher concentrations, worldwide temperature increases may occur. This increased "greenhouse effect" could hasten sea-level rise and damage marine ecological systems. Even considering present erosion and deposition, geologists would like to have us move back from the shore as much as possible and leave it to nature. Engineers, on the other hand, say any coastal region can be stabilized for the lifetime of a development; in that case much public money will be spent stabilizing the coastal property of a few (figure I–2). Perhaps if the sea level does rise, engineers will be too busy protecting present development and will abandon plans to develop new areas!

RATIONAL USE OF TECHNOLOGY

Looking at these issues, we see cause for concern, but certainly not despair. We have a vast resource that hasn't yet been lethally damaged. We have been able to inflict only minor damage here and there along its margin. But as our technology makes us more powerful, the threat of irreversible harm becomes greater—or less, depending on how we use our tools.

Which path will we take? All of you will need to evaluate carefully those you elect to public office, and some of you will have direct responsibility for making the decisions that affect our environment. It is my hope that you, as a new student of the marine environment, will gain enough knowledge from your teacher and this course to help our nation, as well as other nations, make rational use of the oceans in your lifetime.

FIGURE I–1
Coastal Population. Heavily developed bay barrier running from Sea Bright to Monmouth Beach, New Jersey. (Photo © Breck P. Kent.)

FIGURE I–2
Coastal Erosion. Storm surge washes over breakwater during heavy storm. (Photo © Mark Sherman/Bruce Coleman, Inc.)

Chapter 1

History of Oceanography

With each passing year, human interaction with the ocean increases. The fishing effort is stepped up, and the search for petroleum and minerals intensifies. With this increasing rate of ocean exploitation, we are obliged to increase our knowledge of its effect on the marine environment. The following is a summary of events that have brought us to our present state of knowledge and a discussion of how the pressing need for increased knowledge of the oceans may be met through the application of newly developed technology.

THE CONCEPT OF GEOGRAPHY DEVELOPS

Early History

Humankind probably first viewed the ocean as a source of food. At some later stage in the development of civilization, vessels were built to move upon the ocean's surface, thereby making it a wide avenue over which distant societies could begin to interact.

Photo sled ANGUS (Acoustically Navigated Geophysical Underwater System) being lifted aboard the research vessel *Knorr* during the February-March 1977 expedition to a suspected hydrothermal vent area on the Galápagos Rift. Thirteen frames of a 3000 frame "run" above sixteen kilometers of ocean floor revealed a phenomenon previously unknown to science—the deep sea hydrothermal vent biocommunity. (Photo by Al Gunderson. Courtesy of Scripps Institution of Oceanography, University of California.)

The first westerners that are known to have developed the art of navigation were the **Phoenicians,** who as long ago as 2000 B.C. were investigating the Mediterranean Sea, the Red Sea, and the Indian Ocean. The first recorded circumnavigation of Africa in 590 B.C. was made by the Phoenicians, who had also sailed as far north along the European coast as the British Isles.

The Greek view of the world in 450 B.C. is seen in the map of **Herodotus** (figure 1–1). It shows the Mediterranean Sea surrounded by three continents: Europe, Asia, and Libya, bordered by three major seas. The northern and northeastern margins of the continents of Europe and Asia are indicated as unknown, but the Greeks believed that the oceans represented a margin of water that surrounded all three continents.

The Greek astronomer-geographer **Pytheas** sailed northward to Iceland in 325 B.C. and worked out a simple method for determining latitude, which involved measuring the angle between the lines of sight from an observer to the North Star and the northern horizon (appendix VII). Using astronomical measurements, he also proposed that the tides were a product of lunar influence.

Eratosthenes (276–192 B.C.), a Greek librarian at Alexandria, was the first known determiner of the world's circumference. He determined that Earth's meridianal circumference was 40,000 km (24,840 mi) which compares very well with the 40,032 km (24,875 mi) determined by the more precise methods in use today.

The Roman Strabo (63 B.C.–A.D. 24) made a contribution more directly related to the nature of the oceans. He had observed the volcanic activity that is characteristic of the Mediterranean area and concluded that, as a result of this activity, the land periodically sank and rose, causing the sea to invade and then recede from the continents. He further observed that the rivers flowing across the continent erode material from the surface and transport it to the oceans.

In approximately A.D. 150, **Ptolemy** produced a map of the world that represented the Roman knowledge at that time. He introduced lines of longitude and latitude on his map. Mariners could determine the latitude of any point on the surface of Earth using the method introduced by Pytheas, but it was impossible for them to determine longitude accurately. Ptolemy's map indicated, as did the earlier Greek maps, the continents of Europe, Asia and Africa and showed the Indian Ocean to be surrounded by a partly unknown landmass. Unlike Herodotus, Ptolemy considered the major oceans to be seas similar to the Mediterranean, having boundaries defined by unknown landmasses.

The Middle Ages

After the fall of the Roman Empire, the Mediterranean area was dominated by Arab influence, and the writings of the Greeks and Romans passed into the hands of the Arabs, to be forgotten by the Christians in Europe. Subsequently, the Western concept of geography degenerated considerably, and one notion envisioned the world as a disc with Jerusalem at the center.

The Arabs, meanwhile, were trading extensively with East Africa, Southeast Asia, and India and had learned the secret of the monsoons. They took advantage of these winds, making their trade voyages easier. During the summer seasons, when the monsoon winds blew

FIGURE 1–1
The World of Herodotus. The world according to Herodotus, 450 B.C. (From *Challenger* report, Great Britain, 1895.)

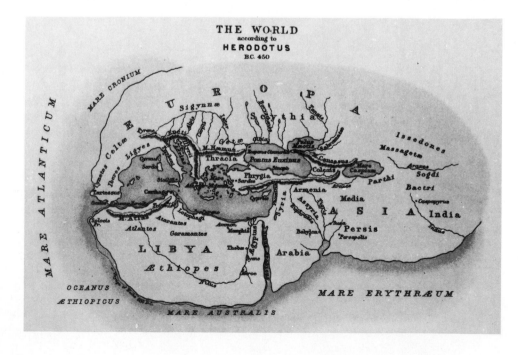

from the southwest, ships laden with goods for trade would leave the Arabian ports and sail eastward across the Indian Ocean. The return voyage would be timed to take advantage of the northeasterly trade winds that occurred during the winter season, making the transit relatively simple for their sailing vessels.

In Europe, the nautical inactivity of the southern Europeans was offset by the vigorous exploration of the **Vikings** of Scandinavia. Late in the 9th century, aided by a period of climatic warming, the Vikings conquered Iceland. In 981 Erik the Red sailed westward from Greenland and discovered Baffin Island. In 995 Erik's son, **Leif Eriksson,** discovered what was then called Vinland, and he spent the winter in that portion of North America we call Newfoundland (figure 1–2).

Age of Discovery

During the 30-year period from 1492 to 1522, known as the **Age of Discovery,** the Western world came to a full realization of the vastness of Earth's water-covered surface. The continents of North and South America were discovered. The globe was circumnavigated, and it was learned that human populations existed elsewhere in the world. Human cultures were found throughout the newly discovered continents and islands, although they were vastly different from those with which the voyagers were familiar.

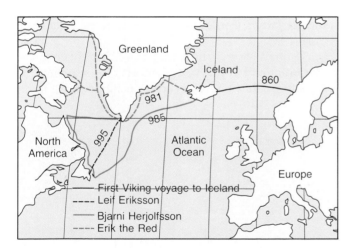

FIGURE 1–2
The Vikings Reach the New World. The expansion of Viking influence spread from Europe and probably reached Iceland by 860. Erik the Red was banished from Iceland and sailed to Greenland in 981. He returned to Iceland in 984 and led the first wave of Viking colonists to Greenland. Bjarni Herjolfsson sailed from Iceland to join the colonists but sailed too far south and is thought to be the first Viking to have viewed what is now called Newfoundland. Bjarni did not land but continued trying to find his way to Greenland. He eventually did land at the settlement Herjolfsness, named for his father, who waited for him there. In 995, Leif Eriksson, son of Erik the Red, bought Bjarni's ship and set out for the land Bjarni had seen to the southwest. Leif set up a camp at Tickle Cove in Trinity Bay and named the land Vinland after the grapes that were found there.

Precipitating these voyages was the capture of Constantinople, the capital of eastern Christendom, in 1453 by Sultan Mohammed II. This event isolated the Mediterranean ports from the riches of the East Indies and caused the Western world to search for a new trading route to this area. As a result of the Arab occupation of Constantinople, the ancient knowledge of the Greeks and Romans was carried out of that city and became available to the powers of southern Europe who were interested in reestablishing trade connections with the East Indies.

Prince **Henry the Navigator,** of Portugal, had established a marine observatory to improve the Portuguese sailing skills, but its attempts to reestablish trade by ocean routes met with failure for years.

One of the greatest obstacles to an alternate trade route was the need to travel around the tip of Africa. Cape Agulhas was finally rounded by **Bartholomeu Diaz** in 1486. He was followed in 1498 by **Vasco da Gama,** who continued his trip around the tip of Africa to India.

The idea for a voyage such as the one **Christopher Columbus** undertook was initiated by the Florentine astronomer Toscanelli. He wrote to the king of Portugal, suggesting that a course be charted to the west in an attempt to reach the East Indies by crossing the Atlantic Ocean. Columbus later contacted Toscanelli and was given a copy of the information indicating that India could be reached at a distance that would have carried him just west of the continent of North America.

After his well-known difficulties in originating the voyage, Columbus set sail with 88 men and three ships August 3, 1492, from the Canary Islands. During the morning of October 12, 1492, the first land was sighted, which is generally believed to have been Watling Island. Because Columbus had greatly underestimated the distance to the East Indies via the Atlantic Ocean, he was convinced that he had arrived at these islands. Upon his return to Spain and the announcement of his discovery, additional voyages were planned, and the Spanish and Portuguese continued to explore the coasts of North and South America.

The Atlantic Ocean became familiar to the European explorers, but the Pacific Ocean was not seen by them until 1513 when Vasco Núñez de Balboa attempted a land crossing of the Isthmus of Panama and sighted the Pacific Ocean from atop a mountain.

The culmination of this period of discovery was the circumnavigation of the globe by **Ferdinand Magellan** (figure 1–3). In September of 1519, Magellan left Sanlucar de Barrameda, Spain, and traveled through a passage to the Pacific at 52°S latitude, now named the Straits of Magellan. After discovering the Philippines on March 15, 1521, Magellan was killed in a fight with the inhabitants of these islands. Juan Sebastian del Caño completed the circumnavigation by taking one of the ships, *Victoria,* across the Indian Ocean and back to Spain in 1522. On

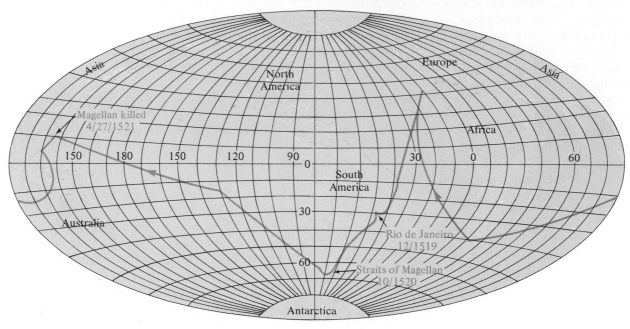

Sanlucar de Barrameda, Spain
Voyage began 9/20/1519
Voyage ended 9/6/1522

FIGURE 1–3
Voyage of Magellan. Searching for a western route to the East Indies, Magellan culminated the Age of Discovery by beginning a voyage that ended for him with death in the Philippines. Juan Sebastian del Caño returned to Spain in the *Victoria,* one of five ships that began the voyage, to complete the first circumnavigation of Earth. (Base map courtesy of National Ocean Survey.)

this trip Magellan had attempted to measure the depth of the Pacific Ocean by a weighted line, but he was unable to reach bottom.

After these voyages, the Spanish initiated many others for the purpose of removing the gold that had been found in the possession of the Aztec and Inca cultures in Mexico and South America. While the Spanish concentrated on plundering the Aztecs and Incas, the English and Dutch plundered the Spanish. The political dominance of Spain came to an end with the defeat of the Spanish Armada by the British in 1588. With this squelching of an attempted invasion of the British Isles, the English became the dominant maritime power, and they remained so until early in the 20th century.

THE SEARCH TO INCREASE SCIENTIFIC KNOWLEDGE OF THE OCEANS

Captain James Cook

The next major focus of interest in the oceans was more scientific, as the English determined that increasing their knowledge of the oceans would help them maintain their maritime superiority. The more successful of the early voyages that were initiated to learn more about the physical nature of the oceans were conducted by the English navigator **James Cook** (1728–79), son of a farm laborer (figure 1–4).

From 1768 until his death in 1779, Captain Cook undertook three voyages of discovery. He searched for the continent Terra Australis and concluded that if it ex-

FIGURE 1–4
Captain James Cook, RN. (From U.S. Navy photograph.)

isted, it lay beneath or beyond the extensive ice fields of the southern oceans. Captain Cook discovered the South Georgia and South Sandwich islands and the Hawaiian Islands, where he was killed after searching for a northwest passage from the Pacific Ocean to the Atlantic Ocean. Cook also developed a diet that prevented his crew from contracting the disease scurvy, which is caused by a vitamin C deficiency.

Cook's expeditions added greatly to the scientific knowledge of the oceans. He determined the outline of the world's largest ocean and was the first person known to cross the Antarctic Circle. Cook also led the way in sampling subsurface temperatures, measuring winds and currents, sounding, and in collecting important data on coral reefs. By proving the value of **John Harrison's** chronometer as a means of determining longitude, Cook made possible the first accurate maps of Earth's surface (see Appendix VII).

An American contemporary of Captain Cook was an early contributor to greater understanding of the ocean's surface currents. **Benjamin Franklin,** deputy postmaster general for the colonies, determined that it took mail ships coming from Europe by a northerly route 2 weeks longer to reach the colonies than it did ships that came by a more southerly route. To find out why this occurred, he asked for information concerning the movement of surface waters from captains who came into port. Franklin inferred that there was a significant current moving in a northerly path along the eastern coast of the United States. The current moved out in a more easterly path across the North Atlantic. He concluded that this eastflowing current, which the ships traveling a northerly route from Europe had to combat, was responsible for increasing the time of their voyage. He subsequently published (in 1777) a map of the Gulf Stream based on these observations (figure 1–5).

Matthew Fontaine Maury

An even greater contribution was made by **Matthew Fontaine Maury** (1806–73). This career officer in the U.S. Navy was placed in charge of the Depot of Naval Charts and Instruments after suffering an injury early in his career. In the depot were log books containing a large amount of information about currents and weather conditions in various parts of the oceans. Maury's systematic analysis of these logs produced a compilation of wind and current patterns that proved very useful to the navigators of the 19th century. Maury helped organize the first International Meteorological Conference in Brussels in 1853 for the purpose of establishing uniform methods of making nautical and meteorological observations at sea. This standardization greatly increased the dependability of such data, which Maury summarized in *The Physical Geography of the Sea* (1855). Maury is often referred to as the father of oceanography (figure 1–6).

FIGURE 1–5
Chart of the Gulf Stream Compiled by Benjamin Franklin in 1777. (Courtesy of U.S. Navy.)

Charles Darwin

In the early 19th century, an English naturalist named **Charles Darwin** (1809–82) entered the scientific scene. Darwin's interest, investigating the whole of nature, led him to make one of the most outstanding contributions

FIGURE 1–6
Lieutenant Matthew F. Maury. A photograph of an engraving by Lemuel S. Punderson after a daguerreotype, autographed and inscribed by Lt. Maury. (Released Naval Historical Center photograph.)

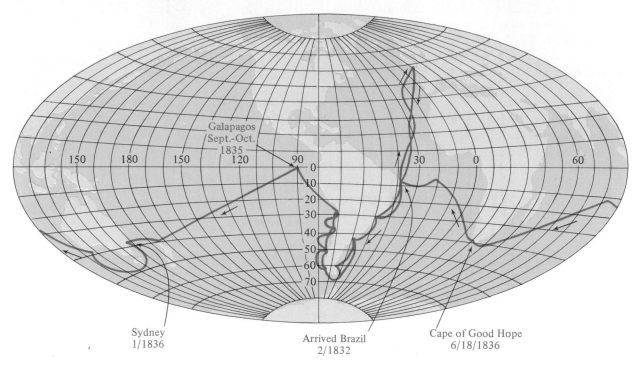

Departure: 12/27/1831
Return: 8/2/1836

Galapagos
Sept.-Oct.
1835

Sydney
1/1836

Arrived Brazil
2/1832

Cape of Good Hope
6/18/1836

FIGURE 1–7
Voyage of the *Beagle*. Sailing as naturalist aboard H.M.S. *Beagle,* Charles Darwin gathered the evidence that enabled him to develop his theory of biological evolution through natural selection. (Base map courtesy of National Ocean Survey.)

to the field of biology. Much of the background on which he based his conclusions concerning evolution by natural selection was gained from observations made during his voyage aboard HMS *Beagle.*

The *Beagle* sailed from Devensport on December 27, 1831, under the command of Captain Robert Fitzroy with the major objective of completing the survey of the coast of Patagonia and Tierra del Fuego.

Darwin spent the next 5 years aboard the *Beagle,* which continued to sail around the world for the purpose of carrying on chronometric measurements (figure 1–7). The voyage allowed the young naturalist the opportunity to study the plants and animals throughout the world. He studied closely the changes that occurred in the animal populations living in different environments and concluded that all animals change slowly over the immense period of time represented by the geologic past.

Darwin's observations led him to conclude that birds and mammals must have evolved from the reptiles and that the similar skeletal framework of the human, the bat, the horse, the giraffe, the elephant, the porpoise, and other vertebrates required that they be grouped together. Darwin felt that the superficial differences that could be

observed between populations were the result of adaptation to different environments and modes of existence.

On his return to England, Darwin published his controversial book, *The Origin of Species,* which dealt not so much with the origin of life but with the evolution of living things into their many forms. Although Darwin's ideas were highly controversial, many can be indisputably proven by an impressive array of scientific facts.

The Rosses—Sounders of the Deep

Two of the earliest successful sounders of deep oceans were Englishmen, **Sir John Ross** and his nephew **Sir James Clark Ross.** Baffin Bay in Canada was explored by Sir John Ross in 1817 and 1818, and he was able to measure depths. Ross also collected samples of bottom-dwelling organisms and sediments with a device of his own design called a *deep-sea clamm.* He recovered starfish and worms in mud samples from a depth of 1.8 km (1.1 mi).

Sir James Clark Ross extended the soundings to greater depths on voyages to the Antarctic from 1839–43 (figure 1–8). A 7-km (4⅓-mi) sounding line was used on

FIGURE 1–8
Ross Expedition. Members of the Sir James Clark Ross expedition to Antarctica after landing on Possession Island. The penguins seem to be at least as curious about the visitors as the members of the expedition are about them. (Reprinted by permission of Bettman Archive.)

these voyages, and even this great length was at times insufficient to reach the bottom of the ocean. Sir James observed that the animals he recovered from the cold waters of the Antarctic were the same species his uncle had recovered from the Arctic area, and they were found to be very sensitive to temperature increase. This discovery led him to the conclusion that the waters making up the deep ocean must be of uniformly low temperature.

Edward Forbes

Also making an important contribution to the study of marine biology was a British naturalist, **Edward Forbes** (1815–54). Forbes was interested in determining the vertical distribution of life in the ocean, and after repeated observations, he divided the sea into specific life-depth zones. He came to the conclusion that plant life is limited to the zone near the surface. Forbes also found that the animal concentration was greater near the surface and decreased with increasing depth until only a small trace of life, if any, remained in the deepest waters.

Forbes's followers either misinterpreted his statement of the condition of life's distribution in the ocean or decided to apply their own logic to it, and they concluded that no life existed in the deep ocean, as life would be impossible under the conditions of high pressure and absence of light and air in this environment. Ignoring the findings of the Rosses in the high southern and northern latitudes, they persisted in their belief that no life existed in the abyssal ocean.

The *Challenger* Expedition

The controversy over the distribution of life in the oceans was one of the factors that helped stimulate interest in the first large-scale voyage with the express purpose of studying the subject.

In 1871, the Royal Society recommended that funds be raised from the British government for an expedition to investigate the following subjects:

1. Physical conditions of the deep sea in the great ocean basins;
2. The chemical composition of seawater at all depths in the ocean;
3. The physical and chemical characteristics of the deposits of the sea floor and the nature of their origin;
4. The distribution of organic life at all depths in the sea, as well as on the sea floor.

A staff of six scientists under the direction of C. Wyville Thompson left England in December 1872 aboard **HMS *Challenger.*** A 2306-tn corvette that had been refitted to conduct scientific investigation, the *Challenger* returned in May 1876, after having traversed large portions of the Atlantic and Pacific oceans (figure 1–9). The achievements of the *Challenger* expedition, which covered 127,500 km (79,223 mi), included 492 deep-sea soundings, 133 bottom dredges, 151 open-water trawls, and 263 serial water temperature observations. The more outstanding aspects of the voyage were the netting and classification of 4717 new species of marine life and the measurement of a water depth of 8185 m (26,850 ft) in the Mariana Trench. Seventy-seven samples of ocean water collected during the *Challenger* expedition were analyzed in 1884 by C. R. Dittmar. This first refined analysis contributed greatly to our understanding of ocean salinity, which will be discussed in chapter 5.

Fridtjof Nansen

One of the most unusual of the voyages was initiated by a Norwegian, **Fridtjof Nansen** (1861–1930). Dr. Nansen developed a great interest in exploring the North Atlantic and the Arctic area. To aid him in his planning, the results

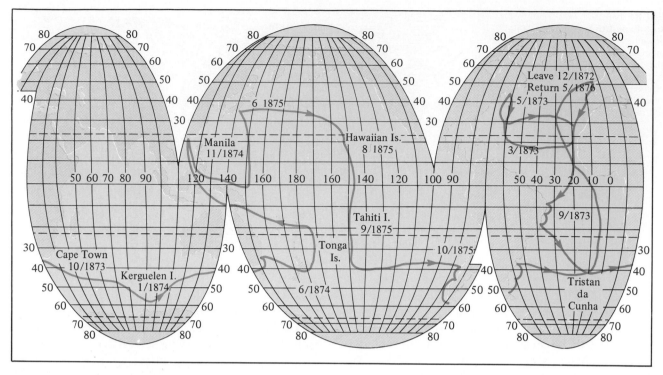

FIGURE 1–9
Voyage of HMS *Challenger*. The route traveled by HMS *Challenger* during the first major oceanographic voyage (December 1872 to May 1876). (Base map courtesy of The National Ocean Survey. Right, HMS *Challenger* from the *Challenger* Report, Great Britain, 1895.)

of earlier attempts to explore the Arctic during the latter part of the 19th century were compiled. Of particular interest to Nansen was the ill-fated American voyage of George Washington DeLong and his crew on the *Jeanette* in 1879. This expedition had attempted to sail through the Bering Strait to Wrangell Island, from which it planned to go overland to the pole. Scientists believed that Wrangell Island was possibly the southern tip of a peninsula extending southward from the great, but yet undiscovered, Arctic continent.

The *Jeanette* became stuck in the ice on September 6, 1879, and drifted north of Wrangell Island, proving that the island was not a peninsula of the Arctic continent. After two years of drifting, the *Jeanette* sank off the New Siberian Islands. DeLong and many of his crew perished in the Lena Delta region of eastern Siberia.

In 1884, a number of articles that could be traced to the *Jeanette* were found frozen in the pack ice off the eastern coast of Greenland. On the basis of this evidence, Nansen considered that the articles must have drifted from the wreck of the *Jeanette* to the observed location by following a path that would have taken them over or near the North Pole. He saw no reason why this ice flow that transported these articles could not be used to transport an expedition across the North Pole as well.

Eventually Nansen was able to raise sufficient funds for building the ***Fram,*** a ship that would be so designed that the expanding ice would not crush it but rather would force it up to the surface, free of the grip of the growing ice sheets.

Provisions for 13 men for 5 years were stored away on the small ship, and the crew set sail from Oslo on June 24, 1893. On September 21, after fighting their way along the northern coast of Siberia, they had not yet sighted the New Siberian Islands when the ice captured them at a latitude of 78°30′N. They were 1100 km (683 mi) from the North Pole. They had now begun what would be a long and lonely endeavor (figure 1–10).

On November 15, 1895, the *Fram* reached its northernmost point, 394 km (244 mi) from the pole. On August 13, 1896, the little *Fram* broke ice and was again

FIGURE 1–10
Voyage of the *Fram*. The course
followed by the *Fram* after becoming
frozen in ice near the New Siberian
Islands, September 21, 1893. On August 13, 1896, the little ship was released by the ice off Spitsbergen.

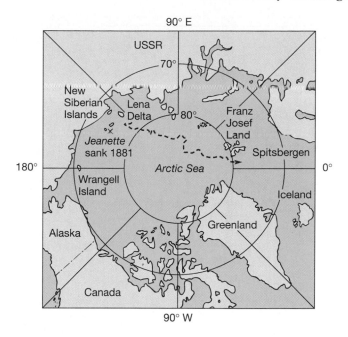

unbound in the open sea. She had drifted a total of
1658 km (1028 mi) during her 3-year entrapment.

The drift of the *Fram* proved that no continent existed by the Arctic Sea and that the ice that covered the
polar area throughout the year was not of glacial origin
but a freely moving pack ice accumulation that had been
formed directly upon the ocean surface. During the voyage, the crew had found that the depth of this ocean ex-
ceeded 3000 m (9840 ft) and that there was a rather surprising body of relatively warm water with temperatures
as high as 1.5°C (35°F) between the depths of 150 and
900 m (about 500–3000 ft). Nansen correctly described
this water as being a mass of Atlantic water that had sunk
below the less saline Arctic water.

Nansen's realization of the need for more accurate
measurements of salinity and temperature led him to
develop the once widely used bottle for sampling the
ocean depths, the *Nansen bottle*. His observations of the
direction of ice drift relative to the wind direction helped
V. Walfrid Ekman, a Scandinavian physicist, to develop
the mathematical explanation of this phenomenon. Nansen, who was awarded the Nobel Peace Prize in 1922,
Ekman, and other Scandinavian scientists led the way in
the development of the field of physical oceanography in
the early 20th century.

20TH CENTURY OCEANOGRAPHY

Voyage of the *Meteor*

Oceanography entered a new era with the voyage of the
Meteor, 1925–27. This German expedition was the first
to use an echo sounder, making it possible to obtain a
continuous depth recording as a vessel proceeded along
its course. This represented a great advancement over
reliance on scattered soundings and interpolation of the
water depth between these points of known depths. The
Meteor gathered data day and night for 25 months and
concentrated on increasing the knowledge of the South
Atlantic Ocean. It was this expedition that first revealed
the true ruggedness of the ocean floor (figure 1–11).

Many contributions from a number of nations have
been made to the understanding of the oceans subsequent to the voyage of the *Meteor*. We will not discuss the
details of these voyages, but we must emphasize that the
knowledge we now possess concerning the world's
oceans is still far from complete. The efforts that will be
needed to bring it to a more complete state will, of necessity, continue for years into the future.

Oceanography in the United States

Although the first large-scale contribution to oceanography by a citizen of the United States resulted from the
efforts of Lieutenant Matthew Maury of the U.S. Navy and
culminated in his book *The Physical Geography of the
Sea,* it was not until later in the 19th century that the
United States became active in organizing voyages for the
purpose of increasing our knowledge of the oceans. This
phase of involvement actually began in 1877 with the voyages of the *Blake* under the direction of **Alexander Agassiz.** This young scientist, the son of the Swiss naturalist
Louis Agassiz, contributed greatly to the development of
oceanographic research in the United States. Following
his voyages, many more have been initiated with the
prime objective of increasing the scientific knowledge of
the world's oceans. Rather than detail these individual
voyages, the following paragraphs will discuss briefly
some projects undertaken by the major institutions in the

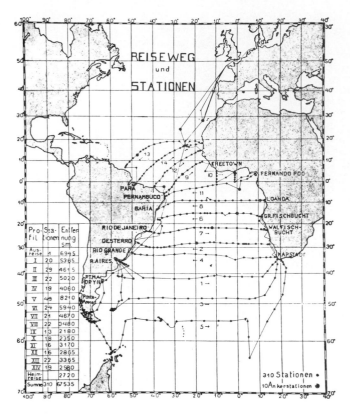

FIGURE 1–11
Voyage of the *Meteor*. This 1925 cruise crisscrossed the South Atlantic, where marine data were sparse, in an effort to make possible a comprehensive study of the water masses and circulation of the Atlantic Ocean. (From Wüst, G., 1935. The stratosphere of the Atlantic Ocean. Scientific results of the German American Expedition of the Research Vessel *Meteor* 1925–27, Vol. VI, sec. 1. W. J. Emery, trans. and ed. Amerind Pub. Co., New Delhi, 1978.)

United States that are concerned with the investigation of the world's oceans.

The U.S. Navy Oceanographic Office has the primary objective of improving the combat readiness of the fleet, and basic research performed toward this objective is a part of its assignment. The results of their research, as well as that of the Office of Naval Research and the U.S. Coast Guard, are of value and basic importance to all oceanographers.

On February 10, 1807, the U.S. Coast and Geodetic Survey was created by an act of Congress after President Jefferson requested a survey of the coast of the United States. This organization, now the National Ocean Survey, has the primary responsibility for bathymetric (ocean depth) surveys in the waters adjacent to the United States and its territories. Although the work done by this agency is primarily that of underwater mapping, the vessels that are involved are outfitted to conduct a wider scope of oceanographic research, gathering a wide range of oceanographic data.

The Commerce Department's National Oceanic and Atmospheric Administration (NOAA) works to ensure wise use of ocean resources through its National Ocean Survey, National Marine Fisheries Service, and Office of Sea Grant.

Not directly under the control of the federal government, but very important in the achievement of research of national interest, are three marine laboratories that have pioneered much of the marine research in the United States. The oldest of these institutions is **Scripps Institution of Oceanography** of the University of California at San Diego. Originated at La Jolla, California, in 1903, Scripps investigates a wide spectrum of marine problems.

Woods Hole Oceanographic Institution originated at Woods Hole, Massachusetts, in 1930 as a private nonprofit organization devoted to the scientific study of the world ocean. The ship *Atlantis II* operated by this institution was one of the first modern ships designed to conduct oceanographic research.

The **Lamont-Doherty Geological Observatory** of Columbia University was founded at Torrey Cliffs Palisades, New York, in 1949 under the direction of **Maurice Ewing.** He was one of the great pioneers of geophysical methods for the investigation of the geology of the ocean bottom (figure 1–12).

These institutions have led the way in 20th-century oceanographic exploration. Their development of new devices for the study of the ocean depths has been sparked by the desire to unravel more of the mysteries hidden there. One of the first of these imaginative devices was the **Floating Instrument Platform** (FLIP)

FIGURE 1–12
Maurice Ewing. Dr. Maurice Ewing (1906–74) pioneered seismic techniques for studying sediments on the ocean floor. He was director of Lamont-Doherty Geological Observatory at Columbia University from 1949 to 1972. (Photo courtesy of Columbia University.)

FIGURE 1–13
Floating Instrument Platform (FLIP). FLIP, a 108-m (356-ft) Floating Instrument Platform, designed and developed by the Marine Physical Laboratory at Scripps Institution of Oceanography. In the vertical position (inset) it gives scientists an extremely stable platform from which to carry out underwater acoustic and other types of oceanographic research. FLIP has no motive power and must be towed to a research site in the horizontal position. Once on station, the ballast tanks are flooded and the vessel flips vertically, leaving 17 m (56 ft) of the platform above water. The work completed, water is forced from the tanks and FLIP resumes the horizontal position. (Courtesy of Scripps Institution of Oceanography, University of California, San Diego.)

shown in figure 1–13. It was designed for making acoustical and other open-ocean measurements that require a stable platform. Camera systems and magnetometer instrument capsules have been developed for free-fall emplacement on the ocean floor. They can be left to operate for days or months. Automatic timers or acoustical signals release these devices to float to the surface for recovery.

The need for cooperative efforts was recognized in the **Deep Sea Drilling Project,** which was being carried out under the leadership of the Scripps Institution of Oceanography and through the cooperation of a number of educational institutions that have developed oceanographic programs. In November 1963, the director of the National Science Foundation announced the establishment of a national program for sediment coring. Scripps Institution of Oceanography and three other major oceanographic institutions—the Rosenstiel School of Atmospheric and Oceanic Studies at the University of Miami, Florida; Lamont-Doherty Geological Observatory of Columbia University; and Woods Hole Oceanographic Institution in Massachusetts—united to form the Joint Oceanographic Institutions for Deep Earth Sampling (JOIDES). They were later joined by the departments of oceanography of the University of Washington, Texas A&M, the University of Hawaii, Oregon State University, and the University of Rhode Island.

A ship capable of drilling the ocean bottom while floating 6000 m (3.7 mi) above at the ocean surface had to be designed specifically to carry on this program. The **Glomar Challenger** was constructed and launched March 23, 1968. The first leg of the Deep Sea Drilling Project commenced on August 11 of that year. The *Glomar Challenger* is 122 m (400 ft) long and 20 m (66 ft) in beam, and its displacement is 10,500 tn loaded. The vessel is a self-sustained unit capable of remaining at sea for 90 days.

Initially the oceanographic research program was financed by the U.S. government, but in 1975 it became international. Financial and scientific support was received from West Germany, France, Japan, the United Kingdom, and the USSR. In 1983 the Deep Sea Drilling Project became the **Ocean Drilling Program** (ODP) with the broader objective of also drilling the continental margins. This new phase of operations is supervised by Texas A&M University. Accompanying this change in emphasis was the decommissioning of *Glomar Challenger* after 15 highly successful years during which it made possible a major advancement in the field of earth science—confirmation of sea-floor spreading, a process in which new ocean floor is created along submarine mountain ranges. Following in the wake of the *Glomar Challenger* as the ODP drill ship is the much larger **JOIDES Resolution,** which conducted its first scientific cruise in January of 1985.

The Geochemical Ocean Sections (GEOSECS) program has been operating since 1972. This project involves sampling and analyzing water from all oceans to study ocean circulation patterns, mixing processes, and biogeochemical cycling. The GEOSECS program is discussed further in chapter 5.

Putting humans into the marine environment to observe it directly has always been a dangerous business. The effects of increased pressure on body cavities and gases dissolved in body tissues limit the depth and duration of dives where the diver interacts directly with the marine environment. The record for such dives in the ocean is 457 m (1500 ft), but the researchers in this field believe humans will eventually be able to work for extended periods of time at depths of more than 600 m (1970 ft). Table 1–1 provides the dates of some significant events in the history of human progress toward deep-sea descent.

The use of submersibles that take humans deep into the ocean housed in chambers in which the pressure remains the same as at the ocean surface (1 atmosphere)

TABLE 1–1
Some Important Achievements in the History of Human Descent into the Ocean.

Date	Event
360 B.C.	Aristotle records the use of air trapped in kettles lowered into the sea by Greek divers in his *Problematum*.
330 B.C.	Alexander the Great descends in a glass berrylike bell during the siege of Tyre.
1620	Dutch inventor Cornelius van Drebbel tests first submarine in the Thames River. Drebbel descended to a depth of 5 m (16.5 ft) with King James I of England on board.
1690	Sir Edmond Halley descends in a lead-weighted diving bell. Air was replenished from barrels lowered into the Thames River.
1715	Englishman John Lethbridge tests a leather-covered barrel of air with viewing ports and waterproof armholes to depth of 18 m (60 ft). This design represented the first armored diving suit.
1776	Colonist David Bushnell's *Turtle* attempted the first military submarine attack on the British ship H.M.S. *Eagle*. The attempt failed, but the British moved their fleet to less accessible waters.
1800	Robert Fulton builds the submarine *Nautilus*. It was the first of several to bear this name and was powered by a hand-driven screw propeller when submerged.
1837	Augustus Siebe invents a prototype of the modern helmeted diving suit.
1913	A German company, Neufeldt and Kuhnke, manufactures an armored diving suit with articulating arms and legs.
1930	William Beebe and Otis Barton descend in a bathysphere to a depth of 923 m (3027 ft) off Bermuda.
1943	Jacques-Yves Cousteau and Émile Gagnan invent the fully automatic, compressed air aqualung.
1960	Jacques Piccard and Donald Walsh descend to a depth of 10,915 m (35,801 ft) in the bathyscaphe *Trieste*.
1962	Hannes Keller and Peter Small make an open-sea dive from a diving bell to a depth of 304 m (1000 ft) using a special gas mixture. Small dies during decompression.
1962	Albert Falco and Claude Wesly live for one week under conditions of saturation in Cousteau's Continental Shelf Station off Marseilles, France.
1964	*Sealab I*—U.S. Navy team led by George F. Bond stays 11 days at 59 m (193 ft) off Bermuda.
1966	Submersibles *Alvin, Aluminaut,* and *Cubmarine,* with the aid of unmanned remote-controlled vehicle *CURV,* locate and recover an atomic bomb from water over 800 m (2,624 ft) deep off Palomares, Spain.
1969	Armored diving suit, *Jim,* is built to operate at depths below 304 m (1000 ft).
1969–70	*Tektite I* and *II* accommodate 12 scientific teams for from 14 to 60 days at a depth of 15 m (50 ft) off the U.S. Virgin Islands.
1976	Jacques Mayol dove to 100 m (328 ft) while holding his breath for 3 minutes and 40 seconds.
1981	Diving scientists at Duke University simulate a dive to 686 m (2250 ft) in a pressure chamber.

has been very successful as a means of directly observing the deep-sea environment. In 1934, zoologist William Beebe descended to a depth of 923 m (3027 ft) off Bermuda in a tethered *bathysphere* to observe deep-sea life. After World War II, the Swiss scientist Auguste Piccard designed his *bathyscaphe* (deep boat), *Trieste,* which was

bought by the U.S. Navy. This vessel, which operates much like a balloon with a gondola, submerged to the record depth of 10,915 m (35,801 ft) in the Mariana Trench off Guam on January 23, 1960 (figure 1–14).

Submersibles that are now widely used in deep-sea research are two untethered vessels. **Alvin** can descend

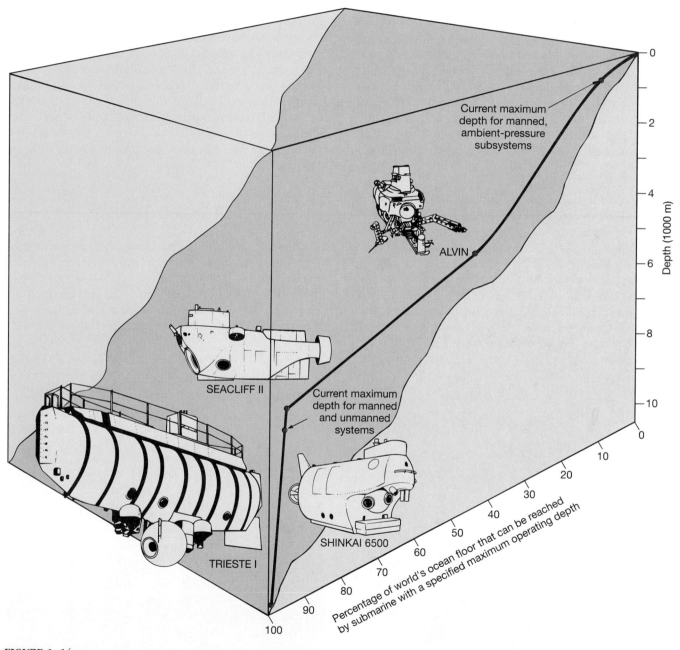

FIGURE 1–14
Manned Submersibles and the Ocean Depths. This figure shows the percentage of ocean floor that is reachable by four submersibles of importance in the history of ocean exploration. *Alvin,* which is the first submersible to be used extensively in marine research, can explore about 44 percent of the ocean floor with its maximum operating depth of 4000 m (13,120 ft). *Sea Cliff II* has an operations limit of 6000 m (20,000 ft). The *Shinkai 6500* is the world's deepest-diving manned research submarine. It is operated by the Japanese government and can explore 97 percent of the ocean floor. The 1960 dive of *Trieste* to a depth of 10,915 m (35,801 ft) shows it to have the capability of exploring any part of the deep ocean. However, given its immobility, it is not practical to use it for deep-sea exploration.

to a depth of 4000 m (13,120 ft), and **Sea Cliff II** can descend to a depth of 6000 m (20,000 ft). The deepest diving manned submersible is *Shinkai* 6500. This Japanese vessel can dive to 6500 m (21,320 ft) (figure 1–14).

A newly developed Remote Underwater Manipulator (RUM III), about the size of a compact car, will provide deep-ocean research capability previously available only through the costly use of manned submersibles (figure 1–15). Weighing 363 kg (800 lb) in water, it is light enough to maneuver across the softest sediments using its canvas tracks. Operated from a computerized console aboard ship, it will be towed by steel-armored coaxial cable. RUM III has a manipulator arm that can lift samples and experimental equipment weighing up to 100 kg (220 lb), thrusters for positioning it above and on the ocean floor, and television cameras. Two side-looking sonar instruments will be able to map the terrain of the ocean floor. RUM III can work at depths of 6000 m (20,000 ft).

Another remote system similar to RUM III is the **Seabeam/Argo-Jason** system (figure 1–16). Seabeam is a complex sonar mapping system operating from a research vessel. *Argo,* which contains an array of sonar and television systems, was being towed above bottom by the R/V *Knorr* on September 1, 1985, when its television cameras discovered the remains of the RMS *Titanic* (figure 1–17). *Jason,* a self-propelled vehicle, can be tethered to *Argo.* With its stereo color television and two manipulator arms, *Jason* can be lowered to the ocean floor to investigate areas of interest identified on *Argo's* television and acoustical images.

Remote sensing from satellites is also becoming an important tool in ocean studies. Instruments aboard satellites can measure the temperature, ice cover, color, and topography of the ocean surface as well as give us much information about the topography of the ocean floor. Figure 1–18 shows *Seasat A,* the first truly oceanographic satellite, launched in 1978. Unfortunately it shorted out after three months but still provided enough data to prove the great benefit to be derived from remote satellite sensing of the ocean's surface.

Many examples of ingenious uses of today's technology in the study of the oceans are discussed in later chapters. They include the use of sound waves to image horizontal slices of ocean temperature by use of **Ocean Acoustical Tomography.** This technique is similar in design to the use of X rays to image slices through the human body with Computerized Axial Tomography, or CAT scans.

LAW OF THE SEA

Because of the imminence of extensive exploitation of the ocean floor beyond the jurisdiction of the nations that border the seas and the developing problems relative to international fisheries and pollution, one of the most urgent problems awaiting international agreement has been the establishment of laws by which these activities may be regulated.

In 1609, Hugo Grotius, the Dutch jurist and statesman whose writings were important in the formulation of

A.

B.

FIGURE 1–15
RUM III. *A,* diagram of the RUM III showing *(A)* thrusters, *(B)* manipulator arm shown picking up a magnetometer capsule, *(C)* canvas tracks, and *(D)* counterweights to help stabilize RUM III when heavy objects are lifted. *B,* RUM III during sea trials in January 1986. (Photo supplied by The Marine Physical Laboratory of the Scripps Institution of Oceanography.)

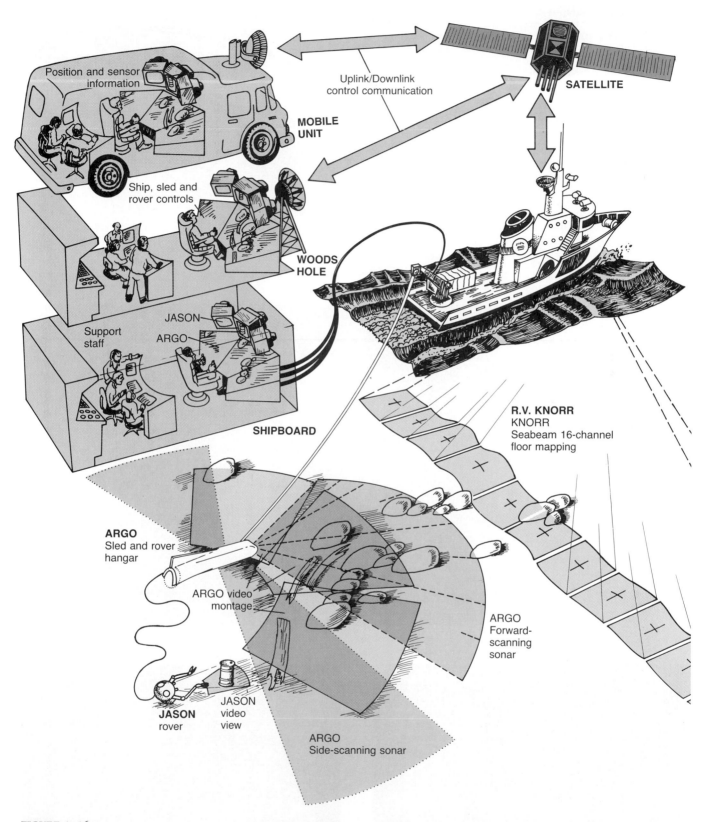

FIGURE 1–16

Seabeam/Argo-Jason. Seabeam, with its computerized system, makes very accurate sonar maps that can be converted to three-dimensional models of the ocean floor. When the *Argo-Jason* system reaches the ocean floor that has been modeled, the shipboard operator will have a view of the ocean floor as if he or she were inside the remote system. Both *Argo* and *Jason* will use television eyes to guide the shipboard operator across the ocean floor. *Argo* will send out sonar signals to help determine its position on the three-dimensional model created by Seabeam.

FIGURE 1–17
RMS *Titanic* is Photographed.
RMS *Titanic,* unseen by human eyes since it sank in 1912 600 km (373 mi) south of Newfoundland, is at last visible in television and photographic images. The ship was discovered during sea trials of the new deep-sea investigative instrument *Argo* on September 1, 1985. Photos *A* and *B* were taken by a towed photographic unit, ANGUS (Acoustically Navigated Geological Underwater Survey System), that takes high-resolution still photographs. *A,* anchor chains, capstans, and windlasses; *B,* loading cranes on the bow of the *Titanic.* Photos *C* and *D* were taken the next year when *Alvin* returned for the sea trials of *Jason. C,* electric winch on boat deck photographed by prototype of *Jason* called *Jason Jr. D, Jason Jr.* photographed from *Alvin. Jason Jr.*'s "garage" on *Alvin* is in foreground. (Courtesy of Woods Hole Oceanographic Institution.)

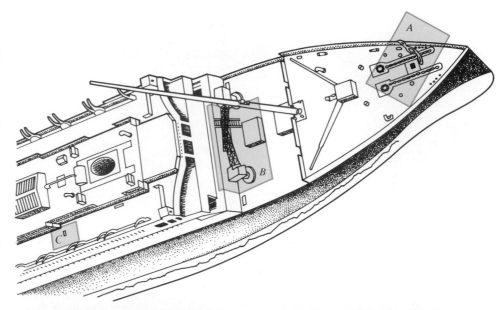

A.

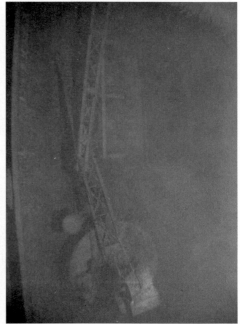

B.

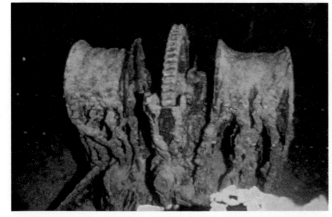

C.

D.

FIGURE 1–18
Seasat A. This satellite, which pioneered the remote sensing of the oceans, was launched July 7, 1978. It shorted out on October 10, 1978, but provided sufficient data to prove the value of satellite remote sensing to the study of the oceans. The ghostly image below *Seasat* is the H.M.S. *Challenger,* which pioneered the study of the oceans from the ocean surface in 1872. (Courtesy Jet Propulsion Laboratory, NASA.)

international law, established in *Mare liberum* the doctrine that the oceans are free to all nations. Although controversy continued over whether nations could control portions of the oceans, Cornelius van Bynkershonk produced a satisfactory solution to this problem in *De dominio maris,* published in 1702. It provided for national domain over the sea for the distance that could be protected by cannons on the shore. The first official dimension of this **territorial sea** was probably established in 1672 when the British determined that they would exercise control over a strip of water 1 league (3 nautical miles) wide along the shores of the Commonwealth—the territorial sea.

Because of the rapidly developing technology for drilling beneath the ocean, the first **United Nations Conference on the Law of the Sea,** held in Geneva in

1958, produced a treaty relating to the continental shelf. It stated that prospecting and mining of minerals on the shelf are under the control of the country that owns the nearest land. Unfortunately, the seaward limit of the continental shelf was not well-defined in the treaty.

The second United Nations Conference on the Law of the Sea was held in Geneva in 1960 and achieved little. The third had its first meeting in New York in 1973. Subsequent meetings were held at least once a year.

The third conference came to a disappointing conclusion in April of 1982, when a large majority—130 to 4, with 17 abstentions—voted in favor of a new Law of the Sea treaty. Most of the developing nations who could benefit significantly from the provisions of the treaty voted for its adoption. The opposition was led by the United States, who, along with Turkey, Israel, and Venezuela, voted against it. These countries support private companies planning seabed mining operations and believe certain provisions of the treaty will make it unprofitable. Among the abstainers were the Soviet Union, Great Britain, Belgium, the Netherlands, Italy, and West Germany, all of whom are interested in seabed mining. Because only 30 nations have ratified the treaty, its ability to facilitate the exploitation of seabed resources will be severely hampered.

The primary features of the treaty are as follows:

1. *Coastal Nations Jurisdiction:* A uniform 12-mi territorial sea and a 200-mi **Exclusive Economic Zone** (EEZ) were established. The coastal nation has jurisdiction over mineral resources, fishing, and pollution within the EEZ and beyond it, up to 350 mi from shore, if the continental shelf (defined geologically) extends beyond the EEZ.

2. *Ship Passage:* The right of free passage on the high seas is maintained; it is also provided for within territorial seas through straits used for international navigation.

3. *Deep-Ocean Mineral Resources:* Private exploitation of seabed resources may proceed under the regulation of the International Seabed Authority (ISA), within which will be a mining company chartered by the United Nations, called Enterprise. Mining sites will be operated in pairs: one chosen by the private mining concern, the other by Enterprise, and both approved by ISA. Revenues from the private concern's chosen site will go to that concern, and those from the other site will be divided among the developing nations.

4. *Arbitration of Disputes:* A Law of the Sea tribunal will perform this function.

SUMMARY

In the Western world, the Phoenicians were the first great navigators. Later the Greeks, Romans, and Arabs made significant contributions that lay hidden behind the curtain of the Middle Ages for over 1000 years. The Age of Discovery renewed the Western world's interest in exploring the unknown. It began with the voyage of Columbus in 1492 and ended in 1522 with the circumnavigation of Earth by a voyage initiated by Ferdinand Magellan.

Captain James Cook, who sailed in the late 18th century, was one of the first navigators to sail the oceans with the primary purpose of learning their natural history. He also made the first accurate maps using the chronometer developed by John Harrison. Matthew Fontaine Maury was the first American to make a large-scale contribution to the study of the oceans with publication of *The Physical Geography of the Sea* in 1855. By the mid-19th century, the English biologists Charles Darwin, Sir John Ross, Sir James Ross, and Edward Forbes had made significant contributions to increasing our knowledge of life in the oceans.

The voyage of HMS *Challenger,* from December 1872 through May 1876, represented the first major full-scale oceanographic voyage. A feel for the significance of this exploration can be gained by contemplating the fact that 4717 new species of marine life were identified during the voyage. An unusual voyage aboard the *Fram,* which spent 3 years entrapped in Arctic ice was initiated by the Norwegian explorer Fridtjof Nansen in 1893. It was with this voyage that the first knowledge of the oceanography of the Arctic Ocean was gained. From 1925 through 1927 during a period of 25 months, the German ship *Meteor* criss-crossed the South Atlantic Ocean, recording the

first large-scale continuous depth record with an echo sounder. This voyage first revealed the true ruggedness of the ocean floor.

Recent oceanographic investigations have been numerous. In the United States, the early ones were conducted by government agencies and three pioneering oceanographic institutions—Scripps Institution of Oceanography, Woods Hole Oceanographic Institution, and Lamont-Doherty Geological Observatory. They have since been joined by many educational institutions. Major projects such as the Deep Sea Drilling Project, Ocean Drilling Program, and Geochemical Ocean Sections have resulted from this widespread cooperation among institutions and nations.

The use of manned submersibles such as *Alvin* and *Sea Cliff II* has contributed significantly to new discoveries in the deep ocean. Remote unmanned systems such as RUM III and the Seabeam/*Argo-Jason* systems will be able to replace more expensive submersibles for much deep-ocean research. Many new techniques have been developed for studying the oceans. Ocean Acoustical Tomography and remote sensing from satellites will become increasingly important.

Because of the imminence of extensive exploitation of the oceans, the United Nations has conducted conferences on the Law of the Sea to develop international agreement on how the resources of the oceans can be shared and managed. These efforts were certainly set back, and may have failed, in April 1982, when developed and developing nations could not agree. The treaty voted on then may not be ratified by enough developed nations to significantly increase current levels of seabed resource exploitation.

QUESTIONS AND EXERCISES

1. Construct a time line that includes the major events of human history that have resulted in a greater understanding of our planet in general and the oceans in particular.
2. List the Mediterranean cultures that appear on your time line in the probable order of their dominance in the field of marine navigation.
3. Using a diagram, describe the method used by Pytheas to determine latitude in the Northern Hemisphere.
4. How did the Greek concept of geography provided by Herodotus differ from that of the Romans as given by Ptolemy?
5. While the Arabs dominated the Mediterranean region during the Middle Ages, what were the most significant events taking place in northern Europe?
6. List some of the major achievements of Captain James Cook.
7. What was Matthew Fontaine Maury's major contribution to an increased knowledge of the oceans?
8. Define in your own words the process of evolution by natural selection.
9. What was the controversy that developed due to the observations of the Rosses and Edward Forbes?
10. List some major achievements of the voyage of HMS *Challenger*.
11. What new knowledge of the Arctic polar region did the voyage of the *Fram* provide?
12. The voyage of which ship first revealed the true relief of the ocean floor?
13. List the names and locations of three pioneering oceanographic research institutions.
14. Describe the functions of the Floating Instrument Platform (FLIP), the Remote Underwater Manipulator (RUM III), and Seabeam/*Argo-Jason*.
15. Name the major ocean research programs designed to drill into the ocean floor and study ocean circulation.
16. Discuss possible reasons that less developed nations might prefer the concept that the open ocean is the common heritage of all nations, while the more developed nations might prefer the concept that the open ocean's resources belong to those who recover them.

KEY TERMS

Agassiz, Alexander (p. 11)
Age of Discovery (p. 5)
Alvin (p. 15)
Challenger, HMS (p. 9)
Columbus, Christopher (p. 5)
Cook, James (p. 6)
Da Gama, Vasco (p. 5)
Darwin, Charles (p. 7)
Deep Sea Drilling Project (p. 13)
Diaz, Bartholomeu (p. 5)
Ekman, V. Walfrid (p. 11)
Eratosthenes (p. 4)
Eriksson, Leif (p. 5)
Ewing, Maurice (p. 12)
Exclusive Economic Zone (p. 20)
Floating Instrument Platform (FLIP) (p. 12)

Forbes, Edward (p. 9)
Fram (p. 10)
Franklin, Benjamin (p. 7)
Glomar Challenger (p. 13)
Harrison, John (p. 7)
Henry the Navigator, Prince (p. 5)
Herodotus (p. 4)
Joides Resolution (p. 13)
Lamont-Doherty Geological Observatory (p. 12)
Law of the Sea (p. 19)
Magellan, Ferdinand (p. 5)
Maury, Matthew Fontaine (p. 7)
Meteor (p. 11)
Nansen, Fridtjof (p. 9)

Ocean Acoustical Tomography (p. 16)
Ocean Drilling Program (p. 13)
Ptolemy (p. 4)
Pytheas (p. 4)
Ross, James Clark (p. 8)
Ross, John (p. 8)
Scripps Institution of Oceanography (p. 12)
Seabeam/*Argo-Jason* (p. 16)
Sea Cliff II (p. 16)
Territorial Sea (p. 19)
Vikings (p. 5)
Woods Hole Oceanographic Institution (p. 12)

REFERENCES

Allmendinger, E. 1982. Submersibles: Past—present—future. *Oceanus* 25:1, 18–35.

Bailey, H. S., Jr. 1953. The voyage of the *Challenger*. *Scientific American* 188:5, 88–94.

Borgese, E. M., and Ginsberg, N. 1991. *Ocean Yearbook 9*. Chicago: The University of Chicago Press.

Brewer, P., ed. 1983. *Oceanography: The present and future*. New York: Springer-Verlag.

Duxbury, A. C. 1971. *The earth and its oceans*. Reading, Mass.: Addison-Wesley.

Knauss, J. A. 1974. Marine science and the 1974 Law of the Sea Conference: Science faces a difficult future in changing Law of the Sea. *Science* 184:1335–41.

Mowat, F. 1965. *Westviking*. Boston: Atlantic-Little, Brown.

Ryan, P. R. 1986. The *Titanic* revisited. *Oceanus* 29:3, 2–17.

——— 1985. The *Titanic*: Lost and found. 1985. *Oceanus* 28:4, 1–112.

Sears, M., and Merriman, D., eds. 1980. *Oceanography: The past*. New York: Springer Verlag.

SUGGESTED READING

Sea Frontiers

Baker, S. G. 1981. The continent that wasn't there. 27:2, 108–14. A history of the search for Ptolemy's *Terra Australis Incognita,* the unfound southern continent.

Charlier, R. H., and Charlier, P. A. 1970. Matthew Fontaine Maury, Cyrus Field, and the physical geography of the sea. 16:5, 272–81. A biography of Maury with emphasis on his accomplishments as superintendent of the Department of Charts and Instruments. His role in laying the first transatlantic cable and a discussion of his most famous work, *The Physical Geography of the Sea,* are included.

Engle, M. 1986. Oceanography's new eye in the sky. 32:1, 37–43. Photographs taken by Paul Scully-Power, a U.S. Navy oceanographer, from the space shuttle add to our knowledge of ocean currents and waves.

Loeffelbein, B. 1983. Law of the Sea treaty: What does it mean to nonsigners (like the United States)? 29:6, 358–66. An overview of the Law of the Sea Convention and a discussion of the implications of the treaty not being signed by all nations.

Maranto, G. 1991. Way above sea level. 37:4, 16–23. A discussion of Navstar Global Positioning System satellites and their potential for helping us map changes in sea level.

McClintock, J. 1987. Remote sensing: Adding to our knowledge of oceans—and Earth. 33:2, 105–113. The role of remote sensing, primarily in aiding us in understanding weather and climate, is discussed.

Rice, A. L. 1972. HMS *Challenger:* Midwife to oceanography. 18:5, 291–305. An interesting account of the achievements recorded during the first major oceanographic expedition.

Schuessler, R. 1984. Ferdinand Magellan: The greatest voyager of them all. 30:5, 299–307. A brief history of the voyage initiated by Magellan to circumnavigate the globe.

Van Dover, C. 1987. *Argo* Rise: Outline of an oceanographic expedition. 33:3, 186–194. A description of the first scientific use of the *Argo* system to survey hydrothermal vent activity on the East Pacific Rise off the coast of Mexico.

———— 1988. Dive 2000. 34:6, 326–33. Events related to the 2,000th dive of DSRV *Alvin* on the East Pacific Rise are recounted.

Scientific American

Bailey, H. S. 1953. The voyage of the *Challenger.* 188:5, 88–94. A summary of the accomplishments of the English oceanographic expedition of 1872.

Borgese, E. M. 1983. The law of the sea. 248:3, 42–49. One hundred nineteen nations sign a convention aimed at establishing a new international scheme of regulating the oceans.

Herbert, S. 1986. Darwin as a geologist. 254:5, 116–123. Before publication of *Origin of species,* Charles Darwin considered himself to be primarily a geologist. This article outlines his contributions in this field with special attention to his theory of coral reef formation.

Chapter 2

The Ocean: Its Origin and Provinces

It can be a wonderful and rewarding experience to learn about the world's oceans. The wonder is heightened by adding this new knowledge to a general understanding of the place of the oceans in the cosmos.

A great deal of speculation is involved in these considerations, but science has not made advances by restricting itself to that which is already known. We should, however, begin our discussion with those aspects of the universe which we can readily observe and describe. Then we will discuss some theories of the origin of the universe and our planet.

The chapter concludes with a discussion of the distribution of Earth's present ocean basins and an overview of oceanic provinces.

THE UNIVERSE WE SEE

As astronomers looked into the sky surrounding the planet Earth, they gradually developed an understanding of Earth's position in what is now called the **solar system** (figure 2–1). The solar system is composed of the

The *Glomar Challenger*. (Photo © Scripps Institution of Oceanography, University of California, San Diego.)

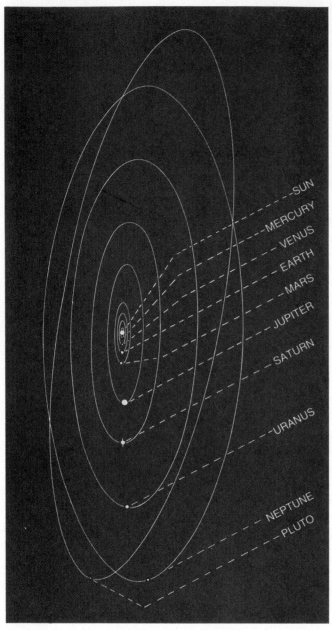

A.

B.

FIGURE 2–1
The Solar System. *A,* the orbits of the planets of the solar system are drawn to scale. *B,* relative sizes of the sun and the planets are shown. The distance scale is not maintained in *B.* (Reprinted by permission from Tarbuck, E. J., and Lutgens, F. K., *Earth science,* 5th ed. (Columbus: Merrill, 1988), figs. 19.1 and 19.2.)

sun and its nine planets, which occupy a disc-shaped portion of the universe about 13 billion kilometers (8 billion miles) in diameter. Although this seems to be an enormous structure, we find that it represents a very minute portion of the total universe. Our sun is one of perhaps 100 billion billion, or 10^{20}, stars that constitute the known universe. In terms of distances with which we are familiar, the solar system is incomprehensibly isolated. The distance from the sun to the nearest star is 40.7 thousand billion kilometers (25 thousand billion miles).

While the planets revolve around the sun, the entire system is moving at about 280 km/s (174 mi/s) around the center of the Milky Way, the **galaxy** or group of stars to which our solar system belongs. Estimates of the dimensions of the Milky Way indicate that it is a disc-shaped accumulation of stars about 100,000 light-years in diameter and 10,000–15,000 light-years thick near the center, thinning gradually toward the edges. A **light-year** is a unit of measure used by astronomers to describe distances within the universe and is equal to the distance traveled by light at a speed of 300,000 km/s (186,000 mi/s) during a period of one year—almost 10 trillion (9.8×10^{12}) km (6.2 trillion miles). Within the Milky Way are some 100 billion (10^{11}) stars, tens of millions of which very probably possess families of planets, millions of which may be inhabited by intelligent creatures, according to the estimates of some astronomers.

All the stars you see with your naked eye belong to the Milky Way (figure 2–2). If you have exceptional vision and live in the Northern Hemisphere, you may see a hazy patch of light in the constellation Andromeda that represents another galaxy within the universe. This galaxy is some 1.5 million light-years away. It appears that galaxies are exceedingly numerous, and within a distance of a billion light-years from our galaxy, at least 100 million galaxies have been observed by our telescopes.

By observing light energy that radiates from these distant galaxies, astronomers have been able to determine that most are moving away from us. The velocity with which the galaxies recede from our system is directly proportional to their distance from it. Velocities of nearly 240,000 km/s (149,000 mi/s) have been calculated for the most distant galaxies as they recede from the Milky Way. This is in excess of three-fourths of the speed of light!

One may be left with the impression that all the galaxies are being repelled by our solar system, but a more reasonable consideration would be that galaxies are moving away from one another as would fragments that result from an explosion. The universe appears to be rapidly expanding, its component galaxies moving ever farther apart. If all these galaxies are moving away from one another in a manner similar to that of fragments created by explosion, we might ask if they all belonged to one large mass. We, of course, do not know. But if that was the case, the time required for them to have reached their present distribution would have been at least 15 billion years.

FIGURE 2–2
The Milky Way. The sun and its planets are located about two-thirds the distance from the center of the Milky Way to the edge. The Milky Way completes a rotation every 200 million years. To maintain its position in the Milky Way, the solar system travels around the center of the galaxy at a velocity of about 280 km/s (174 mi/s). (Reprinted by permission from Tarbuck, E. J., and Lutgens, F. K., *Earth science,* 5th ed. (Columbus: Merrill, 1988), fig. 20.17.)

ORIGIN OF THE EARTH

As insignificant as our planet is in the overall composition of the cosmos, it is nonetheless of interest to us to learn what we can of its origin. Although no complete understanding of the origin of Earth has yet been developed, it is possible to outline the probable sequence that led to its present form. As galactic matter revolved around its center of rotation, the stars began to form as individual glowing masses due to the concentration of particles under the force of gravity. These concentrations may have occurred in local eddies and over a vast period of time took their present form. Many of the resulting stars developed assemblages of planets that revolved around them.

Such a general sequence may have occurred with the formation of our sun, and in its early stages it may have had a size at least equal to the diameter of the solar system that we observe today. As this large accumulation of gas and dust began to contract under the force of gravity, a small percentage of it was left behind in smaller eddies with features similar to the large eddies in which the stars developed in association with the rotation of the galaxies.

As this material was left behind, it flattened itself into a disc that became increasingly dense. The stage was set for the formation of the suite of planets that revolve around the sun. Because of the increase in density of this disc, it became gravitationally unstable and broke into separate small clouds. These were protoplanets, which later consolidated into the present planets.

Protoearth was a huge mass, perhaps 1000 times greater in diameter than Earth today and 500 times more massive. The heavier constituents of the planet Earth, as with all the other protoplanets, were migrating toward the center to form the heavy core surrounded by lighter materials. Similarly, the satellites we observe around the planets, such as our moon, began to form. Our moon is relatively large compared to the size of Earth. In its early formation it was much closer to Earth than it is today.

During this early formation of the protoplanets and their satellites, the sun eventually began to "shine" as it condensed into such a concentrated mass that forces were set up within its interior, creating energy through a process known as the carbon cycle. When temperatures reach tens of millions of degrees, hydrogen atoms, in the presence of carbon, are converted to helium atoms, and large amounts of energy are released. In addition to the

FIGURE 2–3
The Earth's Structure. The cross section shows the three major subdivisions of Earth: core, mantle, and crust. The inner core, outer core, and mantle are drawn to scale, but the thickness of the crust is exaggerated about five times. The boundary between the crust and the mantle is the Mohorovičič discontinuity (Moho). (Reprinted by permission from Tarbuck, E. J., and Lutgens, F. K., *The earth: An introduction to physical geology*, 3d ed. (Columbus: Merrill, 1990), fig. 1.11.)

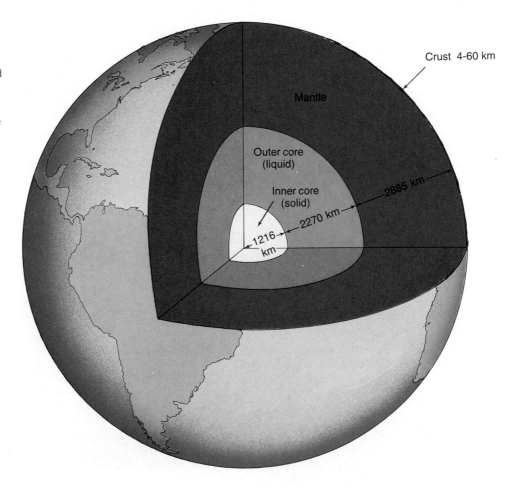

light which is so obviously emitted by the sun, there is an intense emission of ionized, or electrically charged, particles. In the early stages of the development of our solar system, this emission of ionized particles served to blow away the nebular gas that remained from the formation of the planets and their satellites.

Meanwhile, the cold protoplanets nearer the sun were being warmed by the solar radiation, and their atmospheres began to boil away. The combination of emission of ionized solar particles and the internal warming of the planets resulted in a drastic decrease in the size of the planets closer to the sun. Since the gas envelope that surrounded the planets was composed mostly of hydrogen and helium (we may consider this to be the composition of Earth's first atmosphere), the energy that was available easily allowed atoms of this small size to escape from the gravitational attraction of the planets. As the protoplanets continued to contract, the heat produced deep within their cores as a product of spontaneous disintegration of atoms (radioactivity) became more intense.

The evidence indicates that Earth ultimately became molten, allowing the heavier components, primarily iron and nickel, to concentrate in the **core.** The lighter materials segregated according to their densities in concentric spheres around this core. Surrounding the core is a zone composed of heavy iron and magnesium silicates—the **mantle.** The thin **crust** overlying the mantle ranges in thickness from 4 to 10 km (2.5 to 6.2 mi) for the oceanic crust composed of the black volcanic rock, basalt, to a 35- to 60-km (22 to 37-mi) thickness for the lighter granite of the continental crust. The crust is separated from the mantle by a distinct density change called the **Mohorovičič discontinuity (Moho),** after the Yugoslavian seismologist who discovered its existence (figure 2–3). It should be noted that the internal structure of Earth appears to be much more complex than might be gathered from this brief discussion.

ORIGIN OF THE ATMOSPHERE AND OCEANS

The solidification of Earth's crust marks the beginning of geologic history, and the dating of rocks from Earth and the moon indicates that this event occurred about 4.6 billion years ago (figure 2–4). At this time Earth had lost all but a small fraction of its original gas envelope.

The Atmosphere Forms

Where did the material to make up the present atmosphere and ocean come from? Although Earth's surface had become relatively cool by this stage in its development, there was still much volcanic activity according to the geologic record. As volcanic gases bubbled up from Earth's upper mantle, they formed a new atmosphere as

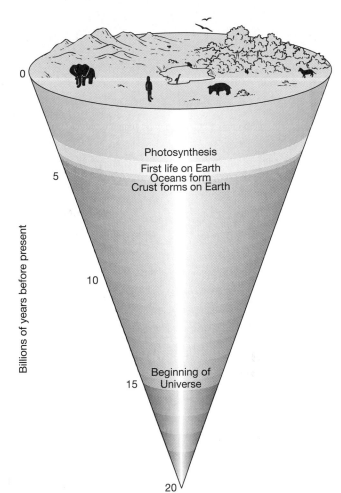

FIGURE 2–4
Major Events in the Evolution of Earth.

soon as Earth had cooled enough to retain one. This atmosphere was far different from the one that Earth had lost and from the one it presently has. It contained little free oxygen and was composed of large percentages of carbon dioxide and water vapor. As Earth cooled to the point at which it could retain these gases, a vast envelope of clouds surrounded the planet, causing Earth to reflect about 60 percent of all the sunlight that was directed toward it. Most of the energy it absorbed from the sun remained in the thick envelope of gases and did not reach Earth's solid surface.

Water vapor released from the surface was captured in the thick envelope and was forced to remain there as a vapor until sufficient cooling of Earth's crust would allow it to condense and begin to accumulate as a liquid. Eventually, cooling of the crust allowed the water vapor to condense and fall under the force of gravity to Earth's surface. Finally, with further decreases in the surface temperature, permanent accumulation of water began in depressions on Earth's irregular surface (figure 2–5).

FIGURE 2–5
Formation of Earth's Oceans. *A,* widespread volcanic activity released H_2O and smaller quantities of CO_2, Cl_2, N_2, and H_2, which produced a water vapor atmosphere that also contained carbon dioxide (CO_2), methane (CH_4), and ammonia (NH_3). *B,* as the earth cooled, the water vapor (1) condensed and (2) fell to Earth's surface. There it accumulated to (3) form the oceans.

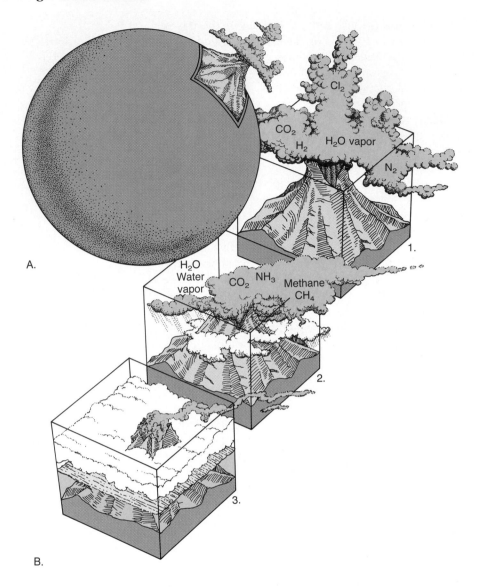

The Oceans Evolve

These accumulations were the initial oceans on planet Earth, and it is assumed that this water has been permanently in existence since its formation. The primary crystalline rocks that made up Earth's crust were weathered and eroded. The inorganic compounds of which they were composed were dissolved in the waters that fell relentlessly upon their surfaces and were carried into the newly forming oceans.

THE CHEMICAL BALANCE SHEET

As chemical weathering of the primary rocks that composed Earth's original crust was made possible by the appearance of water, the process of chemical weathering should be considered. As a result of chemical weathering, the elements that were contained in the primary crystalline rock were freed, dissolved in the water, and carried off to accumulate in the ocean. As the water carried these dissolved elements and compounds into the ocean, it also carried a great load of particulate material that had been weathered and eroded from the crystalline rocks and deposited this material as sediment. The components of the primary crystalline rock that were freed by chemical weathering must be found (1) dissolved in the ocean, (2) as components of Earth's atmosphere, or (3) chemically bound with the sediments that have been carried into the sea by the rivers (see figure 2–6).

The estimates that have been made for these various amounts, of course, will not be exact. The estimate for the ocean is probably one of the most accurate, since the ocean volume is relatively well known and its composition is comparatively uniform because it is well mixed. The estimates are less accurate for crystalline rock, which is quite heterogeneous, and for the sediments, which are of various types. Nevertheless, geochemists have drawn up balance sheets for the various elements, and those

FIGURE 2–6
Geochemical Balances.

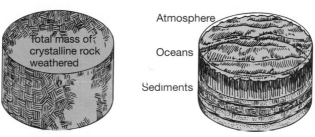

The total mass of crystalline rock components freed by chemical weathering must be found as components of the atmosphere, oceans, or sediments.

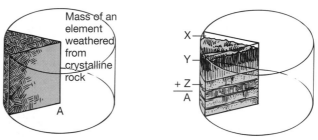

When considering one element represented by the dark wedge in the "pie" to the left (A), the sum of the masses of that element found in the atmosphere (X), oceans (Y), and sediments (Z), should equal the mass weathered from the crystalline rock, A = X + Y + Z.

calculated by different individuals working on the problem agree fairly well. For those elements that are most common in the crystalline rocks, the geochemical equations balance well within the limits set by the accuracy of the estimates.

Excess Volatiles

There are some components that do not balance in any of the attempts that have been made. These elements and compounds appear to be much more abundant in the atmosphere, the oceans, and the sediments than can be accounted for by the volume of crystalline rocks that have been estimated to have undergone chemical weathering. All these substances that are found in excessive amounts compared to the products of chemical weathering are volatile (gaseous) at or slightly above the average temperatures found at Earth's surface and are therefore called **excess volatiles.** Most abundant of the excess volatiles are water vapor and carbon dioxide; also important are quantities of chlorine, nitrogen, sulfur, hydrogen, and fluorine (see table 2–1).

Volcanic activity is presently producing these gases and venting them into the atmosphere. A great percent-

TABLE 2–1
Excess Volatiles. The volatile elements and compounds that are found in the atmosphere, ocean, and sedimentary rocks in far greater amounts than have been made available by chemical weathering of primary crystalline rocks are called *excess volatiles.* They are believed to have been released from the molten interior of the earth by volcanic activity. Note the absence of free oxygen.

	Mass of Elements or Compounds in 10^{20} g					
	H_2O	C as CO_2	Cl	N	S	H, B, Br Ar, F, etc.
Found in:						
Atmosphere and ocean	14,600	1.5	276	39	13	1.7
Buried in sedimentary rocks	2,100	920	30	4	15	15
Total	16,700	921.5	306	43	28	16.7
Supplied by:						
Weathering of crystalline rocks	130	11	5	0.6	6	3.5
Excess volatiles (unaccounted for by rock weathering)	16,570	910.5	301	42.4	22	13.2

Source: Courtesy of the Geological Society of America. After Rubey, 1951.

age of the gases that surface as a result of volcanic activity may simply be recycled volatiles that have been dissolved in groundwater and carried down from Earth's surface, but it has been determined that a small percentage of the total volatiles is always composed of new gases that have for the first time been released from the crystalline rock of the mantle. When **magma** (molten rock) cools, solid minerals are formed and some of the gaseous constituents of the magma are freed. These gases, freed during the crystallization process, represent the new volatiles coming to the surface from the mantle.

DEVELOPMENT OF THE OCEANS AND THEIR BASINS

When Earth began to form its sold crust about 4.6 billion years ago, the mass of the oceans was zero, while at present it is 1.4×10^{24} g. We do not know what the rate of volcanic release of gases from the mantle was throughout this period of 4.6 billion years. However, if we assume it to have been equal to the present rate, we can visualize the formation of Earth's oceans as a gradual process.

Does continual volcanism mean that the oceans are gradually covering more and more of Earth's surface? Not necessarily, since the area that the oceans cover is directly determined by the volume that the basin in which they form is able to accept. The continental crust may not have been present during the initial solidification of Earth's crust and may well have formed gradually as the oceans themselves were formed. Since the volume of the ocean basins is directly related to the difference in thickness between the oceanic and the continental crusts, and if we assume that the continental crust has gradually accumulated, we can see that the capacity of the ocean's basins could have gradually increased along with the volume of water that was being produced over this period of time. Thus, it is quite possible that the oceans have had a distribution very similar in total surface area to that of the present oceans for a long period of geologic time and that the major change in the ocean's character has been an increase in depth. Throughout the geologic time during which the oceans and continents were increasing their volumes, the ocean basins were getting deeper to accommodate the increasing volume of water that was being produced in association with the formation of the continental crust. This process will be discussed further in Chapter 3.

Considering the history of ocean salinity, we might ask if the oceans have possessed a relatively uniform salinity throughout their history or if they are getting more or less saline. By far the most important component of salinity is the chloride ion, Cl^-, which is produced by the same process that produces the water vapor forming the oceans. So our considerations boil down to one question: Has the relative amount of water vapor to chloride ion

production remained constant throughout geologic time, or has it varied? We have no indication that there has been any fluctuation in this ratio throughout geologic time. We must then consider, on the basis of the present evidence, that the oceans' salinity has been relatively constant, while their volume and the volume of the crystalline rock that makes up the continental crust has increased throughout the 4.6 billion years since Earth's crust first began to form.

LIFE BEGINS IN THE OCEANS

We did not include free oxygen in the excess volatiles that were produced at Earth's surface by volcanic activity. If such a gas were produced, it is certain that it could not have remained free in the atmosphere because of the iron found in volcanic rock. Any oxygen that surfaced during volcanic activity would have combined with this iron. As a result of this condition, we may consider that Earth's early atmosphere did not contain this gas that makes up almost 21 percent of our present atmosphere.

We depend on the oxygen in our atmosphere for protection from the ultraviolet radiation that is constantly bombarding our planet. Much of the energy represented by this ultraviolet radiation is exhausted in our upper atmosphere converting the free oxygen (O_2) into ozone (O_3), thus allowing very little of the ultraviolet energy to reach Earth's surface.

Of course, if oxygen was missing from the primitive atmosphere, the ultraviolet radiation would have readily penetrated the atmosphere and reached the surface of the young oceans. These oceans and the atmosphere contained the gases methane (CH_4), ammonia (NH_3), nitrogen (N_2), and carbon dioxide (CO_2).

Laboratory experiments have shown that exposing a mixture of hydrogen, carbon dioxide, methane, ammonia, and water to ultraviolet light and an electrical spark will produce a large assortment of organic molecules (figure 2–7). The exposure of the mixture to ultraviolet radiation causes photosynthesis, carried on in the absence of chlorophyll. **Photosynthesis** is a chemical reaction in which energy from the sun is stored in the products of the photosynthetic reaction—the organic molecules. The production of organic substances in the shallow oceans that developed early in Earth's history must have resulted in a vast amount of "organic" material, as do the laboratory experiments, but this material did not yet represent life on the planet Earth. There is also evidence that some organic molecules reached Earth from outer space via meteorites.

Carbon—The Organizer of Life

Let us now leave these general considerations for a moment and consider more specifically the element that seems to be the most important in allowing the buildup

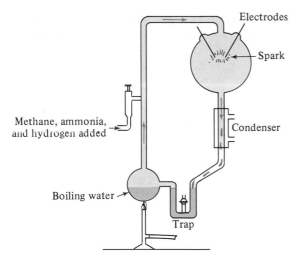

FIGURE 2–7
Synthesis of Organic Molecules. The apparatus used by Stanley L. Miller in the 1952 experiment that resulted in the synthesis of the basic components of life, amino acids. A mixture of water vapor, methane, ammonia, and hydrogen was subjected to an electrical spark that provided the energy for synthesis. This mixture is thought to resemble the composition of the atmosphere and oceans that existed when "organic" molecules first formed on Earth.

of the organic compounds we have just mentioned. This element is carbon. We even refer to the study of the chemistry of carbon compounds as organic chemistry. In studying the chemistry of carbon, we are constantly concerned with molecules which are several times larger than those with which the inorganic chemist is commonly concerned. The greater percentage of inorganic molecules contain fewer than 10 atoms, while in the study of organic chemistry, molecules which may exceed 100 atoms per molecule are not uncommon.

This fact arises out of the "combining power" of the carbon atom. The combining power of an atom is indicated by its **valence.** Carbon has a valence of four (the highest normal valence), which means it can combine with four atoms of hydrogen.

To demonstrate how this ability gives carbon an unusually high combining power, let us compare the carbon atom with other common atoms with valences less than four. If we consider chlorine, which occurs as a gas and has a valence of one, we can see that a single atom of chlorine combines with a single atom of hydrogen to produce a molecule of hydrogen chloride, HCl.

Cl—H

Oxygen, which also occurs as a gas, has a valence of two. An atom of oxygen combines with two molecules of hydrogen in the formation of water, H_2O.

$$O\big\langle {}^{H}_{H}$$

Another gas common in our atmosphere, nitrogen, is tri-

valent. One atom of nitrogen will combine with three atoms of hydrogen to produce the gas ammonia, NH_3.

$$\begin{array}{c} H \\ | \\ N \\ / \ \backslash \\ H \quad\ H \end{array}$$

We return to the carbon atom, with a valence of four, and find that one carbon atom will combine with four atoms of hydrogen to produce the gas methane, CH_4.

$$\begin{array}{c} H \\ | \\ H{-}C{-}H \\ | \\ H \end{array}$$

We can readily see that the higher the valence of an atom of an element, the greater the complexity of the molecules it can form. If we consider that carbon has a valence of four compared to the valence of one for chlorine, then carbon must have a much greater combining power than chlorine or any other element with a valence of one. Consider chlorine as it combines with the metal sodium (Na), which also has a valence of one.

NaCl

We can see that these two elements can combine to form only one compound, sodium chloride (common table salt).

By contrast, methane, an important constituent of Earth's early atmosphere, represents only the simplest of the group of carbon compounds called **hydrocarbons**— those containing only carbon and hydrogen atoms. Somewhat more complex are the **carbohydrates,** which are composed of carbon and water molecules. Belonging to this group are the sugars and starches. Even more complex carbon compounds that are known as **lipids (fats)** and **proteins** may or may not have been present in those early oceans as products of inorganic photosynthesis. Aiding carbon further in forming large molecules is the ability of carbon atoms to attach to one another and produce long chains and rings.

We do not know precisely how the organic material took on the characteristics that allowed it to become living substance. Through some process, however, the organic material became chemically self-reproductive, developed the ability to actively seek light, and began to metabolize and grow toward a characteristic size and shape dictated by forces from within.

Plants and Animals Evolve

These very earliest forms of life must have been **heterotrophs,** organisms that depend on an external food supply. That food supply was certainly abundant in the form of molecules of the nonliving organic material from which the first living single-celled organisms originated.

The **autotrophs,** those organisms that do not depend on an external food supply but manufacture their own, eventually evolved. These first autotrophic forms may well have been organisms similar to our present-day anaerobic bacteria, forms that can exist without free oxygen. They may have been able to oxidize various inorganic compounds that would release energy that could be used to produce their own food internally. This process is called **chemosynthesis.** At some later date, the more complex single-celled plants probably evolved. Having developed a green pigment called **chlorophyll,** plants could capture the sun's energy and, through photosynthesis, produce their own food supply from the carbon dioxide and water that surrounded them.

The autotrophs had developed security in the form of a built-in mechanism to assure them of a food supply. Heterotrophs, not so endowed, were at a considerable disadvantage because they had to search for this necessity of life. To meet this need, animals eventually emerged with their inherent characteristic of mobility and began to develop an awareness or a consciousness that gave them a much more active role in the biological community than that possessed by the plants.

Plants and animals developed in a beautifully balanced environment where the waste products of one filled the vital needs of the other; we can see this expressed chemically by looking at the processes of photosynthesis and **respiration.** In the photosynthetic process, energy is captured and stored in the form of organic compounds. This is an **endothermic** chemical reaction. Respiration is just the opposite—an **exothermic** reaction. It is the reaction by which the energy stored in the foods produced by photosynthesis is extracted by the plants and animals to carry on their life processes.

Photosynthesis → *energy is stored (endothermic)*
$$6H_2O + 6CO_2 + Energy \longleftrightarrow C_6H_{12}O_6 + 6O_2$$
Water + Carbon + Energy   Sugar + Oxygen
dioxide
Respiration ← *energy is released (exothermic)*

Every living organism that inhabits Earth today is the result of an evolutionary process of natural selection that has been going on since these early times and by which various life forms (species) have been able to inhabit increasingly numerous niches within the environment. As these diverse life forms developed in an attempt to exist and adapt to various environments, they also modified the environments in which they lived.

As the plants emerged from the oceans to inhabit the terrestrial environment, they changed the appearance of the landscape from the harsh, bleak panorama we may envision as being typical of a lunar surface to the soft green hues that cover much of the land surface of Earth today. Other changes were manifested in the ocean itself as vast quantities of the hard parts of dead organisms began to accumulate on the bottom. These accumulations of calcium carbonate ($CaCO_3$) and silica (SiO_2) formed the limestone ($CaCO_3$) and diatomaceous deposits (SiO_2 deposits composed mostly of the remains of single-celled plants called diatoms) that we now see exposed on the continents as evidence that these rocks formed from sediments that accumulated on the bottom of the seas. Since over half of the rocks exposed at the surface of continents originally formed on the ocean floor, we can learn much about the oceans of the past by studying these rocks.

Probably the most important environmental modification was that which involved the atmosphere. It was this change that made it possible for most present-day animals to develop. A byproduct of photosynthesis in the early oceans was the free oxygen that makes up 21 percent of our present atmosphere. Carbon dioxide, which at one time must have made up a large portion of our atmosphere, was removed by photosynthesis to produce a concentration of 0.035 percent in Earth's present atmosphere. These changes in the atmosphere must have had a very significant effect on the climates and populations of organisms that have evolved throughout Earth's history (figure 2–8).

Because of human activities over the past 100 years or so, the atmospheric concentration of CO_2 and other gases that help warm the atmosphere has increased. There is concern that Earth is warming to a degree that could cause serious problems for your generation and those that follow you. This phenomenon, referred to as the increased greenhouse effect, is discussed in some detail in chapter 6.

We also see, in petroleum and coal deposits, the remains of plant and animal life that were not completely oxidized but were buried in a reduced (oxygen-free) environment that allowed their energy to remain stored. These deposits provide us with over 90 percent of the energy used for domestic and industrial purposes today. So not only do we, as animals, depend upon the present productivity of plants to supply the energy required by our life processes, but we also depend very heavily on the energy stored by plants during the geologic past.

BATHYMETRY

Throughout the passage of time, the shape of the ocean basins has changed as continents moved across Earth's surface in response to forces related to the release of heat from Earth's interior—a process that will be discussed in chapter 3. The ocean basins as they presently exist reflect the long process of Earth history we have been discussing.

An understanding of the general bathymetric features of the ocean is important because these features will be related ultimately to the origin of the ocean basins

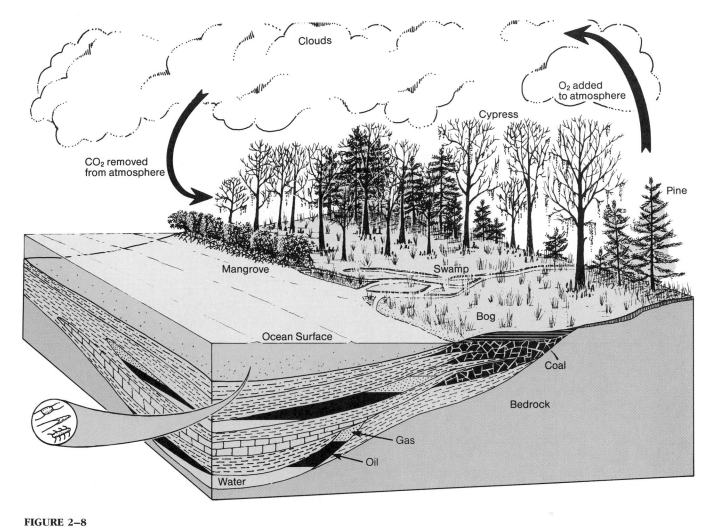

FIGURE 2–8
The Effect of the Evolution of Plants on the Earth Environment. Subsequent to the origin of single-celled plants in the ocean, Earth's atmosphere began to be enriched with O_2 and its CO_2 content was decreased as a result of photosynthesis. As these plants and the microscopic animals that fed off them died and sank to the ocean floor, these remains were incorporated into the marine sediments. Some of the remains were buried before they could be completely decomposed and were ultimately converted to petroleum (oil and gas). When plants moved onto the continents, they evolved into the various forms that support the animals that subsequently evolved there. In the low-lying, highly productive wetlands, plant remains accumulated in the sediment to be converted to peat and ultimately coal.

as well as physical and biological phenomena to be discussed in following chapters. **Bathymetry** in oceanography is analogous to topography in geography. As investigations into the depth of the ocean have proceeded, it has become apparent that a broad shelflike feature that steepens and slopes off into deep basins has generally developed around the continents. Quite commonly, linear mountain ranges run through the basins. In all oceans, but especially around the margin of the Pacific Ocean, deep linear trenches up to 11 km (6.8 mi) in depth separate the slopes from the deep-ocean basin (figure 2–9).

The hypsographic curve showing the distribution of Earth's solid surface at elevations above and below sea level indicates that 71 percent of Earth's surface is to be found beneath the oceans (figure 2–10). The mean elevation of the continents above the sea (840 m; 2755 ft) and mean depth of the oceans (3800 m; 12,460 ft) is the result of different densities of the continental and oceanic crustal rocks. The ocean crust is denser.

ISOSTASY

The basaltic oceanic crust has an average density of about 3.0 g/cm³. The continents are composed of a lighter granitic crustal material ranging in density from 2.67 g/cm³ near the surface to 2.8 g/cm³ deep beneath the continental mountain ranges. The continental crust averages about

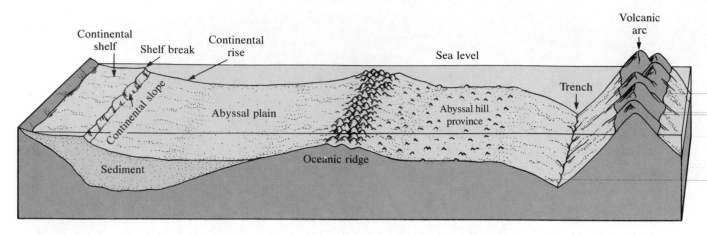

FIGURE 2–9
Marine Physiographic Provinces. A longitudinal profile showing submarine physiographic provinces. To the left of the oceanic ridge is a continental margin typical of the Atlantic Ocean, with a well-developed continental shelf, slope, and rise at the surface of a thick sedimentary deposit. The features to the right are more typical of the Pacific Ocean, where trenches and associated volcanic arcs are common.

35 km (22 mi) in thickness but may reach a maximum thickness of 60 km (37 mi) beneath the highest mountain ranges.

The granitic continental blocks and basaltic oceanic crust "float" on the denser mantle beneath. This concept of crustal flotation, which may be compared with the flotation of ice on water, is called **isostasy** (figure 2–11).

If we take a regularly shaped prism of ice with a density of 0.91 and float it on water with a density of 1.0, the ice will sink into the water until it displaces water equal in mass to the ice. When this displacement, called

buoyancy, occurs, 91 percent of the ice mass will be submerged. Therefore, 91 percent of the length of the prism of ice will be beneath the surface of the water. Similarly, about 55 percent of the mass of the continents is found submerged in the mantle. Actually, we would expect about 91 percent of the continents to be submerged because they are, on the average, about 91 percent of the density of the upper mantle. However, where high heat flow in the mantle heats and reduces the density of the mantle beneath the continents, this low-density mantle provides isostatic compensation. In such situa-

FIGURE 2–10
Elevations of Earth's Crust. The horizontal dashed lines indicate that the average height of the continents is 840 m (2755 ft) above sea level and that the average depth of the oceans is 3800 m (12,460 ft) below sea level. The vertical dashed line marks the division between land and sea. Almost 30 percent or 150 million square kilometers of Earth's surface is above sea level. It is clear that the volume of Earth below sea level is many times greater than the volume of the continents above sea level. (Reprinted with permission from Tarbuck, E. J., and Lutgens, F. K., *The earth: An introduction to physical geology,* 3d ed. (New York: Macmillan, 1990), fig. 1.14.)

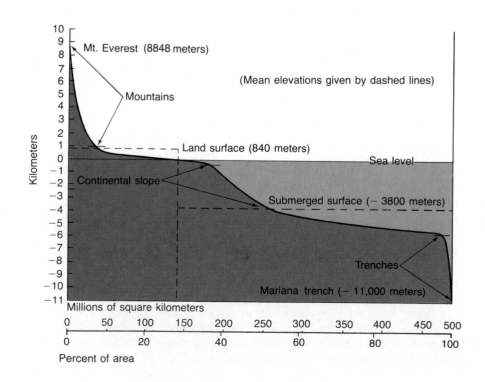

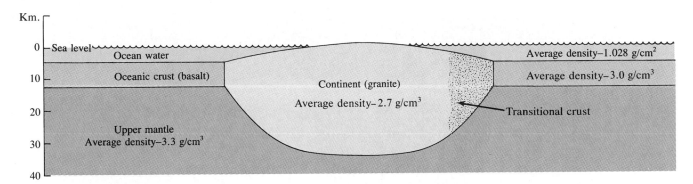

FIGURE 2–11
Isostasy. Isostasy is a state of equilibrium reached by different components of Earth's crust and the mantle. So that the lithosphere (the rock sphere of Earth's structure), which contains the continental and oceanic crustal units as well as the upper mantle, may maintain a uniform average density, continental crustal units that extend up to 14 km (8.7 mi) above the floor of ocean basins must send deep roots into the mantle.

tions as exist beneath the southwestern United States and eastern Africa, the low-density mantle material provides the buoyancy that would normally be provided by the deep root of continental crust.

TECHNIQUES OF DETERMINING BATHYMETRY

Needless to say, our greatest understanding of the ocean floor relates to the shallow shelf areas at the margins of the continents, over which the most important commercial fisheries have been located. They are presently being explored extensively because of their high potential for petroleum accumulation. Because of their economic importance and the fact that we have a greater knowledge of this region than any other ocean region, we will begin our description of the ocean floor with these marginal shelves.

However, before we discuss the various submarine provinces, we should discuss briefly the technique that has been most used to obtain data regarding submarine geological processes, **seismic surveying.** Such surveys are made by emitting a sound signal that will travel through the water, bounce off an object, and return to be picked up by a receiver. If the velocity at which sound will travel in water has been determined, the distance to the object is equal to the velocity times one-half the time required for the sound signal to make the trip from the surface to the object or ocean floor and back to the receiver (figure 2–12).

If one wishes to know only the distance to an object in the water or the ocean floor, relatively weak high-frequency sound signals can be used. Marine geologists, however, need to know the thickness of various rock units beneath the ocean floor, and for this purpose they use stronger low-frequency signals generated by explosions. These sound signals can penetrate the rocks be-

neath the ocean floor, reflect off the contacts between rock units, and produce seismic reflection profiles such as those shown in chapter 4.

A variation of this technique is **side-scan sonar,** which allows a survey ship to map the topography of the bottom along a strip of ocean floor up to 60 km (37 mi) wide. The map is developed with the aid of computers and sound emitters directed away from both sides of the ship (figure 2–13).

Due to the gravitational effects on the ocean surface, sea level rises or falls 4 m (13 ft) with each decrease or increase in ocean depth of 1000 m (3280 ft). Altimeters aboard *Seasat A* detected these differences and provided data to produce a map of the ocean surface. The maps in figure 2–14 show how these changes in ocean surface elevation reflect major bathymetric features of the ocean floor.

CONTINENTAL MARGIN

Continental Shelf

Extending from the shoreline is a shelflike feature called the **continental shelf,** which is geologically part of the continent. During the geologic past much of it was exposed above the shoreline. Since the shoreline has moved back and forth across the shelf, its general bathymetry can usually be predicted by looking closely at the topography of the adjacent coastal region. With few exceptions, this coastal topography can be expected to extend beyond the shore and onto the continental shelf. The continental shelf may be defined as a shelflike zone extending from the shore beneath the ocean surface to a point at which a marked increase in slope occurs. This point where an increase in the rate of slope occurs is referred to as the shelf break, and the steeper slope be-

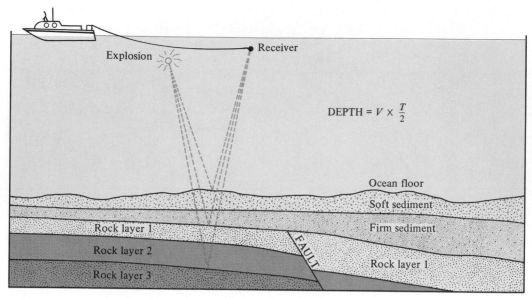

$$\text{DEPTH} = V \times \frac{T}{2}$$

FIGURE 2–12
Seismic Profiling. Low-frequency sound that can penetrate bottom sediments is emitted by an explosion. It reflects off the boundaries between rock layers and returns to the receiver. The depth of each reflecting layer is equal to the velocity of sound travel (V) times one-half the time (T) required for the sound to travel from the source to the reflecting layer and back to the receiver.

FIGURE 2–13
GLORIA System for Side-scanning Sonar Surveys. GLORIA is an acronym for Geological Long-Range Inclined ASDIC. ASDIC is an acronym for Acoustical Side Direction. The neutrally buoyant GLORIA is towed 300 m (985 ft) behind a research vessel. At water depths of 5000 m (16,400 ft), it can map a 60-km (37-mi)-wide strip of ocean floor. At shallower depths, the width of the strip is reduced.

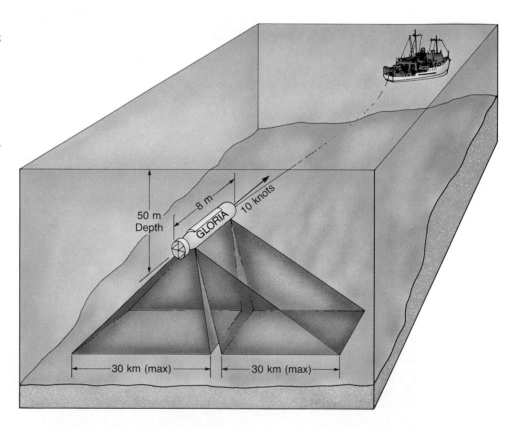

FIGURE 2–14 (opposite)
Ocean Surface Mapped by *SEASAT*. *A,* due to gravitational effects, the ocean surface rises over high volcanic peaks and drops over depressions such as trenches on the ocean floor. An altimeter on the *SEASAT* satellite that operated in 1978 was able to measure these changes in the ocean surface with sufficient precision to map major bathymetric features on the ocean floor. *B,* in the southern Indian Ocean and Pacific Ocean, this remote sensing of the ocean surface was able to identify some new bottom features and provide information that significantly altered the shape or location of previously identified features (circled).

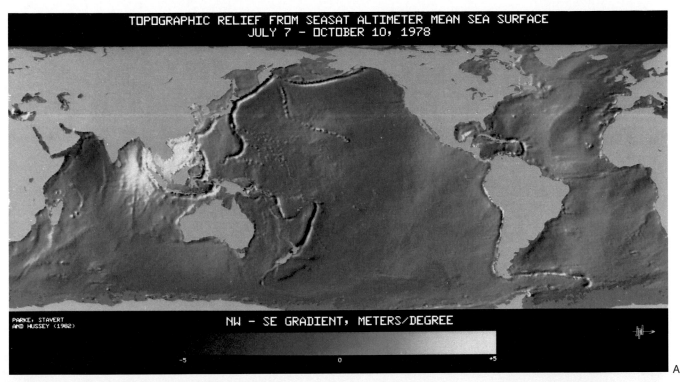

TOPOGRAPHIC RELIEF FROM SEASAT ALTIMETER MEAN SEA SURFACE
JULY 7 — OCTOBER 10, 1978

PARKE, STAVERT
AND HUSSEY (1982)

NW — SE GRADIENT, METERS/DEGREE

-5   0   +5

A.

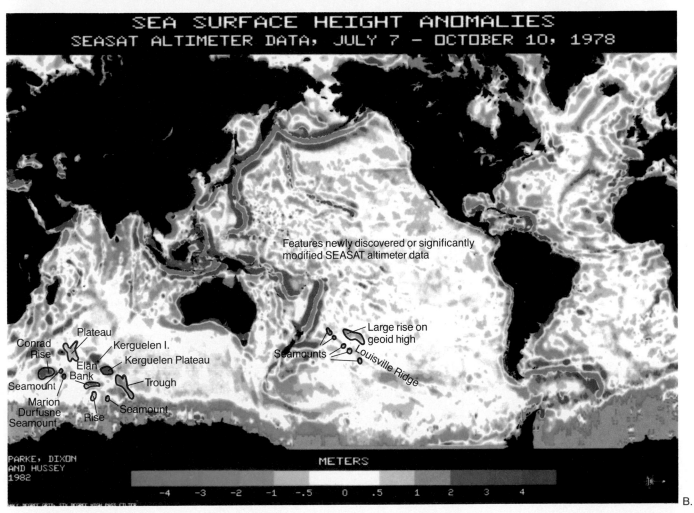

SEA SURFACE HEIGHT ANOMALIES
SEASAT ALTIMETER DATA, JULY 7 — OCTOBER 10, 1978

Features newly discovered or significantly modified SEASAT altimeter data

Plateau
Conrad Rise
Kerguelen I.
Elan Bank
Kerguelen Plateau
Seamount
Trough
Marion Durfusne Seamount
Rise
Seamount

Large rise on geoid high
Seamounts
Louisville Ridge

PARKE, DIXON
AND HUSSEY
1982

METERS

-4   -3   -2   -1   -.5   0   .5   1   2   3   4

B.

yond the break is known as the **continental slope** (shown in figure 2–9).

The width of continental shelves may vary from a few tens of meters to about 1300 km (800 mi). The broadest shelf developments occur off the northern coasts of Siberia and North America in the Arctic Ocean and in the North and West Pacific from Alaska to Australia. The average width of the continental shelf is about 70 km (43 mi), and the average depth at which the break occurs is about 135 m (443 ft). Around the continent of Antarctica, however, the shelf break occurs at a depth of 350 m (2200 ft). The mean slope of the continental shelf if 0°07′ or 1.9 m/km (10 ft/mi).

Sediment is transported across the continental shelf by current flow and mass wasting (submarine sediment slides), which are often generated by earthquakes. Evidence of earthquake-induced sediment failure producing a slide on the gently sloping (0.25°) Klamath River delta off the northern California coast was recorded in 1980. On November 8, 1980, an earthquake of magnitude 6.5 on the Richter scale occurred 60 km (37 mi) off the coast.

During a December survey of the area, using high-resolution seismic reflection and deep-towed side-scan sonar, a slide zone 1 km (0.62 mi) wide and 20 km (12.4 mi) long was discovered. The failure seems to have occurred along the sediment boundary between fine to medium sand and a seaward deposit of mud (clayey silt), shown in figure 2–15.

Figure 2–16 shows seismic profiles of the slide area before and after the earthquake occurred. A continuous gentle slope was recorded in October 1979 prior to the quake. A slide feature characterized by flat terraces and 1 to 1.5 m (3.3 to 5 ft) high scarps at the seaward edge of the terraces appears on the profiles made after the quake.

Continental Slope

The continental slopes beyond the shelf break are features similar in relief to mountain ranges we find on the continents. The break at the top of the slope may be from 1 to 5 km (0.62 to 3 mi) above the deep-ocean basin at its base. In areas where the slope descends into submarine

FIGURE 2–15
Location of Klamath River Delta Failure Zone Resulting from November 8, 1980, Earthquake. The slide zone is 1 km (0.62 mi) wide and about 20 km (12.4 mi) long at an average depth of 60 m (197 ft). It seems to be associated with a sediment contact where nearshore sandy sediment changes seaward to mud. The location of the seismic profiles and side-scan images shown in figure 2–16 are indicated. (Courtesy of Michael Field, USGS.)

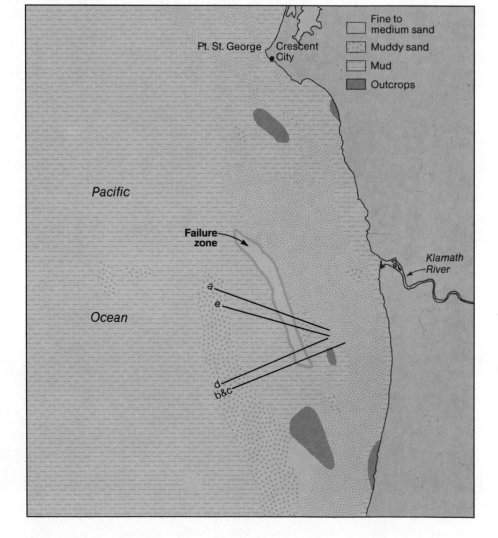

FIGURE 2–16
Comparison of Pre- and Post-earthquake High-resolution Seismic Reflection Records from Klamath River Delta. Vertical scale bars = 10 m (33 ft). Horizontal scale bars = 250 m (820 ft). Solid triangles mark approximate water depth of 60 m (197 ft). Profile locations are shown on figure 2–15. Lines *c* and *d* are uniboom records made with a strong sound source and show the subsurface structure, whereas lines *a*, *b*, and *e* show primarily the condition of the ocean floor. There is clearly no evidence of sediment failure on the pre-earthquake records (*d* and *e*), while the postearthquake records (*a*, *b*, and *c*) show scarps and terraces that have resulted from sediment movement. The unconformity labeled in *d* is an old erosional surface upon which younger sediment has been deposited. (Courtesy of Michael Field, USGS.)

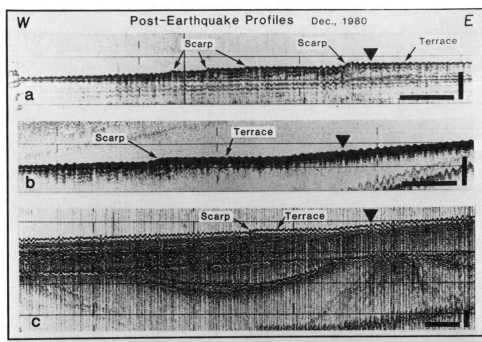

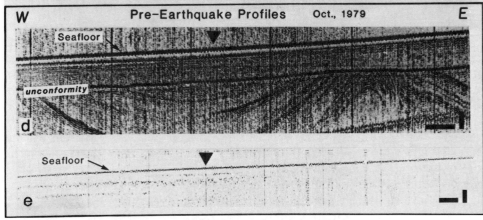

trenches, even greater vertical relief is measured. Off the west coast of South America the total relief from the top of the Andes Mountains to the bottom of the Peru-Chile Trench is about 15 km (9.3 mi).

Continental slopes vary in steepness from 1° to 25° and average about 4°. Around the margin of the Pacific Ocean, where the slope is associated with the processes forming coastal mountain ranges and submarine trenches, the continental slopes are steeper than in the Atlantic and Indian oceans. Slopes in the Pacific Ocean average more than 5°, while those in the Atlantic and Indian oceans are about 3°.

Submarine Canyons

The continental slope and, less commonly, the continental shelf are cut by large **submarine canyons** that resemble the largest of canyons cut on land by rivers. Like the canyons cut by rivers, the submarine canyons have tributaries and steep V-shaped walls that expose a wide range of rock types varying greatly in geologic age. Exposed in the walls of the canyons are rocks ranging from soft shales to quartzite and granite.

The canyons are obviously erosional features; the problem is to explain how the erosion occurred. If the first person to observe a submarine canyon had observed those off the west coast of the French island of Corsica in the Mediterranean, he or she would have concluded that, without doubt, submarine canyons were related to land river systems, since the canyons lead right into the mouths of the rivers that drain the western flank of the island. Sediments recovered during Deep Sea Drilling Project cruises in the Mediterranean Sea indicate this body of water very nearly dried up at least once in its history. Therefore, the submarine canyons off the island of Corsica may well have been created primarily or entirely by stream erosion. The majority of submarine canyons, however, do not tie so nicely with land drainage systems, and many are confined exclusively to the continental slope. The most obvious objection to explaining them as drowned river valleys is the fact that they continue to the base of the continental slope, which averages some 3500 m (11,500 ft) below sea level. Since rivers lose their ability to erode shortly after reaching the ocean, it seems impossible that rivers could have cut canyons this far below sea level.

Side-scan sonar surveys indicate that the continental slope is dominated by submarine canyon topography along the Atlantic coast from Hudson Canyon to Baltimore Canyon (figure 2–17A). Smooth intercanyon areas are particularly rare on the middle and upper surfaces of the slope. There is also a correlation between the spacing of canyons and the continental slope gradient. No canyons are present where the slope is less than 3°. Canyons are from 2 to 10 km (1.2 to 6 mi) apart where the slope is between 3° and 5°, and they are 1.5 to 4 km (0.9 to 2.5 mi) apart where the slope gradient exceeds 6°.

Canyons confined to the continental slope are straighter and have steeper canyon floor gradients than those that extend into the continental shelf. This condition suggests that the canyons are initiated on the continental slope by some marine process and cut their way headward into the continental shelf as they age.

Turbidity Currents

Probably the most widely supported theory concerning the origin of submarine canyons is that which explains the erosion by **turbidity currents,** flows of sediment-laden water that periodically move down the canyon. Turbidity flows have been observed resulting from river input near the heads of some Pacific-coast submarine canyons that extend well onto the continental shelf. However, turbidity currents in many canyons confined to the continental slope apparently are initiated after sediment moves across the continental shelf into the head of the canyon and accumulates there. The actual initiation of the current may be produced by an earth tremor or other disturbance that makes the mass collapse and start to move down the slope under the force of gravity. As it moves, the sediment mixes with the water, producing a high density mass that moves down the canyon, spreads out, and is deposited at the base of the continental slope as part of a deep-sea fan structure found at the mouth of submarine canyons. The flow moves out across the fans somewhat as a river flow moves across the surface of a delta. A pattern of distributaries develops. They are bordered by levees that are deposited as the turbid mass overflows the banks of the relatively shallow distributary channels (figure 2–17B).

The remote sonar recording of a turbidity current generated on August 26, 1976, by the discharge of mine tailings into Rupert Inlet, British Columbia (figure 2–18), may be the only real-time observation of a marine turbidity current moving through its channel. The sonograms show acoustic sounding profiles across the submarine channel in Rupert Inlet before and during the event that lasted for about one and one-half hours.

Another line of evidence that indicates that turbidity currents move across the ocean bottom is associated with trans-Atlantic cable breaks that occurred on the continental shelf and slope south of Newfoundland after an earthquake on November 18, 1929 (figure 2–19). These cables were broken in a pattern that indicated that the cables closest to the earthquake were broken first and those that crossed the shelf at greater depths and distances from the epicenter (point on ocean floor directly above earthquake) were broken later. This phenomenon could be explained by a dense flowing mass, moving away from the epicenter down the slope and snapping the cables as it passed. The velocity required in this particular case would be approximately 27 km/h (16.8 mi/h), which seems quite high. The possibility that such a mass may move with this velocity makes it easier to understand how the submarine canyons may have been formed on the continental slope and continental shelf.

Continental Rise

Deep-sea fans accumulate as deposits at the mouths of submarine canyons (figure 2–17B). The merging of these deep-sea fans along the base of the continental slope is partly responsible for the development of the **continental rise.**

The Amazon Cone is one of the largest deep-sea fans. It extends 700 km (434 mi) northeast of Brazil (figure 2–20). Examination of the fan's distributary system using GLORIA long-range side-scan sonar revealed the following features. The main channel, which extends seaward from the Amazon Submarine Canyon, splits into western and eastern systems that extend across the middle of the fan to a depth of about 4200 m (13,800 ft). Additional branching and development of meanders occur across the middle of the fan. The branching seems to result from channel flow breaching the levees, especially on the outside curves of meanders.

The onset of meandering of the distributary channels across the middle fan corresponds well to the decreased gradient of the middle fan surface as compared to that of the steeper upper fan. The meander patterns, as well as other features—abandoned meander loops, cutoffs, etc.—closely resemble in form and scale those found on the flood plains of mature river systems on land such as the lower Mississippi River.

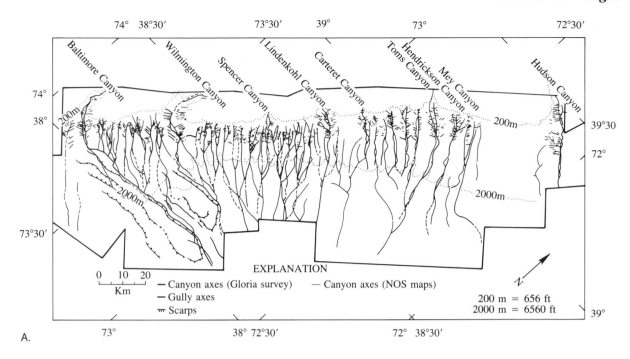

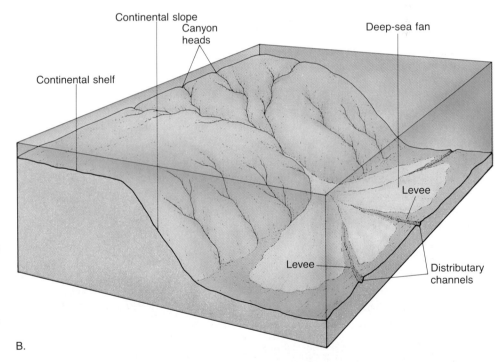

FIGURE 2–17
Submarine Canyons. *A,* submarine canyons along the Atlantic coast. There are no submarine canyons between Hudson Canyon and Mey Canyon where the slope gradient is less than 3°. The distance between adjacent submarine canyons decreases as the slope gradient increases. Between Lindenkohl Canyon and Baltimore Canyon, the gradient exceeds 6° for most of the continental slope. Where the slope is steepest, the submarine canyons are spaced as little as 1.5 km (0.9 mi) apart. *B,* erosional and depositional features associated with the formation of submarine canyons.

It is also clear that mass wasting in the form of slump-debris flows continues to be an important process in distributing sediment across the fan—especially on the steeper upper fan slope.

A major factor in shaping the continental rise is the strong **western boundary undercurrent (WBUC)** or slope current that flows toward the equator at the base of the continental slope along the western boundaries of most ocean basins. This deep boundary current and the

continental rise over which it flows have been extensively studied along the east coast of North America using submersibles, surface ships, and side-scan sonar.

The Coriolis effect (described in chapter 6) forces the WBUC, which originates in the Norwegian Sea and flows into the North Atlantic through deep troughs between Greenland and Scotland, to veer to the right and flow snugly against the base of the continental slope at all times. Picking up volcanic debris from Iceland and sedi-

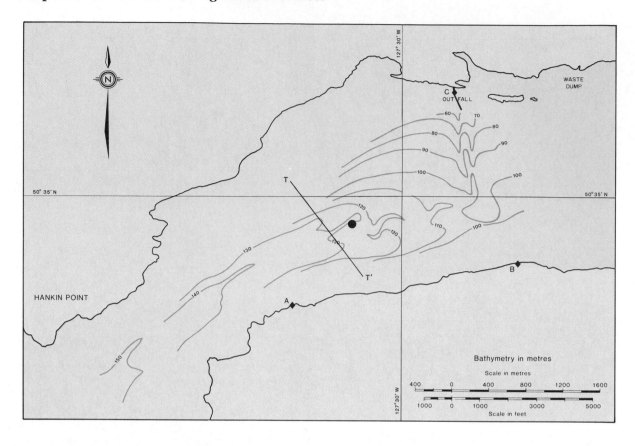

Sonograms

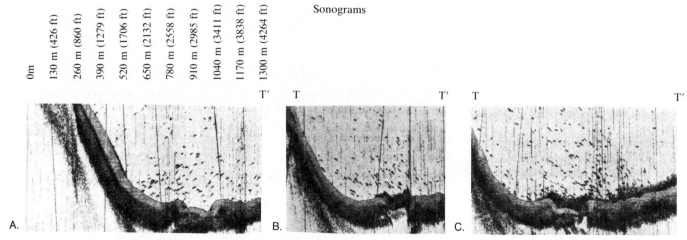

A.   B.   C.

FIGURE 2-18
Rupert Inlet, British Columbia. Top, bathymetry of submarine channel. Line T-T′ is path followed
by the Canadian Survey Ship *Vector* on August 26, 1976, when it recorded the occurrence of a turbidity current. The black circle denotes the approximate location of
a bottom-moored current meter, and the diamonds labeled A, B, and C are locations of
positioning transponders. Bottom, in sonogram A, the ship crossed the channel at about
12:53 P.M., and no turbidity current was flowing in the channel. Note the levees on either side of the
main channel. The discrete spots in the water column above the bottom are thought to be fish. In B,
the ship was over the channel at 1:03 P.M., and the turbidity current had begun to flow in the main
channel. By 1:28 P.M. in C, the turbidity current had overspilled the levees. The turbidity current was
not detectable 1.5 h after its onset. The steep wall is on the north side of the channel. The inclined
vertical lines are the result of transmitted pulses from another sounder. (Reprinted by permission from
Hay, A. E., et al., "Remote acoustic detection of turbidity current surge," *Science* 217: 4562, pp. 833–
835, 27 August 1982. Copyright 1982 by the AAAS. Photos courtesy of Alex E. Hay.)

FIGURE 2–19
Floor of the North Atlantic Ocean. Arrows indicate the path of North Atlantic Deep Water or western boundary undercurrent. Hexagon indicates cable breaks associated with the 1929 Grand Banks earthquake. (North Atlantic Panorama excerpted from *The world ocean floor* by Bruce C. Heezen and Marie Tharp. Copyright © 1977 by Marie Tharp. Reproduced by permission of Marie Tharp, 1 Washington Ave., South Nyack, NY 10960.)

FIGURE 2–20
The Amazon Cone. The Amazon Canyon branches into two major distributary systems, the Western Levee Complex and the Eastern Levee Complex. They are composed of meandering channels bounded by levees. Note the large debris flows originating on the steep upper slope of the fan cone. (Map courtesy of John E. Damuth and Roger D. Flood. Permission granted by the Geological Society of America.)

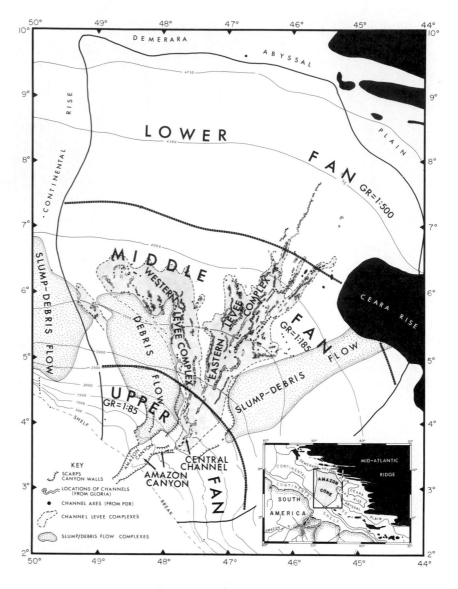

ment from periodic turbidity flows, and scouring sediment from some portions of the continental slope and rise, the western boundary undercurrent, flowing at velocities that reach 40 cm/s (1 mi/h), creates a **benthic nepheloid layer** (**BNL**) of suspended sediment above the ocean floor (figure 2–21). As the current is forced around the bends in the continental slope, its velocity decreases and the rate of deposition increases, producing the deposits called drifts or ridges shown in figure 2–19.

Increased knowledge of these continental-margin features and the processes that create them may be of importance in selecting ocean sites for disposal of toxic waste and for effective exploitation of oil, gas, and mineral resources.

DEEP-OCEAN BASIN

Moving seaward from the base of the continental margin, the ocean floor is underlain by oceanic crust with many

volcanic peaks extending to various elevations above the ocean floor. Those that extend more than 1 km (0.6 mi) above the deep-ocean floor are called **seamounts.** If they have flattened tops, they are called tablemounts or **guyots** after the Swiss scientist, A.H. Guyot. Guyots are less common than seamounts and are found in linear trends in the Pacific Ocean. The tops of the guyots in the Pacific Ocean are between 1800 and 3000 m (5900 and 9850 ft) below the present ocean surface. Since there is evidence to indicate that the flattened surface of the guyots was produced by wave action when they were exposed above the ocean surface, a significant amount of subsidence of the ocean basin in these regions must have occurred since the guyots were at the surface. An explanation of this subsidence is presented in chapter 3.

Volcanic features on the ocean bottom that are less than 1 km (0.6 mi) above the ocean floor are called **abyssal hills** (figure 2–22). They cover a large percentage of the entire ocean basin floor and have an average relief of about 200 m (650 ft). While many such features are found

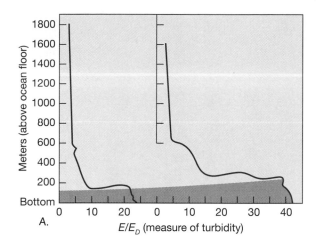

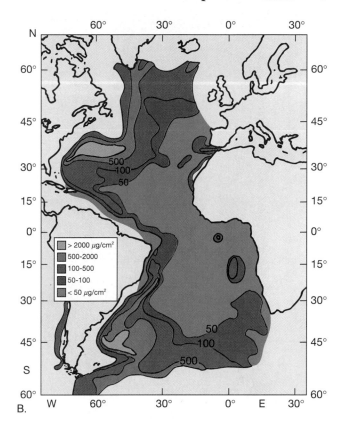

FIGURE 2–21
Benthic Nepheloid Layer. *A,* two light-scattering curves showing the nepheloid layer on the Blake-Bahama Outer Ridge. (See figure 2–19.) *B,* distribution of suspended material in the nepheloid layer in the Atlantic Ocean. Note the high concentrations found only in the western portions of the ocean overlying the continental rise. Compare the concentration pattern in the North Atlantic with the location of the western boundary undercurrent in figure 2–19. (After Biscaye and Eittreim, 1977.)

buried beneath the sediments of abyssal plains of the Atlantic and Indian oceans, the lower rate of sediment deposition in the Pacific Ocean has left extensive regions of ocean floor dominated by abyssal hills. These areas are called **abyssal hill provinces.** The evidence of volcanic activity on the bottom of the Pacific Ocean is particularly widespread—more than 20,000 volcanic peaks are found there. Some marine volcanic activity produces widespread gently sloping surfaces where large volumes of lava flowed out in broad sheets, solidified, and became **archipelagic aprons.** Such features commonly surround volcanic islands.

Abyssal Plains

Extending from the base of the continental rise into the deep-ocean basins, one often finds flat depositional surfaces with slopes of less than 1:1000 that cover extensive portions of the basins (figure 2–22). These flat **abyssal plains** are particularly extensive and flat in the Atlantic Ocean, although they are occasionally interrupted by volcanic peaks. Most abyssal plains lie at depths of between 4500 (14,800 ft) and 6000 m (19,700 ft).

The benthic nepheloid layer that is so well developed over the continental slope is also found (although not so well developed) throughout the deep ocean where bottom current speeds average at least 8 cm/s (approx. 0.15 mi/h). Such currents may play an important role in distributing fine sediment on the deep abyssal plains. Just north of the equator in the eastern Pacific,

average bottom current velocities measured 30 m (100 ft) above the ocean floor reached 18.5 cm/s (0.4 mi/h).

Trenches

Although the continental rise is a feature commonly found at the base of the continental slope where it meets the abyssal plain, the slope sometimes descends into long, narrow steep-sided **trenches.** The deepest portions of the world's oceans are found in these trenches. Such features are characteristic of the margins of the Pacific Ocean along the coast of South America, Central America, and the Aleutian Islands and are also characteristic of the western margin of the Pacific Ocean (figures 2–22 and 2–23). A depth of 11,022 m (36,150 ft) has been recorded in the Challenger Deep of the Mariana Trench. Table 2–2 presents some of the dimensional characteristics of trenches. The landward side of the trench rises as a volcanic arc that may produce islands (Japan) or a volcanic mountain range at the margin of a continent (the Andes).

Oceanic Ridges and Rises

Extending across some 65,000 km (40,400 mi) of the deep-ocean basin are mountainous features that are referred to as **oceanic ridges** if they have steep and irregular slopes, and **oceanic rises** if the slopes are more gentle. The best explored of these features are the Mid-Atlantic Ridge and the East Pacific Rise (figure 2–22). This

FIGURE 2–22
The World Ocean Floor. (By Bruce C. Heezen and Marie Tharp, 1977. Copyright © 1977 Marie Tharp. Reproduced by permission of Marie Tharp.)

Painted by Tanguy de Rémur.

American Geographical Society, Broadway at 56th St. New York, N.Y. 10032.

Aleutian Tr.

Emperor Seamounts

Kurile Tr.

Japan Tr.

Mendocino F. Z.

Philippine Tr.

Mariana Tr.

Hawaiian Chain

Clarion F. Z.

Mid-America Tr.

East

Clipperton F. Z.

Line Is.

Galápagos
Rift

Marquesa Is.

Pacific

Peru-Chile Tr.

Tuamotu Is.

Society Is.

Kermadec-Tonga Tr.

Cook Is.

Rise

Eltanin F. Z.

NORTH AMERICA

SOUTH AMERICA

AUSTRALIA

ANTARCTICA

BERING SEA

OKHOTSK SEA

YELLOW SEA

HUDSON BAY

BAFFIN BAY

120° 140° 160° 180° 160° 140° 120° 100° 80° 60°

70° 60° 40° 20° 0° 20° 40° 60° 70°

Editions Pierre Charron, 51 rue Pierre-Charron. 75008 Paris., Draeger, Imp.

Mercator Projection 1 : 48,000,000 at the Equator.
Depth and Elevations in Meters.

Based on Bathymetric studies by
Bruce C. Heezen and Marie Tharp
of the Lamont Doherty Geological Observatory
Columbia University Palisades, New York, 1964
SUPPORTED BY THE UNITED STATES NAVY
OFFICE OF NAVAL RESEARCH

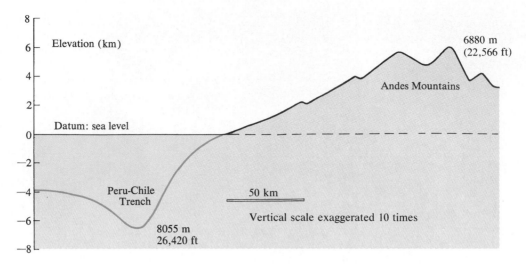

FIGURE 2–23
Profile across Peru-Chile Trench and Andes Mountains off the Coast of Chile. This profile is typical of the deep trenches common around the margin of the Pacific Ocean. Note that over a distance of 200 km (125 mi) a change in elevation of more than 14,900 m (48,870 ft) occurs.

system of mountains is entirely volcanic and composed of lavas with a basaltic composition characteristic of the oceanic crust. The Mid-Atlantic Ridge was a focal point during the development of the theory of global plate tecton-

TABLE 2–2
Dimensions of Trenches

Trench	Depth (m)	Mean width (km)	Length (km)
Peru-Chile	8,100	100	5,900
Aleutian	7,700	50	3,700
Middle America	6,700	40	2,800
Mariana	11,022	70	2,550
Kurile	10,500	120	2,200
Kermadec	10,000	40	1,500
Tonga	10,000	55	1,400
Philippine	10,500	60	1,400
Japan	8,400	100	800

ics because of its position, separating the Atlantic Ocean into equal halves. Recently, the East Pacific Rise has been studied because of the hydrothermal vents and associated biological communities located in its central rift valley.

Oceanic ridges and rises extend an average of 2.5 km (1.5 mi) above the abyssal plains or abyssal hill provinces on either side. Oceanic ridges practically surround Antarctica. A major ridge extending from the southern Indian Ocean to the Red Sea is the Carlsberg Ridge.

Fracture Zones

The oceanic ridges and rises are offset by rugged fault scars called **fracture zones.** (figure 2-22). These linear bands of mountains and troughs intersect all oceanic ridges at intervals at nearly right angles. In the Pacific Ocean, where the scars are less rapidly covered by sediment than in other ocean basins, fracture zones extend for thousands of kilometers and have widths of up to 200 km (120 mi).

SUMMARY

Our solar system, consisting of the sun and nine planets, belongs to the galaxy of stars we call the Milky Way. All the stars we see with the unaided eye belong to this galaxy; yet there are some 100 million galaxies thought to exist in the universe. Most seem to be moving away from us as though all were set in motion by a single large explosion.

The galaxies are thought to be accumulations of the debris from this explosion in large eddies, inside of which are smaller eddies where concentration of matter has produced stars and their associated planets. Stars result from accumula-

tion of masses large enough to produce sufficient internal temperatures and pressures to set off processes causing them to "shine" as they give off energy at high rates. The planets were smaller, so they did not reach this state.

Protoearth was composed mostly of hydrogen and helium, but as it condensed and heated up, these elements were driven off into space. There are indications Earth became molten and developed a layered structure that included a core, mantle, and a crust composed of a thin basaltic unit, which ultimately underlaid the oceans, and the thicker granite crust of the continents. During this period Earth also developed an atmosphere rich in water vapor and carbon dioxide produced by volcanic activity. Methane and ammonia were also present in significant quantities. As Earth's surface cooled sufficiently, the water vapor condensed and accumulated in depressions on the surface to give Earth its first oceans.

It is in these oceans life is thought to have begun. As ultraviolet radiation fell on the oceans with their dissolved carbon dioxide, methane, and ammonia, inorganic molecules may have combined to produce carbon-containing molecules that are now formed naturally on Earth only by organisms. Chance combinations of these molecules eventually produced heterotrophic organisms that were probably similar to present-day anaerobic bacteria. Eventually, chemosynthetic autotrophs evolved, and later some cells included chlorophyll in their makeup. Plants were the result. The photosynthesis of plants extracted carbon dioxide from the atmosphere and produced the first free oxygen at Earth's surface. This produced an oxygen-rich atmosphere in which animals as we know them could survive. Eventually, both plants and animals evolved into forms that could survive in the stark environment of the continents, producing the lush continental environments we enjoy today. Seventy-one percent of Earth is covered by oceans. The continents stand high and bound the oceans because they are composed of granite, which is less dense than the basaltic crust that underlies the oceans. Both of these crustal units "float" on the underlying mantle according to the principles of a concept of crustal flotation called isostasy.

Much of the knowledge we have gained about the nature of the ocean floor has been obtained using seismic surveys and side-scan sonar. Seismic surveys can reveal the topography of the ocean floor and the location of rock unit boundaries beneath it. Side-scan sonar is used to make strip maps of ocean floor topography along the course of the survey ship.

Extending from the shoreline of continents are gently sloping continental shelves that extend out to sea for an average distance of 70 km (43 mi). Sediment is transported across the continental shelves by currents and mass wasting (submarine sediment slides). When the shelves reach an average depth of 135 m they steepen into a feature called the continental slope. The slopes continue to the deep-ocean floor with steepness ranging from 1° to 25° and averaging 4°. Cutting deep into the slopes and reaching well onto the continental shelves, in some cases, are submarine canyons. The origin of the canyons is unknown, but they may be caused in part by periodic flows that are a mixture of sediment and water called turbidity currents. Turbidity currents deposit their sediment load at the base of the continental slope to produce deep-sea fans or cones through a series of meandering distributary channels, which branch out from the mouth of the submarine canyon. The merging of these fans produces a gently sloping continental rise at the base of some continental slopes. A major factor in distributing sediment along the continental slope is the western boundary undercurrent (WBUC) or slope current that flows toward the equator along the western boundary of ocean basins. The sediment suspended and carried above the bottom by these currents creates a benthic nepheloid layer. When the current slows to make turns along the base of the continental slope, it deposits sediment onto features called drifts or ridges that become prominent features of the continental rise.

The continental rises gradually become flat, extensive, depositional features of the deep-ocean basin called abyssal plains. These plains are often penetrated by volcanic peaks. In the Pacific Ocean where sedimentation rates are low, abyssal plains are not extensively developed, and abyssal hill provinces cover broad expanses of ocean floor. In some places, particularly around the margin of the Pacific Ocean, the continental slope does not merge into a continental rise but continues down into deep linear trenches. Trenches are usually bounded on their landward side by volcanic island arcs or continental volcanic mountain ranges if the trench lies near the continent. Oceanic ridges and rises are volcanic mountain ranges running through the deep-ocean basins and rising about 2.5 km (1.5 mi) above the abyssal plains on either side. The volcanic peaks protruding from the abyssal plains are called seamounts if they extend over 1 km (0.6 mi) above the deep-ocean floor and abyssal hills if they are smaller. Some peaks once stood above sea level and had their tops flattened by wave action. They are called guyots or tablemounts. Rugged fault scars called fracture zones cut across vast distances of ocean floor and offset the axes of oceanic ridges and rises.

KEY TERMS

Abyssal hill (p. 44)
Abyssal hill province (p. 45)
Abyssal plain (p. 45)
Archipelagic apron (p. 45)
Autotroph (p. 32)
Bathymetry (p. 33)
Benthic nepheloid layer (BNL) (p. 44)
Buoyancy (p. 34)
Carbohydrates (p. 31)
Chemosynthesis (p. 32)
Chlorophyll (p. 32)
Continental rise (p. 40)
Continental shelf (p. 35)

Continental slope (p. 38)
Core (p. 27)
Crust (p. 27)
Endothermic reaction (p. 32)
Excess volatiles (p. 29)
Exothermic reaction (p. 32)
Fracture zone (p. 48)
Galaxy (p. 25)
Guyot (p. 44)
Heterotroph (p. 31)
Hydrocarbon (p. 31)
Isostasy (p. 34)
Light-year (p. 25)

Lipids (p. 31)
Magma (p. 30)
Mantle (p. 27)
Mohorovičič discontinuity (Moho) (p. 27)
Oceanic ridge (p. 45)
Oceanic rise (p. 45)
Photosynthesis (p. 30)
Protein (p. 31)
Respiration (p. 32)
Seamount (p. 44)
Seismic surveying (p. 35)
Side-scan sonar (p. 35)

Solar system (p. 23)
Submarine canyon (p. 39)
Trench (p. 45)

Turbidity current (p. 40)
Valence (p. 31)

Western boundary undercurrent
(WBUC) (p. 41)

QUESTIONS AND EXERCISES

1. Why is it theorized that the observed motions of galaxies were initiated by an explosion?
2. In what three components of Earth's makeup must we find the total mass of crystalline rock components that were eroded from the primary crystalline rocks of Earth's surface?
3. List the five most abundant excess volatiles. Why are they called excess volatiles?
4. Why does the fact that new water is continually being released to the atmosphere by volcanic activity not necessarily mean the oceans will progressively cover an increasing percentage of Earth's surface?
5. Why do we believe that free oxygen (O_2) was not released into the atmosphere by volcanic activity?
6. How does the presence of oxygen (O_2) in our atmosphere help reduce the amount of ultraviolet radiation that reaches Earth's surface?
7. How did the photosynthesis that produced the first organic molecules in the early oceans differ from plant photosynthesis? How does chemosynthesis differ from photosynthesis?
8. When we say carbon (C) has a greater combining power than oxygen (O), we are referring to their chemical valence. How is valence related to combining power?
9. Describe some basic characteristics of living things.
10. Discuss photosynthesis and respiration and explain their relationship to the chemical processes of endothermic and exothermic reactions.
11. As plants evolved on Earth, great changes in Earth's environment were produced. Describe some of these major changes caused by plants.
12. Compare the concept of isostasy with that of buoyancy, which explains the flotation of ice in water.
13. Describe the major features of the continental margin: continental shelf, continental slope, continental rise, submarine canyon, and deep-sea fans or cones.
14. Why is the suspended sediment layer best developed along the western margin of the ocean basins?
15. In which ocean basin are most ocean trenches found?
16. Describe the following: seamount, abyssal hill, guyot, oceanic ridge, fracture zone.

REFERENCES

Anderson, R. N. 1986. *Marine-geology: A planet earth perspective.* New York: John Wiley.

Biscaye, P. E., and Eittreim, S. L. 1977. Suspended particulate loads and transports in the nepheloid layers of the abyssal Atlantic Ocean. *Marine Geology* 23:155–72.

Damuth, J. E.; Kolla, V.; Flood, R. D.; Kowsmann, R. O.; Monteiro, M. C.; Gorini, M. A; Palma, J. J.; and Belderson, R. H. 1983. Distributary channel meandering and bifurcation patterns on the Amazon deep-sea fan as revealed by long-range side-scan sonar (GLORIA). *Geology* 11:2, 94–98.

Field, M. E.; Gardner, J. V.; Jennings, A. E.; and Edwards, D. E. 1982. Earthquake-induced sediment failures on a 0.25° slope, Klamath River delta, California. *Geology* 10:10, 542–45.

Glaessner, M. F. 1984. *The dawn of animal life: A biohistorical study.* Cambridge: Cambridge University Press.

Gregor, B. C.; Garrels, R. M.; Mackenzie, F. T.; and Maynard, J. B., eds. 1988. *Chemical cycles in the evolution of the earth.* New York: John Wiley.

Hay, A. E.; Burling, R. W.; and Murray, J. W. 1982. Remote acoustic detection of a turbidity current surge. *Science* 217:4562, 833–35.

Holland, J. D. 1984. *The chemical evolution of the atmosphere and oceans.* Princeton, N.J.: Princeton University Press.

Rubey, W. W. 1951. Geologic history of seawater: An attempt to state the problem. *Geologic Society of America Bulletin.* 62:1110–19.

Strom, K. M., and Strom, S. E. 1982. Galactic evolution: A survey of recent progress. *Science* 216:4546, 571–80.

Sverdrup, H. U.; Johnson, M. W.; and Fleming, R. H. 1942. Reprinted 1970. *The oceans: Their physics, chemistry, and general biology.* Englewood Cliffs, N.J.: Prentice-Hall.

Twichell, D. C., and Roberts, D. G. 1982. Morphology, distribution and development of submarine canyons on the United States Atlantic continental slope between Hudson and Baltimore canyons. *Geology* 10:8, 408–12.

Weirich, F. H. 1984. Turbidity currents: Monitoring their occurrence and movement with a three-dimensional sensor network. *Science* 224:4647, 384–87.

SUGGESTED READING

Sea Frontiers

Feazel, C. T. 1986. Asteroid impacts, sea-floor sediments, and extinction of the dinosaurs. 32:3, 169–78. A discussion of the evidence in marine sediments that may support the theory that the demise of the dinosaurs was caused by the impact of a large meteor.

Mark, K. 1976. Coral reefs, seamounts, and guyots. 22:3, 143–49. A discussion of the role of global plate tectonics in explaining the distribution of seamounts, guyots, and the evolution of coral reefs.

Rice, A. L. 1991. Finding bottom. 37:2, 28–33. Depth-sounding devices developed by ingenious early navigators are described.

Schafer, C., and Carter, L. 1986. Ocean-bottom mapping in the 1980s. 32:2, 122–30. The use of SeaMARK (Seafloor Mapping and Remote Characterization), a side-scan sonar device, in mapping the continental margin off the coast of Labrador is described.

Scientific American

Badash, L. 1989. The age-of-the-Earth debate. 261:2, 90–97. A history of the development of knowledge concerning Earth's age.

Barrow, J. D., and Silk, J. 1980. The structure of the early universe. 242:4, 118–28. Although the universe is inhomogeneous on the small scale of the solar system or a galaxy, it is very homogeneous on the scale of the universe as a whole.

Emery, K. O. 1969. The continental shelves. 221:3, 106–25. The nature of the continental shelves and the effect of the advance and retreat of the shoreline across them as a result of glaciation are discussed.

Frieden, E. 1972. Chemical elements of life. 227:1, 52–64. The roles of the 24 elements known to be essential to life are discussed, including some background on how they may have been selected from the physical environment in which life evolved.

Gott, J. R., and Gunn, J. E. 1976. Will the universe expand forever? 234:3, 62–79. Based on data regarding the recession of galaxies, average density of matter, and chemical elements, the authors suggest expansion will be reversed.

Heezen B. C. 1956. The origin of submarine canyons. 195:2, 36–41. Theories explaining the origin of submarine canyons are presented along with data on the location and nature of such features.

Herbert, S. 1986. Darwin as a geologist. 254:5, 116–23. Before publication of *The Origin of Species,* Charles Darwin considered himself to be primarily a geologist. This article outlines his contributions in this field, with special attention being given to his theory of coral reef formation.

Horgan, J. 1991. In the beginning. 264:2, 116–25. An overview of data relating to the validity of the big bang theory of the origin of the universe.

Kasting, J. F.; Toon, O. B.; and Pollack, J. B. 1988. How climate evolved on the terrestrial planets. 258:2, 90–97. A discussion of the possible sequence of events that culminated with the atmospheres that now exist on Mercury, Venus, Earth, and Mars.

McMenamin, M. A. S. 1987. The emergence of animals. 256:4, 94–103. A discussion of how the explosive diversification of animal forms 570 million years ago may have been related to the breakup of a single large continental landmass.

Menard, H. W. 1969. The deep-ocean floor. 221:3, 126–45. A summary of the dynamic effects of sea-floor spreading is presented with a description of related sea-floor features.

Stebbins, G. L., and Ayala, F. J. 1985. The evolution of Darwinism. 253:1, 72–85. New advances in molecular biology and new interpretations of the fossil record add to the knowledge of evolution.

Vidal, G. 1984. The oldest eukaryotic cells. 250:2, 48–57. Cells with a nucleus appear to have evolved as marine plankters 1.4 billion years ago.

Wilson, A. C. 1985. The molecular basis of evolution. 253:4, 164–175. Mutations within the genes of organisms play an important role in evolution at the organismal level.

Chapter 3

Global Plate Tectonics

The fact that we live on a dynamic Earth on which movement is the rule rather than the exception has long been accepted by geologists. Geologic processes responsible for the ever-changing landscape of the continents are believed to require long periods of geologic time to build such features as mountains. However, many geologists were not prepared until recent years to accept as a process associated with the building of mountain ranges a dynamic characteristic of Earth that was of a much broader scope—the horizontal movement of continents. When geologists speak of the movement of continental masses across Earth's surface, they sometimes refer to this phenomenon as **continental drift.**

Marine geologists who have studied the ocean floor and developed theories concerning why the continents are moving relative to one another may use the term **sea-floor spreading.** The name for this theory is derived from evidence indicating that new oceanic crust and rigid upper mantle material is being formed along the axes of a series of oceanic ridges and rises. This newly formed material then moves at right angles

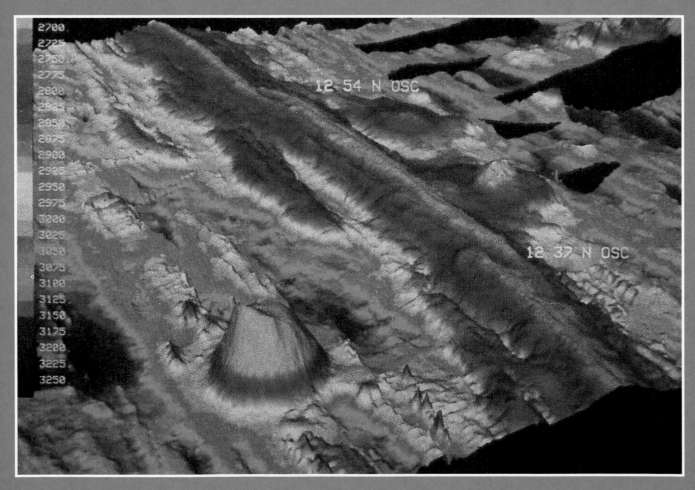

False-color map of the East Pacific Rise between 12°N and 13°N. Black represents areas of no data. Map constructed by S. P. Miller using seismic tomographic data from K. C. Macdonald and P. J. Fox. (Photo Courtesy K. C. Macdonald, University of California, Santa Barbara.)

away from the ridge and rise axes. The continents float on this denser material and are carried along by the moving layer of denser rock.

A third term used to encompass the totality of the process is **global plate tectonics.** Study of the process has indicated that there are a number of major plates into which this rock sphere, or **lithosphere,** can be divided. The interaction of these plates as they move builds the structural features of Earth's crust that can be observed by geologists. Thus, **tectonics** refers to the building of Earth's crustal structure and is derived from the Greek *tektonikos,* which means "to construct."

The possibility that the continents may be moving across Earth's surface is not new. Such a possibility was suggested early in the 19th century, and the first theory attempting to explain the movement was presented in 1912. With such a long history of awareness of the possibility of such movement, why has the acceptance of theories related to the process been so long in coming?

DEVELOPMENT OF THE THEORY

Alfred Wegener (figure 3–1) is considered by most scientists to be the pioneer of the modern continental drift theory. This German scientist originally was drawn to the concept in an attempt to explain the ancient climates that were recorded in the rocks deposited in ocean basins and on landmasses of the past. Wegener's theory was published in 1912 and met a great deal of resistance from the scientific community.

Wegener postulated that about 200 million years ago all the continental mass of Earth was one large continent, **Pangaea.** About 180 million years ago, the continent began to break up, and the various continental masses we know today started to drift toward their present positions. Since it was impossible at that time to present evidence that would support a mechanism for such transportation, the theory did not receive wide acceptance.

Although many Southern Hemisphere geologists had accepted continental drift as a geologic reality, it was not until the 1950s that geologists of the Northern Hemisphere began to give it serious attention. The impetus for the renewed attention arose from the study of Earth's ancient magnetism. The British geophysicist S. Keith Runcorn explained his observations of the magnetic properties of the rocks of Europe and North America in terms of continental movements. The details of these considerations will be discussed later in this chapter.

On the basis of this study of Earth's ancient magnetic fields, convincing arguments could be made for the fact that the continents had drifted relative to one another. As study continued, more evidence was gathered to support this movement, and additional data suggested the mechanism by which the movement might have taken place. In the next section, we will outline in some detail the early observations that led to the initial interest in the theory of continental drift and those more recent findings that led to the development of the theory of global plate tectonics.

FIGURE 3–1
Alfred Wegener. Shown waiting out the 1912–13 Arctic winter during an expedition to Greenland, where he made a 1200-km traverse across the widest part of the island's ice cap. (Courtesy of Bildarchiv Preussischer Kulturbesitz, West Berlin.)

Continental Jigsaw

Before we look at the bulk of scientific evidence that has been gathered in relation to plate tectonics, we should consider how well the continents do fit together. Attempts were made by many of the early investigators to arrange the continents in a manner that would achieve a reasonable fit and support their data. Most of these attempts used the existing shoreline as the margin of the continent. Based on our previous discussion of the geomorphic provinces of the ocean basin, we considered the continental shelf and continental slope to be part of the continental mass. Studies of the magnetic intensity of Earth's crust support such a contention.

There is a distinct increase in the magnetic intensity of oceanic crustal rocks compared to that of the continental crustal rocks. The increased intensity of the magnetism of the rocks is caused by the higher iron content of oceanic crustal rocks.

Sir Edward Bullard, an English geophysicist, constructed a computer fit of all the continents in 1965. The arrangement that produced minimum overlaps and gaps was found to be formed by continents outlined by the 2000-m (6560-ft) depth contour. This contour represents a depth that is approximately halfway down the continental slope (figure 3–2).

CONTINENTAL GEOLOGY

To test the fit of the continents, we may compare the rocks along the margins of the two continents that are thought to have been united. We need to identify rocks of the same type and age on the continents along their common margin. Identification is not easy in some areas, since during the millions of years since their separation, younger rocks may have been deposited and covered those rocks that might hold the key to the past history of the continents. However, there are many areas where such rocks are available for observation, and in these areas we can compare the ages by the use of (1) **fossils,** the remains of ancient organisms preserved in rocks, and (2) **radioactive dating** of the rocks.

The Fossil Record

The use of fossils became an important dating device in the early 19th century when it was realized that a particular layer of sedimentary rock would contain the remains of organisms that, as a group, were unique in time. Rocks laid down earlier and those laid down later would have a distinctly different assemblage of plant and animal fossils. Once these assemblages are recognized, a geologist can tell if rocks exposed in one area are younger or older than in another. The unique character of these assemblages is the result of evolution throughout geologic

FIGURE 3–2
Computer Fit of Continents by Sir Edward Bullard. Sir Edward Bullard's fit of the continents was attempted in 1965 after a convincing fit pattern had been achieved in 1958 by the Australian geologist S. Warren Carey without the aid of computers. Bullard's fit was in complete agreement with that of Carey. (From *Continental drift* by Don and Maureen Tarling, © 1971 by G. Bell & Sons, Ltd. Reprinted by permission of Doubleday & Co., Inc.)

time. Appendix III shows the general pattern of this evolution and the geologic time scale.

Dating by such a method is referred to as **relative dating.** One can tell only whether or not an assemblage is younger or older than another, but not its actual age in years. Such assemblages have been found abundantly in sedimentary rocks of the last 600 million years. Note that this is but a little over one-eighth of the time represented by Earth's existence, which is thought to exceed 4.6 billion years. Therefore, this method will not be effective in comparing the ages of the continental margins composed of rocks in excess of 600 million years of age.

Radioactive Dating

Most of the rocks we find on the continents contain small amounts of radioactive elements such as uranium, thorium, and potassium, which break down into atoms of other elements. Each radioactive **isotope** (atom of an element with an atomic weight different that that of other atoms of the element) has a specific **half-life,** the time it takes for one-half of the atoms present in a sample to decay to atoms of some other element. By comparing the quantities of the radioactive isotope with the quantities of their decay products in given continental rocks, the age of the rocks may be determined within 2 to 3 percent of their actual age. Although in very old rocks this error can be quite significant, this method of age determination is a very powerful tool and represents the first possibility that we have had of giving the actual age of rocks in years. We refer to such dating as **asbolute dating.** As compared to the relative dating possible with the use of fossils, absolute dating tells us not only which rocks are younger or older, but how much younger or older (figure 3–3).

Ancient Life and Climates

The fossil record of plants and animals contained in the sedimentary rocks can tell us much about the environments of the past. For instance, it is relatively simple to determine whether an organism lived in the ocean or on land. This can normally be told also by the characteristics of the sedimentary rock itself. However, finer distinctions in environmental conditions may also be determined.

Upon examining the distribution of some fossil assemblages and other characteristics of rocks that give clues to the climate under which they were formed, we find some that could not have formed under the climatic conditions present in the areas where we find them today. These anomalous locations of fossil assemblages and rock types may be explained by assuming that either (1) the factors that control climatic belts today are different from the factors that controlled climates in the past, or (2) the factors controlling climate throughout geologic time have not changed. The second assumption, which seems more logical, leaves us with the conclusion that these assemblages and rock types in the sediments of the past must have been carried to their present position by the movement of the continental masses.

The dominant factor controlling climatic distribution is latitudinal position on the rotating Earth. Assuming that Earth's axis of rotation has not changed significantly throughout its history, we may conclude that given latitudinal belts have possessed climatic characteristics that have not changed greatly during the evolution of life on Earth.

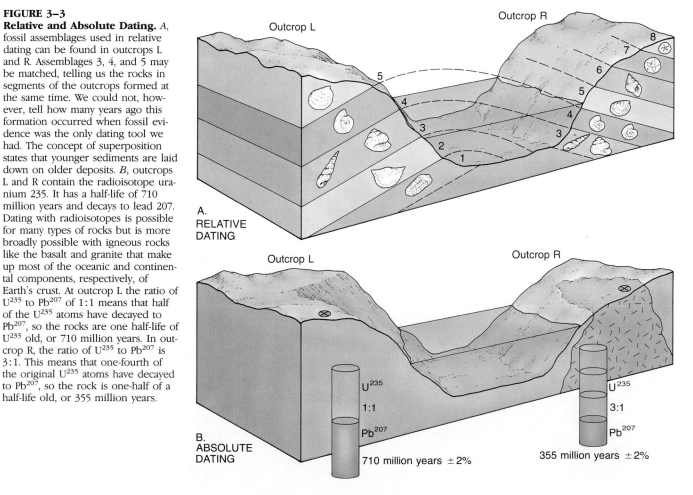

FIGURE 3–3
Relative and Absolute Dating. *A,* fossil assemblages used in relative dating can be found in outcrops L and R. Assemblages 3, 4, and 5 may be matched, telling us the rocks in segments of the outcrops formed at the same time. We could not, however, tell how many years ago this formation occurred when fossil evidence was the only dating tool we had. The concept of superposition states that younger sediments are laid down on older deposits. *B,* outcrops L and R contain the radioisotope uranium 235. It has a half-life of 710 million years and decays to lead 207. Dating with radioisotopes is possible for many types of rocks but is more broadly possible with igneous rocks like the basalt and granite that make up most of the oceanic and continental components, respectively, of Earth's crust. At outcrop L the ratio of U^{235} to Pb^{207} of 1:1 means that half of the U^{235} atoms have decayed to Pb^{207}, so the rocks are one half-life of U^{235} old, or 710 million years. In outcrop R, the ratio of U^{235} to Pb^{207} is 3:1. This means that one-fourth of the original U^{235} atoms have decayed to Pb^{207}, so the rock is one-half of a half-life old, or 355 million years.

Laurasia and Gondwanaland

Modern reef-building corals are known to exist in an environment where the water is clear and shallow and its temperature does not fall below 18°C. Although we cannot be sure such conditions have been required throughout the long history of coral evolution, we may assume that the conditions required were similar.

Exactly the same species of coral are found in rocks 350 million years old in western Europe and eastern North America, as well as throughout the Alps and Himalayas. Other fossil evidence indicates that throughout much of the 600-million-year fossil record, a major ocean existed that separated a large continent composed of North America, Europe, and Asia to the north from a large continent composed of South America, Africa, India, Australia, and Antarctica to the south. This ocean mass has been given the name **Tethys Sea,** and the supercontinents were **Laurasia** to the north and **Gondwanaland** to the south. The record also indicates that the two supercontinents periodically came into direct or close contact across the Tethys Sea. It appears that both continents together formed one large landmass, **Pangaea** (one earth), about 200 million years ago. Pangaea was surrounded by the ocean **Panthalassa** (one sea) (figure 3–4).

Fossils from sediments that were laid down within a land environment also support the existence of the Tethys Sea. From 350 to 285 million years ago there existed two distinct floral assemblages on the two supercontinents. The Laurasian floral assemblage included many species of tropical plants that were incorporated into the sediment to form the extensive coal beds mined throughout the eastern United States and Europe. Throughout Gondwanaland existed an assemblage represented by a few species of plants that were thought to have grown in a cold climate. Supporting this belief are indications of glaciation in South America, Africa, India, and Australia, which at that time must have been very near the southern polar region (figure 3–4A and B).

Continental Magnetism

Although we cannot reconstruct the relative position of continents prior to 200 million years ago with great accuracy, there is enough evidence from that time to the present to give us some idea of the paths followed by the continents after the breakup of Pangaea. The first clues to such movements came from the study of the magnetism of continental rocks.

All igneous rocks contain some particles of magnetite that are natural magnets and align themselves with Earth's magnetic field at the time of the rocks' formation. Volcanic lavas such as basalt are high in magnetite content and solidify from molten material that possesses temperatures in excess of 1000°C (1832°F). As they cool below 600°C (1112°F), the magnetite particles become oriented in the direction of Earth's magnetic field, permanently recording that field relative to the rock location. If Earth's magnetic field changes subsequent to the formation of the igneous rock, the alignment of these particles will not be affected.

Magnetite is also deposited in sediments. While the deposit is in the form of a sediment surrounded by water, magnetite particles have an opportunity to align themselves once more with Earth's magnetic field. This alignment is preserved when the sediment is buried.

Although a number of rock types may be used for the study of Earth's paleomagnetism, the basaltic lavas and other igneous rocks high in magnetite content are best for such studies. The magnetite particles act as small compass needles, as shown in figure 3–5. They not only point in a north-south direction but also point into Earth at an angle relative to Earth's surface called the **magnetic dip,** or inclination, which is related to latitude. At the equator, the "needle" will not dip at all, but will lie horizontally. It will point straight into Earth at the magnetic north pole and straight out of Earth at the magnetic south pole. At points between the equator and the pole the angle of dip increases with increased latitude. It is this

FIGURE 3–4 (opposite)
Fossil and Glacial Evidence Supporting Continental Drift and the Existence of Pangaea. *A,* The present distribution of continents contains three fossil assemblages that help in the reconstruction of past continental distributions. Assemblage A is a tropical assemblage found on Laurasia between 280 and 350 million years ago. This assemblage helps support the idea that North America and Europe were combined at that time to form Laurasia. Assemblage B is of a cold climate flora found in association with evidence of glaciation. This assemblage indicates a large Southern Hemisphere continent, Gondwanaland, which contained present-day South America, Antarctica, Africa, India, and Australia, was in existence between 280 and 350 million years ago. Assemblage C is of fossil corals that are believed to have grown in the shallow Tethys Sea that separated the two large continents. *B,* Laurasia and Gondwanaland may have been in the relative positions shown here during the interval from 250 to 350 million years ago. *C,* By 200 million years ago Laurasia and Gondwanaland combined to produce the single large continent, Pangaea. The ocean that covered the rest of Earth, Panthalassa, may be considered the ancestral Pacific Ocean, which has been decreasing in size since the breakup of Pangaea about 180 million years ago.

GLACIATION
250 – 300 million years ago
Arrows show direction of
glacial movement.

 POLAR GONDWANALAND FLORA
280–350 million years ago

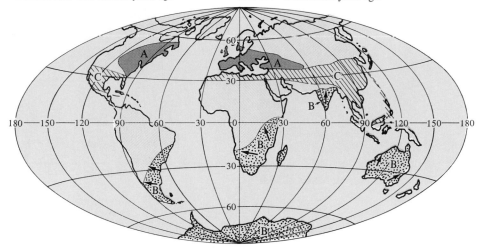 TROPICAL LAURASIAN FLORA
280–350 million years ago

TETHYS CORAL ASSEMBLAGE
350 million years ago

A.

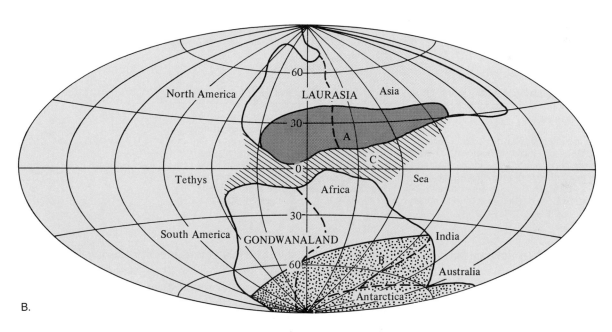

B.

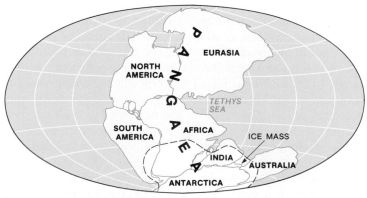

C.

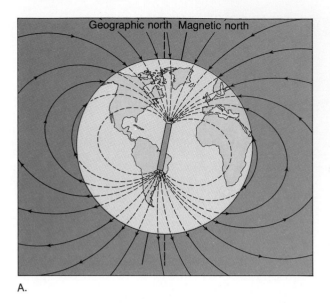

A.

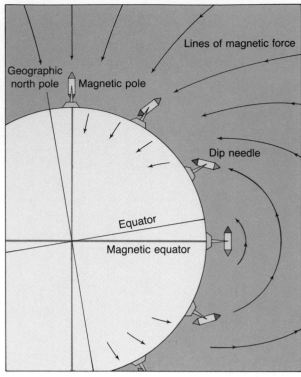

B.

FIGURE 3–5
Nature of Earth's Magnetic Field. *A*, Earth's magnetic field consists of lines of force much like those a giant bar magnet would produce if placed inside Earth. Although Earth's magnetic poles seldom coincide with the geographic poles, over long periods of time the average position of the magnetic poles can be considered to represent the position of the geographic poles. *B*, magnetite particles in newly formed rocks are aligned with the lines of force of Earth's magnetic field. This causes them to dip into Earth. The angle of dip increases uniformly from 0° at the magnetic equator to 90° at the magnetic poles. When the rocks solidify, the magnetite particles are frozen into position and become fossil compass needles that tell today's investigators about the strength and alignment of Earth's magnetic field when the rocks of which they are a part formed. The angle at which they dip into Earth also tells at what latitude they formed. (Reprinted by permission from Tarbuck, E. J., and Lutgens, F. K., 1990, figs. 18.8 and 18.9.)

dip that is retained in magnetically polarized rocks. By measuring the dip angle, we can estimate the latitude at which the rock formed.

Apparent Polar Wandering. Because the present magnetic poles do not coincide with geographical poles related to latitude, it might be felt that determining latitude by magnetic dip would give some incorrect determinations. However, for the last few thousand years the average positions of the magnetic poles have coincided with those of the geographic poles. If we can assume that this is true for the past, we can determine average positions of the magnetic poles during a time interval and consider them to represent the geographic poles.

As these average positions for the magnetic poles are determined for rocks on the continents, it is found that their positions changed with time. It appears the magnetic poles were wandering. The wandering curve of the pole for North America shows an interesting relation-

ship with that determined for Europe. Both curves have a similar shape, but for all rocks older than 70 million years the pole determined from North American rocks lies to the west of that determined by the study of European rocks. This difference implies that North America and Europe have changed position relative to the pole and relative to each other (figure 3–6).

If there can be only one north magnetic pole at any given time and its position must be at or near the north geographic pole, a problem exists that can be solved only by moving the continents. The two wandering curves can be made to coincide by moving the two continents together as we go back in time. Now we have a single wandering curve, which shows the magnetic north pole to be much too far south during the interval of time from 200 million to 300 million years ago. Rotating the merged continents can bring the pole into the proper position. The fact that moving the continents was the only solution to this problem was very strong evidence in support of the movement of continents throughout geologic time.

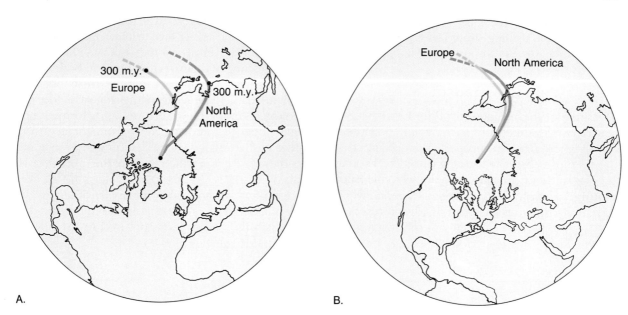

FIGURE 3–6
Apparent Wandering Curves of North Magnetic Pole Determined from Rocks of North America and Europe. *A,* apparent wandering curves of the north magnetic pole as determined from North America and Europe, going back in time from the present to 300 million years ago, follow divergent paths. *B,* if one moves the apparent position of the pole as determined from North America to a position that makes it coincide with the pole as determined from Europe, the direction and distance equal that required to close the Atlantic Ocean between the two continents. This still leaves the apparent north magnetic pole out in the North Pacific Ocean about 40° south of the geographic north pole. Again, movement of the continent is required to solve this problem. Rotation of Laurasia, which is formed by closing the Atlantic Ocean and joining Europe and North America, can bring the apparently wayward north magnetic pole to a position coincident with the north geographic pole. (Reprinted by permission from Tarbuck, E. J., and Lutgens, F. K., 1990, fig. 18.10.)

Estimating the latitude of formation for rocks from many areas throughout the continents on the basis of paleoclimatic and paleomagnetic evidence produces similar results. The most logical explanation for the changes that have occurred throughout the past in both climate and magnetic dip of the rocks at a given location is that the continents have moved.

Magnetic Polarity Reversals. Not only have the magnetic poles as determined from continental rocks seemed to wander throughout geologic time, but also the polarity, the direction of the magnetic field, seems to have reversed itself periodically. Reversals in polarity can be described in the following way. A compass needle that would point to the magnetic north today would, during a period of reversal, point south. It is not known why these reversals occur, but for the last 76 million years they have occurred once or twice each million years. The period of time during which a change in polarity occurs lasts for a few thousand years. It is identified on the basis of magnetic properties of rocks by a gradual decrease in the intensity of the magnetic field of one polarity until it disappears, followed by the gradual increase in the intensity of the magnetic field with the opposite polarity. The time during which a particular paleomagnetic condition existed can be determined by the radiometric dating of the

igneous rock from which the paleomagnetic measurements were taken.

Earth's present magnetic field has been weakening for the last 150 years, and some investigators believe that the present polarity will disappear in another 2000 years.

MARINE GEOLOGY

We have so far considered only data obtained from the continents in considering the theory of moving continents. This continental evidence did not convince many geologists that the continents had moved. For many years no other data were available since extensive sampling of the deep-ocean bottom did not become technologically feasible until the 1960s.

The Deep Sea Drilling Project, which began in 1968, and the **Ocean Drilling Program,** which followed it in 1983, have added greatly to our knowledge of the ocean floor. The impetus for such programs was provided in great part by the need for observations of oceanic sediment and crust to check the theory of sea-floor spreading. Extensive geophysical studies, including those related to the paleomagnetic characteristics of the crust, were the first to be carried out. The results of these studies gave the planners of the Deep Sea Drilling Project clues as to where the drilling should be concentrated in order to gain the greatest amount of new knowledge.

Paleomagnetism

A detailed study of the magnetism of the Pacific Ocean floor by Scripps Institution of Oceanography identified narrow strips where magnetic properties of oceanic crust differed from those currently forming (magnetic anomalies). The **magnetic anomalies** ran parallel to the **Juan de Fuca Ridge.** Each anomaly represents a period when the polarization of Earth's magnetic field was reversed compared to today's. The anomalies were separated by bands of ocean crust that displayed present-day polarity.

It was observed that the sequence of polarity changes on one side of an oceanic ridge is identical to the sequence of reversals on the opposite side of the ridge. During dating of the reversal points on both flanks of the ridge, it became apparent that the rocks became older with increased distance from the ridge axis. The evidence indicated new oceanic crust was being formed at the oceanic ridges and moving away from them on opposite sides of the ridges. Having determined from continental studies the dates at which many of the more recent reversals occurred, it was possible to determine the age of the ocean floor at each strip boundary. Dividing the width by the number of years that polarity lasted produced the rate at which the ocean floor appeared to be moving away from the ridge (figure 3–7).

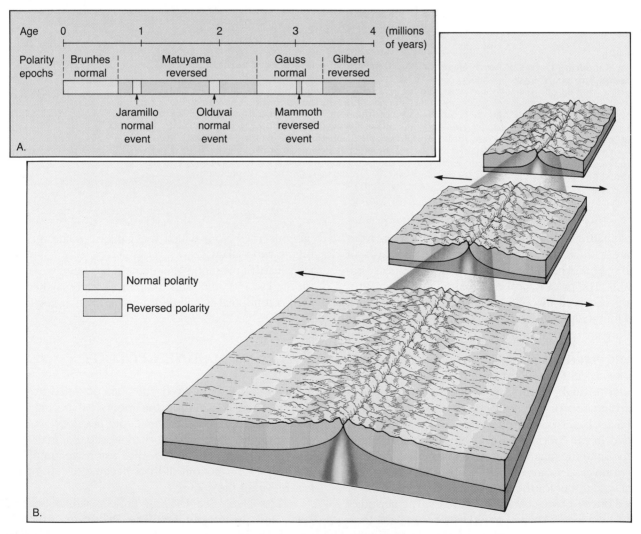

FIGURE 3–7
Magnetic Evidence in Support of the Sea-Floor Spreading Hypothesis. *A,* time scale of Earth's magnetic field in the recent past. This time scale was developed by establishing the magnetic polarity for volcanic lavas of known ages. *B,* new sea floor records the polarity of the magnetic field at the time it formed. Hence it behaves much like a tape recorder, as it records each reversal of Earth's magnetic field. (Data from Allan Cox and G. B. Dalrymple. Reprinted by permission from Tarbuck, E. J., and Lutgens, F. K., 1990, fig. 18.12.)

Confirmation of the spreading of the oceanic crust away from the oceanic ridges required obtaining actual samples of the crust at various locations for radiometric dating. In addition, since sediment could not be laid down on crustal material until it formed at the oceanic ridges, it would be expected that the fossil assemblages observed in sediments immediately overlying the oceanic crust should be representative of organisms that existed at the time of the crustal formation. Also, it should be expected that the sediment thickness would be greater on older sea floor than on younger sea floor. A significant task of the Deep Sea Drilling Project was to check the age of the ocean bottom by drilling through sedimentary sections into the oceanic crust. Although all attempts to do this did not meet with success, enough data based on fossil assemblages immediately overlying the crust and radiometric age determinations of the crust itself confirmed that the oceanic floor is moving in the manner suggested by paleomagnetic data. The youngest oceanic crust was found at the axes of the oceanic ridges, with the age of the crust increasing at greater distances from the ridge crests.

Sea-Floor Spreading

New ocean floor appears to be forming at the oceanic ridges and rises. It then is carried away from the axes of such features, and volcanic processes fill the void with yet a younger strip of ocean floor. Thus, the axes of oceanic ridges and rises are referred to as **spreading centers.** The term **sea-floor spreading** is applied to this theory.

It is, however, more than the sea floor that is moving. Although the details are far from clear, the mechanism of spreading involves two fundamental units of Earth's structure confined to the uppermost 700 km (435 mi) (figure 3–8). The lithosphere (rock sphere) is a relatively cool, rigid shell that includes the crust and upper mantle. The lithosphere is broken into a dozen or so lithospheric plates (figure 3–9) that we observe moving across Earth's surface. Underlying the lithosphere is a high-temperature plastic layer within the mantle. This layer, called the **asthenosphere** (weak sphere), can flow slowly, allowing the rigid lithospheric plates resting at its upper surface to move.

The ultimate fate of a lithospheric plate, if mass is to be conserved, is its **subduction**—a process by which it descends beneath another plate and is ultimately resorbed into the mantle.

A Possible Mechanism for Spreading. An absolute answer to the mechanism that might cause the formation of lithosphere at the oceanic ridges and its movement away from those ridges at right angles has not been determined. However, the lithosphere thickens from a few kilometers near the ridge to over 200 km (124 mi) beneath some continental regions. The increasing thickness

of the lithosphere correlates well with increasing depth of the ocean and decreasing heat flow as the lithosphere ages (see figure 3–10).

Within Earth, radioactive atoms are breaking down and releasing energy that must find its way to the surface as heat. It is conceivable that this heat moves to the surface through **convection cells** that carry the heat up in the regions of the oceanic spreading centers. If this is the case, there must be some regions on Earth where the cooler portions of the mantle descend to complete the convection cell.

Heat-flow measurements taken throughout Earth's crust show that the quantity of heat flowing to the surface along the oceanic ridges is as much as eight times in excess of the average for Earth's crust (1.25 μcal/cm^2/s). In addition, areas where ocean lithosphere subducts at trenches have heat flow as little as one-tenth the average (figure 3–11).

The lithosphere at the spreading centers would be thin because of the proximity of high temperatures immediately beneath the ocean floor. This heat would also cause expansion of the upper mantle rocks, raising the oceanic ridges and rises to elevations well above those of the ocean floor on either side of the zones of upwelling heat. As the lithosphere moves away from the spreading centers toward zones of convection downwelling beneath the trenches, it will encounter decreasing rates of heat flow through the ocean floor from the underlying mantle. This would result in increasing thickness of the lithosphere as cooling asthenosphere at the base of the lithosphere is converted to rigid lithosphere. The thermal contraction of hot, fluffy asthenosphere into cold, brittle lithosphere accounts for the increasing depth of the ocean away from the spreading centers.

It was once believed that most of the heat driving global plate tectonics was generated primarily by radioactivity in the upper mantle. However, now there is evidence that a significant amount of heat is derived from the core of Earth. If heat from the core is a significant source of energy driving mantle convection, a study of three-dimensional spherical models of convection in Earth's mantle shows that upwelling occurs as cylindrical **plumes** arising from the base of the mantle. Such features are known to exist, and their surface expressions are referred to as **hot spots.** The high rates of volcanic activity associated with hot spots have produced such features as the Hawaiian Islands and Iceland.

In these models, the downwelling occurs in sheets that form a ringlike pattern around the upwelling plumes. The overall pattern is similar to what we now see occurring in the Pacific Ocean. Because seismic data indicate the magma chambers that provide the lavas for the spreading centers extend no deeper than 350 km (220 mi) into the mantle, it may well be that they are passive features that result from the lithosphere being

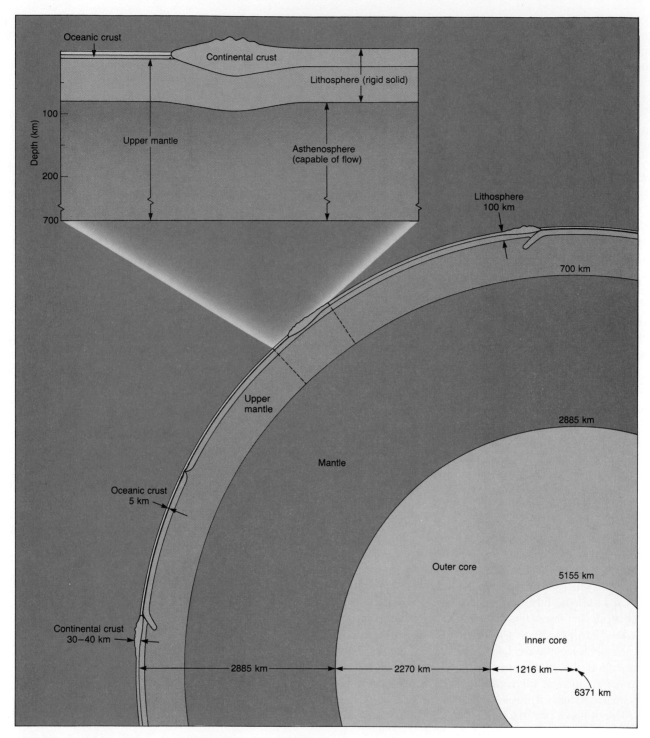

FIGURE 3–8
Lithosphere and Asthenosphere. The lithosphere is the rigid outer shell of Earth's structure that includes the crust and cooler upper mantle. It is supported and allowed to move across Earth's surface by flow within the warmer, plastic underlying portion of the mantle called the asthenosphere. This weaker, possibly partially molten layer is thought to extend to a depth of approximately 700 km (435 mi). The three major subdivisions of the earth—core, mantle, and crust—are shown to clarify the position of the lithosphere and asthenosphere. (Reprinted by permission from Tarbuck, E. J., and Lutgens, F. K., 1991, fig. 5.19.)

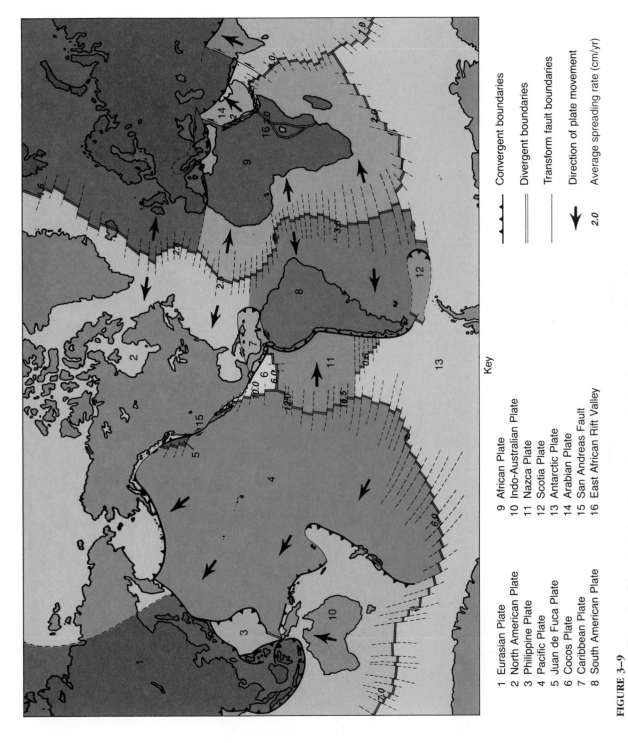

Key

1 Eurasian Plate	9 African Plate
2 North American Plate	10 Indo-Australian Plate
3 Philippine Plate	11 Nazca Plate
4 Pacific Plate	12 Scotia Plate
5 Juan de Fuca Plate	13 Antarctic Plate
6 Cocos Plate	14 Arabian Plate
7 Caribbean Plate	15 San Andreas Fault
8 South American Plate	16 East African Rift Valley

─────── Convergent boundaries

═══════ Divergent boundaries

─────── Transform fault boundaries

↓ Direction of plate movement

2.0 Average spreading rate (cm/yr)

FIGURE 3–9

Lithospheric Plates. The lithospheric plates form at the axes of submarine mountain ranges (divergent boundaries) and dive back into Earth's interior beneath deep-ocean trenches (convergent boundaries). The divergent plate boundaries are offset by transform faults, along which the plates slide past one another in opposite directions. The lithospheric plates move at rates ranging from 1.5 cm (0.8 in.) per year for the American plates to more than 10 cm (5.4 in.) per year for the Pacific Plate.

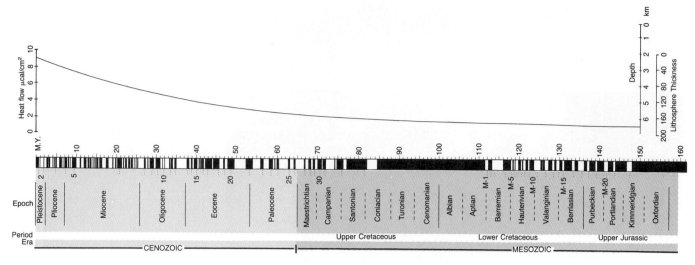

FIGURE 3-10

Magnetic Time Scale, Heat Flow, Ocean Depth, and Lithosphere Thickness. Black bands represent periods when the magnetic polarity of Earth was the same as at present, and white bands represent periods when it was reversed. The curve shows that the ocean gets deeper and the lithosphere thicker with increasing age while the rate of heat flow to Earth's surface through the lithosphere decreases as the lithosphere ages. Although this curve could not be accurately applied to any particular region of the ocean basins, it indicates that, in general, heat flow decreases while ocean depth, thickness of the lithosphere, and age of the ocean floor increase with increasing distance from the spreading centers. Time is the independent variable for all of these changes. The geologic time scale is shown at the bottom (*see* Appendix III).

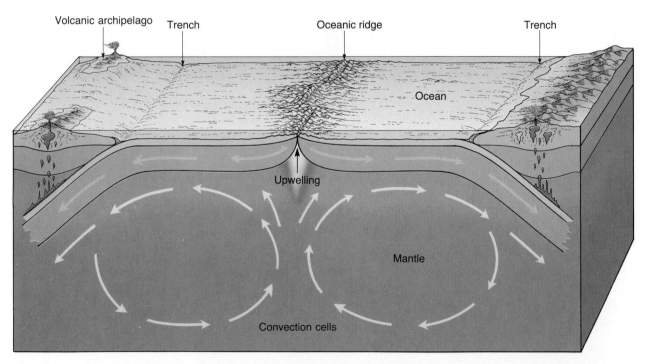

FIGURE 3-11

Movement in the Asthenosphere. Lateral flow within the asthenosphere is a possible mechanism for the lateral movement of the lithosphere, including the crust. The asthenosphere is a low-strength plastic zone within the mantle where pressure is low enough and temperature sufficiently near the melting point of the material to allow it to flow slowly. Some form of thermal convection within the asthenosphere appears to create new lithosphere at the oceanic ridges and rises and may carry old lithosphere back into the mantle to be subducted beneath oceanic trench systems. (Reprinted by permission from Tarbuck, E. J., and Lutgens, F. K., 1990, fig. 18.11.)

pulled apart by the subducting slabs at the edges of these sheets (figure 3–12).

PLATE BOUNDARIES

The process of sea-floor spreading sets up stresses within the lithosphere as plates move across the spherical surface of Earth and interact at relative velocities of up to 16 cm (5 in.)/yr. There are seven major and a half-dozen

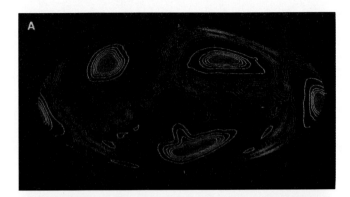

FIGURE 3–12
Three-Dimensional Spherical Model of Convection in Earth's Mantle with Half of the Heating Occurring from within the Mantle and Half from the Core/Mantle Boundary. *A*, contours of vertical velocity of convection at a level midway between the base and top of the mantle. Red and yellow represent high and moderate rates of upwelling, respectively. Green represents no vertical motion, and light blue and dark blue represent low and high rates of downwelling, respectively. Note that upwelling appears as circles where the upwelling plume cuts through the spherical surface. Downwelling is seen as rings around the upwelling plumes where sheetlike downwellings intersect the sphere. *B*, in this longitudinal cross section of the model mantle, contours represent entropy values. Colors can be thought of as representing hot (red), warm (yellow), cool (cyan), and cold (blue). Red regions represent the centers of upwelling plumes that rise to the surface from the core/mantle boundary. Blue contours are found in the areas of convective downwelling. (Images courtesy of D. Bercovici. From Bercovici, D., et al. "Three-Dimensional Spherical Models of Convection in the Earth's Mantle," *Science* 244, pp. 950–55, 26 May, 1989. Used with permission of the author and publisher. Copyright 1989 A.A.A.S.)

or more minor plates, some of which are shown in figure 3–19. Our first clue as to the locations of plate boundaries was that most tectonic events such as earthquakes and volcanic activity were confined to linear belts (figure 3–13). Closer examination of these active regions identified three fundamental types of boundaries. The plate boundaries where new lithosphere is being added along oceanic ridges are called constructive or **divergent boundaries.** Boundaries where plates collide and one subducts beneath the other are called destructive or **convergent boundaries.** Lithospheric plates also move past one another along shear or **transform boundaries** (figure 3–14).

Divergent Boundaries

Because deepening of the ocean floor is correlated directly with the amount of time the lithosphere has cooled, the rate of sea-floor spreading has a significant effect on the steepness of the flanks of the mountain ranges associated with the spreading.

The term **spreading rate** applies to the total widening rate of an ocean basin resulting from the motion of both plates away from a spreading center where they are created. These rates are shown for the major spreading centers in figure 3–19. The faster the spreading rate, the broader the mountain range associated with the spreading. These gently sloping features are the oceanic rises; the best example is the **East Pacific Rise** between the Pacific Plate and the Nazca Plate, where spreading rates are as high as 16.5 cm (6.5 in.)/yr. The **Mid-Atlantic Ridge,** with spreading rates in the 2–3 cm (0.8–1.2 in.)/yr range, displays the steeper slopes characteristic of slow spreading.

There are significant differences in the nature of features found at the axes of fast- and slow-spreading divergent plate boundaries. These differences appear to result from the fact that a much greater magma supply is available at fast-spreading oceanic rises than is available at slow-spreading oceanic ridges. Figure 3–15 shows a comparison of the sizes of axial rift valleys, which are much larger on the slow-spreading ridges. Project FAMOUS, the French-American Mid-Ocean Undersea Study, conducted in 1974, made observations in the rift valley running along the axis of the Mid-Atlantic Ridge. These observations provide a clue to the spreading process (figure 3–16).

Fissures oriented parallel to the ridge axis range from hairline size near the center of the rift valley to more than 10 m (32.8 ft) in width near both margins of the valley. This evidence supports the theory that the plates are being continuously pulled apart rather than being pushed apart by upwelling of material beneath the ridges. Possibly, the upwelling of magma beneath the oceanic ridges is more that of filling in the void left by the

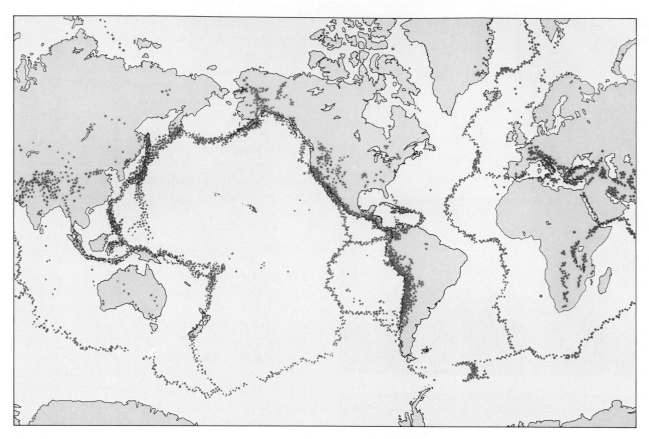

FIGURE 3–13
World Distribution of Earthquakes over a Nine-year Period. Note that the pattern of
eathquake distribution matches that of the lithospheric plate boundaries shown in figure
3–9. (Data from NOAA. Reprinted by permission from Tarbuck, E. J., and Lutgens, F. K., 1991, fig. 5.11.)

separating plates of lithosphere. Whatever the mechanism of emplacement, a large mass of magma or viscous igneous material must exist near the surface beneath the ridges. This is manifested in the oceanic ridges and rises being elevated to mountainlike proportions above the deep-ocean basins. This process produces 20 km³ (4.8 mi³) of new ocean crust per year.

The magnitude of energy released by earthquakes along the divergent plate boundaries is closely related to spreading rate. As measured by the Richter scale, the maximum magnitude recorded to date is 8.9 (not, however, on a divergent plate boundary). Earthquakes occurring in the rift valley of the slow-spreading Mid-Atlantic Ridge reach a maximum magnitude of about 6, while those occurring along the axis of the fast-spreading East Pacific Rise seldom exceed 4.5. Each unit increase of Richter scale magnitude represents an increase of energy release of about 30 times. All spreading-center earthquakes are confined to the lithosphere, so the difference in observed magnitude may result from the fact the slow-spreading centers have thicker and colder lithosphere at the plate boundaries.

Hydrothermal Vents at Spreading Centers. Near the axes of oceanic ridges and rises, where magma cham-

bers may be less than 1 km (0.6 mi) beneath the ocean floor, seawater seeps down fractures in the ocean crust and is heated by the hot underlying magma. It then rises back toward the surface to produce **hydrothermal vents** (figure 3–17A). This rising water may exit into the ocean as warm-water vents (10°–20°C), white smokers (30°–330°C) containing white particles of barium sulfate, or black smokers with temperatures around 350°C (662°F) and containing black metal sulfides (figure 3–17B). The entire volume of ocean water is recycled through this hydrothermal circulation system in about 3 million years, so the chemical exchange between ocean water and the basaltic crust has a major influence on the nature of ocean water. The deposition of metallic sulfides around these vents is likely a major source of continental sulfide ore deposits on Earth as they may be transferred to the continents during subduction of the oceanic plate.

The oceanic hydrothermal vents support unusual biological communities first discovered in 1977 in the **Galápagos Rift** between South America and the Galápagos Islands (figure 3–17C). Large clams, mussels, tube worms, and a large variety of other animals depend on bacteria that oxidize the hydrogen sulfide gas dissolved in the vent water to provide energy to chemosynthetically produce food for the community.

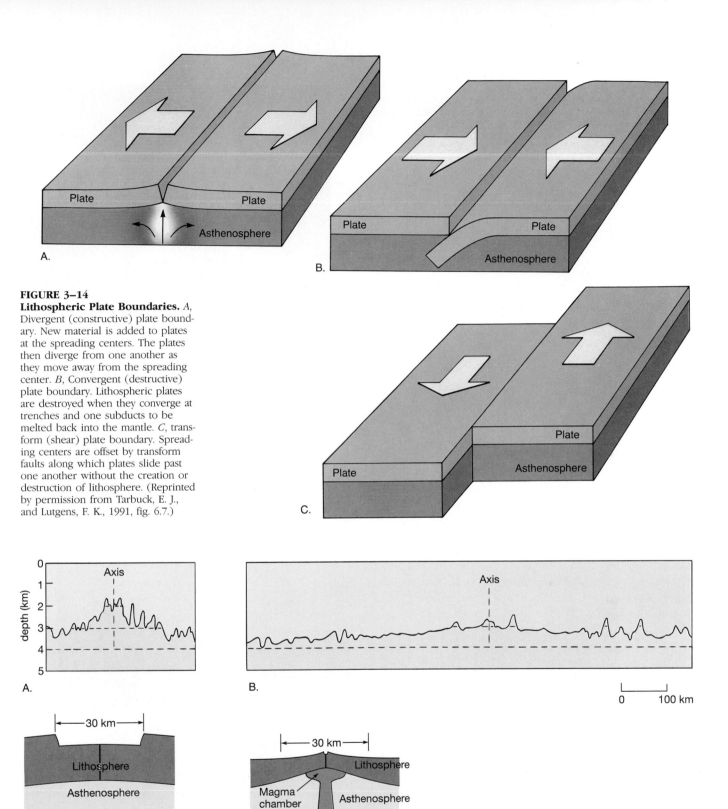

FIGURE 3–14
Lithospheric Plate Boundaries. *A*, Divergent (constructive) plate boundary. New material is added to plates at the spreading centers. The plates then diverge from one another as they move away from the spreading center. *B*, Convergent (destructive) plate boundary. Lithospheric plates are destroyed when they converge at trenches and one subducts to be melted back into the mantle. *C*, transform (shear) plate boundary. Spreading centers are offset by transform faults along which plates slide past one another without the creation or destruction of lithosphere. (Reprinted by permission from Tarbuck, E. J., and Lutgens, F. K., 1991, fig. 6.7.)

FIGURE 3–15
Oceanic Ridges and Rises. *A*, the Mid-Atlantic Ridge has a steeper and more irregular topography as a result of a low spreading rate. *B*, the high spreading rate of the East Pacific Rise has produced a more gently sloping and smoother bottom topography. Vertical exaggeration 50 ×. (*The ocean basins: Their structure and evolution,* The Open University and Pergamon Press.) *C*, the axial valley of a slow-spreading ridge is typically 30 km wide and may be from 100 to 3,000 m deep. The lithosphere may be from 3 to 10 km thick at the plate boundary near the center of the axial valley. *D*, at fast-spreading rises an axial ridge a few hundred meters high and from 2 to 10 km in width typically overlies the plate boundary above a magma chamber that may be about 4 km wide and 1 km thick. An axial valley less than 1 km wide and up to 100 m deep may run along the axis of the axial ridge. The thickness of the lithosphere at the rise axis may be less than 2 km.

67

FIGURE 3–16
Rift Valley Fissures. *A*, a fissure in the rift valley of the Mid-Atlantic Ridge photographed from the submersible *Alvin* during Project FAMOUS, 1974. (Courtesy of Woods Hole Oceanographic Institution.) *B*, a more readily observable fissure above sea level in the rift valley of Iceland. Here the Mid-Atlantic Ridge rises above the ocean surface as an island because of the high rate of volcanic activity in the Iceland hot spot. (Courtesy of Br. Robert McDermott, S. J.)

A.

B.

Transform Boundaries

Geophysical studies of divergent plate boundaries have indicated that a segmentation of the oceanic ridges and rises is associated with the emplacement of separate magma chambers beneath their axes (figure 3–18). Each segment may range from 10 to 80 km (6 to 50 mi) in length and is centered over a magma chamber. The lithosphere is thicker at the boundary between adjacent segments, and the segments may be offset at this boundary. The offset, called a **transform fault,** may be very minor, or it may be a major offset of hundreds of kilometers.

Transform Faults. As newly formed lithosphere moves away from the axis of an oceanic ridge, it travels perpen-

dicular to the axis. On a spherical surface like that of Earth, all of this new lithosphere will converge toward two points. For instance, if the equator were an oceanic ridge, the points of convergence would be the North and South poles. Obviously, all this new matter cannot be accommodated at a single point, and stresses are set up within the lithosphere. The stresses help cause the lithosphere to break along lines perpendicular to the ridge axis. Movement along the breaks offsets the ridge axis, producing a transform fault (figure 3–19). Along such faults one plate slides past another, producing shallow earthquakes in the lithosphere. Since the lithosphere is thicker at transform faults than along the spreading axes, these earthquakes may be larger. Magnitudes of 7 have

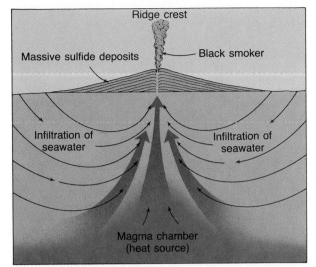

A.

B.

C.

FIGURE 3–17
Hydrothermal Vents along Spreading Center Axes. *A,* cross section across the axis of a spreading center to illustrate the hydrothermal circulation in oceanic crust. Seawater percolates down along a broad zone of fractured crustal rock, is heated as it approaches the underlying magma chamber, and rises in a narrow zone near the axis of the spreading center. The resulting vents usually lie within 200 m (656 ft) of the ridge or rise axis. (Reprinted with permission from Tarbuck, E. J., and Lutgens, F. K., 1988, fig. 10.17.) *B,* black smoker. During a 1979 expedition to 21°N on the East Pacific Rise, this "black smoker" spewing out 315°C (600°F) water was observed. In the foreground is *Alvin's* sample basket. *C,* large tube worms observed at East Pacific Rise warm-water vent. The tubes are about 1 m (3.2 ft) long and the vestimentiferan worms that inhabit them have no mouth or digestive tract. Symbiotic chemosynthetic bacteria oxidize the hydrogen sulfide gas dissolved in the vent water and use the energy released to produce food for themselves and the worms. (Photos courtesy of Scripps Institution of Oceanography, University of California, San Diego; Photo *B* by Dr. Fred N. Spiess.)

been recorded along oceanic transform faults. An example of such a fault that has come ashore is the **San Andreas Fault,** which runs across California from the head of the Gulf of California to the San Francisco area. Since the San Andreas Fault runs through continental crust, the lithosphere is much thicker than at oceanic sites. Earthquakes along this fault have had magnitudes of up to 8.5 on the Richter scale.

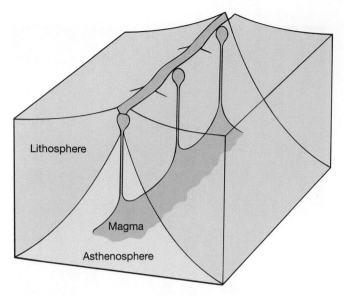

FIGURE 3–18
Segmentation of Oceanic Ridges and Rises. Beneath the ridge axis, the heat-bearing asthenosphere (yellow) rises between the separating lithospheric plates (green). A zone of partially molten asthenosphere (red) is less dense than the cold lithosphere above, and diapirs of magma rise from it to form magma chambers at intervals along the ridge axis. The ridge axis rises over the magma chambers and is deepest at the segment boundaries midway between magma chambers.

Fracture Zones. Continuing along the alignment of the transform fault beyond the offset axes are fracture zones. The transform faults represent active displacements of the axes; the fracture zones are evidence of past transform-fault activity. On opposite sides of the transform fault, the lithospheric plates are moving in opposite directions, whereas there is no relative plate motion occurring along a fracture zone. In other words, the transform faults serve as plate boundaries, while the fracture zones are embedded in a single plate (figure 3–19). Earthquakes with foci above a depth of 10 km are common along the transform faults, whereas the fracture zones are aseismic.

Because the age of the ocean floor on one side of a fracture zone is greater than that of the ocean floor on the opposite side, the older ocean floor will have subsided more from thermal contraction. This results in a vertical escarpment that faces the older ocean floor and increases in height with increased difference in the age of ocean floor on opposite sides of the fracture zone.

Convergent Boundaries

Island Arc Trench System. Oceanic ridges, oceanic rises, and trench systems represent significant discontinuities in Earth's crustal structure and are characterized by earthquakes and volcanic activity. There is a significant difference in the earthquake activity of the two regions, however. Shallow quakes, usually less than 10 km (6 mi)

in depth, are associated with the constructive ridge boundaries where the lithosphere is thin. Movements that cause earthquakes in the trench regions may occur at depths down to 670 km (415 mi).

The earthquake focus, the point at which the movement that causes the quake occurs, is usually shallow in the immediate area of the trench. Earthquake activity in the subduction zone deepens and reaches a maximum value of 670 km. The Wadati-Benioff seismic zone is a band 20 km (12.5 mi) thick that dips from the trench region under the overriding plate, within which all of the earthquake foci are located. The angle of dip is approximately 45°, becoming steeper at greater depths (figure 3–20A). Because the descending lithosphere would become too ductile below depths of 70–100 km to produce earthquakes, some other mechanism must be found to explain them to depths of 670 km. It is possible that these deep-focus earthquakes result from the changing of the mineral, olivine, to a denser form, spinel. Olivine is a mineral that makes up a large percentage of the descending lithospheric mantle. This process is considered a likely cause of deep-focus earthquakes because it should occur between the depths of 400 and 670 km.

This band within which the earthquakes originate must lie within the slab of the lithosphere that is descending. The fact that the lithosphere generated at the oceanic rises and ridges is subducted at the trenches is supported by the fact that no oceanic crustal material more than 175 million years old has been recovered from the ocean bottom. Continental crust, which is very unlikely to be subducted because of its low density, remains at Earth's surface, and continental rocks as old as 3.8 billion years have been found.

As the lithosphere is carried into the asthenosphere, it is heated by being subducted to greater depths. Water and other volatiles are freed, producing a low-density mixture that rises to the surface on the overriding plate side of the trenches to produce basaltic volcanoes; these volcanoes may become island arcs, which usually develop about 100 km (62 mi) above subducting lithosphere (figure 3–20B).

Continental Arc Trench System. Should an oceanic plate subduct beneath a plate with a continent at its leading edge, the melting of the subducting oceanic plate occurs beneath the continent. Consequently, the rising basalt melt passes through and mixes with the granite of the continental crust. This results in a continental arc of volcanoes along the edge of the continent (figure 3–20A). The volcanoes are composed of a rock called andesite, which is of a composition intermediate between those of basalt and granite and is thought to form from the mixing of the two basic crutal rock types.

If the spreading center producing the subducting plate is far enough from the subduction zone, an oceanic

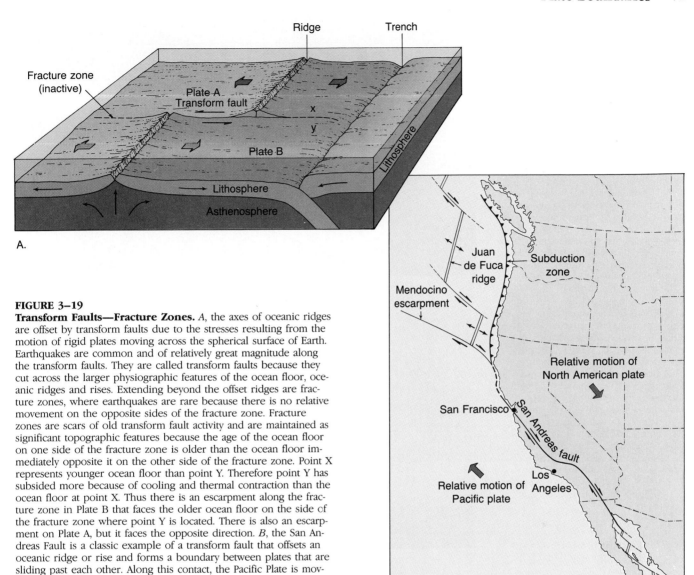

A.

B.

FIGURE 3–19
Transform Faults—Fracture Zones. *A*, the axes of oceanic ridges are offset by transform faults due to the stresses resulting from the motion of rigid plates moving across the spherical surface of Earth. Earthquakes are common and of relatively great magnitude along the transform faults. They are called transform faults because they cut across the larger physiographic features of the ocean floor, oceanic ridges and rises. Extending beyond the offset ridges are fracture zones, where earthquakes are rare because there is no relative movement on the opposite sides of the fracture zone. Fracture zones are scars of old transform fault activity and are maintained as significant topographic features because the age of the ocean floor on one side of the fracture zone is older than the ocean floor immediately opposite it on the other side of the fracture zone. Point X represents younger ocean floor than point Y. Therefore point Y has subsided more because of cooling and thermal contraction than the ocean floor at point X. Thus there is an escarpment along the fracture zone in Plate B that faces the older ocean floor on the side of the fracture zone where point Y is located. There is also an escarpment on Plate A, but it faces the opposite direction. *B*, the San Andreas Fault is a classic example of a transform fault that offsets an oceanic ridge or rise and forms a boundary between plates that are sliding past each other. Along this contact, the Pacific Plate is moving north relative to the North American Plate. (Reprinted by permission from Tarbuck, E. J., and Lutgens, F. K., 1990, figs. 18.22 and 18.23.)

trench is well developed along the margin of the continent. The Peru-Chile Trench is an example. It is associated with the Andes Mountains continental arc of volcanoes from which the rock, andesite, takes its name. No trench is visible where the Juan de Fuca Plate subducts beneath the North American Plate (NAP) off the coasts of Washington and Oregon to produce the Cascade Mountains continental arc. Here, the Juan de Fuca Ridge is so near the NAP that the subducting lithosphere is less than 10 million years old and hasn't cooled enough to produce deep-ocean depths. Also, the large amount of sediment carried to the ocean by the Columbia River has filled what trench structure may have devloped. Many of the Cascade volcanoes have been active within the last 100 years. Most recently, Mount St. Helens erupted in May of 1980 and killed 62 people (figure 3–21).

Continental Mountain System. If two lithospheric plates collide and both contain continental crust near their leading edges, the surface expression will be a mountain range composed of the folded sedimentary rocks derived from sediments deposited in the sea that previously separated the continental blocks. For the most part, these sedimentary rocks do not subside in a zone of subduction because of their relatively low density, having been derived from the continents. The oceanic crust itself may, however, subside beneath such mountains. The exact nature of the subsidence of oceanic crustal material in these regions has not been determined (figure 3–20C).

Some degree of folded mountain formation is frequently associated with plate collisions, but ocean trenches are absent in some areas.

FIGURE 3–20
Effects of Plate Collisions. *A*, when an oceanic lithospheric plate subducts beneath a plate with continental crust at its leading edge, an andesitic continental arc develops. The volcanic rock andesite is named for the Andes Mountains, where a continental arc exists. Andesite has a composition between that of granite and basalt. It is thought to form as melted basalt from the descending oceanic plate rises through the continental granite to produce the volcanic continental arc. The sloping belt in which earthquake foci are located is the Wadati-Benioff seismic zone. *B*, classic oceanic trench systems develop where oceanic lithosphere subducts beneath a plate with oceanic crust at its leading edge and a continent trailing some distance behind. A basaltic island arc-trench system develops. *C*, as two plates carrying continents near their leading edges converge, trenches do not develop because sediments that accumulated in the sea between the continents are too light to be subducted. The shortening of Earth's crust in such regions is accommodated by the folding of the sediments into mountain ranges such as the Appalachians, Alps, Himalayas, and Urals. The lower lithosphere and asthenosphere may be involved in low rates of subduction beneath the mountains. (Reprinted by permission from Tarbuck, E. J., and Lutgens, F. K., *Earth Science,* 5th ed. [New York: Macmillan], 1988, fig. 6.11.)

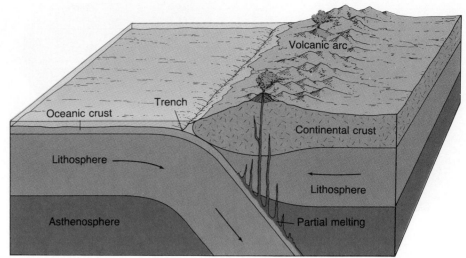

A.

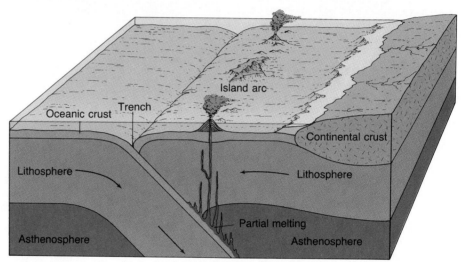

B.

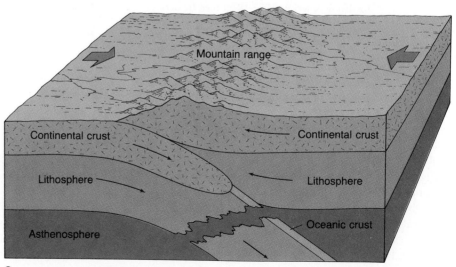

C.

FIGURE 3–21
Mount St. Helens. *A,* tectonic features of the Mount St. Helens region. The mountains shown are volcanoes within the Cascade Mountain Range. *B,* the eruption of Mount St. Helens resulted from the subduction of the Juan de Fuca Plate beneath the North American Plate.

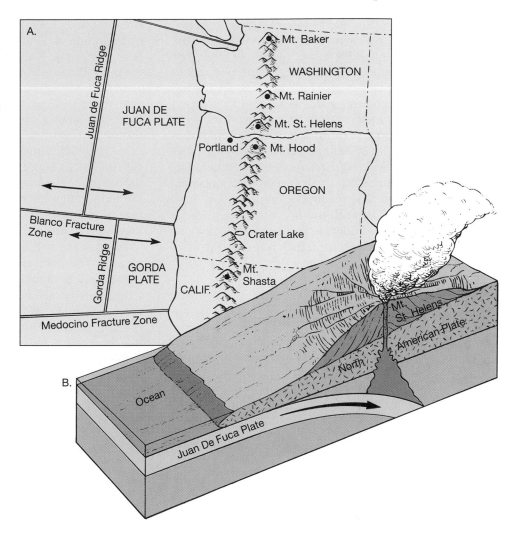

INTRAPLATE FEATURES

Seamounts and Tablemounts

Highly characteristic of the floor of the Pacific Plate are seamounts and tablemounts. A theory related to the formation of such features in association with the oceanic ridge systems could account for the existence of many seamounts and tablemounts (figure 3–22). Considering the oceanic ridge system on a worldwide basis, active oceanic volcanoes are characteristic of the axes of the ridges. Moving away from the ridge axis, the volcanic activity gradually decreases as the long cracks in Earth's crust serving as conduits for rising lava near the axis of these ridges are sealed off as they move away. Active volcanoes over 30 million years of age are rare.

FIGURE 3–22
Formation of Seamounts and Guyots. Volcanic activity develops near the crest of the ridge because of tensional fractures that develop parallel to the ridge. These fractures serve as conduits for lava but usually seal within 30 million years.

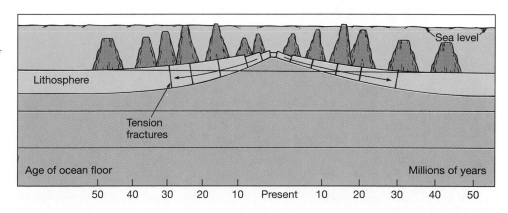

Let us consider the life history of a theoretical volcano that is carried away from an oceanic ridge by the lithospheric plate. A volcano may originate near the axis of the ridge and gradually increase in size as it moves downslope. By the time it is 10 million years old, if it has been a very active volcano, it may develop into an island that may exist for another 10 to 15 million years if the volcano remains active. The volcano will probably become dormant within 30 million years of its origin, and as it continues to move down the flank of the oceanic ridge, the top will be flattened by wave action and ultimately will be submerged. The flat-topped structure will sink deeper and deeper beneath the ocean surface as it descends the slope of the ridge. It then becomes a guyot or tablemount, most of which are at least 30 million years of age.

Coral Reef Development

Although the development of a comprehensive theory of global plate tectonics was not to be accomplished until over 100 years later, Charles Darwin developed a theory on the origin of coral reefs that depended on the subsidence of volcanic islands. Darwin published his theory in the book *The Structure and Distribution of Coral Reefs* in 1842. Figure 3–23 illustrates Darwin's theory of reef development. Drilling into reefs before and after World War II provided the evidence to prove the validity of Darwin's theory.

There are three basic types of reefs—fringing, barrier, and atoll (figure 3–23). **Fringing reefs** develop at the margin of any landmass where the temperature, salinity, and turbidity of the water are suitable for reef-building corals. Due to the close association between the landmass and the reef, there will be inevitable periods of time when runoff from the landmass will carry too much sediment for the coral in a particular area to survive. Very commonly, the amount of living coral in a fringing reef at any given time will be relatively small. The greatest concentration of live reef will be on the seaward margin that is more protected from sediment and salinity changes due to runoff. Without some change in sea level relative to the landmass during the existence of the fringing reef,

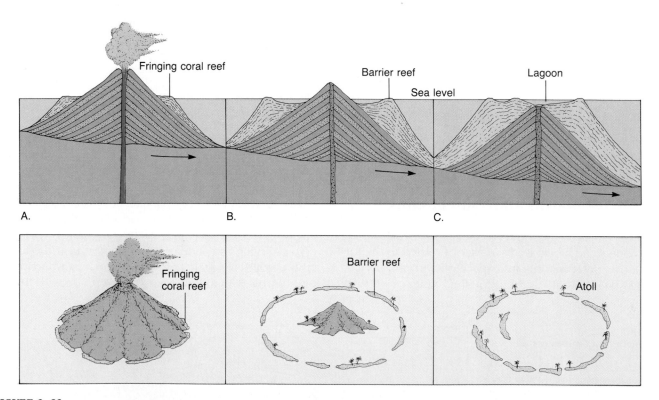

FIGURE 3–23
Darwin's Theory of Development of Reef Types. A volcanic island forms in warm-water latitudes. *A*, a fringing coral reef forms in the shallow, sunlit waters adjacent to the island shore. *B*, as the dormant volcano gradually subsides due to thermal contraction of the lithosphere, the reef maintains its position near the surface by growing upward. This produces a barrier reef where the actively growing reef is separated from the volcanic island by a well-developed lagoon. *C*, continued subsidence submerges the volcano, but the reef may continue to grow upward to produce an atoll with active reef growth confined mostly to the outer edge of the reef. Deposition of reef debris may produce low-lying islands inside the protective ring of coral reef. Within the atoll ring lies a broad, shallow lagoon. (Reprinted by permission from Tarbuck, E. J., and Lutgens, F. K., *Earth Science,* 5th ed (New York: Macmillan), 1988, fig. 10.19.)

it is not likely that the other two types of reefs we have mentioned, the barrier reef or the atoll, will form.

Barrier reefs are linear or circular developments separated by a lagoon from the continent or island on whose margin they are growing. As the island or margin of the continent subsides, the fringe reef maintains its position at the optimum water depth by growing upward, producing the barrier reef. Studies of reef growth rates indicate most have grown 3–5 m (10–16 ft) per 1000 years during the recent geologic past. There is some evidence, however, that reefs in the Caribbean have grown at rates of over 10 m (33 ft) per 1000 years.

The most outstanding example of a barrier reef is the Great Barrier Reef, which is 150 km (93 mi) wide and extends for more than 2000 km (1240 mi) along the northeast coast of Australia. The effects of global plate tectonics are clearly visible in the structure of the Great Barrier Reef. It is oldest (±25 million years) and thickest at its northern end. The reef thins to the south, and the basal reef material becomes younger. This is clearly the result of Australia's moving north from colder high-latitude waters into warmer low-latitude waters as a result of the northward movement of the Indo-Australian Plate (figure 3–24). Smaller barrier reefs are found around the island of Tahiti and other islands in the western Atlantic and Pacific oceans.

Atolls may form either on a subsiding continental shelf or around the margin of a sinking volcanic island in the open ocean. Most atolls are of the oceanic variety, and the greatest number occur in the equatorial Pacific. Atolls may take a circular or irregular shape and surround a lagoon that is usually 30–50 m (98–165 ft) deep at the deepest part. The fringing reef will generally have many channels that allow circulation between the lagoon and the open ocean. The reef flat is commonly broader on the windward side of the atoll, and channels are more numerous on the leeward side (figure 3–25).

Islands large enough to inhabit may develop as deposits of coral debris from the windward reef flat, which is characterized by a well-developed *lithothamnion ridge* on the windward edge of the reef. The lithothamnion ridge, built by calcium-secreting algae, can extend above the low tide level, but beyond it on the reef front will be found a profuse growth of coral below the low water line.

The floor of the lagoon will be filled with coral debris. Oval-shaped reefs called *pinnacles* or *knolls* maintain active reef growth where the lagoon is in good communication with the open ocean through the channels.

Hot Spots

Throughout the Pacific Plate are many northwest-southeast-trending island chains. The most intensely studied of these is the chain of islands between Midway Island and

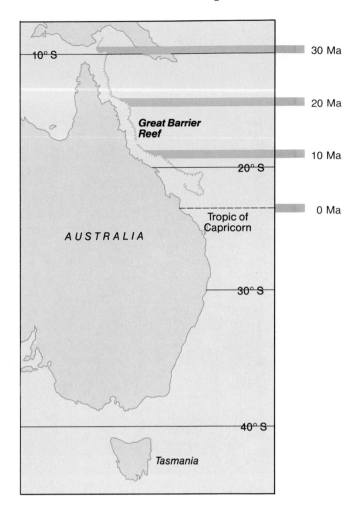

FIGURE 3–24
Australia's Great Barrier Reef and Global Plate Tectonics. The Great Barrier Reef extends for more than 2000 km (1240 mi) along the northeastern coast of Australia. It extends between latitudes 9° and 24° S. The reef first began to develop about 25 million years ago when northern Australia moved into tropical waters from colder Antarctic waters to the south. The southern end of the Great Barrier Reef has moved into tropical waters only recently.

Hawaii. There are no active volcanoes on Midway Island; the volcanic activity appears to have ceased 20 million years ago. The ages of the islands decrease toward Hawaii, where the only active volcanism in the chain exists. It is manifested in the volcano Kilauea, which is less than 1 million years old (figure 3–26). An active submarine volcano, Loihi, presently exists southeast of Hawaii and may either become part of that island or with increased growth create a new island.

Continuing northwest from Midway Island, submerged seamounts have the same alignment as the island chain from Midway to Hawaii. At about 188°W longitude and 30°N latitude, an abrupt change in the alignment of the seamounts occurs. From this point on, the alignment is slightly west of true north. The age of the seamounts at this juncture of the two trends is thought to be between 30 and 40 million years, while the oldest seamounts in

FIGURE 3–25
Atolls. View from space of a group of atolls in the Pacific Ocean. (Photo courtesy of NASA.)

the northward-trending Emperor Seamount Chain formed some 70 million years ago. If the Emperor Seamount Chain and the Hawaii seamounts and islands were generated by the same hot spot, there must have been an abrupt change in the direction of the movement of the Pacific Plate between 30 and 40 million years ago.

The Line Islands to the south of the Hawaiian Islands do not show as distinct a pattern of progressive formation as the Hawaiian Islands and the Emperor Seamount Chain. Beginning with Christmas Island near the southern end of the Line Islands, just north of the equator, an age of approximately 75 million years seems to be the minimum that could be determined. At the northern end of the chain, the age appears to be no more than 85 million years. Although the islands at the southern end are younger, the difference in age is not so great as might be expected had the hot spot theory accounted for this alignment.

South of the equator are other northwest-southeast-trending volcanic island chains—the Marquesas Islands, Tuamotu Islands, Socity Islands, and Cook Islands. It appears that groups of islands in a given chain may have

formed more or less contemporaneously. Although there are indications these chains of islands may have a hot spot origin, the evidence is not so clearly developed as for the Hawaiian Islands.

THE GROWTH OF OCEAN BASINS

So far in this chapter we have discussed some specific features associated with construction, transport, and destruction of tectonic plates in light of what they can tell us about the specifics of these processes. In this section we concentrate on the broader aspect of changes that have occurred since the breakup of Pangaea 180 million years ago. The study of the historical changes of shapes and positions of the continents and oceans is **paleogeography,** while the term **paleoceanography** applies to the study of the changes in the physical and biological character of the oceans brought about by paleogeographical changes.

Figure 3–27A and B show the changing positions of the continents we know today during the last 150 million

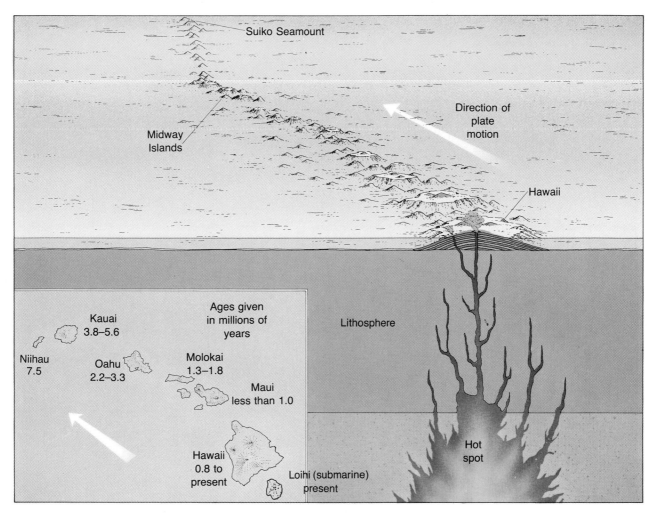

FIGURE 3–26
Hawaiian Islands—Emperor Seamount Chain. The chain of islands and seamounts that extends from Hawaii to the Aleutian trench results from the movement of the Pacific Plate over an apparently stationary hot spot. Radiometric dating of the Hawaiian Islands shows that the volcanic activity decreases in age toward the island of Hawaii. (Reprinted with permission from Tarbuck, E. J., and Lutgens, F. K., 1990, fig. 18.28.)

years for the Northern Hemisphere and Southern Hemisphere, respectively. Observing the changes in the Northern Hemisphere, it is clear that the major change is the separation of North America and South America from Europe and Africa to produce the Atlantic Ocean. During this process, Europe, Africa, and North America also moved a bit northward. Many investigators of plate motion observe that the evidence indicates rates of spreading have historically shown a pattern of higher rates at lower latitudes. This shows up in the pattern of formation of the North Atlantic Ocean. Europe and North America have separated at a much lower rate than have Africa and North America, which began separating 180 Ma (million years ago). It was not until 60 Ma that a noticeable separation between North America and Europe began, and the island of Greenland appeared to be moving on the same plate as Europe. By 30 Ma, the north end of the Mid-Atlantic Ridge shifted to the east of Greenland, putting it back on the North American Plate. North America and South America were not fully connected by the Isthmus of Panama until about 5 Ma. This event had a marked effect on the patterns of ocean circulation and biological species distribution.

In the Southern Hemisphere, the early stages of continental breakup show South America and a continent composed of India, Australia, and Antarctica separating from Africa. By 120 Ma, there is a clear separation between South America and Africa, and India had moved north away from the Australia-Antarctica mass, which began moving poleward. As the Atlantic Ocean continued to open, India moved rapidly north and collided with Asia about 35 Ma. Subsequent to this collision, the spreading center that ran between India and Australia foundered and was replaced by one that began to sepa-

FIGURE 3–27A
Paleogeographic
Reconstruction—Northern Hemisphere. These paleoceanic projections represent an attempt to determine the relative positions of the continents over the last 150 million years. Ages are quoted in Ma (millions of years ago) at intervals of 30 million years.

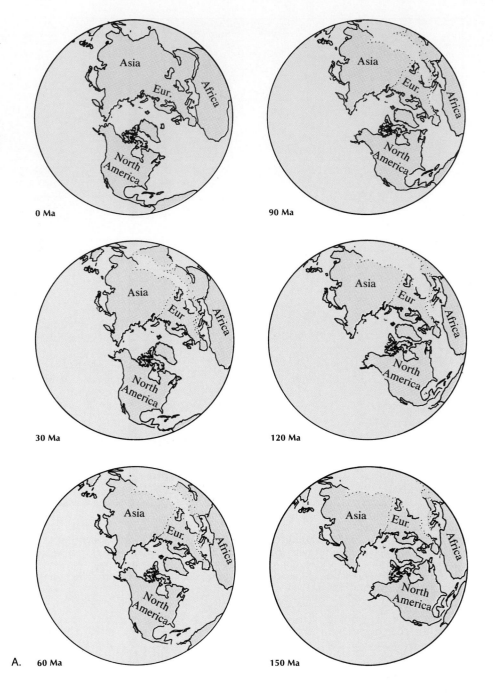

0 Ma

90 Ma

30 Ma

120 Ma

A.   60 Ma

150 Ma

rate Australia from Antarctica. The scars representing shear-plate boundaries during the northward movement of the Indian Plate are the Chagos-Laccadive Ridge to the west and the Ninety East Ridge to the east. These features are clearly visible in figure 2–22.

The overall effect of global plate tectonic events over the past 180 million years is the creation of the new Atlantic Ocean, which continues to grow, and the division of Panthalassa into the Pacific Ocean and Indian Ocean. The fate of the Indian Ocean in terms of its changing size is questionable, but it is clear that the Pacific Ocean con-

tinues to shrink as oceanic plates produced within it subduct into the many trenches that surround it and continent-bearing plates bear in from east and west.

Age of the Ocean Floor

Much can be learned of the history of ocean basins from mapping the pattern of age distribution for the ocean crust. Observing figure 3–28, it can be seen that the Atlantic Ocean has the simplest and most symmetrical pattern of age distribution. As Pangaea was rifted by the in-

FIGURE 3–27B
Paleogeographic
Reconstruction—Southern Hemisphere. The projections are based on data taken from the first 60 legs of the Deep Sea Drilling Project, conducted during a decade of deep-ocean drilling from August 1968 to August 1978. (Courtesy of P. L. Firstbrook, B. M. Funnell, A. M. Hurley, and A. G. Smith in association with Scripps Institution of Oceanography, University of California, San Diego.)

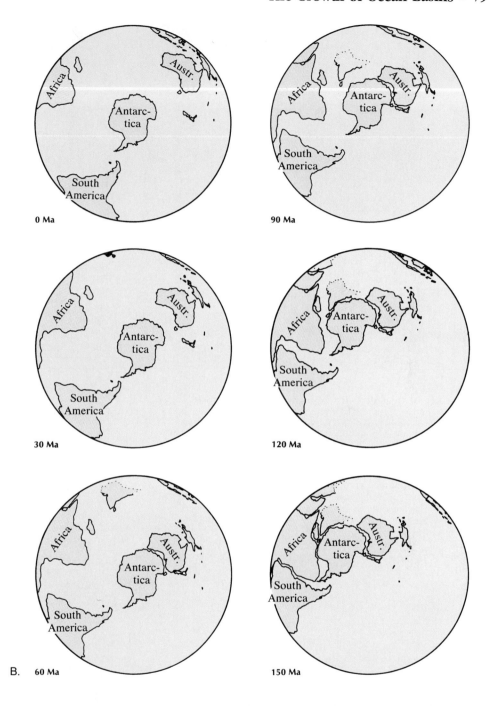

cipient Mid-Atlantic Ridge, the first separation involved North America and Africa over 175 Ma. The Pacific Ocean has the least symmetrical pattern because of the large amount of subduction that surrounds it. The ocean floor east of the East Pacific Rise that is older than 40 Ma has been subducted, whereas ocean floor in the northwestern Pacific is over 170 Ma. The Rise itself has even disappeared under North America. The broader age bands on the Pacific Ocean floor are clear evidence that the spreading rate throughout this history has been greater in the Pacific Ocean than in the Atlantic or Indian oceans.

Metallic Ores and Plate Tectonics

The origin of metallic ores that contain rich deposits of copper, lead, zinc, and silver associated with the destructive margins of lithospheric plates is beginning to be understood. Their origin is related to the process of plate destruction (figure 3–29). Copper sulfide ores associated with sections of oceanic crust that have been pushed up onto continental plates, *ophiolites,* are part of this process. A good example of this kind of ore is the Troodos Massif along the southern coast of Cyprus, which has

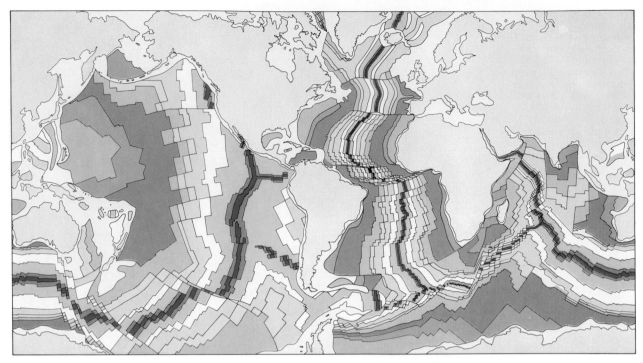

FIGURE 3–28
Relative Age of Oceanic Crust beneath Deep-Sea Deposits.
Notice that the youngest rocks (bright red areas) are found along
the oceanic ridge crests and the oldest oceanic crust (brown areas)
is located adjacent to the continents. When you observe the Atlantic
basin, a symmetrical pattern centered on the Mid-Atlantic ridge crest
becomes apparent. This pattern verifies the fact that sea-floor
spreading generates new oceanic crust equally on both sides of a
spreading center. Further, compare the widths of the yellow stripes
in the Pacific basin with those in the South Atlantic. Because these
stripes were produced during the same time period, this comparison verifies that the rate of sea-floor spreading was faster in the Pacific than in the South Atlantic. (Reprinted from Tarbuck, E. J., and
Lutgens, F. K., 1990, fig. 19.16, after *The Bedrock Geology of the
World,* by R. L. Larson et al. Copyright © 1985 by W. H. Freeman.)

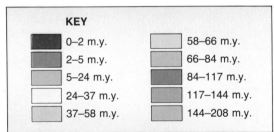

KEY

▨ 0–2 m.y.	58–66 m.y.
2–5 m.y.	66–84 m.y.
5–24 m.y.	84–117 m.y.
24–37 m.y.	117–144 m.y.
37–58 m.y.	144–208 m.y.

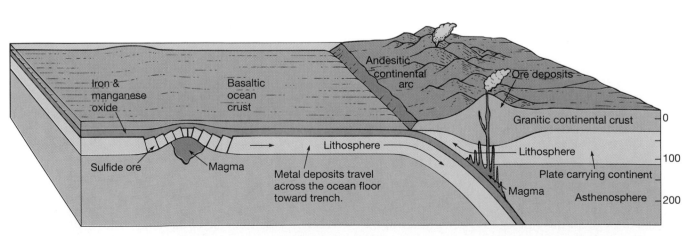

FIGURE 3–29
Theory of Metallic Ore Production.

been mined since early in the development of the Mediterranean civilizations.

The constructive margins of plates associated with oceanic ridges may be the source of this metallic enrichment. The metals found here are predominantly iron and manganese, with significant amounts of copper, nickel, cobalt, zinc, and barium. Analysis of such sediments in a 200-km^2 (77-mi^2) area of the Atlantis Deep in the Red Sea have shown that they contain more than 3 million tons of zinc, 1 million tons of copper, almost 1 million tons of lead, and 5000 tons of silver. Similar iron- and manganese-rich deposits were recovered by submersibles diving along the Mid-Atlantic Ridge during Project FAMOUS in 1974.

SUMMARY

Scientific investigation of evidence suggesting continents are moving relative to one another began with Alfred Wegener at the start of the 20th century. After investigations into the way continents might be fit together and the stronger magnetic field associated with the ocean crust compared to the continental crust, many geologists and geophysicists who had resisted such beliefs throughout the first half of the 20th century began to believe the continents were indeed moving. The evidence gathered by the late 1960s indicated all the present continents were combined into one large continent, Pangaea, about 200 million years ago. Just prior to the formation and possibly also after the breakup of Pangaea, two large continents appear to have existed, Laurasia in the north and Gondwanaland to the south.

The discovery that Earth's magnetic field has changed polarity throughout Earth's history and that the record of these changes was permanently recorded in the rocks of the ocean crust made it possible to develop a hypothesis of sea-floor spreading as the process by which the continents could be moved. According to the hypothesis, new rock is added to rigid lithospheric plates by intrusion and extrusion of molten material along oceanic ridges and rises. These plates are composed primarily of dense mantle rocks with a thin crust of basaltic lava beneath the ocean basins. As material is added to the plates near the axes of the oceanic ridges and rises, it moves away down the slopes of the mountain ranges to make room for new material surfacing at the axes. Thus, the crustal rocks beneath the oceans should increase in age as one moves from the axes of the mountain ranges down their slopes into the deep-ocean basins. Continents located on a lithospheric plate are carried passively along as the plate moves away from the oceanic ridge or rise. The sea-floor spreading hypothesis was confirmed when the *Glomar Challenger* began the Deep Sea Drilling Project to sample the sediments and crust beneath the oceans in 1968. The drilling proved the ocean crustal rocks were indeed older as one moved away from the oceanic ridges and rises into the deep-ocean basins. This movement of the lithosphere appears to be made possible by the existence of a plastic region of mantle material with temperatures near the melting point of the mantle, the asthenosphere, immediately beneath the lithosphere.

As new mass is added to the lithosphere at the oceanic rises and ridges (constructive or divergent boundaries), the leading edges of the plates are subducted into the mantle at ocean trench systems or beneath continental mountain ranges such as the Himalayas (destructive or convergent boundaries). Additionally, there are plate boundaries called transform faults along which oceanic ridges and rises are offset and plates slide past one another—shear plate boundaries. Although Earth is over 4.6 billion years old, no rocks have been found on the ocean floor that are as much as 200 million years old. It is becoming clearer that sinking at the leading edges of the plates rather than upwelling of new material at the ridges drives the convectionlike movement, and it is obvious that the process produces recycling of material between the lithosphere and the deeper mantle.

As the lithospheric plates move away from the high-heat-flow oceanic ridges and rises toward low-heat-flow trenches, they cool and thicken. The age and depth of the ocean floor increase with increasing distance from the spreading centers.

Along the axes of oceanic ridges and rises, seawater percolates down through fractures to be heated and returned to the surface as hydrotheral vents with temperatures up to 350°C (662°F). This process has produced most of Earth's metal sulfide ore deposits, and new biological communities were discovered in 1977. These vent biocommunities depend on chemosynthesis by hydrogen sulfide–oxidizing bacteria rather than the photosynthetic plankton that support most marine life.

Additional intraplate features are seamounts and tablemounts (guyots) that formed as volcanic peaks and islands along the spreading centers. As they moved away from the spreading centers on the cooling lithospheric plates, the volcanoes became inactive within 30 million years and submerged to their present positions. Where average monthly temperatures exceed 18°C and deposition of sediment is minimal, coral reefs develop around volcanic peaks. With continued subsidence, they will pass from fringing reefs to barrier reefs and atolls. There is evidence that reefs can survive subsidence rates of over 10 m per 1000 years. Another feature that appears to have its classic development in the Pacific Ocean is the hot spot, a region of local upwelling of molten material. The volcanic activity presently producing the island of Hawaii also produced all the volcanic peaks of the Hawaiian Island chain and the Emperor Seamounts during the past 70 million years.

The paleogeography, or changing shape and location of ocean basins and continents, has produced changes in the physical and biological character of the oceans. The study of these changes is called paleoceanography. Since the breakup of Pangaea, which began 180 Ma, the Atlantic Ocean has come into existence and Panthalassa has been reshaped into the Pacific Ocean and Indian Ocean. There is evidence that metallic ore deposits associated with continental mountain ranges originated as part of the global plate tectonics process.

KEY TERMS

Absolute dating (p. 55)
Asthenosphere (p. 61)
Atoll (p. 75)
Barrier reef (p. 75)
Bullard, Edward (p. 54)
Continental drift (p. 52)
Convection cell (p. 61)
Convergent plate boundary (p. 65)
Divergent plate boundary (p. 65)
East Pacific Rise (p. 65)
Fossil (p. 54)
Fringing reef (p. 74)
Galápagos Rift (p. 66)
Global plate tectonics (p. 53)
Gondwanaland (p. 56)

Half-life (p. 54)
Heat flow (p. 61)
Hot spot (p. 61)
Hydrothermal vent (p. 66)
Isotope (p. 54)
Juan de Fuca Ridge (p. 60)
Laurasia (p. 56)
Lithosphere (p. 53)
Magnetic anomaly (p. 60)
Magnetic dip (p. 56)
Mid-Atlantic Ridge (p. 65)
Ocean Drilling Program (p. 59)
Paleoceanography (p. 76)
Paleogeography (p. 76)
Pangaea (p. 53)

Panthalassa (p. 56)
Plume (p. 61)
Radioactive dating (p. 54)
Relative dating (p. 54)
San Andreas Fault (p. 69)
Sea-floor spreading (p. 52)
Spreading center (p. 61)
Spreading rate (p. 65)
Subduction (p. 61)
Tectonics (p. 53)
Tethys Sea (p. 56)
Transform fault (p. 68)
Transform plate boundary (p. 65)
Wegener, Alfred (p. 53)

QUESTIONS AND EXERCISES

1. Why couldn't Alfred Wegener convince most scientists that the continents were indeed moving across Earth's surface?
2. What evidence was developed in the mid-20th century that supported the possibility of continental movement?
3. How do the magnetic properties of the continents and ocean basins differ? Why?
4. Describe how fossil evidence was used in relative dating of rock units found on Earth.
5. What discovery made absolute dating of rock units possible? Why was this technique a significant improvement over relative dating?
6. What is the age of a rock that has a ratio of ^{235}U to ^{207}Pb of 1 to 3?
7. Describe the evidence found on the present continents that suggests Laurasia and Gondwanaland existed between 250 and 350 million years ago.
8. How does the magnetic dip of magnetite particles found in igneous rocks tell us at what latitude they formed?
9. How does continental movement account for the two apparent wandering curves of the north magnetic pole as determined from Europe and North America?
10. What property of the magnetic record found in oceanic crustal rocks gave rise to the idea of sea-floor spreading?
11. Describe a possible mechanism responsible for moving the continents over Earth's surface. Include a discussion of the lithosphere and asthenosphere.
12. Describe the general relationships that exist among distance from the spreading centers of the oceanic ridges, heat flow, age of the ocean crustal rock, and ocean depth.
13. Discuss the three types of plate boundaries, spreading centers of the oceanic ridges, trenches, and transform faults.

Explain why earthquake activity is usually confined to depths less than 10 km (6 mi) along the spreading centers and transform faults while it may occur as deep as 670 km (415 mi) near the trenches. Construct a plan view and cross section showing each of the types of boundary and direction of plate movement.
14. What evidence indicates that the plates are more likely being pulled down in the trench areas than being pushed away from the oceanic ridge spreading centers?
15. Describe the general process of hydrothermal circulation associated with spreading centers and their relationship to mining ore deposits and chemosynthetically supported biocommunities.
16. Explain why seamounts and guyots increase in age and depth with increased distance from the oceanic ridges and rises. Discuss the length of time they probably were active volcanoes.
17. Discuss the possible relationship of coral reef development to sea-floor spreading. Include the effects of both vertical and horizontal motions.
18. How are the alignments and age distribution patterns of the Emperor Seamount and Hawaiian Island chains explained by the hot spot theory?
19. When does evidence for the following events first show up on the paleogeographic reconstructions of figure 3–27?
   a. Greenland separates from Europe
   b. South American begins movement away from Africa
   c. India separates from Antarctica
   d. Australia separates from Antarctica
20. Describe the relationship of metallic ore deposits in the lithosphere to the plate tectonics process.

REFERENCES

Apperson, K. D. 1991. Stress fields of the overriding plate at convergent margins and beneath active volcanic arcs. *Science* 254:5032, 670–78.

Bercovici, D.; Schubert, G.; and Glatzmaier, G. A. 1989. Three-dimensional spherical models of convection in the earth's mantle. *Science* 244:4907, 950–54.

Bullard, Sir Edward. 1969. The origin of the oceans. *Scientific American* 221:66–75.

Burnett, M. S.; Caress, D. W.; and Orcutt, J. A. 1989. Tomographic image of the magma chamber at 12°50′N on the East Pacific Rise. *Nature* 339:206–8.

Carrigan, C. R., and Gubbins, D. 1979. The source of the earth's magnetic field. *Scientific American* 240–2:118–33.

Davies, P. J.; Symonds, P. A.; Feary, D. A.; and Pigram, C. J. 1987. Horizontal plate motion: A key allocyclic factor in the evolution of the Great Barrier Reef. *Science* 238:1697–1700.

Dietz, R. S., and Holden, J. C. 1970. The breakup of Pangaea. *Scientific American* 223:30–41.

Hamilton, W. B. 1988. Plate tectonics and island arcs. *Bulletin Geological Society of America* 100:1503–1526.

Hess, H. H. 1962. *History of ocean basins. Petrologic studies: A volume to honor A. F. Buddington,* Engel, A. E. J.; Lames, H. L.; and Leonard, B. F., eds., 559–620. New York: Geological Society of America.

Liu, M.; Yuen, D. A.; Zhao, W.; and Honda, S. 1991. Development of diapiric structures in the upper mantle due to phase transitions. *Science* 252:1836–39.

Mcnard, H. W. 1986. *The ocean of truth: A personal history of global tectonics.* Princeton, N. J., Princeton University Press.

Mid-ocean ridges. 1992. *Oceanus* 34:4, 1–111.

Olson, P.; Silver, P. G.; and Carlson, R. W. 1990. The large-scale structure of convection in the Earth's mantle. *Nature* 344:209–14.

Sclater, J. G.; Parsons, B.; and Jaupart, C. 1981. Oceans and continents: Similarities and differences in the mechanisms of heat loss. *Journal of Geophysical Research* 86:11, 535–52.

Shepard, F. P. 1977. *Geological oceanography.* New York: Crane, Russak.

Tarbuck, E. J., and Lutgens, F. K. 1990. *The earth: An introduction to physical geology,* 3d. ed. New York: Macmillan.

———. 1991. *Earth science.* 6th ed. New York: Macmillan.

van der Hilst, R.; Engdahl, R.; Spakman, W.; and Nolet, G. 1991. Tomographic imaging of subducted lithosphere below northwest Pacific island arcs. *Nature* 353:37–42.

SUGGESTED READING

Sea Frontier

Burton, R. 1974. Instant islands. 19:3, 130–36. A description of the 1973 eruption on the island of Heimaey south of Iceland and its relationship to plate tectonics processes.

Dietz, R. S. 1976. Iceland: Where the mid-ocean ridge bares its back. 22:1, 9–15. A description of the rift zone of Iceland and its relationship to the Mid-Atlantic Ridge and sea-floor spreading.

———. 1977. San Andreas: An oceanic fault that came ashore. 23:5, 258–66. A discussion of the San Andreas Fault and its relationship to global plate tectonics.

———. 1971. Those shifty continents. 17:4, 204–12. A very readable and informative presentation of the crustal features and the possible mechanism of plate tectonics.

Emiliani, C. 1972. A magnificent revolution. 18:6, 357–72. A discussion of advances in studying earth science from the sea, including climate cycles, plate tectonics, and the economic potential of resources lying within marine sediments.

Mark, K. 1974. Earthquakes in Alaska. 20:5, 274–83. A discussion of the origin, nature, and effects of earthquakes in Alaska.

———. 1972. Ocean fossils on land. 18:2, 95–106. A discussion of the significance of marine fossils found in rocks now many miles inland is centered on the work of an early American paleontologist, James Hall.

Rona, P. 1984. Perpetual seafloor metal factory. 30:3, 132–141. A discussion of how metallic mineral deposits form in association with hydrothermal vents located on oceanic ridges and rises.

Scientific American

Bloxham, I., and Bubbins, D. 1989. The evolution of the earth's magnetic field. 261:6, 68–75. The theory of how Earth's magnetic field is associated with motions in Earth's liquid outer core is discussed.

Bonatti, E., and Crane, K. 1984. Oceanic fracture zones. 250:5, 40–51. The role of oceanic fracture zones in the plate tectonics process is related to the spherical shape of Earth.

Brimhall, G. 1991. The genesis of ores. 264:5, 84–91. After being emplaced into Earth's crust at oceanic ridges, metal deposits undergo a number of processes before they become mineable ores embedded in continental rocks.

Macdonald, K. C., and Fox, P. J. 1990. The mid-ocean ridge. 262:6, 72–95. An interesting and comprehensive overview of present knowledge of the oceanic ridges and rises is presented.

Chapter 4

Marine Sediments

Over one-half of the rocks exposed at the surface of the continents above the shoreline were formed by the lithification of sediment laid down in past ocean environments. From this observation, it appears that there is a significant relationship between the formation of the continental masses and the surrounding ocean basins, but the exact nature of this relationship has for years eluded the inquiring geologist. Geologists once believed that if they could examine the entire sedimentary column in the deep-ocean basin, which they assumed to have been permanently in existence since the initial formation of Earth's oceans, a very great portion of the history of Earth might well be recorded in these sediments.

Because of recent advances in technology, it has been possible to sufficiently examine the sediments in the deep-ocean basins to make it seem certain such is not the case. It seems unlikely that sediment representing more than 200 million years of Earth's 4.6-billion-year history will be found in the deep-ocean basins. This, however, does not dim the marine geologist's interest in the study of these sediments, since they will

The drill crew working aboard the D/V *Glomar Challenger* during Deep Sea Drilling Project operations, conducted from 1968 to 1983. (Photo © Scripps Institution of Oceanography, University of California, San Diego.)

help determine the details of this shorter history of the present ocean basins. Clues to past climates, movements of the ocean floor, ocean circulation patterns, and nutrient supply for the marine plant population are embedded in the sedimentary deposits throughout the ocean basins.

An important source of information, however, is rocks formed from sediments that once were deposited in ancient ocean basins. Such rocks cover over half of the continental area of our planet. Most of our knowledge of Earth's ancient history has been obtained by studying these rocks, some of which date back 3.8 billion years. The farther back we go, the more difficult it is to obtain information from these deposits. We may have to climb mountains to elevations kilometers above sea level to observe this record. Yet, with the expanding technology that can be used to extract information from these rocks, we hope to develop an increasingly precise understanding of the processes that have shaped our Earth throughout its history.

In the following pages, we will consider some general properties of sediments and then some specific sedimentary deposits found in today's oceans.

SEDIMENT TEXTURE

Sediment texture, which is determined primarily by grain size, is indicative of the energy condition under which a deposit is laid down. The abbreviated Wentworth scale presented in table 4–1 classifies particles in categories ranging from boulders to colloids. Between these extremes are cobbles, pebbles, granules, sand, silt, and clay-sized particles. Deposits that are laid down in areas where wave action is strong (areas of high energy) may be composed primarily of the larger particles—cobbles and boulders. The deposition of clay-sized particles occurs in areas where the energy level is low and the current speed is minimal.

Sediments composed of particles that are primarily within the same size classification are *well sorted*. A beach sand is usually a well-sorted deposit. Glacial till that may be found on the continental shelf is an example of a poorly sorted deposit because it contains particles ranging in size from the colloid to boulders that have been carried by glaciers and dropped out as the glacier melted.

As particles are carried from the source to their point of deposition, they increase in *maturity*. This results from their association with moving water, which has the capacity to carry particles of a certain size away from a deposit and leave behind other larger particles. As the time in the transporting medium increases, particles of sand size or larger become more rounded through chemical weathering and abrasion. Increasing **sediment maturity** is indicated by (1) decreasing clay content, (2)

TABLE 4–1
Wentworth Scale of Grain Size for Sediments.

Size Range (mm)	Particle	
256	Boulder	
64	Cobble	
4	Pebble	
2	Granule	
1	Very coarse sand	SAND
1/2	Coarse sand	
1/4	Medium sand	
1/8	Fine sand	
1/16	Very fine sand	
1/32	Coarse silt	SILT
1/64	Medium silt	
1/128	Fine silt	
1/256	Very fine silt	
1/640	Coarse clay	
1/1024	Medium clay	CLAY
1/2360	Fine clay	
1/4096	Very fine clay	
	Colloid	

Source: Wentworth (1922), after Udden (1898)

increasing degree of sorting, and (3) increasing rounding of the grains within the deposit.

The poorly sorted glacial till, which contains relatively large quantities of clay-sized and larger particles that have not been well rounded, is an example of an immature sedimentary deposit. The beach sand we used as an example of a well-sorted sediment contains very little clay-sized material and is usually composed of well-rounded particles that have undergone considerable transportation. The beach sand is an example of a mature sedimentary deposit. Figure 4–1 illustrates the nature of sediments of varying degrees of maturity based on these criteria.

SEDIMENT TRANSPORT

The sediment derived from the continents and carried to the margins of the ocean by the rivers and glaciers flowing from the continents either settles out to fill bays, becomes a part of a delta, is spread across the continental shelf by currents, or is carried beyond the shelf to the deep-ocean basin. The greatest volume of sediment transport in the ocean is achieved by the longshore currents near the margins of the continent. Lower-energy currents distribute the finer components of the sediment

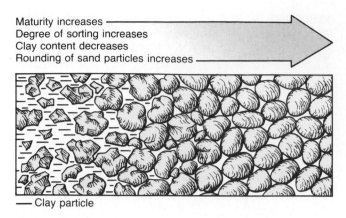

Maturity increases ⟶
Degree of sorting increases
Clay content decreases
Rounding of sand particles increases ⟶

— Clay particle

FIGURE 4–1
Sediment Maturity.

along the margin of the continental shelf and even into the deep-ocean basin.

Figure 4–2 shows the relationship between average horizontal current velocity and the erosion, transportation, and deposition of particles ranging in size from 0.001 to 100 mm (0.00004 to 4 in.). As might be expected, the curve that separates transportation of particles from their deposition shows that larger particles will settle out at higher current velocities than smaller ones.

Observing the curve that separates the process of erosion from transportation, we can see that it takes higher velocity currents to erode (pick up and carry away) larger particles from the sand-sized particles up. The surprising fact that clay-sized particles require a higher velocity current than that needed to erode the larger sand-sized particles results from the fact that clay-sized particles are flat rather than round like typical sand-

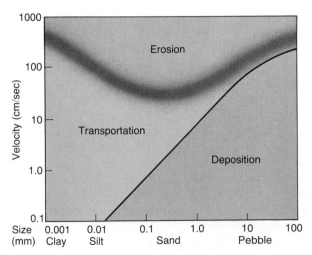

FIGURE 4–2
Hjulstrom's Curve. Horizontal current velocity vs. erosion —deposition of sediment by particle size. (Reprinted with permission from Tarbuck, E. J., and Lutgens, F. K., *The earth: An introduction to physical geology,* 3d ed. (New York: Macmillan, 1990), fig. 10.15.)

sized particles. Thus they have a greater surface area per unit of mass in contact with one another when they are laid down in a deposit. This causes a great cohesive force in clay sediments. To overcome this cohesive attraction and pick up these particles once they are deposited requires a high velocity current.

Particles less than 20 μm (1μm = 1 millionth meter) in diameter can be carried far out over the open ocean by prevailing winds. Such particles either settle out as the velocity of the wind decreases or find their way into the ocean during precipitation. They commonly serve as nuclei around which raindrops and snowflakes form. Some particles that reach high altitudes can be carried very rapidly by the jet stream that exists in the mid-latitudes.

COMPOSITION OF MARINE SEDIMENT

Clues to the origin of sediments can be found in their mineral composition. Certain minerals such as quartz and feldspar are characteristic rock-forming minerals that, when found in a marine sediment, indicate it was derived from rocks. Although silica (SiO_2) and calcite ($CaCO_3$) both may be found in other types of deposits, they are the most abundant compounds in sediments derived from organisms. The minerals that precipitate from ocean water are numerous but, as will be discussed later in the chapter, are characteristic of such origin. Within the following paragraphs we will describe the various origins of marine sedimentary particles.

Lithogenous Sediment

Lithogenous means "derived from rocks." Most of the rock mass that supplies lithogenous sediment is that which makes up the continents. The volcanic islands found in the open ocean are also important sources of lithogenous sediment. All parts of Earth's crust originally formed from the solidification of molten material into **igneous** rocks (*ignis* is Latin for "fire"). Igneous rocks solidified at temperatures and pressures well above those of atmospheric conditions and in an environment in which free oxygen and water were scarce.

Igneous rocks are composed of discrete crystals of naturally occurring compounds called **minerals.** These minerals may be grouped into two basic categories: **ferromagnesian** and **nonferromagnesian.** Ferromagnesian minerals are relatively high density, dark in color, and contain significant amounts of the elements iron and magnesium. Common varieties are olivine, augite, hornblende, and biotite. The nonferromagnesian minerals quartz, feldspar, and muscovite contain no iron or magnesium and have a lower density and lighter color than the ferromagnesian minerals. All these minerals are—to varying degrees—unstable under the conditions that prevail at Earth's surface and begin to undergo chemical and physical breakdown known as **weathering.** As they break down, the particles are carried to the ocean, where they are deposited. By far the greatest quantity of lithogenous material is found around the margins of the continents, but there are no parts of the ocean basins where traces of this sediment cannot be found.

Figure 4–3 shows the major types of igneous rocks and their general mineral composition and texture. **Intrusive rocks** such as granite cool and solidify slowly beneath the surface of Earth. The individual crystals grow to large size and give the rocks a coarse-grained texture. **Extrusive rocks** cool more rapidly at or near Earth's surface, so the mineral crystals don't have time to grow large. They have a fine-grained texture.

Clays. The products of the decomposition of feldspars (aluminum- and silica-rich) and ferromagnesian (iron- and magnesium-rich) minerals found in igneous rocks comprise four basic groups of **clays** (hydrated aluminum silicates) identifiable in marine sediment. *Chlorite* is a greenish platy mineral that forms from the decomposition of ferromagnesian minerals and contains magnesium and iron. *Montmorillonite* contains sodium and potassium, as well as calcium, magnesium, and iron. The chemical composition of *illite* is usually restricted to the metallic ions of potassium, iron, and magnesium. *Kaolinite* is a product of the weathering of potassium feldspar and muscovite and contains only the metallic ions aluminum and potassium. In addition, **zeolites** are a group of hydrated aluminum silicates resulting primarily from the weathering of feldspars and possessing varying quantities of potassium, sodium, and calcium. Phillipsite is one of the most abundant zeolites found in marine sediments.

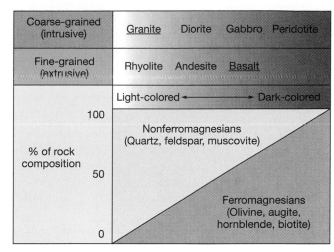

FIGURE 4–3
Igneous Rock Classification. This table shows the most common types of igneous rocks. The coarse-grained ones cool slowly deep beneath Earth's surface, and granite, the typical intrusive rock of the continents, is the most common variety. Peridotite is found in Earth's mantle. The fine-grained rocks, which are associated with volcanic activity, cool rapidly at or near Earth's surface. The most common variety is basalt, which is characteristic of the oceanic crust. The rocks on the left of the table are light colored and contain mostly nonferromagnesian minerals. Moving to the right, ferromagnesian minerals replace nonferromagnesian minerals, and the rocks become darker in color.

Wind Transport. A large percentage of the lithogenous particles that find their way into the deep-ocean sediments far from the continents are transported by the prevailing wind systems that remove small particles from the subtropical desert regions of the continents. Figure 4–4 shows the content of small shards of quartz that are present in the surface sediments of the ocean floor. It shows a

FIGURE 4–4
Lithogenous Quartz (SiO_2) in Surface Sediments of the World's Oceans. Occurring as chips and shards mostly within the 5–10 μm (0.0002–0.0004 in.) range, most quartz found in deep-sea sediments is believed to have been transported by winds. Desert areas of Africa, Asia, and Australia are important sources of eolian (windblown) quartz. Turbidity flows have been significant sources of quartz in the equatorial, northwestern, and northern Atlantic Ocean and the Bay of Bengal, where deep-bottom currents have also modified sediment distribution. The distribution of lithogenous clays transported by winds may show a pattern similar to that shown here. (After Leinen et al. 1986.)

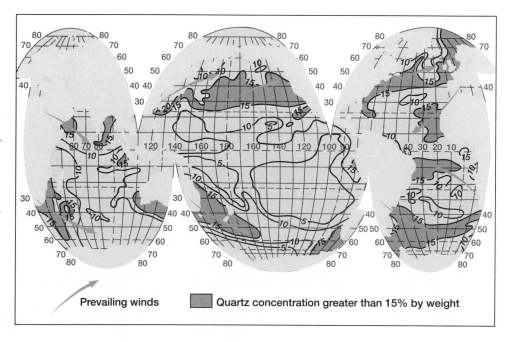

Prevailing winds

Quartz concentration greater than 15% by weight

close relationship between the location of strong prevailing wind systems and high concentrations of quartz shards in ocean sediment. The pattern of clay particle distribution in these sediments would likely be similar.

Biogenous Sediment

Biogenous means "derived from organisms." The insoluble remains of organisms, such as bones and teeth of animals and the protective coverings of minute plants and animals, are deposited on the ocean bottom. The most common chemical compounds found in biogenous sediment are **calcium carbonate** ($CaCO_3$) and **silica** (SiO_2).

Contributing most of the silica component of biogenous particles, which rarely exceed 100 μm in diameter, are microscopic plants (**diatoms**) and animals (**radiolarians**). **Foraminifera,** close relatives of radiolarians, are responsible for most of the calcium carbonate component, but other significant calcareous contributors are the larger pteropods and microscopic algae, **coccolithophores** (figure 4-5).

Massive, undisturbed, and detrital reef material is also included within the biogenous classification. Composed primarily of the calcareous skeletal structure secreted by corals and algae, coral reefs have been deposited as sedimentary rock, although by biological rather than physical processes.

Hydrogenous Sediment

Hydrogenous means "derived from water." These deposits are formed by a chemical reaction that occurs within seawater. Manganese deposits, phosphorite, and glauconite are minerals that form by chemical precipitation from water. The rate of accumulation of this type of sediment is very slow.

Manganese Nodules. Among the most important sediments on the ocean floor in terms of economic potential are the **manganese* nodules** (polymetalic nodules) found on the deep-ocean floor (figure 4–6). Such nodules have been known to be relatively abundant in all of the major ocean basins since the voyage of the *Challenger.* The major components of these nodules are MnO_2 and Fe_2O_3, with the average manganese dioxide content ranging around 30 percent and the average iron oxide content around 20 percent by weight. Also occurring in economically significant concentrations in the nodules are copper, cobalt, and nickel. Although these concentrations are usually less than 1 percent, they can exceed 2 percent by weight.

Since iron and manganese both occur in hydrogenous deposits in concentrations greater than those they reach in igneous rocks, there exists a problem of explaining the concentration of these elements in the marine deposits. This problem has not yet been successfully solved, but it has been concluded that there are three possible sources for manganese and iron. It is thought that they are derived from (1) the weathering of volcanic material produced by volcanic activity on the ocean floor; (2) concentration in hydrothermal springwater found at the axes of oceanic ridges and rises; and (3) runoff from the continents that carries the iron and manganese minerals as soluble compounds from the continents. There seems to be little variation in the chemistry of deep-ocean water throughout the world ocean. Studies on the relationship between manganese nodules and organisms suggest that the presence of microorganisms, primarily bacteria, may be a determining factor in the formation of nodules. The Manganese Nodule Project (MANOP), which began in 1977, uses sediment traps to catch particles falling to the ocean floor and a Bottom Lander to carry out seabed experiments designed to explain how nodules form and why they have the observed range of composition.

Phosphorite. Phosphorus in the form of P_2O_5 is found abundantly as a precipitate in nodules, as a thin crust on the continental shelf, and on banks at depths above 1000 m (3280 ft). Concentrations of phosphates in such deposits commonly reach 30 percent by weight. Again, the exact process of formation is not determined, but the deposits seem to be associated with areas of upwelling water rich in phosphorus.

Glauconite. This generally greenish hydrous silicate has a complex and variable composition which includes ions of potassium, magnesium, and iron. **Glauconite** may form by submarine weathering of ferromagnesian minerals, although the exact mode of formation is not known. Glauconite, which is found to depths of 2500 m (8200 ft) but more commonly on topographically high areas near the coastline, is considered to form only in the marine environment. Forming as casts in organic remains and as encrustations, glauconite is more commonly found in grains of silt or sand size in mud and sand deposits. Deposits in which the glauconite is sufficiently abundant are frequently referred to as *green sands* or *green muds* due to the coloration that results from the presence of glauconite.

Carbonates. The two most important carbonates in marine sediment are **aragonite** ($CaCO_3$) and **calcite** ($CaCO_3$). Aragonite and calcite have the same chemical

---

*The element manganese (Mn) has an atomic weight (54.94) that is more than twice the atomic weight of magnesium (Mg) (24.31). Manganese falls next to iron (Fe), which has an atomic weight of 55.85, on the periodic table of elements (Appendix IV).

A.

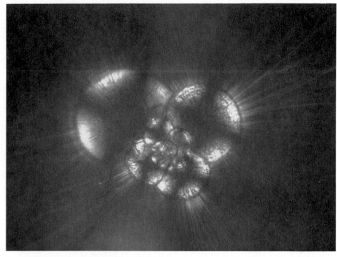

C.

B.

D.

FIGURE 4–5
Microscopic Skeletons from the Deep Sea. *A,* magnified 160 times by scanning electron microscopy, the skeletons are foraminifers and radiolarians found in sediment cores taken at Site 55 in the Caroline Ridge area of the west Pacific Ocean by the Deep Sea Drilling Project. They were recovered in water 2843 m (9315 ft) deep, with further penetration of 130 m (426 ft) into the ocean bottom. Rounded globose forms are foraminifers, whose skeletons are calcareous. Forms with reticulate skeletons are radiolarians and are siliceous. In life, these tiny organisms lived near the ocean surface, but upon death their skeletons fell to the sea bed to become entombed in the sediments. By carefully studying them, scientists can determine how long ago they lived and can learn much about the history of the ocean. (Courtesy of the Deep Sea Drilling Project, Scripps Institution of Oceanography, University of California, San Diego.) *B,* coccoliths magnified about 10,000 times. (Photo courtesy of Deep Sea Drilling Project, Scripps Institute of Oceanography.) *C,* foraminifer, *Orbulina universa,* is magnified 300 times. (Photo courtesy of Dr. Howard Spero.) *D,* pteropod, *Corolla spectabilis.* (Photo courtesy of James M. King, Graphic Impressions, P. O. Box 21626, Santa Barbara, CA 93121.)

A.

B.

FIGURE 4–6
Hydrogenous Sediment. *A,* photograph of the surface of a manganese nodule, magnified 156 times by a scanning electron microscope. The small tube-and-dome structures were built by organisms for shelter, according to Dr. Jimmy Greenslate at Scripps Institution of Oceanography. He has studied 71 nodules from the Pacific Ocean and found biologically derived structures both buried inside them and covering large portions of their surfaces. Nodules may owe their existence to organisms that participate in their formation. *B,* manganese nodules on the Pacific Ocean sea floor are formed as chemical precipitates. They contain quantities of manganese, iron, copper, nickel, and cobalt. Although the manganese is inferior to manganese ores mined on land, it is still a valuable commodity, making harvesting of the lumps a probability in the future. (Courtesy of Scripps Institution of Oceanography, University of California, San Diego.)

composition but different crystalline structures. Aragonite is the less stable form and over a period of time changes to calcite. Carbonate precipitation may occur directly from ocean water through physiochemical precipitation when water conditions are right. The most common forms of precipitated carbonate are aragonite crystals less than 2 mm (0.08 in.) in length and **oolites,** onionlike spheres that precipitate around a nucleus and reach diameters of less than 2 mm (0.08 in.).

Since ocean water is essentially saturated with calcium (Ca^{2+}), the only factor preventing precipitation of calcium carbonate (the most common carbonate) is the absence of carbonate ions (CO_3^{2-}). Carbon dioxide (CO_2) is important in carbonate chemistry. To see how it influences the precipitation of carbonates, we need to look at the various forms in which CO_2 is stored in ocean water. The following reactions tell us that when carbon dioxide combines with water in the ocean, carbonic acid (H_2CO_3) is formed. Carbonic acid may break down and produce a hydrogen ion (H^+) and a bicarbonate ion (HCO^-_3). On further breaking down of the bicarbonate ion, an additional hydrogen ion may be formed along with carbonate ion (CO_3^{2-}).

H_2O	+	CO_2	$\rightleftharpoons$	H_2CO_3	$\rightleftharpoons$	H^+	+ $HCO_3 \rightleftharpoons H^+ +$	H^+	+	CO_3^{2-}
Water		Carbon dioxide		Carbonic acid		Hydrogen ion	Bicarbonate ion	Hydrogen ions		Carbonate ion
				Low pH	$\longleftarrow\qquad\longrightarrow$			High pH		
				7.5				9.0		

The presence of carbon dioxide and the carbonate ion in ocean water is mutually exclusive. If there is a high concentration of CO_2, there will be little or no CO_3^{2-} in the water. Removal of carbon dioxide raises the concentration of CO_3^{2-}, so we might look to areas where photosynthesis is occurring at rates that remove significant amounts of carbon dioxide from the water for areas of carbonate precipitation. Heating water also reduces its ability to dissolve carbon dioxide and decreases carbon dioxide concentration. Both factors are at work in areas where precipitation is known to occur. For instance, in the Bahama Banks, water that is moving north from the deep flow of the Florida Straits moves up over the shallow bank, where it is heated and exposed to high plant productivity. Both changes tend to reduce carbon dioxide content of the water and enhance precipitation of carbonates.

Cosmogenous Sediment

Typical particles of cosmic origin found in marine sediment are microscopic magnetic spherules rich in iron and rocky chondrules (figure 4–7). They had long been thought to have been shed from meteors entering Earth's atmosphere. Recent studies have, however, resulted in a different conclusion. Based on the comparison of their shape, size, and composition with that of meteorites, in-

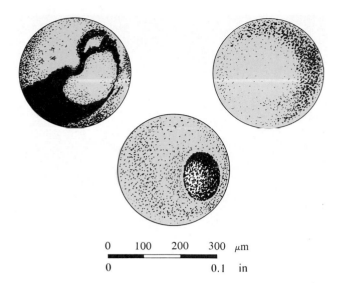

0   100   200   300 μm

0   0.1   in

FIGURE 4–7
Cosmogenous Sediment. Cosmic spherules from the *Challenger* expedition.

vestigators believe they form in the asteroid belt as sparks produced when asteroids collide. They then rain down on Earth as a general component of space dust.

Recent investigations have shown that a 170-km (106-mi) diameter crater along the north coast of Yucatán peninsula, Mexico, may well be the site of a meteor impact 65 Ma that could have brought about the extinction of the dinosaurs and other species. Analysis of cores

taken in the Antarctic Basin about 1400 km (870 mi) west of Cape Horn indicates that an asteroid meteorite struck the area 2.3 million years ago. This is the only known impact of a meteor into a deep-ocean basin (figure 4–8).

The evidence is in the form of three types of particles—all of which are thought to represent meteoric material. The particles are a nonterrestrial basalt that shows signs of impact stress, vesicular particles that formed from melted meteoric material, and a few pieces of metal. The basaltic particles and vesicular particles have similar compositions (figure 4–9). Commonly the basaltic particles are embedded in or are in contact with vesicular material. This indicates that vesicular material, while in a plastic state, collided with the solid basaltic particles in the atmosphere during impact.

The metallic element iridium (Ir) is found in greater concentrations in meteoric material than on Earth, so layers of sediment that contain unusually high concentrations of Ir are usually accepted as evidence of meteoric impact if there is no evidence of certain types of volcanic events that might have produced them.

Based on the Ir concentrations shown in figure 4–8, the amounts and distribution pattern suggest that the meteorite that hit the Pacific Ocean was between 0.5 and 1 km (0.31 and 0.62 mi) in diameter. It hit the ocean without producing a crater in the ocean floor.

This event occurred near the start of the Pleistocene ice age. Although it surely didn't cause the climatic change, it could have contributed to a more rapid de-

FIGURE 4–8
Evidence of Meteorite Impact in Antarctic Basin. The impact of an asteroid ranging in diameter between 0.5 and 1.0 km (0.31 and 0.62 mi) occurred after an approach from a direction a little south of west according to the distribution of Ir-rich particles recovered in piston cores labeled on the map. The greatest concentrations of Ir-rich particles occur in cores 13-3 and 13-4, which are starred. (Data courtesy of Frank T. Kyte, University of California, Los Angeles.)

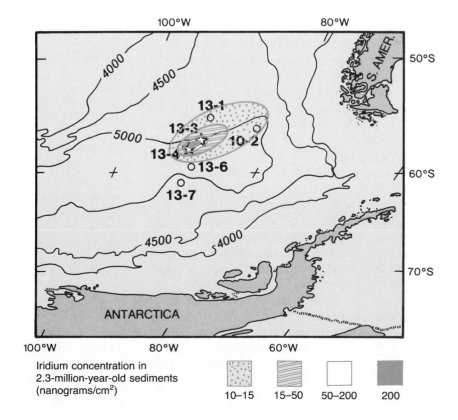

Iridium concentration in
2.3-million-year-old sediments
(nanograms/cm²)

10–15   15–50   50–200   200

A.

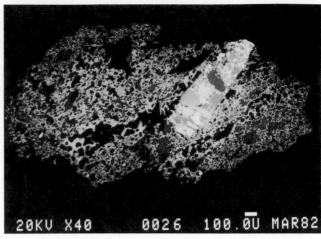

B.

FIGURE 4–9
Iridium-Rich Meteoric Particles. *A,* scanning electron micrograph (SEM) of a vesicular particle. Up to 50 percent of the volume of these particles is pore space. Bar length is 100 μm. *B,* SEM of a vesicular particle with a basaltic inclusion. Sharp boundary between basalt and vesicular material indicates the solid, rigid basalt collided with a plastic mass of vesicular material as they were propelled through the atmosphere. Bar length is 100 μm. (SEMs courtesy of Frank T. Kyte, University of California, Los Angeles.)

crease in global temperature resulting from the meteoric dust and water that the event could have sent into the stratosphere.

TYPES OF MARINE SEDIMENTARY DEPOSITS

As the result of marine sediment deposition, the relative amounts of lithogenous, biogenous, hydrogenous, and cosmogenous particles found in marine sedimentary deposits vary considerably. The two largest categories into which marine sedimentary deposits can be divided are neritic deposits, found near the continents, and oceanic deposits, characteristic of the deep-ocean basins.

Neritic sediment refers to material with a wide range of particle size that is composed primarily of lithogenous particles derived from the continents. It accumulates rapidly on the continental shelf, slope, and rise. Neritic sediment also contains biogenous, hydrogenous, and cosmogenous particles. These constitute only a minor percentage of the total sediment mass because of the rapid deposition rate of lithogenous particles near the continents. Neritic sediment is distributed along the continental shelf by surface currents and carried down the submarine canyons of the continental slope to the continental rise by turbidity currents. Here the sediment is further distributed by deep boundary currents.

Oceanic sediment is the very fine-grained material that accumulates at a slower rate on the deep-ocean-basin floor. It is possible for a greater variety of oceanic sediment types to accumulate because the rate of deposition of lithogenous particles is greatly reduced from that observed on the continental margins. Although the lithogenous component dominates the sediment found in most of the deeper basins of the ocean floor, biogenous and hydrogenous components are abundant in many areas of the ocean bottom.

Sediments of the Continental Margin (Neritic Sediments)

Due to the rise in sea level that occurred with the melting of glaciers at the end of the last ice advance, many rivers of the world today deposit their sediment in estuaries rather than carry it onto the continental shelf as they did at various times during the geologic past. In many areas the sediments that cover the continental shelf, **relict sediments,** were deposited from 3000 to 7000 years ago and have not yet been covered by recent deposits. Such sediments presently cover about 70 percent of the world's continental shelf.

Turbidites.   Although it seems unlikely that wave action and ocean current systems could carry coarse material beyond the continental shelf into the deep-ocean basin, there is evidence that much neritic sediment has been deposited at the base of the continental slope forming the continental rise. These accumulations thin gradually toward the abyssal plains. Such deposits are called **turbidites** and are thought to have been deposited by turbidity currents that periodically move down the continental slopes through the submarine canyons, carrying loads of neritic material that spreads out across the continental rise (figure 4–10).

Turbidites are characterized by graded bedding. This means that each deposit that is laid down has coarser material at its base and decreasing particle size toward the top of the deposit. This results from particles of different sizes settling out of a moving water mass as the velocity decreases. The coarser material settles out first, while the finer material stays in suspension longer. The seaward end of a turbidite deposit may be composed mostly of fine clay-sized particles, which represent a contribution to the sediment of the abyssal plains beyond the continental rise.

During the millions of years that sediments accumulated at the margins of continents, the continental shelf, slope, and rise developed as the surface features of a sedimentary wedge more than 10 km (6 mi) thick that contains over 75 percent of the sediment to be found on the ocean floor. Less than 25 percent of marine sediment will be found in the deep-ocean basins that cover in excess of 80 percent of the ocean floor.

Glacial Deposits. Poorly sorted deposits containing particles of all sizes, from boulders to clay, may be found in the high-latitude portions of the continental shelf. These glacial deposits were laid down by melting glaciers that covered the continental shelf area during the Pleisto-

FIGURE 4–10
Turbidites and Graded Bedding.
A, turbidity currents flow down submarine canyons and deposit sediment as deep-sea fans. *B,* examination of the deep-sea fans shows that they are composed of a series of deposits called turbidites. Each turbidite represents the sediment deposited by one turbidity current event. Each turbidite displays graded bedding in which the sediment particles grade from coarse at the bottom to fine at the top. (Reprinted by permission from Tarbuck, E. J., and Lutgens, F. K., *The earth: An introduction to physical geology,* 3d ed. (New York: Macmillan, 1990), figs. 19.7 and 19.8.)

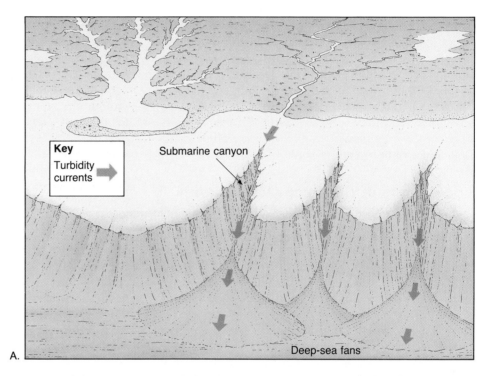

A.

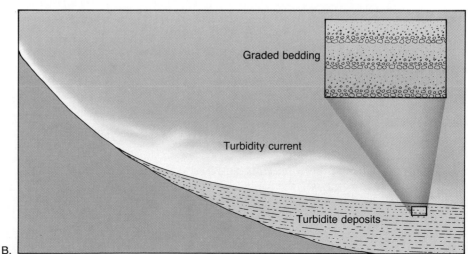

B.

cene epoch, when glaciers were more widespread than they are today and sea level was lower. Glacial deposits are still forming around the continent of Antarctica and around the island of Greenland by ice-rafting, a process by which rock particles trapped in glaciers are carried out to sea by icebergs that break away from glaciers as they push into the coastal ocean. As the icebergs melt, the lithogenous particles are released and settle to the ocean floor.

Carbonate Deposits. During the geologic past, deposits of limestone (CaCO₃) in the marine environment appear to have been widespread. Some contain fossil evidence indicative of a biogenous origin, while others appear to have formed as chemical precipitates. There appear to be very few places where nonbiogenous carbonates are presently forming. All these deposits are found in low-latitude shallow waters of continental margins or adjacent to islands. The Bahama Banks is the largest region where significant carbonate deposition is presently occurring, although deposits are also forming on the Great Barrier Reef, in the Persian Gulf, and in other local areas in the low latitudes. Of course, coral reefs are biogenous deposits of calcium carbonate found extensively on the continental margins of low-latitude continents and around oceanic islands.

Sediments of the Deep-Ocean Basins (Oceanic Sediments)

As was previously discussed, most of the neritic sediment is deposited on the continental margin as part of the continental shelf, continental slope, or the continental rise. Since the continental rise is a transitional feature between the continental margin and the deep-ocean basin, we may well conclude that neritic material is deposited in significant quantities in the deep-ocean basin as part of the deep-sea fan deposits making up the continental rise. This is true, but when we consider the total surface area represented by the deep-ocean basin, we find that more commonly no neritic material will be found in the sedimentary section. Covering most of the deep-ocean floor are two types of oceanic sediments, abyssal clay and oozes.

Abyssal Clay. Accumulating at a rate of approximately 1 mm (0.04 in.) per 1000 years, **abyssal clays** cover most of the deeper ocean floor. They are commonly red-brown or buff in color due to phillipsite and oxidized iron. Abyssal clay consists predominantly of clay-sized particles derived from the continents and carried by winds or ocean currents to the open-ocean regions. These particles, along with some cosmic and volcanic dust, settle slowly to the ocean floor.

A significant component of the abyssal clay is represented by the minerals montmorillonite and phillipsite,

which are thought to form from the interaction of ocean water with volcanic and other lithogenous particles.

Oozes. At somewhat shallower depths, biogenous material makes up a significant portion of the sediment that reaches the ocean bottom. It consists primarily of the minute protective hard coverings of microscopic plants and animals that accumulate on the ocean floor. If sediment contains 30 percent or more skeletal material by weight, that deposit is called an **ooze.** Oozes may be classified as **calcareous** (CaCO₃) or **siliceous** (SiO₂), depending on the chemical composition of the dominant type of biogenous material making up the deposit.

Oozes are not present on the continental margin. This is not because the skeletal remains of plants and animals do not accumulate there but because the rate of accumulation of lithogenous sediment is so great that organically derived material never makes up 30 percent or more of the sediment. Oozes can form only where the deposition of material other than the remains of plants and animals occurs at a very low rate.

More specific names may be given to oozes that indicate the types of organic remains most abundant in the deposits. *Diatom oozes* are siliceous oozes that contain mostly the remains of diatoms. Other descriptive names commonly used on this same basis are *radiolarian ooze, foraminifera ooze,* and *pteropod ooze.* The predominant type of ooze is *Globigerina ooze,* which is especially widespread in the Atlantic and South Pacific oceans. *Globigerina* is an abundant and widespread genus of foraminifera.

The rate of accumulation of biogenous material on the ocean floor depends upon three fundamental processes—productivity, destruction, and dilution. *Productivity* of the planktonic* forms that contribute most of the biogenous material in oceanic sediments is greater in the areas of upwelling along the margins of continents and associated with the diverging water masses in equatorial regions.

Destruction of skeletal remains occurs primarily through solution. The ocean is undersaturated with silica at all depths. So for siliceous shells to last until they are incorporated in the sediment, they must be thick, as thin fragments will be dissolved into the ocean water before they can settle out and be covered by the sediment. The solubility of calcium carbonate varies with depth in the ocean. At the surface, where the water temperatures are relatively high, the water is generally saturated with calcium carbonate. At greater depths, the temperature decreases and the carbon dioxide content increases, allowing the calcareous fragments to be readily dissolved. Below 4500 m (14,760 ft) the carbon dioxide content in-

---

*In this instance, primarily floating microscopic plants and animals.

creases to the point that calcium carbonate dissolves quite readily. Carbonate oozes are generally rare below a depth of 5000 m (16,400 ft). The depth at which calcium carbonate is dissolved as fast as it falls from above is the **carbonate compensation depth (CCD).** It may be as deep as 6000 m (19,680 ft) in portions of the Atlantic Ocean, but may be above 3500 m (11,480 ft) in regions of low biological productivity in the Pacific Ocean.

Dilution is illustrated by the fact that calcareous oozes are not found on the continental margins where rapid rates of deposition of lithogenous sediment prevail. Even in areas of upwelling where productivity is high, oozes do not occur on the continental shelf because the concentration of biologically derived silica or calcium carbonate remains well below 30 percent by weight as a result of the high rate of deposition for lithogenous materials in these areas. The rates of deposition for biogenous sediment range between 1 and 15 mm (0.04–0.6 in.) per 1000 years, depending on the net effect of productivity and destruction. Figure 4–11 shows the relationships among carbonate compensation depth, sea-floor spreading, productivity, and destruction in determining what type of sediment will accumulate on the ocean floor.

Figure 4–12 shows the percentage by weight of calcium carbonate accumulation in the ocean basins. It shows that high concentrations (in excess of 80 percent) are found high on the oceanic ridges, whereas little is found in deep-ocean basins like the North Pacific, where the bottom lies beneath the carbonate compensation depth.

DISTRIBUTION OF OCEANIC SEDIM

Oceanic sediments cover about 75 percent of the ocean bottom. Table 4–2 shows that calcareous oozes cover almost 48 percent of the deep-ocean floor. Abyssal clay covers 38 percent and siliceous oozes 14 percent of the total area. Calcareous oozes are the dominant oceanic sediment in the Indian and Atlantic oceans, while abyssal clay is the principal oceanic sediment in the Pacific Ocean (figure 4–13). This is undoubtedly related to the fact that the Pacific Ocean is deeper, so that more of its bottom lies beneath the carbonate compensation depth. Siliceous oozes cover a smaller percentage of the ocean bottom in all the oceans because regions of high productivity of diatoms and radiolarians, the major components of these deposits, are restricted to areas of high biological productivity along the equator and in the Antarctic. Biogenous silica can be differentiated from lithogenous silica because biogenous silica contains molecular water; it is called **opal** ($SiO_2 \cdot nH_2O$). The distribution of opal in the surface sediments of the oceans is shown in figure 4–14.

Fecal Pellets

One major problem that puzzled marine geologists studying oceanic sediment is that sediments on the deep-ocean floor very closely reflect the particle composition of the surface water directly above. This was difficult to understand because it would typically take these very tiny

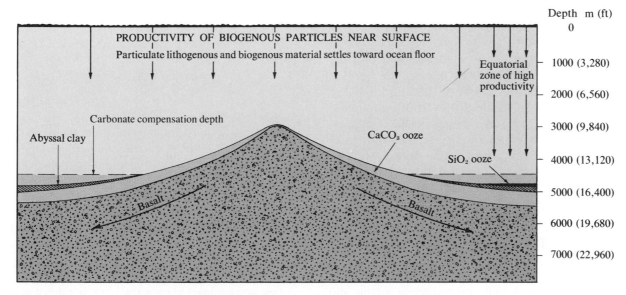

FIGURE 4–11
Sea-Floor Spreading and Sediment Accumulation. New basaltic oceanic crust is forming at ridges that run across the ocean floor and rise to depths of about 3000 m (9840 ft). As the crust moves away from the ridges and sinks to greater depths, the type of sediment that accumulates on it changes. Until it sinks below the carbonate compensation depth, the typical sediment that forms is $CaCO_3$ ooze. Below this depth $CaCO_3$ is dissolved, and abyssal clays dominate. If the crust passes beneath a region of high biological productivity, such as the Pacific equatorial region, oozes could again accumulate.

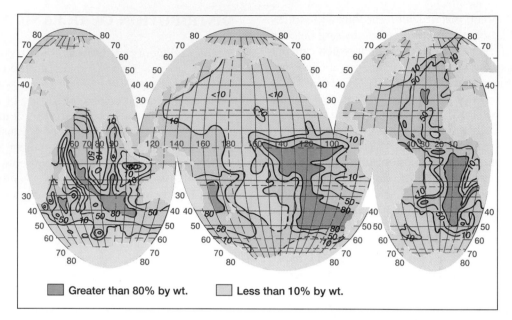

Greater than 80% by wt.   Less than 10% by wt.

FIGURE 4–12
Calcium Carbonate in Surface Sediments of the World's Oceans. Concentrations of calcium carbonate ($CaCO_3$) in ocean sediments is negatively correlated with ocean depth. The deeper the ocean floor, the less calcium carbonate found in the sediment. It is also low in sediments accumulating beneath cold, high-latitude waters. Extensive areas of concentrations of calcium carbonate in the sediment in excess of 80 percent are associated with relatively shallow ocean floor on the Carlsberg Ridge in the Indian Ocean, the East Pacific Rise, and the Mid-Atlantic Ridge. Note that the deepest ocean basin, the North Pacific, lies for the most part beneath the carbonate compensation depth and has very little calcium carbonate in its accumulating sediment. (After Biscaye et al., 1976; Berger et al., 1976; and Kolla et al., 1976.)

particles from 10 to 50 years to sink from the ocean surface to abyssal depths. During this interval, a horizontal ocean current of only 1 cm/s (about 0.02 mi/h) would carry them from 3000 to 15,000 km (1800–9300 mi) laterally before they reached the deep-ocean floor.

However, study of GEOSECS samples shows that 99 percent of particles that fall to the ocean floor do so as part of **fecal pellets** produced by tiny animals living in the water column above. These pellets (figure 4–15), though still small (50–100 μm [0.002–0.004 in.]) in their smallest dimension, are large enough to allow the particulate matter to reach the abyssal ocean floor in 10 to 15 days. If this process is important in transporting particles from surface water to abyssal depths, it could readily explain the similarity in particle composition of the surface waters and the sediment immediately below them.

HOW FAST ARE THE CONTINENTS GROWING?

After we have considered all of the volcanic activity and subduction associated with global plate tectonics, the erosion of continents, and deposition of marine sediment, we may ask, "How fast are continents growing?"

A study conducted by the U.S. Geological Survey has created a possible budget for the growth and denudation of continents. By calculating the yearly flux of various components of marine sediment, seamounts, and island arc volcanoes, a somewhat tentative rate of growth for continents since the breakup of Pangaea has been determined.

It was determined that:

1. Ocean crust is recycled every 110,000,000 years, and its average age is 55,000,000 years.
2. Lithogenous particles accumulate in the deep ocean at a rate of 1.27 km³/yr. Biogenous particles, the chemical components of which come ultimately from continents, accumulate at the rate of 0.38 km³/yr. This totals to a rate of continental denudation of 1.65 km³/yr. (Sediment deposited on the continental shelf, slope, and rise is not included because, while derived from the continent, it is still considered to be part of the continent.)
3. Adding to this the 0.05 km³/yr of volcanic sediment that accumulates gives a total yearly rate of sediment accumulation of 1.7 km³/yr.
4. Seamounts add 0.2 km³/yr of mass to the ocean crust, and volcanic island arcs add 1.1 km³/yr.

TABLE 4–2
Distribution of Oceanic Sediment. This graph shows that the percentage of an ocean basin floor covered by calcareous ooze decreases with increasing mean depth of the basin. This is probably because the deeper an ocean basin is the greater the percentage of its floor that lies beneath the carbonate compensation depth. The mean depths calculated for this table exclude the shallow adjacent seas where little oceanic sediment accumulates. Notice that the dominant oceanic sediment found in the deepest ocean basin, the Pacific Ocean, is abyssal clay, while calcareous ooze is the most widely deposited abyssal sediment in the shallower Atlantic and Indian oceans.

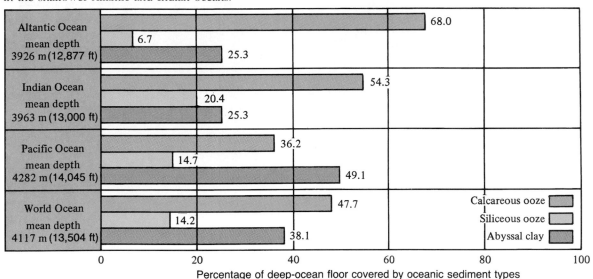

Source: After Sverdrup, Johnson, and Fleming, 1942.

5. Total material available to accrete to continents during subduction is 1.7 + 0.2 + 1.1, or 3 km³/yr.
6. Subtracting the rate of continental denudation (1.65 km³/yr) from this total leaves 1.35 km³/yr of possible continental growth.
7. Whatever amount of deep-sea sediment or seamounts is subducted will have to be subtracted from this growth rate. The numbers for these variables are not yet known.

OCEAN SEDIMENTS AS RESOURCES

Many potential mineral and organic resources are contained within the deposits on the ocean floor. Because of the high costs involved, few will probably be exploited in the near future. Some resources are presently being exploited, however, and a few have been explored sufficiently for us to discuss their nature and future potential.

Petroleum

The remains of microscopic plant and animal cells buried within marine sediments before they can decompose are the source of our most valuable marine resource. More than 95 percent of the value of nonliving resources exploited from the oceans is in the form of oil and gas.

Petroleum provides most of the energy on which the world economy depends. There is believed to be enough oil to meet our needs until about the year 2020. Beyond that date, major changes in the energy supply will be needed. We would then need to convert to unconventional sources such as extra-heavy oil, tar sands, and other high-cost sources. Converting to natural gas would extend the fuel supply only a few years.

With almost no hope of finding major undiscovered reserves on land and only a slightly increased likelihood of finding them offshore, future oil exploration will still be intense in offshore waters. The percentage of world oil production coming from offshore has increased from a trace in 1930 to over 26 percent in 1987 (figure 4–16).

Major discovered offshore reserves exist in the Persian Gulf, the Gulf of Mexico, off southern California, the North Sea, and in the East Indies. Additional reserves are expected to be found off the north coast of Alaska, the Canadian Arctic, Asian seas, Africa, and Brazil, but there is little likelihood they will be able to reverse the existing trends.

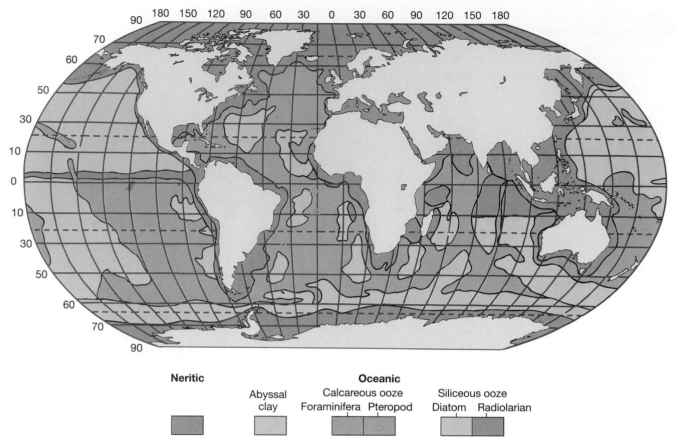

FIGURE 4–13
World Distribution of Neritic and Oceanic Sediments. Beyond the neritic deposits of the continental margins lie the oceanic marine deposits. Abyssal clays are found in the deeper ocean basins, where the bottom lies beneath the carbonate compensation depth. Calcareous oozes are found best developed on the relatively shallow deep-ocean environments of the oceanic ridges and rises. Siliceous oozes are found beneath areas of unusually high biological productivity like the Antarctic and equatorial Pacific Ocean.

Sand and Gravel

The offshore sand and gravel industry is second in value only to that of petroleum. This resource, which includes rock fragments and shells of marine organisms, is mined by technology similar to that used on land. It is used primarily for beach fill, land fill, and concrete aggregate.

Offshore deposits are a major source of sand and gravel in New England, New York, and throughout the Gulf Coast. Many European countries, Iceland, Israel, and Lebanon also depend heavily on such deposits.

Some gravel deposits are rich in valuable minerals. Diamonds have been found in gravel deposits in South Africa and Australia. Sediments rich in tin have been mined for years from Thailand to Indonesia. Platinum and gold have been found in deposits in gold mining areas throughout the world.

One of the major concerns of mining sand and gravel deposits is that mining not be conducted too near shore. Nearshore mining operations may result in damage to beaches as beach material moves offshore into the mined areas.

Sulfur

Bacteria use the sulfate ion in gypsum and anhydrite deposits associated with salt domes as an oxidizing agent for organic matter. Gypsum and anhydrite are minerals associated with the formation of halite (NaCl) by the evaporation of seawater. When the Atlantic Ocean began to form, isolated basins of seawater were evaporated, and a thick bed of halite was laid down under the denser sediments that have subsequently accumulated in areas such as the present Gulf Coast. The salt rises in diapirs through the sediments to produce salt domes. In time, the salt on the top of the dome is dissolved by water, and the more insoluble gypsum and anhydrite remain (figure 4–17). The bacteria convert these minerals that cap the salt dome to free **sulfur.**

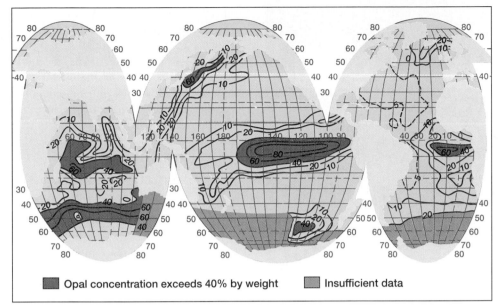

Opal concentration exceeds 40% by weight   Insufficient data

FIGURE 4–14
Biogenous Silica (SiO$_2$·nH$_2$O—Opal) in Surface Sediments of the World's Oceans. The distribution pattern of opal in surface sediments of the ocean shows maximum concentrations in areas of highest biological productivity. These particles are produced by diatoms and radiolarians in the surface waters. In the equatorial and northwest Pacific, opal is produced predominantly by radiolarians. Elsewhere, diatom remains are more abundant. The highly productive waters south of the Antarctic Convergence show up well in the south Indian Ocean and southeast Pacific Ocean. In the southern Atlantic Ocean and southwest Pacific Ocean there are insufficient data to determine the relationship between highly productive surface waters and opal concentrations in the sediment. (After Leinen et al., 1986.)

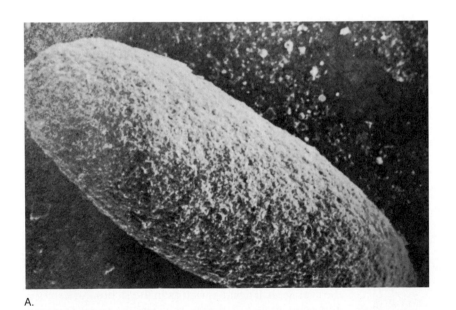

A.

B.

FIGURE 4–15
Fecal Pellets. *A,* fecal pellet produced by zooplankton. The pellet is 200 μm (0.008 in.) long. *B,* view of the surface of the fecal pellet showing the remains of small phytoplankton and detritus. (Photos courtesy of Susumu Honjo, Woods Hole Oceanographic Institution.)

FIGURE 4–16
Offshore Drilling Rig. This rig in the Gulf of Mexico off the Texas coast is drilling for gas and oil. (© Walter Frerck/Odyssey/Chicago.)

Using the Frasch process, one can mine the sulfur by pumping hot water and air into the sulfur to melt it and force it back to the surface through the space between the hot water pipe and another larger pipe that surrounds it. Recently, with the recovery of large amounts of sulfur by pollution control equipment, most mining operations have been shut down.

Phosphorite

Although no commercial phosphorite mining from the oceans is presently occurring, the marine reserve is estimated to be more than 50 billion tons. **Phosphorite** occurs at depths less than 300 m (1000 ft) on the continental shelf and slope. Deposits can be used to produce

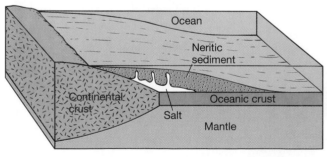

FIGURE 4–17
Sulfur Deposits. Salt, anhydrite, and gypsum were deposited on the floor of the newly formed Atlantic Ocean 170 to 180 million years ago when the ocean began to form. Evaporation of ocean water in the early shallow sea produced these deposits which were subsequently covered by denser sediments of lithogenous origin. The less dense salt flowed upward through the denser sediments as diapirs to form salt domes.

phosphate fertilizer if economic conditions make them recoverable. Some shallow sand and mud deposits contain up to 18 percent phosphate. Some phosphorite deposits occur in nodule form. A hard crust forms around a nucleus and may be as small as a sand grain or up to a meter in diameter. Nodules may contain over 25 percent phosphate. Most land sources have had their phosphate concentration enriched to more than 31 percent by groundwater leaching.

Polymetallic Crusts and Nodules

Manganese nodules have been known to exist in the deep ocean since the voyage of the HMS *Challenger* (1872–76). Containing significant concentrations of manganese and iron and smaller concentrations of copper, nickel, and cobalt, these nodules caught the attention of mining companies, which began in the 1960s to assess the potential for mining them. Explorations found the richest metallic content in deposits in the eastern Pacific Ocean between Hawaii and Mexico (figure 4–18).

Recently, interest has faded in the face of a depressed metals market. A secondary source of concern by the mining companies is the uncertainty of ownership of mining claims growing out of disagreement concerning provisions of the United Nations Law of the Sea Convention. This Convention has still not been signed by most of the developed nations with mining interests.

Cobalt is an important strategic metal that is not mined in the United States. It is required to produce dense, strong alloys with other metals for use in high-speed cutting tools, powerful permanent magnets, and jet engine parts.

Major enrichments of cobalt in Earth's crust are confined to central southern Africa and deep-ocean nodules and crusts. Considering the unstable political situation in Africa, the United States has been looking to the ocean floor as a more dependable source. However, the political and economic uncertainties associated with the United Nations Law of the Sea Convention have made mining interests back away from the mining of deep-sea nodules beyond the 200-mile-wide Exclusive Economic Zone (EEZ). In 1981 **cobalt-rich manganese crusts** were found on the upper slopes of islands and seamounts that lie within the EEZs of the United States and its allies. The cobalt concentrations in these crusts are on the average half again as rich as in the richest African ores and at least twice as rich as in deep-sea manganese nodules.

The greatest concentrations seem to be associated with depths between 1000 and 2500 m (3280 and 8200 ft) on the flanks of islands and seamounts in the central Pacific Ocean. Additional sampling needs to be done to assure that crust deposits are extensive enough to provide a dependable source of cobalt. At least one international

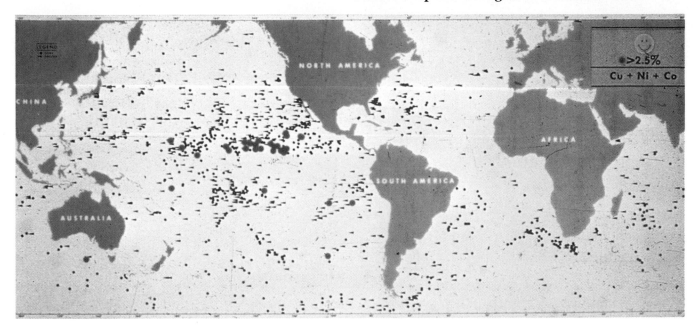

FIGURE 4–18
Distribution of Deep-water Manganese Nodules Containing More Than 2.5 Percent Combined Copper, Nickel, and Cobalt. The manganese (ferromagnesian) nodule deposits with the greatest economic potential are those that have the highest concentrations of Cu, Ni, and Co. They are found between the Clarion and Clipperton fracture zones in the eastern Pacific Ocean. See figure 2–22. (Courtesy Scripps Institution of Oceanography, University of California, San Diego.)

consortium, International Hard Minerals Company, is moving into the study of the feasibility of crust mining. If the results of these investigations are positive, the first deep-sea mining operations may be for cobalt and other metals found in these crusts that lie within the EEZs of the United States or one of its allies. There are at least 100 of these seamounts in the EEZs of the Line Islands and Hawaiian Islands, and each could yield up to 4 million metric tons of ore (figure 4–19).

SUBSEA DISPOSAL OF HIGH-LEVEL NUCLEAR WASTE

For nearly four decades, high-level nuclear waste has been accumulating from the production of nuclear weap-

ons and from commerical power generation. The U.S. alone has more than 75 million gallons of waste from weapons production and 12,000 metric tons of spent reactor fuel that must be safely disposed of until it is no longer dangerous.

This material will be radioactive for more than one million years, but a "safe" disposal system would likely require a shorter time of highly secure confinement. The wastes are composed of over 50 isotopes, each of which has different chemical, half-life, abundance, and radioactive emission characteristics. Considering such factors, investigators believe confinement that allows no more release of radiation into the atmosphere than that of the natural uranium ore from which it was generated may be considered safe and is possible with today's technology (figure 4–20).

FIGURE 4–19
Cobalt-Rich Ocean Crusts. Cobalt-rich manganese crusts, composed primarily of iron and manganese oxides, contain up to 2.5 percent cobalt. This high cobalt concentration in the crusts, which are up to 2 cm (0.8 in.) thick, has aroused interest in their economic potential.

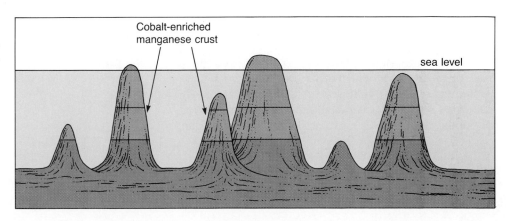

FIGURE 4–20
Radiation Doses from Various
Sources in Millirems per Year.
The Environmental Protection Agency
is responsible for setting the minimum performance requirements (levels of radiation emitted from high-
level disposal sites). It is believed
that they will be no higher than that
of natural uranium ore.

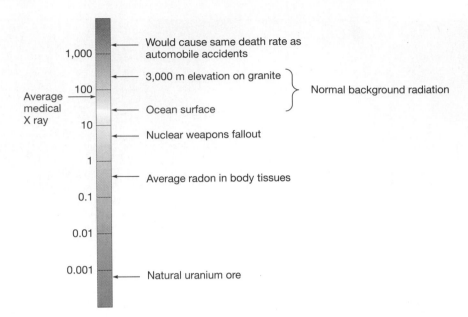

The major emphasis at present is on disposal at land sites because the materials remain accessible. However, since 1974, research has been conducted into the possibility of disposal of high-level nuclear waste in the deep sea. This research was terminated by the United States in 1986 because of budget constraints and on an international basis in 1987 when international agreement was reached to pursue land-based disposal. However, the seabed disposal research did produce encouraging results. Given the political realities that may be faced as work proceeds toward land-based disposal, deep-sea disposal may be considered again in the future. The major features of the seabed disposal program are described below.

Given the requirements that the waste must be kept out of the way of human activities, safe from exposure by natural erosion, and away from seismically active regions, the centers of oceanic lithospheric plates might be ideal disposal sites. Such regions are beneath at least 5000 m of water far from the marine activities of humans. Fishing, petroleum production, and mining activities are now conducted primarily on the continental shelves. Only a potential area of manganese nodule mining between Mexico and the Hawaiian Islands would need to be avoided.

The lithospheric plates are covered with up to 1000 m of fine sediment that has accumulated uninterrupted for over 100 million years in some regions. This pattern of sediment accumulation indicates long periods of stability as the plates move across Earth's surface, and the pattern can be expected to continue for millions of years. In addition, the sediment could serve as an ideal medium in which to place the high-level nuclear waste. Many of the radionuclides, after being released from the canisters (which may fail within 1000 years), would naturally adhere to the clay particles in the sediment. This would slow diffusion of radioactivity away from the burial site.

A number of mid-plate, mid-gyre sites (MPGs) away from ocean currents have been identified in the North Atlantic and North Pacific oceans (figure 4–21). The initial plan is to place the canisters under 30 to 100 m of sediment at intervals of at least 100 m. How to emplace the canisters has not been decided yet, but it may be possible to simply drop them from an appropriate distance above the ocean floor (figure 4–22).

Models indicate that heat release from the waste will be transferred almost entirely by conduction, so upward convection of radiation carried by sediment pore water may be negligible. This is a crucial consideration. Although these results are promising, yet to be completed are models of how well the burial holes will reseal themselves and the means by which radiation will be transported through the water column once it reaches the sediment surface. Should this option ever be used, the present thinking is that it will not be for at least another 50 years. Much additional testing will be conducted before such a form of disposal would be put into operation.

FIGURE 4–21
MPG Regions. Mid-plate, mid-gyre (MPG) regions in northern oceans where the environment can be expected to be stable for millions of years.

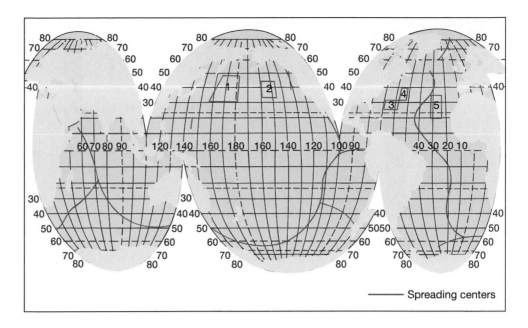

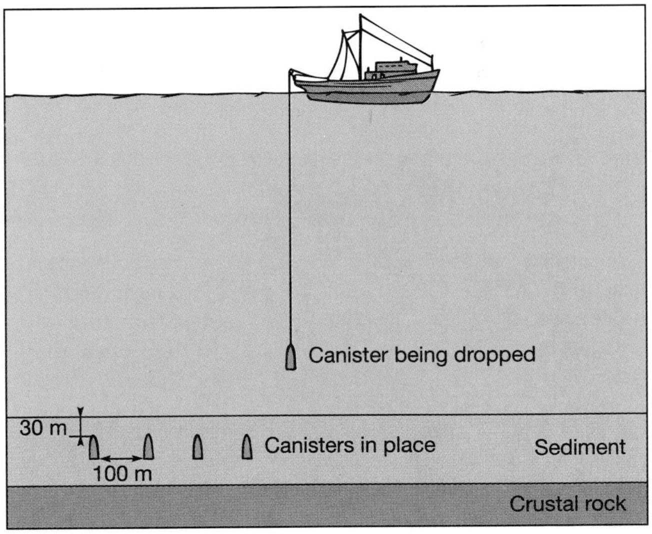

——— Spreading centers

FIGURE 4–22
Canister Placement. Plan of canister emplacement shows them buried at least 30 m (98 ft) beneath ocean floor spaced 100 m (328 ft) apart. Because of the softness of the sediment, it may be possible to implant the canisters by simply dropping them from an appropriate distance above the sediment surface. (Figure not to scale.)

Canister being dropped

30 m

100 m

Canisters in place   Sediment

Crustal rock

SUMMARY

Sediments that accumulate on the ocean floor are classified by origin as lithogenous (derived from rock), biogenous (derived from organisms), hydrogenous (derived from water), and cosmogenous (derived from beyond Earth's atmosphere). Sediment texture, determined in part by the size and sorting of particles, is affected greatly by the type of transportation that brought it to the deposit (water, wind, or ice) and the energy conditions under which it was deposited.

Lithogenous sediment is composed of fragments of rocks. Biogenous sediment is composed primarily of the compounds calcium carbonate ($CaCO_3$) from the remains of foraminifera, pteropods, and coccolithophores and silica (SiO_2) from the remains of diatoms and radiolarians. Hydrogenous sediment includes a wide variety of materials that precipitate from the water or form from interaction of substances dissolved in the water with materials on the ocean floor. Manganese nodules, phosphorite, glauconite, carbonates, and zeolites are examples. Cosmogenous sediment is composed of nickel-iron spherules and silicate chondrules that probably represent fragments resulting from asteroid collision.

Neritic sediment accumulates rapidly along the margins of continents. It is dominated by sediment of lithogenous origin. Due to the recent rise in sea level from melting glaciers, many rivers throughout the world, including those flowing into the Atlantic along the east coast of the United States, are not now depositing sediment on the continental shelves but rather in their estuaries. About 70 percent of the sediment covering the continental shelves is relict sediment 3000 to 7000 years old.

Turbidity currents are thought to transport shelf sediment down submarine canyons and deposit it as turbidite on the deep-ocean floor. More than 75 percent of the sediment mass found on the ocean floor is part of the thick sediment wedge underlying the continental shelves, slopes, and rises. In high latitudes this accumulation includes poorly sorted glacial deposits.

Oceanic sediment accumulates at low rates on the floor of the deep-ocean basins, far from the continents. In the deeper basins, where most biogenous sediment is dissolved before reaching bottom, abyssal clay deposits predominate. Oozes composed of over 30 percent by weight of biogenous sediment are found in shallower basins on ridges and rises. The rate of biological productivity measured against the rates of destruction and dilution of biogenous sediment determines whether abyssal clay or oozes will form on the ocean floor. Much of the destruction of calcium carbonate occurs because the carbon dioxide content of the ocean water increases with depth. At the carbonate compensation depth, ocean water dissolves calcium carbonate at a rate equal to the rate it settles from above, thus preventing accumulation on the ocean floor below that depth.

Although it would take from 10 to 50 years for individual sediment particles in the ocean surface waters to settle to the bottom, most appear to be combined into larger aggregates as fecal pellets of small marine animals and reach the ocean floor in 10 to 15 days. This process of particle transport explains the similarity in particle composition of the surface waters and sediment on the ocean floor immediately below them.

It is estimated that the rate of continental growth is 1.35 km³/yr as a result of an annual rate of continental denudation of 1.65 km³ and a rate of continental accretion of 3 km³. The annual rate of sediment accumulation in the ocean basins from all sources is 1.7 km³.

The most valuable resource taken from the ocean today is petroleum, which accounts for more than 95 percent of the value of nonliving resources presently being exploited. Sand and gravel are removed from the coastal ocean to be used as beach fill, land fill, and concrete aggregate. Sulfur was once a significant mineral recovered from the caps of salt domes but is now being replaced by sulfur recovered by pollution control equipment. Phosphorite is deposited in shallow coastal waters and has the potential to be used to produce phosphate fertilizer. Manganese nodules and cobalt-rich manganese crusts have created interest but will not likely be mined in the near future.

Although present plans call for the disposal of high-level nuclear waste on land, a deep-sea disposal program has been studied. It involves burying the waste in deep-sea sediments.

KEY TERMS

Abyssal clay (p. 94)
Aragonite (p. 88)
Biogenous (p. 88)
Calcareous (p. 94)
Calcite (p. 88)
Calcium carbonate (p. 88)
Carbonate compensation depth (p. 95)
Clay (p. 87)
Cobalt-rich manganese crust (p. 100)
Coccolithophore (p. 88)
Cosmogenous (p. 90)
Diatom (p. 88)
Extrusive rocks (p. 87)
Fecal pellet (p. 96)

Ferromagnesian (p. 86)
Foraminifera (p. 88)
Glauconite (p. 88)
Hydrogenous (p. 88)
Igneous (p. 86)
Intrusive rocks (p. 87)
Lithogenous (p. 86)
Manganese nodules (p. 88)
Mineral (p. 86)
Neritic sediment (p. 92)
Nonferromagnesian (p. 86)
Oceanic sediment (p. 92)
Oolite (p. 90)

Ooze (p. 94)
Opal (p. 95)
Petroleum (p. 97)
Phosphorite (p. 100)
Radiolarian (p. 88)
Relict sediment (p. 92)
Sediment maturity (p. 85)
Silica (p. 88)
Siliceous (p. 94)
Sulfur (p. 98)
Turbidite (p. 92)
Weathering (p. 87)
Zeolite (p. 87)

QUESTIONS AND EXERCISES

1. What characteristics of marine sediment indicate increasing maturity?
2. Why is a higher velocity current required to erode clay-sized particles than larger, sand-sized particles?
3. List the four basic sources of marine sediment.
4. Referring to figure 4–3, describe how the rocks granite, andesite, and basalt differ in texture, color, mineral composition, and density.
5. List the four common clays produced by weathering of common rock-forming minerals.
6. List the two major chemical compounds of which most biogenous sediment is composed and the organisms that produce them.
7. What is the chemical composition of hydrogenous deposits of manganese nodules, phosphorite, and glauconite? In which environments do they form? What is the relationship between a deposit's formation and its environment?
8. What are the two most common forms in which calcium carbonate precipitates from ocean water? Describe the environment in which such precipitation is most likely to occur. Include a discussion of biological processes and water temperature conditions that enhance precipitation.

9. Describe the most common types of cosmogenous sediment and give the probable source of these particles.
10. Describe the basic differences between neritic sediment and oceanic sediment.
11. Explain why many rivers are not now carrying sediment to the continental self, but depositing it in their estuaries.
12. Discuss the processes by which sediments are carried to and distributed across the continental margin.
13. How do oozes differ from abyssal clay? Discuss how productivity, destruction, and dilution combine to determine whether an ooze or abyssal clay will form on the deep-ocean floor.

14. Refer to figure 4–11 and explain why abyssal clay deposits on the floor of deep-ocean basins are commonly underlain by calcareous ooze.
15. How do fecal pellets help explain why the particles found in the ocean surface waters are closely reflected in the particle composition of the sediment directly beneath? Why would one not expect this?
16. Discuss the present importance and the future prospects for the production of petroleum, sand and gravel, sulfur, phosphorite, and polymetallic crusts and nodules.
17. What areas of concern must be addressed before high-level nuclear waste could be deposited in deep-sea sediments?

REFERENCES

Berger, W. H.; Adelseck, C. G., Jr.; and Mayer, L. A. 1976. Distribution of carbonate in surface sediments of the Pacific Ocean. *Journal of Geophysical Research* 81:15, 2617–29.

Biscaye, P. E.; Kolla, V.; and Turedian, K. K. 1976. Distribution of calcium carbonate in surface sediments of the Atlantic Ocean. *Journal of Geophysical Research* 81:15, 2592–2602.

Broadus, J. M. 1987. Seabed minerals. *Science* 235:853–60.

Emery, K. O., and Uchupi, E. 1984. *The geology of the Atlantic Ocean.* New York: Springer-Verlag.

Hildebrand, A. R., and Boynton, W. V. 1990. Proximal Cretaceous-Tertiary boundary impact deposits in the Caribbean. *Science* 248:4957, 843–46.

Howell, D. G., and Murray, R. W. 1986. A budget for continental growth and denudation. *Science* 233:4762, 446–49.

Inter-University Program of Research on Ferromanganese Deposits of the Ocean Floor. Phase 1 Report. Unpublished.

Kolla, V., and Biscaye, P. E. 1976. Distribution of calcium carbonate in surface sediments of the Atlantic Ocean. *Journal of Geophysical Research* 81:15, 2602–16.

Kyte, F. T.; Zhou, L.; and Wasson, J. T. 1988. New evidence on the size and possible effects of a late Pliocene oceanic asteroid impact. *Science* 241:4861, 63–65.

Leinen, M.; Cwienk, D.; Heath, G. R.; Biscaye, P. E.; Kolla, V.; Thiede, J.; and Dauphin, J. P. 1986. Distribution of biogenic silica and quartz in Recent deep-sea sediments. *Geology* 14:3, 199–203.

Masters, C. D.; Root, D. H.; and Attanasi, E. D. 1991. Resource constraints in petroleum production potential. *Science* 253:146–152.

Sverdrup, H. U.; Johnson, M. W.; and Fleming, R. H. 1942. Renewal 1970. *The oceans: Their physics, chemistry, and general biology.* Englewood Cliffs, N.J.: Prentice-Hall.

Udden, J. A. 1898. Mechanical composition of wind deposits. *Augustana Library Pub.* 1.

Weaver, P. P. E., and Thomson, J., eds. 1987. *Geology and geochemistry of abyssal plains.* Palo Alto: Blackwell Scientific.

Wentworth, C. K. 1922. A scale of grade and class terms for clastic sediments. *Journal of Geology* 30, 377–92.

SUGGESTED READING

Sea Frontiers

Dietz, R. S. 1978. IFOs (Identified Flying Objects). 24:6, 341–46. The source of Australasian tektites and microtektites is discussed.

Dudley, W. 1982. The secret of the chalk. 28:6, 344–49. An informative discussion of the formation of marine chalk deposits.

Dugolinsky, B. K. 1979. Mystery of manganese nodules. 25:6, 364–69. The problems of origin, growth, and the environmental implications of manganese nodule mining.

Feazel, C. T. 1986. Asteroid impacts, seafloor sediments, and extinction of the dinosaurs. 32:3, 169–78. A discussion of the possibility that high concentrations of iridium and osmium in a marine clay deposited at the time dinosaurs and many other species died out 65,000,000 years ago may have resulted from the collision of Earth with a meteor 6 mi in diameter.

Feazel, C. T. 1989. Inner space: Porosity of seafloor sediments. 35:1, 49–52. A petroleum geologist discusses how porosity forms in marine sediments and the significance of the pores.

Mark, K. 1988. Ancient ocean rocks: High in the Sangre de Cristo Mountains. 34:1, 22–29. The history of an ancient ocean is revealed in the sediments of the mountains of Colorado and New Mexico.

Prager, E. J. 1988. Curious nodules on Florida's outer shelf. 34:6, 354–58. The origin of calcareous nodules found near Molasses Reef south of Key Largo is discussed.

Shinn, E. A. 1987. Sand castles from the past: Bahamian stromatolites discovered. 33:5, 334–43. The formation of stromatolites, sand domes trapped by algal growth, in the Bahamas is discussed.

Victory, J. J. 1973. Metals from the deep sea. 19:1, 28–33. The formation and economic potential of mining manganese nodules are discussed.

Wood, J. 1987. Shark Bay. 33:5, 324–33. The history of Shark Bay, Australia, since its discovery by Dirk Hartog in 1616 is summarized. The stromatolites and "tame" dolphins that attract tourists are also discussed.

Scientific American

Nelson H., and Johnson, K. R. 1987. Whales and walruses as tillers of the sea floor. 256:2, 112–18. The 200,000 walruses and 16,000 gray whales that feed by scooping sediment from the Bering Sea continental shelf suspend large amounts of sediment that is transported by bottom currents.

Chapter 5

Properties of Water

Before we can go further in our investigation of the ocean, we must establish a background that will allow us to understand the various properties of the substance that makes up 96.5 percent of the mass of the oceans—water. We at first may not consider water to be an unusual substance, inasmuch as it is probably the one we would mention if we were asked to name the most common or prevalent substance that exists on Earth. Water most certainly is abundant, but it does have unique properties; we find that it is a very uncommon substance.

It is the presence of water on our planet that makes life possible. Water controls the distribution of heat over Earth's surface as well as other conditions for life.

THE WATER MOLECULE

When two hydrogen atoms combine with one oxygen atom to form one water molecule, they join in such a manner that the hydrogen atoms are separated by an

Water makes possible life on Earth. (Photo © Fred Bavendam/Peter Arnold, Inc.)

angle of 105° rather than being at opposite ends of a linear molecule and separated by 180° (figure 5–1A). As the electrons move around within the structure, this arrangement of atoms within a molecule produces a greater concentration of electrons around the nucleus of the oxygen atom than around the hydrogen nuclei. As a result, a positive charge associated with the unshielded proton in the nucleus of each hydrogen atom is concentrated at the location of the hydrogen atoms. This situation produces polarity of electrical charge, in which the end of the molecule represented by the oxygen atom is slightly more negatively charged and the end with the hydrogen atoms is slightly more positively charged: The water molecule is **dipolar,** and water is known as a polar substance.

SOLVENT PROPERTIES

The polarity in the distribution of electrical charge does not make the water molecule behave as an ion. Pure water is actually a very poor conductor of electricity, since the molecule will not move toward either the negatively charged pole or the positively charged pole in an electrical system. Instead, the water molecule simply becomes oriented in an electrical field with its positively charged hydrogen nuclei toward the negative plate and its negatively charged oxygen end toward the positively charged plate. This orientation of water molecules in an electrical field tends to neutralize the field.

Owing to their polar nature, water molecules form bonds between one another. These **intermolecular bonds** are referred to as **hydrogen bonds.** The positively charged hydrogen areas of the water molecule attract the negatively charged oxygen portion of the neigh-boring water molecules and bond together by electrostatic forces. When polar molecules are bound together in complexes, as are the water molecules, their ability to reduce the intensity of an electrical field acting on the water and, therefore, the electrostatic attraction between ions of opposite charges introduced into water, is greatly increased. In fact, these forces can be reduced to 1/80 their value out of water.

If we consider sodium chloride, a compound containing **ionic bonds,** we can see that simply by placing that substance in water, we have reduced the electrostatic attraction between the sodium and chloride ions by 80 times. This reduced attraction makes it much easier for the ions to be dissociated. Once they do become dissociated, the sodium ions are attracted to the negative pole of the water molecule and the chloride ions are attracted to the positive pole of the water molecule (figure 5–1).

As more and more ions of sodium and chlorine are freed by the weakening of the electrostatic attraction that is holding them together, they become surrounded by the polar molecules of water, or **hydrated.** Sodium ions are surrounded by water molecules with their negative poles toward the ion, and chloride ions are surrounded by water molecules oriented so that their positively charged poles are directed toward the ion.

THERMAL PROPERTIES

Freezing and Boiling Points of Water

The Celsius (centigrade) temperature scale was constructed on the basis of the characteristics of water at one atmosphere of pressure, with the **freezing point** of water, the temperature at which it changes from a liquid to a solid, representing 0° and the **boiling point** of

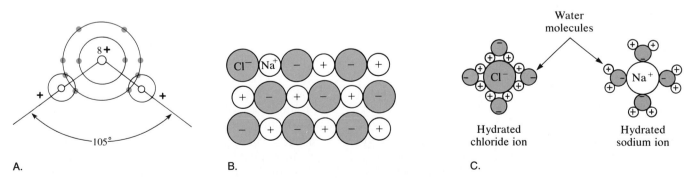

A.   B.   C.

FIGURE 5–1
Water as a Solvent. *A,* water molecule. With the angle of 105° between the two hydrogen atoms in the water molecule, a dipolar molecule results. The oxygen end of the molecule is negatively charged, and the hydrogen regions exhibit a positive charge. *B,* sodium chloride solid state. Ionic bonds hold sodium ions (Na^+) and chloride ions (Cl^-) together in the compound sodium chloride (NaCl). *C,* sodium chloride in solution. The positively charged hydrogen end of the dipolar water molecule is attracted to the negatively charged Cl^- ion; the negatively charged oxygen end is attracted to the positively charged Na^+ ion. The ions become hydrated when surrounded by water molecules that are attracted to them. In this state they have become solute ions in a water solution.

water, the temperature at which it changes from a liquid to a gas, representing 100°. If we compare these values to those of some compounds similar to water in makeup, we will see that the freezing and boiling points of water are uniquely high. The other compounds occur in nature as gases, while water occurs in the gaseous, liquid, and solid states of matter within the narrow range of atmospheric conditions.

We should now consider the nature of the forces that must be dealt with to change the state of a compound. There exists between all molecules of any compound a relatively weak attraction that results from the electrostatic attraction between the nuclear parts of one molecule and the electrons of another. This attraction slightly exceeds the electrostatic repulsion that exists between the nuclear parts of the same molecules and the electrons of the same molecules.

This intermolecular force is known as **van der Waals force,** and it becomes significant only when molecules are very close together, such as in the solid or liquid states. Generally, the heavier the molecule, the greater the van der Waals attraction between individual molecules of the compound. Therefore, with increasing molecular weight, a greater amount of energy is needed to overcome that attraction and allow a change of state, say from a solid to a liquid or a liquid to a gas, to occur. Consequently, the melting and boiling points of compounds generally increase as the molecular weight increases.

What form of energy is used to change the state of matter? We should first define a few terms before we go further in our discussion. **Heat,** a form of energy associated with and proportional to the energy level of molecules, is the total **kinetic energy** of a sample. It may be generated by combustion, chemical reaction, friction, or radioactivity. So that we can measure the amount of energy that is being added to or removed from the molecules, we must introduce the term **calorie** as a unit for measuring quantity of heat. A calorie is defined as the amount of heat required to raise the temperature of one gram of water one degree Celsius. So that we can measure the amount of energy that the molecules within the substance have, we will need to understand the meaning of the word **temperature.** Temperature is defined as the direct measure of the average kinetic energy of the molecules that make up a substance. Kinetic energy is energy of motion, so the higher the temperature, the greater the velocity of the molecules of the substance for which the temperature is being measured.

Let us return now to the van der Waals force that causes molecules of a given substance to be attracted to one another. We can see that if this attraction is to be broken, energy must be given to the molecules so that they can move more rapidly, in the ways described in figure 5–2, to overcome this force. It is this attraction that

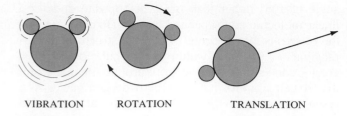

VIBRATION ROTATION TRANSLATION

FIGURE 5–2
Motion of Molecules. *Vibration* is the typical movement of molecules in a crystalline solid; *rotation* is the movement in liquids and gases; and all gas molecules move randomly by *translation.*

must be broken if we are going to produce a change of state in any substance from solid to liquid or liquid to gas.

We find that in a crystalline **solid state,** although bonds are constantly being broken and reformed, the predominant relationship between molecules is that of a rather firm attachment produced by the nearness of the molecules and the great effect the van der Waals force has upon molecules that are close together. Vibration is the dominant type of **molecular motion** in the solid state as the molecules vibrate but remain in relatively fixed positions.

In the **liquid state** the molecules have gained enough energy to overcome many of the van der Waals forces that bound them together in the solid state. The molecules have enough freedom to move relative to one another. In this state we see all forms of molecular movement—vibration, rotation, and translation. The molecules are free to move relative to one another but are still attracted by one another. Bonds are being formed and broken at a much greater rate than in the solid state.

In the **gaseous state,** translation has become the dominant type of motion. Molecules are now moving at random, and there exists no attraction between individual molecules. The only effect that one molecule will have on another is that which is produced by collision during random movement.

Returning to our comparison of the properties of water and compounds of similar composition containing two hydrogen atoms and an atom of another element, we will look at H_2S, H_2Se, and H_2Te. The molecular weights of the four compounds are H_2O (18), H_2S (34), H_2Se (80), and H_2Te (129). Figure 5–3 shows that, as was predicted in the discussion of the van der Waals force, the freezing points and boiling points for H_2S, H_2Se, and H_2Te increase with increased molecular weight. When we insert H_2O into the scale, the freezing point and boiling point based on its molecular weight should be about −90°C (−130°F) and −68°C (−108.4°F), respectively. The fact that water freezes at 0°C (32°F) and boils at 100°C (212°F), seems to be a violation of natural order. Here we see again the great significance of the polarity of the water molecule and the hydrogen bond that this structure produces. The high freezing and boiling points of

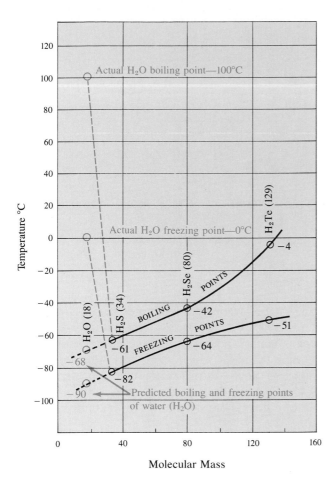

FIGURE 5–3
Molecular Mass—Freezing and Boiling Points. Because of the hydrogen bonds that must be broken in addition to overcoming the van der Waals force in achieving changes of state, the freezing and boiling points of water are higher than would be expected for a compound of its molecular makeup.

water are the manifestations of the additional kinetic energy required to overcome not only van der Waals bonds but also the hydrogen bonds to achieve a change of state.

Heat Capacity

Another direct result of the hydrogen bond is the high heat capacity of water. As noted, a calorie is the amount of heat required to raise the temperature of 1 g (0.035 oz) of water 1°C (1.8°F). The heat capacity of water compared to that of most other substances is great, and we use the capacity of water to absorb heat as a standard against which we compare the heat capacities of other substances. The amount of heat that is required to raise the temperature of 1 g of any substance 1°C is the **heat capacity** of the substance. It is 1 cal for water and less than 1 cal for most other substances. It is this property that produces the mild climates in coastal regions. Great amounts of heat can be gained or lost by the coastal water without causing extreme changes in temperature.

Latent Heats of Melting and Vaporization

Closely related to water's unusually high heat capacity are its high latent heat of melting and latent heat of vaporization. The continuous addition of heat to a substance in the solid or liquid state will bring about a change of state in the substance. A solid converts to a liquid at a temperature called its freezing (melting) point and a liquid is changed to a gas at a temperature defined as its boiling (condensation) point.

When changing the state of any substance, there may be no increase of temperature at that point where a change of state occurs even though heat is continuously being added. The heat energy is being used entirely to break all the intermolecular bonds required to complete the change of state. While the bonds are being broken, a mixture of the substance in both states is in equilibrium. Only upon completion of the change of state will the temperature again rise. The heat that is added to 1 g of a substance at the melting point to break the required bonds to complete the change of state from solid to liquid is the **latent heat of melting.** The heat applied to effect a change of state at the boiling point is the **latent heat of vaporization.**

The word *latent* is used in describing the heats of vaporization and melting because the heat that must be added to a given mass of ice or water to convert it to a higher energy state, water or water vapor, is held in reserve or "hidden" by that mass of water or water vapor. When water vapor returns to a liquid, it **condenses** and heat is released into the surrounding air. Release of heat also occurs with the freezing of water, which is the change of state from liquid to solid. During condensation and freezing, the same amount of heat that is added to change the state of that water from a liquid to a gas or from a solid to a liquid is released.

A practical application of these principles of heat transfer can be seen in the use of ice for the purpose of refrigeration. A block of ice set in a closed container among food articles will lower their temperature because heat energy is drawn from the food articles and added to the molecules of ice to change them to the liquid state as the ice melts. The principle of cooling air with water is similar: In arid climates, the hot dry air that passes through a surface coated with liquid water will lose heat to the water. The water is converted to a vapor. Thus, after passing through or across the water-covered surface, the air is considerably cooler.

To help clarify this phenomenon, we will discuss the heat transfer and the change of state of H_2O from a solid at −40°C (−40°F). Observe the heat-temperature graph, shown in Figure 5–4, which has a vertical temperature scale ranging from −40° to 120°C (248°F) and a horizontal scale which begins at 0 and continues through 800 cal.

FIGURE 5–4
Change of State Graph—Water.
The latent heat of melting (80 cal) is much less than the latent heat of vaporization (540 cal) because only a few hydrogen bonds must be broken to convert 1 g of ice to a liquid, while all remaining hydrogen bonds must be broken to convert 1 g of liquid water to a gas.

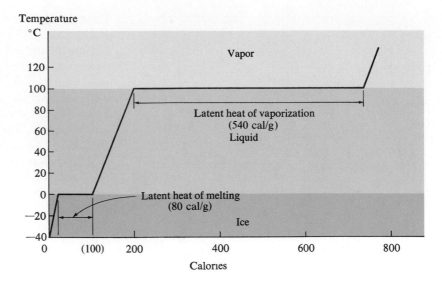

For 1 g of ice, the addition of 20 cal of heat raises the temperature 40°, from −40°C to 0°C. Thus, the heat capacity of ice is 0.5 cal/g, half that of liquid water.

Once the temperature has been raised to 0°C (32°F), the continuous addition of heat does not result in an increase in temperature until 80 cal have been added. We do not observe an increase in temperature during the addition of these 80 cal of heat because all the heat energy added was used to break the intermolecular bonds that were holding the water molecules in the solid state. The temperature will remain unchanged until all of the necessary bonds are broken and the mixture of ice and water has been changed to 1 g of water. The amount of heat required to convert 1 g of ice to 1 g of water, 80 cal, is termed the latent heat of melting, and it is higher for water than for any other commonly occurring substance.

The bonds that are broken in converting most substances from a solid to a liquid are the van der Waals bonds. In water, however, not only the van der Waals bonds but some of the hydrogen bonds must also be overcome. It is necessary only to break the ice structure down into numerous small clusters surrounded by individual water molecules to the extent that these remaining ice clusters can move relative to one another and allow the mass to assume a liquid state. Liquid water, particularly at low temperature near the freezing point, could well be described as a pseudocrystalline liquid, as there are still many small clusters of ice crystals contained within it. Once enough hydrogen bonds have been broken to allow freedom of movement among the clusters and individual water molecules, the temperature of the water will again rise with the addition of heat.

After the phase change from ice to liquid water has occurred at 0°C, the addition of heat to the water causes the temperature to rise again. As it does, note that it requires 1 cal of heat to raise the temperature of water 1°C (1.8°F). We must therefore add another 100 cal or a total

of 200 cal, before one gram of water reaches the boiling point of 100°C (212°F). At this temperature we note the development of another plateau, far more prominent on the graph than that which represents the latent heat of melting, 80 cal/g. This plateau at 100°C represents an addition of 540 cal, the latent heat of vaporization, to the gram of water before complete conversion to the vapor state.

Why is so much more heat energy required to convert 1 g of water to water vapor than was required to convert 1 g of ice to water? We defined a gas as a substance in which the molecules were moving at random, free from the influence of other molecules, except when they collide. To make the conversion from ice to water, not all the hydrogen bonds had to be broken, but only enough to allow freedom of movement among the various ice clusters that remained and the individual molecules that were also present in the system. To convert water to water vapor, every molecule must be freed from the attraction of other water molecules. Therefore, every hydrogen bond must be broken (figure 5–5). To do this requires a great amount of heat energy.

We can now see that large amounts of heat energy are absorbed or released by water as it changes state, particularly if the change of state is from a liquid to a vapor or vice versa. Of course, we do not have temperatures of 100°C at the surface of the ocean where this phenomenon of conversion of water to vapor occurs in nature. Sea-surface temperatures average about 20°C (68°F) or less.

The conversion of a liquid to a gas below the boiling point is called **evaporation.** At ocean-surface temperatures, individual molecules that are being converted from the liquid to the gaseous state have a lower amount of energy than do the molecules of water at 100°C. Therefore, to gain the additional energy necessary to break free of the surrounding water molecules, an individual mole-

FIGURE 5–5
Hydrogen Bonds in H₂O.

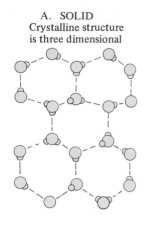

A. SOLID
Crystalline structure
is three dimensional

B. LIQUID

C. GAS

cule must capture energy from its neighbors. This phenomenon explains the cooling effect of evaporation. The molecules left behind have lost heat energy to those that escape. To produce 1 g of water vapor from the ocean surface at temperatures less than 100°C requires more than the 540 cal of heat that are required to make this conversion at the boiling point. For instance, the **latent heat of evaporation** is 585 cal/g at 20°C. This higher value is due to the fact that more hydrogen bonds must be broken at this lower temperature for a gram of water molecules to enter the gaseous state.

We can readily see the significance of the evaporation-condensation cycle in Earth's surface temperatures. Evaporation removes heat energy provided by the sun and stored in the oceans. This energy is carried high into the atmosphere as the water vapor rises and is released there when the vapor condenses and falls as precipitation. It is water's latent heat of evaporation that accounts for the removal of great quantities of heat from the low-latitude oceans by evaporation. It is later released in the heat-deficient higher latitudes after the vapor is transported through the atmosphere and condenses there as rain and snow (figure 5–6).

SURFACE TENSION

Next to mercury, water has the highest **surface tension** of all commonly occurring liquids. We observe a surface tension phenomenon in filling a container with water to the brim and even beyond (figure 5–7A). You will note that the water can be piled up above the rim of the container, forming a convex surface that is the water interface with the atmosphere. Water drops are also a very common manifestation of surface tension. These phenomena result from the tendency of water molecules to attract one another or to cohere at the surface of any accumulation of water. Because of this cohesive ten-

dency, it is possible to float on water objects that are much heavier than water. A razor blade carefully laid on the surface will float, although it is normally five times as dense as water. Many insects use the surface of water as if it were a solid surface, moving across it at will.

Surface tension is a manifestation of the presence of the hydrogen bond. Those molecules of water that are at the surface are strongly attracted to the molecules of water below them by their hydrogen bonds. The air above the surface has a very low density of molecules as compared to the water itself, and although water molecules are attracted to molecules of other substances, the attraction of the hydrogen bond holds the water molecules in the surface layer to those below (figure 5–7B).

Water clings to the surface of many substances; we refer to this phenomenon as wetting. Water will adhere strongly to glass, organic substances, and inorganic material such as rocks and soil. When water is poured into a container made of a substance to which it strongly adheres, the adhesive force, or the force of attraction between water molecules and molecules of another substance, will cause the surface tension layer to take on a form unlike the convex one described previously. In this case the surface tension layer is concave, being drawn up on the container's sides (figure 5–7C).

If the container is glass, the positively charged portions of the water molecules are attracted to unbonded electrons in the oxygen atoms that are part of the makeup of the glass container. Being strongly attracted by these oxygen atoms, the water molecules will "climb" up the side of the container and will be held back only by the hydrogen bond attraction that exists between individual water molecules. In fact, if the diameter of the container is decreased to a very fine bore, the combination of cohesion, which holds the water molecules together, and the adhesive attraction between the water molecules and the glass container will pull the column of water to great heights. This phenomenon is known as **capillarity.**

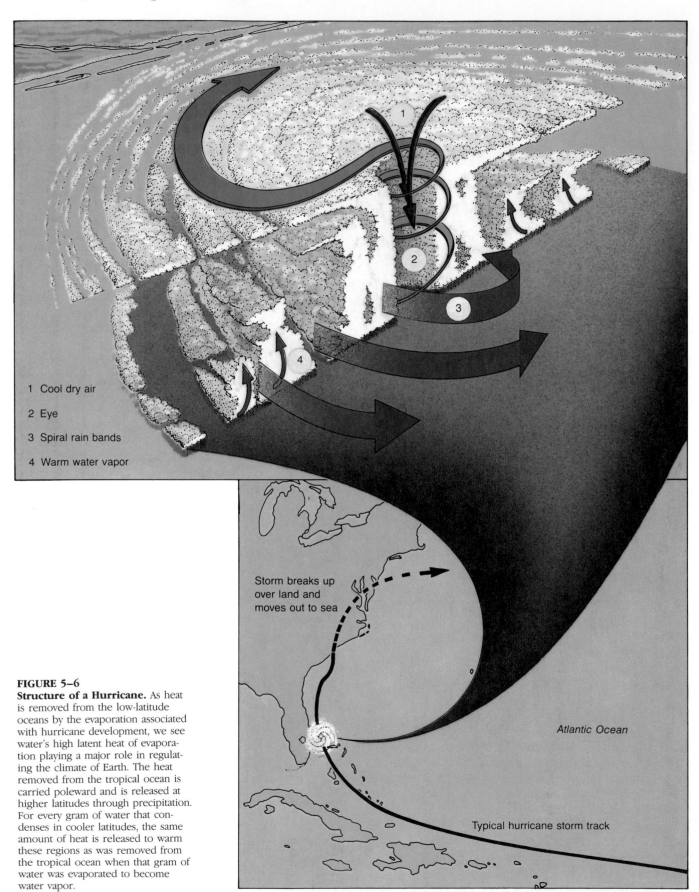

1 Cool dry air

2 Eye

3 Spiral rain bands

4 Warm water vapor

Storm breaks up over land and moves out to sea

Atlantic Ocean

Typical hurricane storm track

FIGURE 5–6
Structure of a Hurricane. As heat is removed from the low-latitude oceans by the evaporation associated with hurricane development, we see water's high latent heat of evaporation playing a major role in regulating the climate of Earth. The heat removed from the tropical ocean is carried poleward and is released at higher latitudes through precipitation. For every gram of water that condenses in cooler latitudes, the same amount of heat is released to warm these regions as was removed from the tropical ocean when that gram of water was evaporated to become water vapor.

A. A container can be filled above its rim due to the strength of the surface layer.

C. Capillarity-water climbs the walls of a tubular container. The smaller the diameter, the higher it will climb.

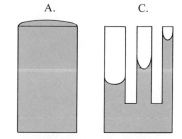

B.   Hydrogen bonds

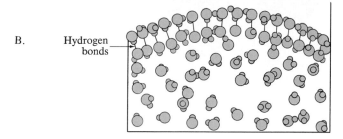

FIGURE 5–7
Surface Tension Phenomena. A surface tension layer forms as hydrogen bonds form a strong attraction between the top layer of water molecules and the underlying molecules.

SALINITY OF OCEAN WATER

Although the proportion of water to dissolved salts may vary within the ocean, the major component ions are distributed in ocean water in relatively constant proportions. This **rule of constancy of composition** was first established by Dittmar in his analysis of the *Challenger* expedition water samples discussed in chapter 1. Water makes up, on the average, 96.53 percent of the ocean's mass, so it determines most of the physical properties that we observe in ocean water.

It seems probable that when the techniques are developed for making such minute measurements, all of the known elements will be found to be dissolved in ocean water. Merely six elements, however, account for over 99 percent of the dissolved solids in ocean water. These major components are chlorine, sodium, magnesium, sulfur (as SO_4), calcium, and potassium (table 5–1 and figure 5–8).

Salinity is defined as the total amount of solid material dissolved in a kilogram of seawater when all the carbonate has been converted to oxide, all bromine and iodine replaced by chlorine, and all organic matter completely oxidized. Given this definition, it may appear that salinity is a very complex property and very difficult to measure. However, since the oceans are so very well mixed and the relative abundance of the major constituents is essentially constant, we have a condition that makes the chemical measurement of salinity relatively simple.

TABLE 5–1
Ocean Salinity. Considering the average salinity of the world ocean, 34.7 by weight, a 1000 g sample of this water would contain 34.7 g of dissolved solids. Six ions account for 99.28 percent of the dissolved solids.

Major Constituents (over 100 parts per million)	
Ion	Percentage
Chloride, Cl^-	55.04
Sodium, Na^+	30.61
Sulfate, SO_4^{-2}	7.68
Magnesium, Mg^{+2}	3.69
Calcium, Ca^{+2}	1.16
Potassium, K^+	1.10
	99.28

Minor Constituents (1–100 parts per million)	
	p.p.m.
Bromine, Br	65.0
Carbon, C	28.0
Strontium, Sr	8.0
Boron, B	4.6
Silicon, Si	3.0
Fluorine, F	1.0

Trace Elements (less than 1 part per million)	
Nitrogen, N	Iodine, I
Lithium, Li	Iron, Fe
Rubidium, Rb	Zinc, Zn
Phosphorus, P	Molybdenum, Mo

Because of constancy of composition, it is necessary to measure the concentration of only one of the major constituents in order to determine the salinity of a given water sample. The constituent that occurs in the greatest abundance and is the easiest to measure accurately is the chloride ion, Cl^-. The portion of the weight of a given sample of water that is the direct result of the presence of this ion is called chlorinity and is usually expressed in the terms grams per kilogram of ocean water (g/kg) or parts per thousand (‰). The chloride ion always accounts for 55.04 percent of the dissolved solids in any sample of ocean water. Therefore, by measuring its concentration, we can determine the total salinity in parts per thousand by the following relationship:

$$\text{Salinity (‰)} = 1.80655 \times \text{Chlorinity (‰)}$$

For example, given the average chlorinity of the ocean, 19.2‰, the ocean's average salinity can be calculated as $1.80655 \times 19.2‰$, or 34.7‰. Although 1/0.5504 equals 1.81686 (not 1.80655), oceanographers have agreed on 1.80655 because the constancy of composition has been found to be more of an approximation than an absolute condition.

**FIGURE 5–8
Constituents of Ocean
Salinity.**

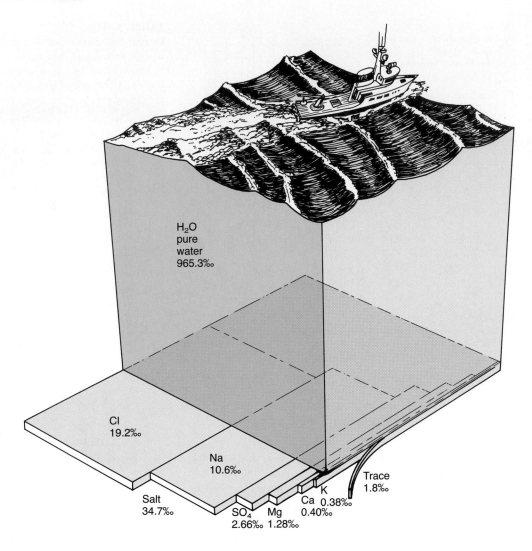

From 1902 to 1975, the Hydrographic Laboratory at Copenhagen, Denmark, provided Standard Seawater samples to assure that salinity determinations made throughout the world would be based on the same reference. In 1975, this duty was taken over by the Institute of Oceanographic Services located in Wormly, England. **Standard Seawater** consists of ocean water analyzed for chloride ion content to the nearest ten-thousandth of a part per thousand. It is then sealed in ampules and labeled to be sent to laboratories throughout the world (figure 5–9). The chlorinity of water samples taken and measured in other parts of the world can then be compared if the titration solution or conductivity instrumentation is calibrated with the standard water sample.

The chemical determination of the chloride content in seawater can be made with a high degree of accuracy but requires a great deal of time and care. It is tedious to carry out the chemical titration with silver nitrate aboard ship. With the advancements in instrumentation in the field of oceanography, this task has been greatly simplified. Since the electrical conductivity of ocean water is known to increase with increased temperature and salinity, salinity is now most commonly determined by measuring the electrical conductivity of ocean water (at a constant temperature of 15°C, or 59°F) with a modern conductance-measuring instrument, the **salinometer.** Salinity can be determined to better than 0.003‰ by this method.

WATER DENSITY

The density of water in its various states and at different temperatures is of great importance in considering the movement of water in the ocean. Anything that is more dense than water will sink into it, and a substance that is less dense will float on the surface. We define **density** as mass per unit of volume, usually grams per cubic centimeter (g/cm^3).

Density is affected by temperature, salinity, and pressure. With most substances we observe that a decrease in temperature produces an increase in the den-

FIGURE 5–9
An Ampule of Standard Seawater.

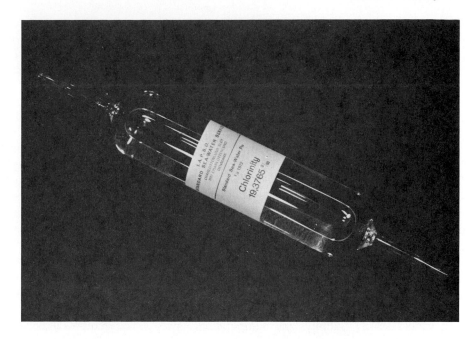

sity of the substance. The increase in density is the result of the same number of molecules occupying less space as they lose energy. This condition of thermal contraction is also found in water. As the temperature of water decreases, the density increases as long as this temperature decrease occurs above 4°C (39.2°F). If we consider the change in the density of water below 4°C, we encounter considerations that must include the hydrogen bond. As the temperature of water is lowered from 4° to 0°C (32°F), we observe that its density decreases owing to expansion of the water.

This anomalous behavior can be explained only by considering the molecular structure of water and the hydrogen bond. As we lower the temperature of water

from 20°C (68°F) and reduce the amount of thermal agitation of the water molecules, the unbonded water molecules occupy less volume because of their decreased energy. But, as we approach the freezing point below 4°C, this reduction in volume owing to the decreased energy of unbonded molecules is not sufficient to compensate for another phenomenon that is occurring. Ice crystals, open six-sided structures in which water molecules are widely spaced, are becoming more abundant. The greater rate of increase of ice crystals as the temperature approaches the freezing point accounts for the decreased density of water below 4°C (figure 5–10).

The **temperature of maximum density** for fresh water, 4°C, is lowered by increasing pressure or adding

FIGURE 5–10
Formation of Ice Clusters in Water. Note that the density ranges from 0.9982 g/cm³ at 20°C to a maximum of 1.000 g/cm³ at 4°C. The density of ice (solid water) is only 0.9170 g/cm³, so it floats on water.

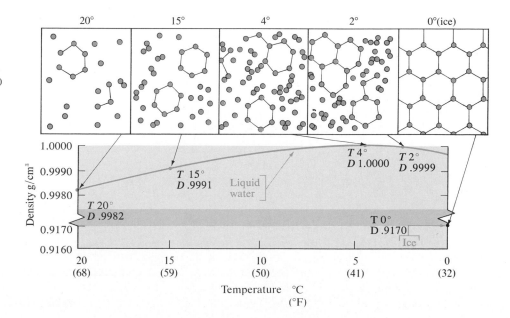

solid particles such as salt, both of which will inhibit the formation of the ice clusters. Thus, more energy must be removed to produce crystals equal in volume to those that could be produced at 4°C in fresh water, causing a reduction in the temperature of maximum density. This interference with the formation of ice crystals also produces a progressively lower freezing point for water as more solids are added. The freezing point and temperature of maximum density converge as they drop and coincide when the salinity reaches 24.7 parts per thousand (‰) at a temperature of −1.33°C (29.61°F) (figure 5–11). When salinity is higher than 24.7‰, the density of water increases with decreasing temperature until the water freezes. Thus, average seawater (34.7‰) has no maximum density anomaly.

Because the components of salt have a density greater than the density of water, the density of ocean water increases with increasing salinity. Physical oceanographers are greatly concerned with the relationships among water density, salinity, and temperature since density distribution is an important determinant of ocean circulation. Density of ocean water cannot easily be measured directly, so to determine density distribution within a region of study, oceanographers measure temperature and salinity and then use these values to calculate density.

Table 5–2 summarizes the physical and biological significance of the properties of water.

RESIDENCE TIME

The **residence time,** or the average length of time an atom of an element will spend in the ocean, can be calculated by the following equation:

$$\text{Residence time} = \frac{\text{Amount of element in the oceans}}{\text{Rate at which element is added to or removed from the oceans}}$$

The oceans are considered to be well mixed and in a steady state; that is, the rate at which an element is removed from the oceans must be equal to the rate at which it is being added. For that reason, we could use either of these values as the denominator in the equation.

The residence time of an element in the ocean depends on how reactive it is with the marine environment. Those elements that are more reactive have a shorter residence time. The reactive elements aluminum and iron have residence times of 100 and 140 years, respectively, while less reactive sodium has a residence time of 260 million years (see table 5–3).

The short residence time for aluminum shown in table 5–3 (100 years) is the average for particulate and dissolved aluminum. Recent investigation has shown that dissolved aluminum has a residence time of 1400 years, indicating that particulate aluminum must be removed from ocean water at a very rapid rate. It is believed that comparisons of the residence times of soluble fractions of elements in the oceans must be known to understand their chemical reactivity in the oceans since the residence time of particles is affected primarily by physical mixing processes.

Elements are removed from ocean water through biological reactions, or they are incorporated into sediment. Each year a 1-m (3.28-ft) layer of the ocean's surface is evaporated. This water is returned directly by precipitation or indirectly by runoff from the continents. The ocean, which has an average depth of 4000 m (13,120 ft), would completely evaporate in 4000 years, but because the water is returned, this is the time required to completely recirculate the ocean. Thus, the residence time of water in the oceans is 4000 years.

The process by which water is recycled among the ocean, atmosphere, and continents is called the **hydrologic cycle** (figure 5–12). Of the total yearly evaporation of water from Earth's surface, 83 percent is removed from the oceans and 17 percent is removed from the surface of the continents. Seventy-six percent of this water is re-

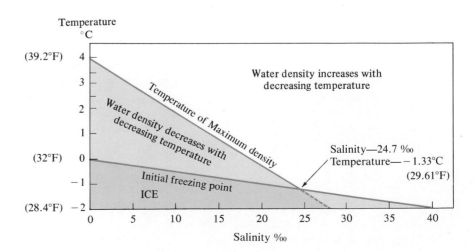

FIGURE 5–11
Salinity, Freezing Point, and Temperature of Maximum Density.

TABLE 5–2
Properties of Water.

Property	Physical and Biological Significance
Physical States. Water is the only substance that occurs as a gas, liquid, and solid within the temperature range found at Earth's surface.	As a gaseous (vapor) component of the atmosphere, water helps transfer heat from warm low latitudes to cold high latitudes. Liquid water also contributes to this process through ocean currents. Additionally, liquid water runs across the continents, dissolving minerals from the rocks and carrying them to the oceans. It is this form of water that accounts for over 85 percent of the mass of most marine organisms and serves as the medium in which the chemical reactions that support life occur. The freezing of water into a solid (ice) during the winter season at the higher latitudes increases surface salinity and makes possible the sinking of dense surface water. This water is the only source of oxygen for the deep ocean.
Solvent Property. Water can dissolve more substances than any other common liquid.	Ocean water carries dissolved within it the nutrients required by marine plants and the oxygen needed by animals. It is this property that has produced the "saltiness" of the oceans.
Heat Capacity. The quantity of heat required to change the temperature of 1 g of a substance 1°C. Water has the highest heat capacity of all common liquids. The heat capacity of water is used as the unit of heat quantity, *calorie*.	Heat capacity is a major factor in making water the most important moderator of climate. It accounts for the narrow range of temperature change found at any location in the ocean. Water can gain or lose a relatively great amount of heat while undergoing a much smaller temperature change than would occur in other substances.
Latent Heat of Melting. The quantity of heat gained or lost per gram by a substance changing from a solid to a liquid or from a liquid to a solid without a change in temperature. For water, it is 80 cal—the highest of all common substances.	When ice forms, most of the heat energy lost is released to the heat-deficient atmosphere. When ice melts, the energy gained by the water is manifested as molecular energy of the liquid water. This prevents the high-latitude ocean from becoming much warmer or colder than the −1.8°C (28.7°F) freezing temperature of ocean water.
Latent Heat of Vaporization. The quantity of heat gained or lost per gram by a substance changing from a liquid to a gas or from a gas to a liquid without a change in temperature. It is higher for water, 540 cal at 100°C (212°F), than for any common substance. It is 585 cal at 20°C (68°F), at which much of the evaporation from the ocean surface occurs.	A great amount of excess heat energy is removed from the low-latitude ocean by evaporation and released through precipitation into the atmosphere at heat-deficient higher latitudes. It is this property that contributes greatly to the fact that the polar regions do not get increasingly colder and the equatorial region does not get increasingly hotter.
Surface Tension. Highest of all common liquids. Cohesive attraction of hydrogen bonds causes a molecule-thick "skin" to form on a water surface and makes capillarity possible.	Some organisms such as Halobates use this "skin" as a walking surface. Others such as Glaucus hang from its undersurface.
Density. Mass per Unit Volume (g/cm³). The density of ocean water is increased by decreasing the temperature and increasing salinity and pressure. For pure water, the temperature of maximum density is 4°C (39.2°F), but ocean water with a salinity greater than 24.7‰ will get denser as the temperature is lowered to the freezing temperature.	Plankton that stay near the surface through buoyancy and frictional resistance to sinking are greatly influenced by the effect of temperature on density. In low-density warm water, plankton must be smaller or more ornate to obtain the increased ratio of surface area to body mass necessary to remain afloat.

TABLE 5–3
Residence Time of Some Elements in the Oceans.

Element	Amount in Ocean (g)	Residence Time in Years
Sodium, Na	147×10^{20}	2.6×10^8
Potassium, K	5.3×10^{20}	1.1×10^7
Calcium, Ca	5.6×10^{20}	8.0×10^6
Silicon, Si	5.2×10^{18}	1.0×10^4
Manganese, Mn	1.4×10^{15}	7.0×10^3
Iron, Fe	1.4×10^{16}	1.4×10^2
Aluminum, Al	1.4×10^{16}	1.0×10^2

Source: Data from Goldberg and Arrhenius, 1958

moved from the atmosphere by precipitation directly back into the oceans. Twenty-four percent falls as precipitation onto the continents. During a year, the atmosphere transports water vapor equivalent to 8.5 percent of that which is being evaporated from the oceans to the continents. This water is returned to the oceans by stream runoff and as groundwater. At any given time, 97 percent of Earth's water is contained in the ocean basins, 2 percent is in glaciers, 0.6 percent is groundwater, 0.3 percent is in the atmosphere, and about 0.1 percent is in rivers and lakes.

The **World Climate Research Program (WCRP)** has proposed a study of the world hydrologic cycle called

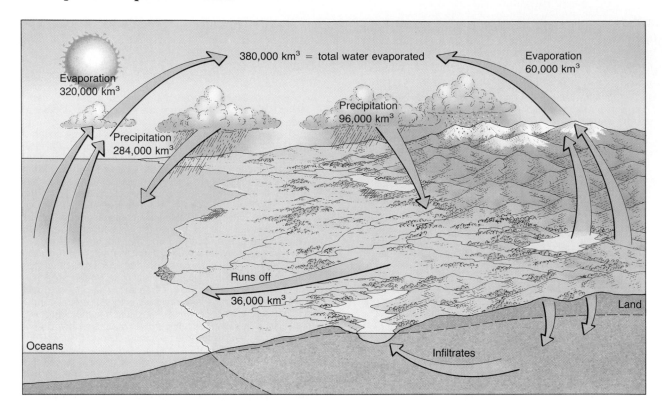

FIGURE 5–12
Hydrologic Cycle. Ninety-seven percent of Earth's water is in the world ocean, 2 percent is in glaciers, 0.6 percent is ground water, 0.3 percent is in the atmosphere, and about 0.1 percent is in rivers and lakes. Although there is some evidence that indicates the glaciers are entering a stage of melting, we will consider them to be in a steady state. Water is removed from the liquid reservoirs by evaporation and returned to them by precipitation. There is a net atmospheric transfer of water vapor from the oceans to the continents, where it is deposited as precipitation and returns to the oceans by stream flow or as groundwater. (Reprinted by permission from Tarbuck, E. J., and Lutgens, F. K., *The earth: An introduction to physical geology,* 3d ed. (New York: Macmillan), fig. 10.1.)

the **Global Energy and Water Cycle Experiment (GEWEX).** Planned for the 1990s, it will use satellite data to help improve long-range weather forecasts and to increase our understanding of hydrology and climate trends.

GEOSECS (GEOCHEMICAL OCEAN SECTIONS)

The GEOSECS project mentioned in chapter 1 is a huge systematic program designed to increase understanding of circulation patterns and mixing processes in the oceans. All of the major ocean basins have been sampled throughout the depth of the water column. The sampling system includes electronic sensors that telemeter temperature, salinity, dissolved oxygen, pressure, and particulate content to a console aboard ship. As rosettes of 30-liter (7.9-gal) sampling bottles are lowered into the ocean, oceanographers have the option of deciding to sample at any depth where an interesting value for any of the properties shows on the console (figure 5–13).

The collected water is analyzed for some 23 chemicals, 15 isotopes, and the amount and types of particulate matter. From the results thus far obtained, oceanographers think the GEOSECS program will not only provide a much improved understanding of physical processes but also will supply much information about biological cycles in the ocean (figure 5–14).

DESALINATION

With great expanses of the continents being arid and the available freshwater supplies throughout the world being used in greater volumes each year, it is becoming urgent that practical methods of **desalination,** or salt removal, be developed so we can take advantage of the greatest water supply on Earth—the oceans. Some of the methods that have been investigated to determine their potential usefulness toward this end are described below.

Distillation involves boiling salt water and passing the water vapor that is produced to a condenser, where it is collected as fresh water. This simple procedure is ex-

FIGURE 5–13
GEOSECS Sampling Bottles. One of the two rosettes carried aboard Scripps Institution of Oceanography's research vessel *Melville* during GEOSECS Pacific operations shows nonmetallic bottles used in collecting seawater samples. The ends of each bottle can be triggered to close by the scientist operating the shipboard console. (Photo from Scripps Institution of Oceanography, University of California, San Diego.)

pensive and requires the use of large amounts of heat energy. Increased efficiency will be required before it will be practical on a large-scale basis.

Another process that requires large amounts of energy is **electrolysis.** In this method, two volumes of

fresh water—one containing a positive electrode and the other a negative electrode—are located on either side of a volume of seawater. The seawater is separated from each of the freshwater reservoirs by semipermeable membranes that are permeable to salt ions but not to water molecules. When an electrical current is applied, positive ions such as sodium ions are attracted to the negative electrode, and negative ions such as chloride ions are attracted to the positive electrode. In time, enough ions are removed through the membranes to convert the seawater to fresh water.

Another application that is limited to small-scale use is **freezing.** This process that naturally produces sea ice may not be feasible on a large-scale basis, but imaginative thinkers have proposed the towing of icebergs to the coastal waters of arid lands. Here, the fresh water that is produced as the icebergs melt could be captured and pumped ashore.

Solar humidification does not require supplemental heating and has been successfully used in large-scale agricultural experiments in Israel, West Africa, and Peru. It involves the evaporation of water in a covered container placed in direct sunlight. Salt water in the container is subjected to evaporation, and the water vapor that condenses on the cover runs into trays from which it is collected.

A method that has much potential for large-scale projects is **reverse osmosis.** In natural osmosis, a net flow of water molecules will pass through a water-permeable membrane from a freshwater solution into a saltwater solution. In reverse osmosis, pressure is applied to the saltwater solution, forcing it to flow through the water-permeable membrane into the freshwater solution (figure 5–15). It is the reverse of natural osmosis. A significant problem with this method is that the membranes must be replaced frequently. At least 30 nations located in arid climates have built reverse osmosis units.

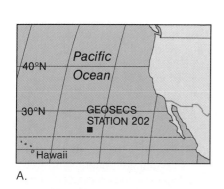
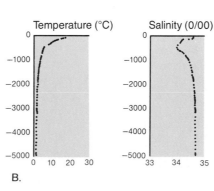
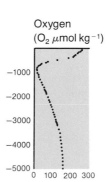
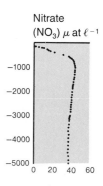

A.   B.

FIGURE 5–14
GEOSECS Data. *A,* the map shows the location of GEOSECS Station 202 at latitude 26°N, longitude 139°W. *B,* the four curves show the plot of temperature, salinity, dissolved oxygen, and nitrate measured from the surface to a depth of 5000 m (16,400 ft) at Station 202. They show that ocean water properties may change significantly from the surface to a depth of 1000 m (3280 ft). The rate of change below that depth is much less. (Plots courtesy of Woods Hole Oceanographic Institution.)

FIGURE 5–15
Reverse Osmosis Process Used to Produce Fresh Water from Salt Water. This fundamentally simple process requires that pressure be applied to a reservoir of salt water to force it through a water-permeable membrane into a reservoir of fresh water, from which it is collected.

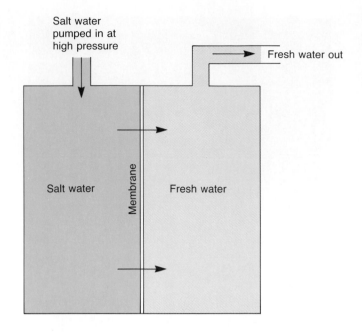

SUMMARY

The hydrogen bond, the bond between water molecules resulting from the dipolar nature of the molecule, plays the major role in giving water its many unusual properties.

Water is a great solvent because its dipolar molecules can attach themselves to charged particles, ions, that make up many substances such as sodium chloride, hydrate them, and put them into solution.

The presence of the hydrogen bond also accounts for the unusual thermal properties of water such as its high freezing point and boiling point, high heat capacity, and high latent heat of melting and latent heat of vaporization.

The surface tension phenomenon that makes water form drops and allows it to be poured into a container until it stands well above the sides of the container results from the water molecules being strongly attracted by the hydrogen bond to those water molecules beside and beneath them. This cohesive attraction combined with the adhesive attraction of water to glass, organic substances, rock, and soil produces the capillarity that pulls water to great heights in small tubular openings.

Salinity is the amount of dissolved solids in ocean water, averaging about 34.7 g of dissolved solids per kg of ocean water. Salinity is usually expressed in parts per thousand (‰). Over 99 percent of the dissolved solids in ocean water are accounted for by six ions, in order of decreasing abundance: chloride, sodium, sulfate, magnesium, calcium, and potassium. In any sample of ocean water these ions will be in the same relative proportions, making it possible to determine salinity by measuring the concentration of only one of them, usually the chloride ion.

The density of pure water increases with decreased temperature, as does the density of most substances, to a temperature of 4°C (39.2°F). Below 4°C, its density decreases with decreased temperature. This change is due to the fact that the rate of formation of open ice crystals increases dramatically below 4°C, the temperature of maximum density for fresh water. By the time water freezes, the density of the ice is only about 0.9 that of water at 4°C. Density increases with increased salinity and pressure. Due to the fact that the components of salinity have a greater density than water, increased salinity increases the density of the ocean water. The residence time of elements in the ocean water ranges from 100 years for aluminum to 260,000,000 years for sodium. The time required to totally recirculate the water in the oceans, its residence time, is 4000 years.

The ocean contains 97 percent of Earth's water. Water is recycled among the ocean, atmosphere, and continents by the hydrologic cycle, which involves evaporation, precipitation, and surface and groundwater runoff from the continents.

The GEOSECS program has helped give oceanographers an increased understanding of physical and biological processes of the oceans. It features an advanced system of sampling and chemically analyzing the properties and content of ocean water.

Desalination of ocean water to provide fresh water for domestic, home, and agricultural use may be achieved by distillation, electrolysis, freezing, solar humidification, and reverse osmosis of seawater. The last two methods show the greatest potential for practical applications.

KEY TERMS

Boiling point (p. 107)
Caloric (p. 108)
Capillarity (p. 111)
Condensation (p. 109)
Constancy of composition, rule of
 (p. 113)
Density (p. 114)
Desalination (p. 118)
Dipolar (p. 107)
Distillation (p. 118)
Electrolysis (p. 119)
Evaporation (p. 110)
Freezing (method of desalination)
 (p. 119)
Freezing point (p. 107)

Gaseous state (p. 108)
Global Energy and Water Cycle
 Experiment (GEWEX) (p. 118)
Heat (p. 108)
Heat capacity (p. 109)
Hydrated (p. 107)
Hydrogen bond (p. 107)
Hydrologic cycle (p. 116)
Intermolecular bond (p. 107)
Ionic bond (p. 107)
Kinetic energy (p. 108)
Latent heat of evaporation (p. 111)
Latent heat of melting (p. 109)
Latent heat of vaporization (p. 109)
Liquid state (p. 108)

Molecular motion (p. 108)
Residence time (p. 116)
Reverse osmosis (p. 119)
Salinity (p. 113)
Salinometer (p. 114)
Solar humidification (p. 119)
Solid state (p. 108)
Standard Seawater (p. 114)
Surface tension (p. 111)
Temperature (p. 108)
Temperature of maximum density
 (p. 115)
Van der Waals force (p. 108)
World Climate Research Program
 (WCRP) (p. 117)

QUESTIONS AND EXERCISES

1. Describe the condition that exists in the water molecule to make it dipolar.
2. Discuss how the dipolar nature of the water molecule makes it such an effective solvent for ionic compounds.
3. Define freezing point and boiling point.
4. There is a fundamental difference between the intermolecular bonds that result from the dipolar nature of the water molecule (hydrogen bond) and the van der Waals force as compared to the chemical bonds (covalent and ionic). What is it?
5. Why are the freezing and boiling points of water higher than would be expected for a compound of its molecular makeup?
6. How does the heat capacity of water compare with that of other substances? Describe the effect this characteristic of water produces in climate.
7. Why does the heat energy added as latent heats of melting and vaporization for water not produce an increase in temperature? Why is the latent heat of vaporization so much greater than the latent heat of melting?
8. Describe how excess heat energy absorbed by Earth's low-latitude regions could be transferred to the heat-deficient higher latitudes through a process that makes use of water's latent heat of evaporation.
9. How does hydrogen bonding produce the surface tension phenomenon of water?

10. As water cools, two distinct changes in the behavior of molecules take place. One tends to increase density while the other decreases density. Describe how the relative rates of their occurrence cause pure water to have a temperature of maximum density at 4°C (39.2°F) and make ice less dense than liquid water.
11. As water becomes more saline, the temperature of maximum density and the freezing temperature of water decrease and converge. At what salinity does water cease to have a temperature of maximum density above its freezing temperature?
12. What condition of salinity makes it possible to chemically determine the total salinity of ocean water by measuring the concentration of only one constituent, Cl^-?
13. If there is an estimated 18×10^{20} g of magnesium (Mg) in the ocean, and it is being added (or removed) at the rate of 56.5×10^{12} g per year, what is the residence time of magnesium in the ocean? (*See* appendix 1.)
14. List the reservoirs of water on Earth and the percentage of Earth's water each holds. Explain how the hydrologic cycle moves water among these reservoirs.
15. Briefly describe how the following methods of desalination would produce fresh water: distillation, electrolysis, freezing, solar humidification, and reserve osmosis.

REFERENCES

Davis, K. S., and Day, J. S. 1961. *Water: The mirror of science*. Garden City, N. Y.: Doubleday.

Hammond, A. L. 1977. Oceanography: Geochemical tracers offer new insight. *Science* 195:164–66.

Harvey, H. W. 1960. *The chemistry and fertility of sea waters*. New York: Cambridge University Press.

Kuenen, P. H. 1963. *Realms of water*. New York: Science Editions.

MacIntyre, F. 1970. Why the sea is salt. *Scientific American* 223:5, 104–15.

Pickard, G. L. 1975. *Descriptive physical oceanography*. 2nd ed. New York: Pergamon Press.

Revelle, R. 1963. Water. *Scientific American* 209:93–108.

Yokoyama, Y.; Guichard, F.; Reyss, J-L; and Van, N. H. 1978. Oceanic residence times of dissolved beryllium and aluminum deduced from cosmogenic tracers ^{10}Be and ^{26}Al. *Science* 201:1016–17.

SUGGESTED READING

Sea Frontiers

Friedman, R. 1990. Salt-free water from the sea. 36:3, 48–54. The economics and nature of the various processes used in desalination are discussed.

Gabianelli, V. J. 1970. Water: The fluid of life. 16:5, 258–70. The unique properties of water are lucidly described and explained. Topics covered include hydrogen bond, capillarity, heat capacity, ice, and solvent properties.

Smith, F., and Charlier, R. 1981. Saltwater fuel. 27:6, 342–49. The potential for using the salinity difference between river water and coastal marine water to generate electricity is considered.

Scientific American

Baker, J. A., and Henderson, D. 1981. The fluid phase of matter. 245:5, 130–39. The structure of gases and liquids is modeled using hard spheres.

MacIntyre, F. 1970. Why the sea is salt. 223:5, 104–15. A summary of what is known of the processes that add and remove the elements dissolved in the ocean.

Chapter 6

Air-Sea Interaction

In this first chapter on physical oceanography we will consider some properties of ocean water, the Coriolis effect, the heat budget of the ocean, and the process by which solar energy is converted into kinetic and heat energy of moving masses of ocean water. All will be discussed in terms of their relationship to the interaction between the atmosphere and oceans.

This interaction has received increasing attention over the years. One of the first major phenomena to spur the quest for greater understanding of how the oceans and atmosphere interact was the recognition in the 1920s of the **El Niño–Southern Oscillation** as an event that was tied to worldwide changes in climate. (This phenomenon is discussed in chapter 7.) Then, in 1988, studies that confirmed that the world's average temperature had risen over the past 100 years led to much investigation into the possibility that the **"greenhouse effect"** of Earth's atmosphere was increasing because of the growing amounts of carbon dioxide, methane, chlorofluorocarbons, and other gases that increase the atmosphere's ability to absorb heat. This topic will be discussed later in this chapter.

Hurricane Gilbert, the strongest hurricane ever to develop in the western hemisphere, moves onto the Yucatán peninsula, Mexico. (Courtesy of NOAA.)

A key factor in all of this investigation is the role of the oceans. Do changes in the ocean produce changes in the atmosphere that lead to the El Niño phenomenon—or vice versa? Or does there exist a complex feedback system that involves both to a more or less equal degree? Because the atmospheric carbon dioxide has increased only half as much as would be predicted from the activities of humans during the past century, where did the rest of the carbon dioxide go? Did it go into the oceans or maybe the biosphere? Many questions are being asked, and answers are probably going to be a long time in coming. Increased understanding of some of the phenomena we will discuss in this chapter will surely be required before the answers are found.

PHYSICAL PROPERTIES OF OCEAN WATER

Light

The color of ocean water ranges from a deep indigo blue in tropical and equatorial regions of little biological productivity to a yellow-green in the coastal waters of high-latitude areas where biological productivity occurs seasonally at a very high rate. The blue of the tropical waters where there is little particulate matter is due to the molecular scattering of solar radiation. This process is also responsible for the blue color of the sky. Higher concentrations of particulate matter, especially where phytoplankton is abundant, result in a greater amount of scattering and absorption and decreased transmission of solar radiation. This produces the greenish color characteristic of such waters.

The many forms of electromagnetic energy radiated by the sun may be seen in figure 6–1, the **electromagnetic spectrum.** The very narrow segment of this spectrum designated as visible light may be broken down on the basis of wavelength into violet, blue, green, yellow, orange, and red energy levels. Combined, these different wavelengths of light produce white light. The shorter wavelength forms of energy to the left of visible light represent highly dangerous forms of electromagnetic energy including X rays and gamma rays. To the right of the visible segment of the electromagnetic spectrum, we find longer wavelength forms of energy that are the basis for the technological fields dealing with heat transfer and communication. We will further consider the electromagnetic spectrum when we examine the role of solar energy in setting the ocean masses in motion. We now consider only that portion of the spectrum that is manifested as visible light.

The reason we see things in color is because objects reflect wavelengths of light that represent the colors of the visible spectrum. If those wavelengths of light are not present in the light that falls upon an object, colors cannot be seen. In the ocean the absorption of visible light is greater for the longer wavelength colors, thus the shorter wavelength portion of the visible spectrum is transmitted to greater depths. As a result of this condition, the red wavelengths are absorbed within the upper 15–20 m (49–66 ft), the yellow disappears before a depth of 100 m (328 ft) has been reached, but green light can still be perceived down to 250 m (820 ft). Only the blue and some green wavelengths extend beyond these depths, and their intensity becomes low. It is because of this pattern of absorption that objects in the ocean usually appear blue-green. Only in the surface waters can the true colors of objects be observed in natural light, since only in the surface waters can all wavelengths of the visible spectrum be found.

To measure the transmission of visible light in the ocean, a simple device about 30 cm (12 in.) in diameter, the *Secchi disc,* is attached to a rope marked off in meters. After the disc is lowered into the ocean, the depth at which it can last be seen gives an indication of the *turbidity*—the amount of suspended material in the water. Increased turbidity increases the rate of absorption, thus decreasing the transmission of visible light.

Density

Density ranges from 1.02200 to 1.03000 g/cm³ in the open ocean. It is an important property of ocean water because it determines the positions and motions of water masses in the ocean. In a stable system, low-density water rests at the top of a column of water in which the density increases with increasing depth. The **in situ density (σ or sigma)** of ocean water is its density *in place,* that is, at whatever depth it is in the ocean. In situ density in the deeper trenches is about 5 percent greater than at the ocean surface because of the effect of pressure.

For many oceanographic applications, it is useful to remove the **adiabatic** effects of pressure on water temperature. Adiabatic refers to the increase in temperature that results from compression of water at increasing depths because of increasing pressure. The density value that has the adiabatic temperature effect removed is called the **potential density (σ_θ, or sigma theta).** Potential density is always less than in situ density except at the surface, where the adiabatic heating of water is zero. Potential density is a useful property of ocean water for oceanographers to know because subsurface currents flow along lines of equal potential density.

Oceanographers usually use only the last four digits in a water density expression and eliminate the units, g/cm³. To derive this simplified form, water density is first converted to specific gravity by dividing it by the density of pure water at 4°C (39.2°F). Suppose we have a water sample with an in situ density (σ) of 1.02567 g/cm³. We

FIGURE 6–1
The Electromagnetic Spectrum and Transmission of Visible Light in Water.

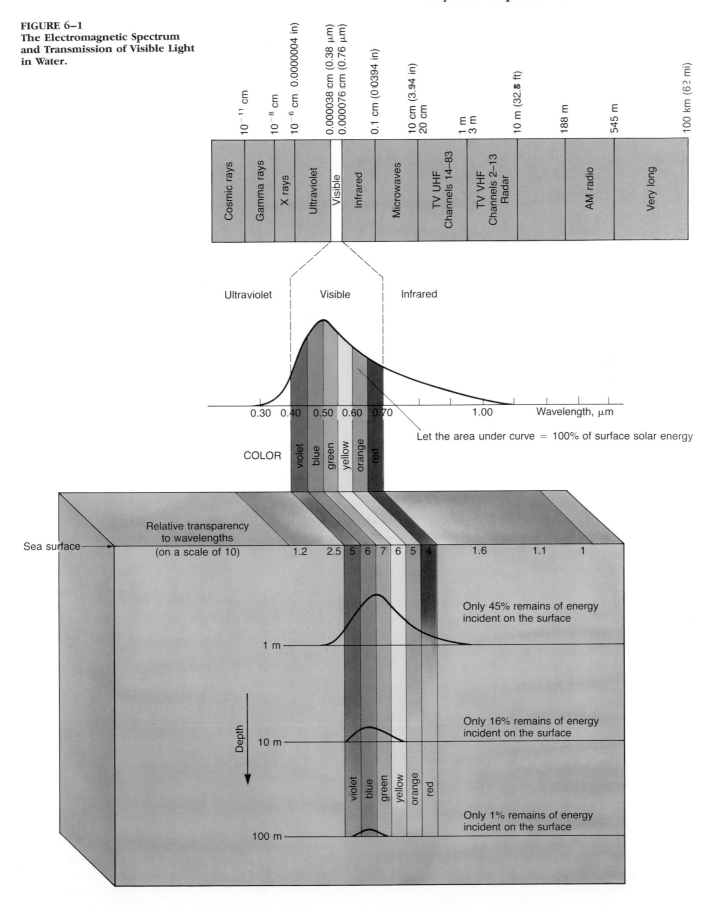

would convert this value to the simplified form by first converting it to specific gravity.

$$\text{Specific gravity} = \frac{\text{density of ocean water}}{\text{density of pure water}}$$
$$= \frac{1.02567 \text{ g/cm}^3}{1.00000 \text{ g/cm}^3}$$
$$= 1.02567$$

This eliminates the density units, g/cm^3. The specific gravity of ocean water is then inserted into the following equation to convert in situ density to the simpler form.

$$\text{In situ density } (\sigma) = (\text{Specific gravity} - 1) \times 1000$$
$$25.67 = (1.02567 - 1) \times 1000$$

It can be determined by observing the temperature-salinity-density relationships in figures 6–2 and 6–3 that temperature change can be expected to have a much greater effect on density in the high-temperature, low-latitude areas than in polar regions. It can be noted that the σ isopleths (lines of constant density) are more nearly parallel to the temperature axis for low temperatures than for high temperatures, showing a greater density change per unit of temperature change in the high-temperature range.

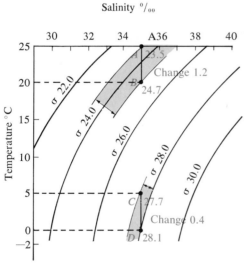

FIGURE 6–2
Temperature-Salinity-Density Relationship. In high-latitude areas characterized by low-temperature water, temperature has less effect on density (sigma, σ) than in the high-temperature, low-latitude areas. Measured at a salinity of 35‰, σ values determined for points *A, B, C,* and *D* show the density change is greater over a 5° temperature span at a higher temperature range. The change in σ across a temperature range of 20–25°C (68°–77°F) is 1.2 compared to a change of only 0.4 across an equal range at the lower temperatures of 0–5°C (32–41°F). Thus, a change in the temperature of warm, low-latitude water has about three times the effect on density that an equal change in temperature occurring in colder, high-latitude waters has. (After G. L. Pickard, *Descriptive physical oceanography* © 1963.)

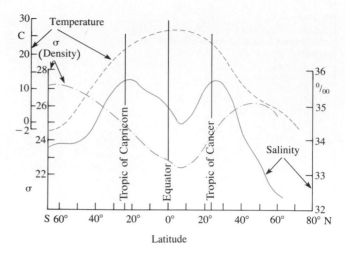

FIGURE 6–3
Average Surface Temperature, Salinity, and Density of World Ocean by Latitude. Density (sigma) increases from about 22 near the equator to maxima of 26–27 near 50°–60° latitude in the surface water of the ocean. At higher latitudes it decreases slightly. Salinity maxima in the tropics do not seem to affect density, pointing out the importance of temperature in controlling density in low latitudes. (After G. L. Pickard, *Descriptive physical oceanography* © 1963.)

Density and the variables that most affect it, temperature and salinity, are important water characteristics defined as **conservative properties** used in the identification of water masses. Conservative properties of water are those that are not significantly altered by any processes other than mixing and diffusion once the water sinks beneath the surface.

A **nonconservative property** of water is changed by some process other than mixing and diffusion after the water sinks beneath the ocean surface. An example is dissolved oxygen, which will be changed by biological processes.

Since the gain or loss of heat energy at the ocean surface is of primary importance in determining density characteristics of the ocean water, we find, as with most water properties, that the rate of change in density is greater in a vertical direction than in a horizontal direction.

The density of surface water is affected primarily by temperature changes in the open ocean. In the extreme high-latitude areas of the ocean, however, where temperatures remain relatively constant, salinity changes can have a significant effect on density. Figure 6–4 shows the vertical distribution of density in various regions of the ocean. In the equatorial region, there is a shallow zone of low-density water near the surface, and below this lies a zone where the increase in density is very rapid. This zone of rapidly increasing density is called the **pycnocline** (density slope) and represents a very stable barrier to the mixing of the low-density water above and the high-density water below. The pycnocline is considered

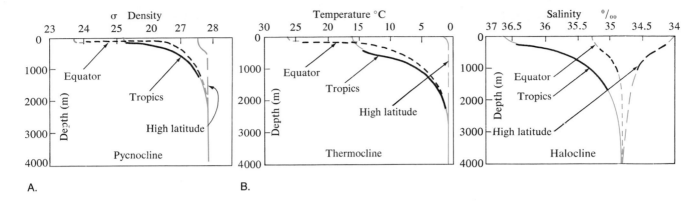

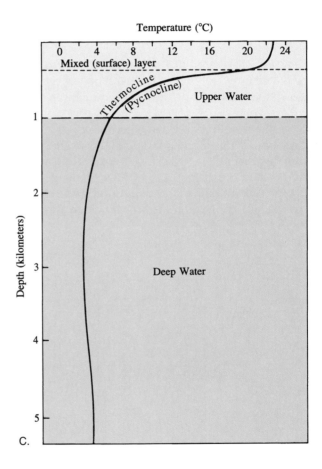

FIGURE 6–4
Vertical Density Profiles for Various Latitudinal Regions. *A,*
because of natural tendencies toward stability, density can normally
be expected to increase with depth. In low-latitude equatorial and
tropical regions a thin layer of low-density surface water is separated
from high-density deep water by a zone of rapid density change, the
pycnocline. The pycnocline is poorly developed or missing in high-
latitude waters. It is the absence of the stable pycnocline that facili-
tates vertical mixing in some high-latitude areas. *B,* the gravitationally stable pycnocline described above is primarily the result of and
coincides with a zone of rapid vertical decrease in temperature, the
thermocline. The thermocline is also more highly developed in low
latitudes. A zone of rapid vertical change in salinity, the halocline,
may also affect density. The halocline usually represents a decrease
in salinity with increasing depth in low latitudes but may represent
the opposite condition in high-latitude areas (note that in *B* the temperature increases to the left. Although this is unconventional, it was
done to make the curves easier to compare). *C,* extending from the
surface down through the mixed, or surface, layer and the pycnocline is the relatively light *upper water.* It is well developed
throughout the low- and midlatitudes. It is underlain by the denser,
cold *deep-water* mass that extends to the deep-ocean floor. The
mixed layer represents water of uniform temperature resulting from
wave mixing. (*A* and *B* after G. L. Pickard, *Descriptive physical
oceanography* © 1963.)

to have a high gravitational stability because it would require a greater amount of energy to move a given mass of
water from some point in the pycnocline either up or
down than would be required to move an equal mass an
equivalent distance in regions where density change occurred very slowly with increasing depth.

The pycnocline develops as a result of the combined effects of zones of rapid vertical changes in temperature, the **thermocline** (temperature slope), and
salinity, the **halocline** (salinity slope). The interrelation-
ship of these three zones, which determine the degree of

separation between the **upper-water** and **deep-water**
masses, is shown in figure 6–4.

The pycnocline is lacking in the high latitudes. It
can be seen that there is not a great amount of difference
in the density of the surface waters as compared to the
bottom waters in these high-latitude areas, thus these
water columns would be considered to be less stable
than the columns possessing a pycnocline that developed
in the low latitudes. We will consider the importance of
the low-stability columns of water in the high latitudes to
mixing the waters of the world ocean in the next chapter.

Sound

Sound can be transmitted much more efficiently through water than through air. This has made it possible to develop systems for determining the position of objects in the ocean as well as determining the distance to the bottom of the ocean by *sonar* (sound navigation and ranging). The average velocity of sound in the ocean is 1450 m (4756 ft)/s, compared with 334 m (1095 ft)/s in a dry atmosphere at 20°C. Velocity of sound increases with increases in temperature, salinity, and pressure. Therefore, the velocity with which sound travels through water in a particular area will vary on a seasonal basis. To determine the exact velocity of sound for any point in the ocean, these three variables—temperature, salinity, and pressure—must first be determined, or a *velocimeter* can measure it directly.

At a depth of around 1000 m (3280 ft) there exists a layer of ocean water where the combined effects of the values of these three variables result in relatively low velocity for the transmission of sound. Sound originating above and below this layer will be refracted, or bent, into the low-velocity layer and be trapped there. Thus, by following this channel, sound can be transmitted for unusually great distances (figure 6–5). This channel is called the **sofar** (sound fixing and ranging) **channel** because of its practical application in distance determination.

SOLAR ENERGY

Distribution of Solar Energy

Essentially all of the energy available to Earth comes from the sun. Solar energy strikes Earth at an average rate of 2 cal/cm²/min. Although energy radiated by the sun covers the full electromagnetic spectrum, most of the energy that reaches Earth's surface is contained in wavelengths close to and including the visible portion of the spectrum.

In the upper atmosphere, molecular oxygen (O_2) is broken into individual atoms of oxygen (O) that recombine with molecular oxygen to produce ozone (O_3). This process absorbs most of the solar radiation with wavelengths shorter than 0.29 μm. Water vapor absorbs a high percentage of solar radiation with wavelengths greater than 0.8 μm leaving most of the solar radiation within the visible spectrum, 0.38–0.76 μm, to penetrate the atmosphere and reach Earth's surface. After scattering by atmospheric molecules and reflection by clouds in the atmosphere, about 47 percent of the solar radiation that is directed toward Earth is absorbed by the oceans and continents. Nineteen percent is absorbed by the atmosphere and 34 percent is lost to space.

If it is true that Earth has maintained a constant average temperature over the years, Earth must be radiating

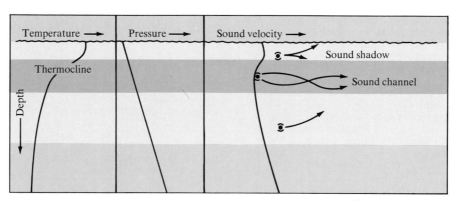

Sound source 💲

FIGURE 6–5
Transmission of Sound in the Ocean. Velocity of sound in the ocean increases with increases in the temperature, salinity, and pressure. Generally salinity changes are not as important in affecting the velocity of sound transmission as are the changes in temperature and pressure. Pressure increases steadily with depth, which increases the velocity of sound. However, the rapid decrease in temperature represented by the thermocline offsets the pressure increase and produces a low-velocity sound channel. Wave phenomena, such as sound, bend (refract) into low-velocity areas. This produces shadow zones when velocity maxima exist and channels that trap sound energy in low-velocity zones. The sound source beneath the sofar channel illustrates how refraction bends sound. If we take a band of the sound that is traveling to the right from the sound source, the sound at the bottom of the band will travel faster than the sound at the top of it. If the sound band is moving as a coherent unit, it will be bent upward into the sofar channel. As it passes through the sofar channel, the top of the band will be moving faster than the bottom and bend it back down into the channel—trapping the sound in the sofar channel.

energy back to space at the same rate it is absorbing solar energy. The energy radiating from Earth falls within the infrared range (0.76 μm–0.1 cm). Figure 6–6 shows the intensity of energy radiated by the sun peaks at 0.48 μm in the visible spectrum. It shows also that the radiation from Earth peaks at 10 μm in the infrared range. While the atmosphere absorbs only 19 percent of the short-wave-length solar radiation, the water vapor and carbon dioxide, as well as other gases, in the atmosphere absorb a large amount of the longer-wavelength infrared radiation emitted by Earth. It is this change of wavelength that is the essence of the greenhouse effect. Some of the infrared energy absorbed in the atmosphere will be reabsorbed by Earth (the rest being lost to space) to continue the process. Therefore, the solar radiation received is retained for a time within our atmosphere to moderate the temperature fluctuations between night and day and between seasons.

The greenhouse gases known to be increasing in their concentrations within the atmosphere as a result of human activities are listed in order of their potential contribution to increased greenhouse effect in table 6–1. Although trace gases such as methane (CH_4), nitrous oxide (N_2O), tropospheric ozone (O_3), and chlorofluorocarbons (CFCs) are present in atmospheric concentrations two to six orders of magnitude lower than carbon dioxide, they are important because, per molecule, they absorb infrared radiation much more strongly than does carbon dioxide (table 6–2).

TABLE 6–1

Estimated Contribution of Various "Greenhouse" Gases to Increased Greenhouse Effect Based on their Present Concentration and Observed Rate of Increase in the Atmosphere.

Species	Concentration (ppbv)	Rate of Increase (% per year)	Relative Contribution (%)
CO_2	353×10^3	0.5	60
CH_4	1.7×10^3	1	15
N_2O	310	0.2	5
O_3*	10–50	0.5	8
CFC-11	0.28	4	4
CFC-12	0.48	4	8

*In the troposphere.

ppbv = parts per billion by volume.

Source: After H. Rodhe, 1990.

Although the average temperature of Earth has risen by 0.5°C (0.9°F) in the last 100 years, there is no proof it is the result of the 25 percent increase in atmospheric CO_2 and the increase in other greenhouse gases (methane, CFCs, etc.) that began with the industrial revolution. Some climate modelers believe the greenhouse warming will produce more cloud cover to block out the sun's rays and significantly reduce the warming effect. What is very clear is that our agricultural and industrial activities are producing changes in the environment. Although we cannot eliminate modifications of the environment by human activities, we can begin a serious at-

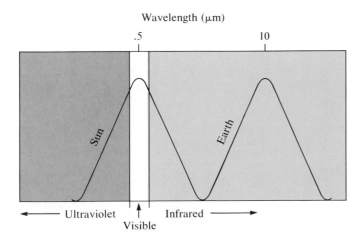

Wavelength (μm)

.5   10

Ultraviolet   Visible   Infrared

FIGURE 6–6

Comparison of Spectrums of Energy Radiated by the Sun and Earth. Most of the energy coming to Earth from the sun is within the visible spectrum and peaks at 0.48 μm (0.0002 in.). The atmosphere is transparent to most of this radiation, and it is absorbed at Earth's surface. When Earth reradiates this energy back toward space, it does so as infrared radiation with a peak at a wavelength of 10 μm (0.004 in.). The CO_2 and H_2O in the atmosphere absorb this infrared radiation to produce the greenhouse effect, which keeps the atmosphere's temperature higher than it otherwise would be.

TABLE 6–2

Comparison of Increased Greenhouse Effect for Known "Greenhouse" Gases. The ability of one molecule of each of these gases to absorb infrared radiation is compared to one molecule of carbon dioxide. It is clear that all of these gases absorb infrared radiation much more efficiently than does carbon dioxide. Table 6–1 shows that their smaller overall contribution to increased greenhouse effect in the atmosphere is due to their low concentrations in the atmosphere compared to carbon dioxide.

Species	Mass Basis (kg^{-1})	Mole Basis (mol^{-1})
CO_2	1	1
CH_4	70	25
N_2O	200	200
O_3*	1,800	2,000
CFC-11	4,000	12,000
CFC-12	6,000	15,000

*In the troposphere.

Source: After H. Rodhe, 1990.

tempt to reduce the magnitude of combustion of fossil fuels and the spread of agriculture. We must preserve as much of Earth's plant communities as possible and replace much of what has been removed.

We must also increase our understanding of the basic cycles of materials within Earth's ecosystems. One of the critical problems in this area is the exchange of gases between the atmosphere and the ocean. But we cannot wait until we have proof that our activities are having a negative effect on Earth's ecosystem before we act. Our activities that are known to change the environment the most must be modified while we continue to strive for understanding of the details of how the system works. Otherwise we are basing our future on a game of Russian roulette.

The amount of atmospheric heating varies with latitude because of the unequal distribution of solar radiation on Earth's surface and water vapor in the atmosphere.

Considering Earth as a sphere, figure 6–7 shows that radiation reaching its surface will strike at an angle of 90° near the center of the sphere that is lit by this radiation, while it will be striking at 0° at the edge of this circle (solar rays will be tangent to Earth's surface). If we consider a 1 km² (0.4 mi²) cross-sectional area of solar radiation that falls near the center of the illuminated portion of the sphere, this ray of light will cover exactly 1 km² of Earth's surface. By contrast, if we consider the same cross-sectional area falling on a high-latitude portion of Earth's surface, where it does not strike Earth at a right angle, this square-kilometer beam will be spread out over considerably more than 1 km² of Earth's surface.

The intensity of radiation available per unit of surface area will be greatly decreased compared to that available in the lower latitudes.

The angle at which direct sunlight strikes the ocean surface is important in determining how much of the solar energy is absorbed and how much is reflected. If the sun shines down on a flat sea from directly overhead, only 2 percent of the radiation will be reflected; 40 percent will be reflected if the sun is only 5° above the horizon (table 6–3).

As Earth rotates on its axis, which is inclined 23.5° from being perpendicular to the ecliptic (plane of Earth's orbit), there will be significant portions of its surface above 66.5°N latitude, the **Arctic Circle,** and 66.5°S latitude, the **Antarctic Circle,** that will spend up to six months in darkness. The direct rays of the sun will migrate back and forth across the equator between the **Tropic of Cancer** and the **Tropic of Capricorn** 23.5° north and south of the equator. The belt between the two tropics will receive a much greater amount of radiation per unit of surface area during a year than will the portions of Earth's surface north of the Arctic Circle and south of the Antarctic Circle.

Oceanic Heat Flow and Atmospheric Circulation

Because of the different angles at which solar radiation strikes Earth's surface and the very highly reflective characteristics of the ice cover found in high-latitude areas of Earth, more energy is absorbed than is radiated back into space in latitudes between 35°N and 40°S, while less en-

FIGURE 6–7
Solar Radiation.

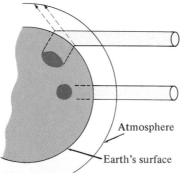

The amount of heat received at higher latitudes is less than that at lower latitudes for three reasons: (1) a ray of solar radiation that strikes the earth at a high latitude is spread over more area than an equal ray that is perpendicular to the earth's surface at a lower latitude, (2) the high latitude ray also passes through a greater thickness of atmosphere, and (3) more of its energy is reflected due to the low angle at which it strikes the earth's surface.

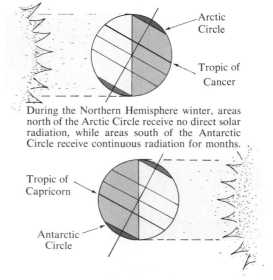

During the Northern Hemisphere winter, areas north of the Arctic Circle receive no direct solar radiation, while areas south of the Antarctic Circle receive continuous radiation for months.

The reverse of the above situation is true during the Northern Hemisphere summer. Throughout the year, the sun is directly over some latitude between the tropics.

TABLE 6–3
Reflection and Absorption of Solar Energy Resulting from Different Angles of Incidence on a Flat Sea.

Elevation of Sun Above Horizon	90°	60°	30°	15°	5°
Percent of radiation reflected	2	3	6	20	40
Percent of radiation absorbed	98	97	94	80	60

ergy is absorbed than is lost to space in latitudes higher than 35°N and 40°S. Figure 6–8 shows how this phenomenon is manifested in the average daily heat flow of the oceans of the Northern Hemisphere.

As a direct result of this condition, one might expect that the equatorial zone would continually get warmer as the years pass and the polar regions would become progressively cooler. Such is not the case. Although the polar regions are considerably colder than the equatorial zone, the temperature difference does not appear to be increasing with time. In order to explain this condition, we must conclude that the excess heat that is absorbed in the low latitudes is transferred by some mechanism into the higher latitudes. Thus a more or less stable climatic difference is maintained between the two regions. The mechanism by which such transfer is achieved involves both the oceans and the atmosphere.

Figure 6–9 shows graphically that the greater heating of the atmosphere over the equator causes air to decrease in density and rise. As it rises, it cools by expan-

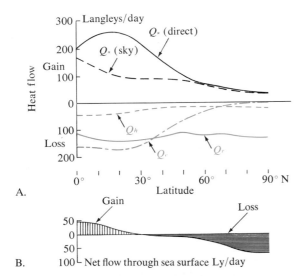

A.

B.

FIGURE 6–8
Heat Flow: Ocean-Atmosphere. *A,* heat flow through the ocean surface is represented as a function of latitude. At all latitudes there is a gain from solar radiation (Q_s) and a loss from back radiation (Q_r), conduction (Q_n), and evaporation (Q_e). The decrease in Q_s (direct) at the equator results from cloud cover associated with the equatorial low-pressure belt. *B,* the amount of heat lost from the ocean by back radiation, conduction, and evaporation is less than that gained from solar radiation below 30° latitude. The heat loss exceeds heat gain above 30° latitude. The langley (ly) is a calorie of heat flow through 1 cm² of the ocean surface. (After G. L. Pickard, *Descriptive physical oceanography* © 1963.)

sion, and the water vapor contained in the rising air mass condenses and falls as precipitation in the equatorial zone. After losing its moisture, this dry air mass descends. The descents occur in the subtropical regions (30° latitude) of the Northern and Southern hemispheres. As the descending air approaches Earth's surface, it is warmed by compression. Upon reaching the surface it moves away from the tropics, toward either the equator or higher latitudes. A high-pressure belt develops in the subtropics owing to the dense air that descends in these regions. Between the subtropical high-pressure belts, the horse latitudes, lies an equatorial low-pressure belt resulting from the low density of the rising air column above the equator.

The masses of air that move across Earth's surface from the subtropical high-pressure belts toward the equatorial low-pressure belt constitute the **trade winds.** Some of the air that descends in the subtropical regions moves along Earth's surface to higher latitudes as the **westerly wind belts.** These masses rise over the dense, cold air moving away from the polar high-pressure caps at the subpolar low-pressure belts located near 60°N and 60°S latitude. The air moving away from the poles produces the **polar easterly wind belts.** The air that rises at the 60° latitudes cools, releases precipitation in these regions, and ultimately descends in the polar regions or in the subtropics. These idealized latitudinal patterns are significantly altered by the uneven distribution of land and ocean over Earth's surface. The general effects of this idealized system are, however, clearly visible on a very broad scale. In the following paragraphs, we will examine how these air masses are affected by Earth's rotation as they move across Earth's surface.

CORIOLIS EFFECT

Any freely moving object on Earth's surface moving horizontally through a long distance for a relatively long period of time will veer to the right in the Northern Hemisphere and to the left in the Southern Hemisphere. The magnitude of this effect will increase with the velocity and the latitude of the object. This phenomenon is called the **Coriolis effect.** It is zero at the equator and maximum at the poles.

We will, for the sake of the following discussion, consider only the Northern Hemisphere. The air masses moving away from the subtropics toward the equator and

FIGURE 6–9
Air Masses. Low-pressure belts develop along the equator and the 60° latitude regions because of rising columns of warm air. Descending cool dry air produces high-pressure belts in the 30° latitude regions (subtropics). Air movements are primarily vertical in these belts. Strong lateral movements of air between the belts produce the westerlies and trade winds. Because of cold air masses overlying the polar regions, high-pressure conditions exist there also. (Reprinted by permission from Tarbuck, E. J., and Lutgens, F. K., *Earth science,* 6th ed. (New York: Macmillan, 1991), fig. 14.11.)

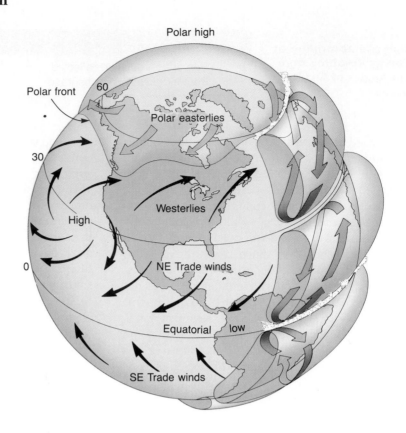

high latitudes are moving basically south in the first case and north in the second. In the case of the trade winds, they do not appear to blow directly out of the north but out of the northeast, while the westerlies, as their name implies, blow out of the southwest. Why do we observe this east-west component of motion in air masses that are moving north and south?

Figure 6–10A shows that as Earth rotates on its axis, points at different latitudes rotate at different velocities. The rotational velocity is proportional to the distance from Earth's surface to the axis of rotation and ranges from 0 km/h at the poles to over 1600 km/h (994 mi/h) at the equator.

The trade winds blow out of the northeast because as the air mass moves south from the subtropical region near 30°N latitude, it is also moving with the rotating earth in an easterly direction at about 1400 km/h (870 mi/h). The air mass starts moving toward a point on the equator at the same longitude that is moving east at 1600 km/h (994 mi/h). The distance that the air mass must cover to reach the equator is about 3200 km (2000 mi). If it moves to the south at 32 km/h (20 mi/h), it will require 100 h to arrive at the equator (fig. 6–10B).

The air mass is moving south at 32 km/h and east at 1400 km/h. For every hour that the air mass moves in the southerly direction, the point on the equator toward which it started moves east 200 km (124 mi) farther than does the air mass (1600 km/h–1400 km/h). In the period of 100 h it takes the air mass to make the trip, the point on

the equator toward which it started will be 20,000 km (100 h × 200 km/h) (12,420 mi) east of the air mass when it reaches the equator. While the air mass was moving the 3200 km in a southerly direction, it appears, as a result of Earth's rotation, to have moved 20,000 km in a westerly direction. Stated another way, as the air mass covered 30° of latitude from north to south, it appeared to move across 180° of longitude in a westerly direction. Certainly, if we were to encounter this air mass aboard ship at some point between the equator and 30°N latitude, it would be coming out of the east-northeast.

A westerly air mass moves in a northerly direction from 30° latitude to 60° latitude. This mass also is moving in an easterly direction at 1400 km/h. The point on the same longitude at 60° latitude toward which the air mass is headed is moving in an easterly direction at only 800 km/h (497 mi/h). Unlike the situation discussed in regard to the trade winds, the westerly air mass is moving in an easterly direction at a velocity greater than that of the point at 60°N latitude toward which it started. For every hour that this mass moves in a northerly direction, it moves 600 km (373 mi) farther east than the point at 60° latitude toward which it started. If we were to encounter this air mass at some point between 30° and 60°N latitude, it would appear to come out of the west-southwest.

If the air mass moves north at the same velocity as a trade wind mass moves south, the amount of deviation to the right in the case of the westerlies will be greater than

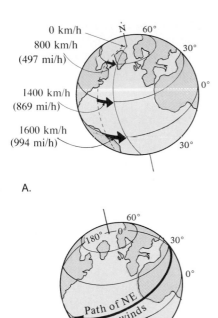

A.

B.

FIGURE 6–10
Coriolis Effect. A, points at different latitudes on Earth's surface rotate at different velocities, ranging from 0 km/h at the poles to 1600 km/h (994 mi/h) at the equator. B, an air mass that leaves the 30° latitude where it is rotating with Earth at 1400 km/h (870 mi/h) and heads south toward the equator at 32 km/h will not reach the equator until 100 hours later. Since the equator is rotating east at 200 km/h (124 mi/h) faster than the air mass, it will meet the equator at a point 20,000 km (12,420 mi) west of the point on the equator that was directly south of the air mass when it left its position on 30° latitude. This air mass could represent the northeast trade winds, which follow a similar path.

for the trade winds. This results from the fact that the condition responsible for the intensity of the Coriolis effect is the different rates at which points at different degrees of latitude rotate about Earth's axis. The rotational velocity of these points ranges from 0 km/h at the poles to more than 1600 km/h at the equator, but the rate of change per degree latitude is not constant. As we approach the pole from the equator, the rate of change of rotational velocity per degree of latitude change increases. As a comparison, we can note that there is a difference of 200 km/h in velocity across 30° of latitude from the equator to 30°N latitude, while there is a difference of 600 km/h in the velocity of rotation from 30°N to 60°N latitude. If we carry this consideration over another 30° of latitude from 60°N latitude to the pole, where the velocity is zero, the difference is over 800 km/h. This explains the fact that the Coriolis effect increases with increased latitude.

The primary factor affecting the amount of deflection resulting from the Coriolis effect is not, however, the magnitude of the effect since it is very small, even at high latitudes. It is the length of time a particle is in motion

that will have the greatest influence on how much it is deflected. Thus, even at low latitudes, a large Coriolis deflection is possible if an object is in motion for a long time.

Although we will not attempt to discuss it here, a complete mathematical treatment of the rotating Earth shows that the effects described for these north–south-moving air masses are valid for movements in any direction. Freely moving objects veer to the right of their intended path in the Northern Hemisphere and to the left in the Southern Hemisphere.

HEAT BUDGET OF THE WORLD OCEAN

As a result of variations in the absorption of solar energy, ocean temperatures will vary from place to place and from time to time. We previously discussed the fact that the temperature difference between polar and equatorial regions remains constant as a result of a transfer of heat energy from the equator to the higher latitudes by moving air and ocean masses. Driven by the moving air masses, large surface current systems are set up in the world's oceans and play an important role in transfer of heat energy. The temperatures at various places in the ocean will depend upon the rate at which heat flows in and out of these areas. Quantitative considerations of heat transfer in ocean water masses constitute the **heat budget** of these masses (figure 6–11).

The heat budget for any *locality* in the ocean is expressed by the following equation:

Rate of heat gain − Rate of heat loss =
Net rate of heat loss or gain

$$(Q_s + Q_c) - (Q_r + Q_e + Q_b) = Q_t$$

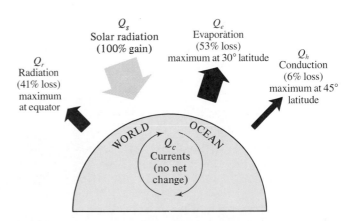

FIGURE 6–11
Avenues of Heat Flow between the Ocean and the Atmosphere. Solar radiation (Q_s) provides heat to the oceans. It is circulated within the ocean by currents (Q_c). Heat is lost from the ocean through evaporation (Q_e), radiation (Q_r), and conduction (Q_b). Heat loss equals heat gain.

In which

Q_s = rate of heat gain from solar radiation

Q_c = rate of heat gain (or loss) through ocean current

Q_r = rate of heat loss through radiation into space

Q_e = rate of heat loss through evaporation

Q_b = rate of heat loss through conduction into the atmosphere

Q_t = net rate of heat loss or gain in the locality

If Q_t is a positive value, it means the temperature of the ocean in that locality is rising. If it is a negative value, the temperature in that locality is falling, and if it is zero, the temperature of the mass of water is not changing.

Theoretically, Q_c and Q_t should be zero when we consider the world ocean over a long period of time. Current flow is internal, and seasonal increases and decreases in temperature should average out so that Q_t becomes zero. Therefore, the rate at which heat is gained in the world ocean through solar radiation should equal the rate at which heat is lost through radiation, evaporation, and conduction into the atmosphere (figure 6–11). Symbolically the heat budget of the *world ocean* is expressed using average values as:

$$Q_s = Q_r + Q_e + Q_b$$
$$100\% = 41\% + 53\% + 6\%$$

The amount of heat lost by radiation, primarily of infrared wavelengths, is greatest near the equator, where the greatest amount of solar energy is absorbed. The greatest heat loss by evaporation is in dry subtropics, while the heat lost through conduction reaches a maximum along the western margins of ocean basins where warm water currents carry water into regions where the atmosphere is much colder than the water.

THE OCEANS, WEATHER, AND CLIMATE

The idealized pressure belts and consequent wind systems previously discussed are significantly modified as a result of two factors. First, the tilt of Earth's axis of rotation produces seasons, and second, the air over continents gets colder in winter and warmer in summer than the air over adjacent oceans. As a result, continents usually develop atmospheric high-pressure cells that reflect the weight of the cold air centered over them during winter and low-pressure cells during the summer (figure 6–12). As air moves away from the high-pressure cells and toward the low-pressure cells, the Coriolis effect produces a cyclonic (counterclockwise) flow of air around low-pressure cells and an anticyclonic (clockwise) flow of air around high-pressure cells in the Northern Hemisphere (figure 6–13). Thus, wind patterns associated with

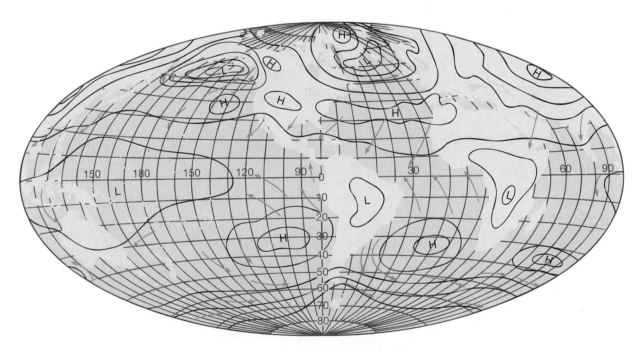

FIGURE 6–12
Sea-Level Atmospheric Pressures and Winds. The idealized high- and low-pressure belts shown in figure 6–9 are modified by the seasons and distribution of continents. Pressure patterns shown in this figure are the averages for the month of January. Because continental rocks have a lower heat capacity than the water of the oceans, continents get hotter in the summer and colder in the winter than the adjacent oceans. The cold air over the continents of the Northern Hemisphere and warm air over the continents in the Southern Hemisphere produce the high- and low-pressure cells centered over the continents.

FIGURE 6–13
High- and Low-Pressure Cells and Air Flow. As air moves away from high-pressure cells (A) and toward low-pressure cells (B), the Coriolis effect causes the air masses (winds) to veer to the right in the Northern Hemisphere. This results in anticyclonic clockwise winds around high-pressure cells and cyclonic counterclockwise winds around low-pressure cells.

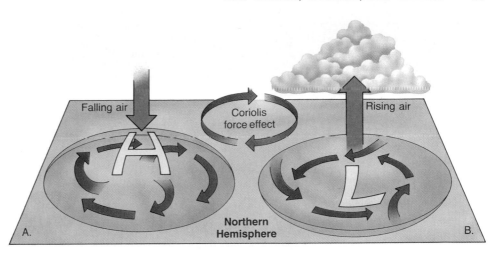

continents may reverse themselves on a seasonal basis as winter high-pressure cells are replaced by summer low-pressure cells.

Day-to-day weather may change little at high and low latitudes. Polar regions are usually cold and dry regardless of the season. Near the equator, the air is warm, damp, and still as the dominant direction of air movement in the doldrums belt is up. Midday rains are common. It is at the midlatitudes that weather gets interesting. Due to the seasonal change of pressure systems over continents, air masses from the high and low latitudes may move into the midlatitudes, meet, and produce severe storms. In the United States, we may be invaded by three major polar air masses and two tropical air masses (figure 6–14).

As polar and tropical air masses move into the midlatitudes, they are also gradually moving in an easterly direction. A **warm front** is the contact between a warm air mass moving east into an area occupied by cold air. A **cold front** is the contact between a cold air mass moving east into an area occupied by warm air (figure 6–15). These confrontations are brought about by the movement of the **jet stream,** an easterly moving air mass centered at about 10 km (6 mi) elevation above the midlatitudes. It usually follows a wavy path and may cause unusual weather by pulling a polar air mass far south or a tropical air mass far to the north. Regardless of whether the colliding air masses represent a cold front or a warm front, the warm air rises above the denser cold air, cools, and the moisture in it condenses as precipitation. A cold front is usually steeper, and the temperature differences across it greater. Therefore, rainfall associated with a cold front is usually heavier and of shorter duration than that resulting from a warm front.

Climate Patterns in the Oceans

The open ocean can be divided into climatic regions with relatively stable boundaries that run generally east-west.

Temperature and salinity of surface waters are determined by the amount of solar radiation and the amounts of evaporation and precipitation (figure 6–16).

In *equatorial* regions the major air movement is vertical as air rises, so winds are weak. Surface waters are warm, and the air is saturated with water vapor (figure 6–17). Heavy precipitation keeps salinity relatively low. Sailors once referred to this region as the **doldrums** because their sailing ships were becalmed by the lack of winds. Meteorologists refer to it as the **intertropical convergence zone** because it is the region where the trade winds converge.

Tropical regions are characterized by strong north-easterly trade winds in the Northern Hemisphere and southeasterly trade winds in the Southern Hemisphere (figure 6–18). These winds push the equatorial currents and create moderately rough seas. Relatively little precipitation falls at higher latitudes within tropical regions, but precipitation increases toward the equator. Hurricanes and typhoons that carry large quantities of heat into higher latitudes are initiated at the doldrums and pass through the tropics as tropical storms.

The belts of high pressure previously described are centered in the **subtropical** region. The dry air descending on the subtropics results in little precipitation and a high rate of evaporation, which produces the highest surface salinities in the open ocean (figure 6–19). Winds are weak in the open ocean, as are currents. However, strong boundary currents flow north and south, particularly along the western margins of the subtropical oceans.

The **temperate** regions are characterized by strong westerly winds blowing from the southwest in the Northern Hemisphere and from the northwest in the Southern Hemisphere. Severe storms are common, especially during winter, and precipitation is heavy.

The **subpolar** ocean is covered in winter by sea ice that melts away, for the most part, in summer. Icebergs are common, and the surface temperature seldom exceeds 5°C (41°F) in the summer months.

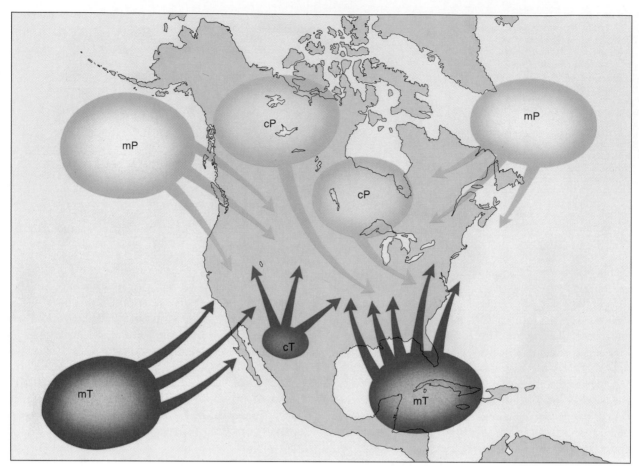

FIGURE 6–14
Air Masses That Affect Weather in the United States. The polar air masses are more likely to invade the United States during winter, and the tropical air masses tend to move in from the south during summer. Air masses are classified on the basis of their source region. The designation continental (c) or maritime (m) gives an indication of moisture content, whereas polar (P) and tropical (T) indicate temperature conditions. (Reprinted by permission from Tarbuck, E. J., and Lutgens, F. K., *Earth science,* 6th ed. (New York: Macmillan, 1991), fig. 15.1.)

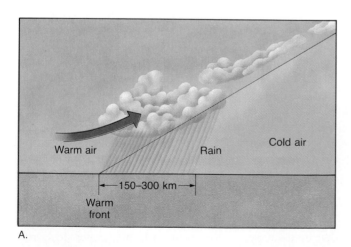

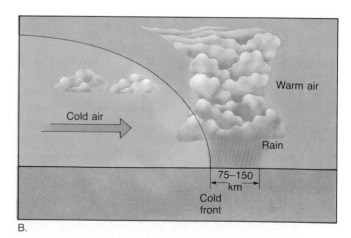

FIGURE 6–15
Warm and Cold Fronts. Cross sections through a warm front *(A)* and cold front *(B)*.

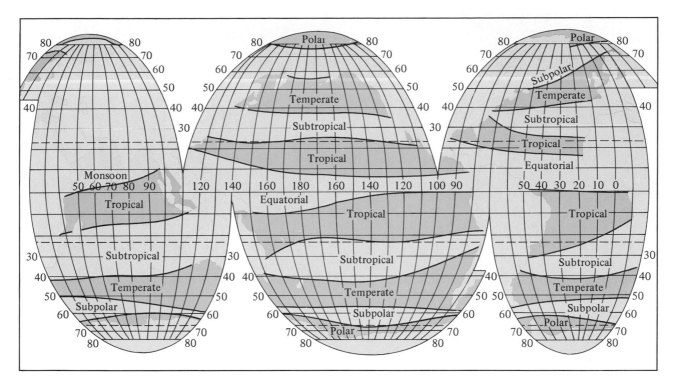

FIGURE 6–16
Climatic Patterns of the Open Ocean.

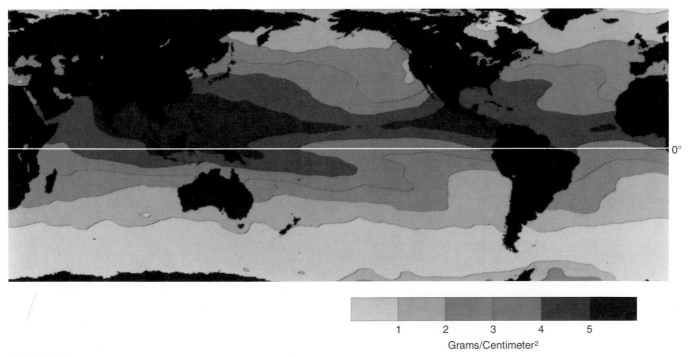

FIGURE 6–17
***SEASAT* SMMR Water Vapor, July 7–October 10, 1978.** Because the ability of the atmosphere to hold water vapor increases with increased temperature, the greatest amount of atmospheric water vapor is found near the equator. This illustration reflects the great amount of heat being removed from the ocean by evaporation in the low latitudes. This belt of maximum atmospheric water vapor content is called the doldrums. It is a region where warm air slowly rises and there are minimal horizontal air movements (wind) to propel sailing vessels or cool their crews. Because it was late summer–early fall in the Northern Hemisphere when this data was obtained, the doldrums are shifted significantly north of the equator.

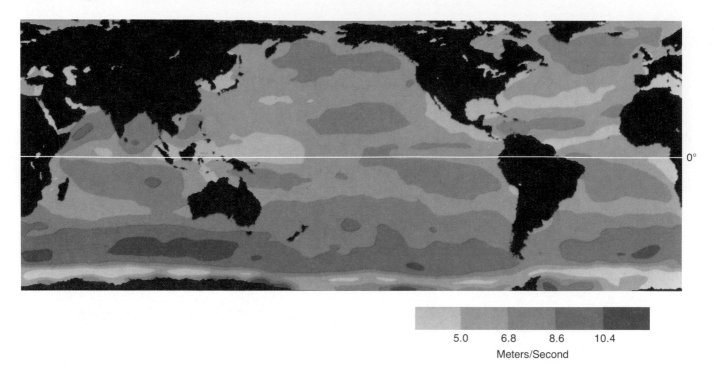

5.0   6.8   8.6   10.4
Meters/Second

FIGURE 6–18
***SEASAT* Altimeter Wind Speed, July 7–October 10, 1978.** Two of Earth's prevailing wind systems are the trade winds that blow from 30° latitudes toward the equator and the westerlies that blow from 30° toward 60° latitudes. Because there are fewer continents in the Southern Hemisphere, the westerly wind belt there contains the strongest year-round winds. The lowest wind speeds occur at the doldrums, where the air is rising, and the horse latitudes (30°N and S latitudes), where air descends. Wind speeds gradually increase away from these belts of minimum wind speed.

Surface temperatures remain at or near freezing in the **polar** areas, which are covered with ice throughout most of the year. In these areas, which include the Arctic Ocean and the ocean adjacent to Antarctica, there is no sunlight during the winter and no night during the summer.

Sea Ice

Resulting directly from the low-temperature conditions that are characteristic of high-latitude areas is the development of a permanent to nearly permanent ice cover at the sea surface. The term **sea ice** is used to distinguish

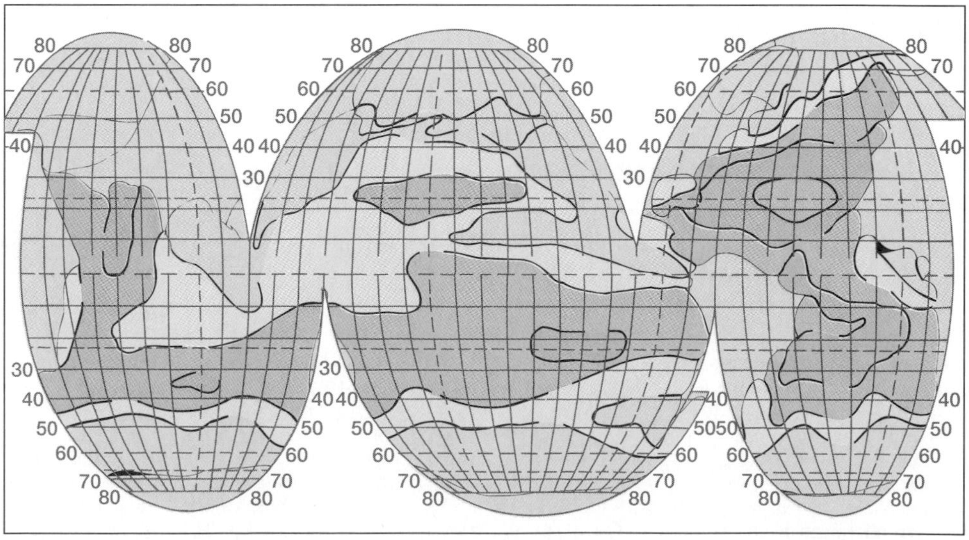

FIGURE 6–19
Surface Salinity of the Oceans in August (‰). (After Sverdrup et al., 1942.)

these masses of ice from **icebergs,** which may also be found at sea but originate by breaking away from glacial masses of ice that form on continental portions of Earth's surface.

The freezing point of water with a salinity of 35‰ is −1.91°C (28.6°F). As water freezes at the ocean surface, the dissolved solids do not fit into the crystalline structure of the ice and are left behind, so that the salinity of the surrounding water increases. This greater salinity tends to lower the freezing point of the remaining water. However, the low-temperature, high-density water that is excluded from the ice tends to sink and be replaced by warmer, less dense water at the surface. This circulation enhances the formation of sea ice, since freezing is aided by low salinity and calm water conditions. Sea ice is found throughout the year around the margin of Antarctica, within the Arctic Sea, and in the extreme high-latitude region of the North Atlantic Ocean.

Sea ice begins to form as small needlelike crystals of hexagonal shape that eventually become so numerous that a slush develops. As the slush begins to form into a thin sheet, it is broken up by wind stress and wave action into disc-shaped pieces called **pancake ice** (figure 6–20A). As further freezing occurs, the pancakes coalesce to form **ice floes.** The rate at which sea ice forms is closely tied to temperature conditions, and large quantities of ice form in relatively short periods of time during which the temperature is very low, for example, −30°C (−22°F). Even at low temperatures, the rate of ice formation will slow as ice thickness increases, due to the poor heat conduction of ice, which is a good insulator. Newly formed sea ice contains a significant quantity of brine that is trapped during the freezing process.

Depending upon the rate of freezing, newly formed ice, *new ice,* may have a total salinity that ranges from 4 to 15‰. The more rapidly it forms, the more brine will be captured and the higher the salinity. After a period of time, the brine will trickle down through the coarse structure of the sea ice, and the salinity of the sea ice will decrease. Normally, by the time it is a year old (old ice), sea ice has become relatively pure.

In the Arctic Sea, ice that forms at the sea surface can be classified into one of three categories—pack ice, polar ice, or fast ice. **Pack ice** forms around the margin of the Arctic Sea, extending through the Bering Strait into the Bering Sea and as far south as Newfoundland and Nova Scotia in the North Atlantic. Pack ice reaches its maximum extent during the month of May and breaks up to cover its least area in September (figure 6–21). This ice, which can be penetrated by icebreakers, reaches a maximum thickness of about 2 m (6.5 ft) in the winter period. The pack ice is driven primarily by the winds, although it also responds to surface currents, which produce stresses that continually break and reform its struc-

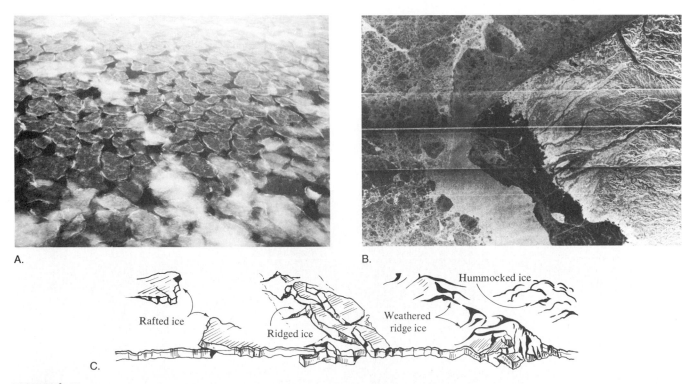

A.

B.

C.

FIGURE 6–20
Sea Ice. *A,* pancake ice. (Photo from Oceanographer of the Navy, Public Affairs Office.) *B,* Banks Island, Canada, to the right, is bounded by fast ice (black). A strip of open water (light) separates the fast ice from pack ice along the left margin. (Photo courtesy of NASA.) *C,* rafted ice.

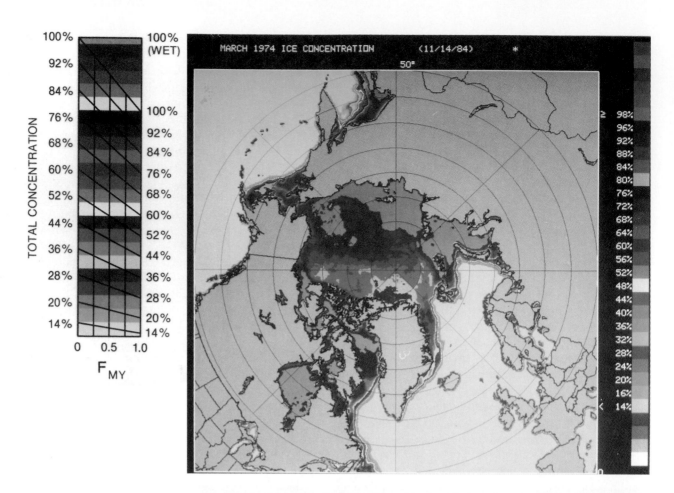

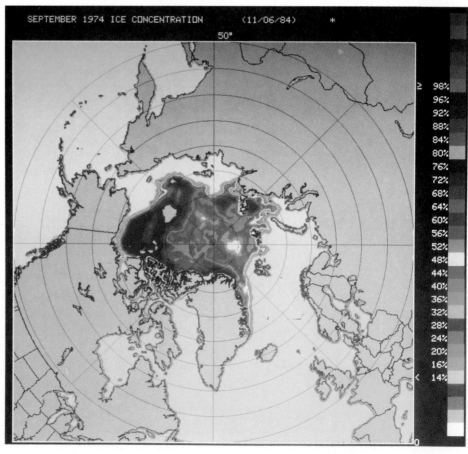

FIGURE 6–21
Extent of Ice in Arctic Sea. These 1974 images showing percent of sea-ice concentration are accurate to within 15 percent for first-year ice (pack ice) and 25 percent for multi-year-ice (polar ice cap). The scales of ice concentration along the right margins of each image are for first-year ice. The scale above can be used for first-year ice (use left side of scale) and multiyear-ice (use right side of scale). The images were developed from the ESMR (Electrically Scanning Microwave Radiometer) aboard the *Nimbus* 5 satellite. (Courtesy of NOAA.)

ture. Its formation is achieved by the expansion of floes that begin to raft onto one another as they expand to cover the sea's surface.

The *polar ice* that covers the greatest portion of the Arctic Sea, including the polar region, attains a maximum thickness in excess of 50 m (164 ft). During the summer months, melting may produce enclosed bodies of water called **polynyas,** although the polar ice never totally disappears. Its average thickness during the summer months is over 2 m (6.5 ft). The polar ice is constantly being exchanged as floes from the pack ice are carried into the polar region during the winter season, and floes break out of the polar ice and reenter the pack ice during the summer. Circling in a clockwise direction around the Arctic Ocean, about one-third of the pack and polar ice is carried into the North Atlantic by the East Greenland Current each year.

Developing in the winter from the shore out to the pack ice is the **fast ice,** which completely melts during the summer. The fast ice, which is firmly attached to the shore, attains a winter thickness in excess of 2 m (6.5 ft).

In the Southern Hemisphere, where the polar region is covered by a continental mass, we might characterize all the sea ice that forms around the margin of the Antarctic continent as pack ice and fast ice that have a rather temporary existence. The pack ice rarely extends north of 55°S latitude and breaks up rather completely from October to January except in the very quiet bays. Winds that are frequently very strong help prevent the formation of a greater pack ice accumulation around the Antarctic continent.

Icebergs

Icebergs are derived from land ice in the form of glaciers that cover the continent of Antarctica and most of the island of Greenland and part of Ellesmere Island in the Arctic (figure 6–22). These vast sheets of ice that grow from the snow accumulation on the landmasses spread outward toward the margins until they push their edges out into the marginal sea. There the ice sheets are buoyed up by the water and break up under the stress of current, wind, and wave action. This breakup that produces icebergs is called *calving*.

In the Arctic, the icebergs that are produced originate primarily from the ice that follows narrow valleys into the sea along the western coast of Greenland. Icebergs are also produced along the east coast of Greenland and the east coast of Ellesmere Island. The East Greenland Current and the West Greenland Current carry the icebergs at rates of up to 20 km (12.4 mi) per day into the North Atlantic, where the Labrador Current may move them into North Atlantic shipping channels. They are seldom carried south of 45°N latitude, but during some seasons icebergs will move as far south as 40°N latitude and become shipping hazards.

Such an accumulation of icebergs developed when the *Titanic* sank in April 1912. After receiving repeated warning of the existence of such a hazard, the 46,000-ton ship containing 2224 passengers proceeded at an excessive speed of 41 km/h (25.5 mi/h) until it came to a grinding halt after hitting an iceberg with its starboard bow. The sinking of the *Titanic* at 41°46'N latitude, 50°14'W longitude near the Grand Banks cost the lives of 1517 people and brought about the formation of the ice patrol that has prevented further loss of life from such accidents. The U.S. Navy began this patrol immediately following the *Titanic* disaster; it became international in 1914. The patrol, now maintained by the U.S. Coast Guard, concentrates its efforts between 40°30' and 48°N latitude and 43° and 54°W longitude.

The most recent major iceberg ramming occurred in the early hours of June 20, 1989. The Soviet cruise liner *Maxim Gorky* rammed an iceberg well north of the Arctic Circle between Greenland and Spitzbergen. Quick action by the crew and the Norwegian Coast Guard prevented any loss of life among the 953 passengers and 378 crew members.

The **shelf ice** that represents the edges of glaciers pushing into the marginal seas of Antarctica produces icebergs that have received less attention than those of the North Atlantic, owing to the fact that they interfere less with shipping. Vast tabular bergs that break from the edges of the shelf ice have lengths over 100 km (62 mi) (figure 6–22C). They may stand as much as 200 m (656 ft) above the ocean surface, though most are probably less than 100 m (328 ft) above sea level. Most of the calving occurs, as it does in the Arctic region, during the summer months. When the sea ice breaks up and allows the swells driven by strong winds to reach the edge of the shelf ice, large icebergs are calved and move to the north into the warmer and rougher water, in which they disintegrate. Carried by the strong West Wind Drift, these icebergs move in an easterly direction around Antarctica, rarely moving farther north than 40°S latitude. In August 1991, an iceberg the size of Connecticut (13,000 km² or 5000 mi²) broke loose from the ice shelf of the Weddell Sea.

RENEWABLE SOURCES OF ENERGY

The potential of extracting energy from the motions and heat distribution patterns of the atmosphere and ocean, which are maintained by the sun, is attractive for the following reasons:

1. Work can be achieved without significant pollution.
2. The amount f energy available at any time is far greater than that in fossil and nuclear fuels.

A.

B.

C.

D.

FIGURE 6–22
Icebergs. *A,* North Atlantic icebergs off Cape York, Greenland. They are generally smaller and more irregular in shape than Antarctic icebergs. *B,* the USS *Wyandotte* moves past a tabular iceberg in the Weddell Sea, Antarctica. Many of the icebergs are kilometers in length and have been mistaken for land. (Photographs *A* and *B* from Oceanographer of the Navy, Public Affairs Office.) *C,* B-9 Iceberg. This infrared image shows the huge iceberg that broke from the Ross Ice Shelf in October 1987. The longitude and latitude lines intersecting near the east end of B-9 are 160°W and 78°S. This monster is 136 km (85 mi) long, but two larger icebergs formed the year before in the Weddell Sea. (Infrared image in *C* courtesy of Robert Whritner, Manager, Antarctic Research Center, Scripps Institution of Oceanography, University of California, San Diego.) *D,* a map showing North Atlantic currents and icebergs (△).

3. Such forms of energy are renewable and will not be depleted.

Considered in order of decreasing energy potential, the sources of renewable energy are (1) heat stored in the oceans, (2) kinetic energy of the winds, (3) potential and kinetic energy of waves, and (4) potential and kinetic energy of tides and currents. In later chapters we will consider the use or potential use of winds, waves, tides, and currents. Here we will consider only the renewable source with the greatest store of potential energy: the surface layers of the tropical oceans.

Ninety percent of Earth's surface between the Tropic of Cancer and Tropic of Capricorn is ocean. What makes this warm tropical surface water such an important source of energy is the presence of much colder water beneath the thermocline. With a temperature difference as small as 17°C (30.6°F), useful work can be done by

A.

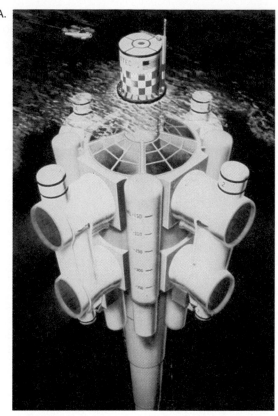

B.

FIGURE 6–23
Proposed Uses of Ocean Thermal Power. *A,* Ocean Thermal Energy Conversion (OTEC) system with crew quarters and maintenance facilities. Attached around the outside are turbine-generators and pumps. It is over 75 m (246 ft) in diameter, 485 m (1595 ft) long, and weighs about 300,000 tn. This unit is designed to generate 160 million watts of power, which is enough to meet the needs of a city with a population of 100,000. (Photo courtesy of Lockheed Missiles and Space Co., Inc.) *B,* OTEC plant for production of ammonia. Moving slowly through tropical waters, this plant could produce 1.4 percent of the ammonia requirements for the United States each year and save 22.6 billion cubic feet of natural gas. (Courtesy of U.S. Department of Energy.)

Ocean Thermal Energy Conversion (OTEC) systems (figure 6–23).

The system works in the opposite way from a typical refrigeration system and on a much larger scale. The warm surface water heats a fluid such as propane or ammonia that is under pressure in evaporating tubes. The fluid is vaporized and passes through a turbine that drives an electrical generator. After passing through the turbine, the fluid is condensed by cold water that has been pumped up from the deep ocean. It is again ready for heating by warm surface water that will cause it to vaporize and pass through the turbine.

The only region with such potential along the conterminous United States coast is a strip about 30 km (18.6 mi) wide and 1000 km (621 mi) long. It extends north from southern Florida along the western margin of the Gulf Stream. The National Science Foundation is supporting an experimental OTEC system to be established 25 km (16 mi) east of Miami, Florida.

Mini-OTEC, a unit mounted on a U.S. Navy barge, was tested off Ke-Ahole Point, Hawaii, in 1979. Designed primarily to test equipment under ocean conditions, it generated only 15 net kilowatts of electricity. A much larger unit, OTEC 1, designed to generate 1 megawatt, tested heat exchanger designs and biofouling systems under ocean conditions for 3 months in 1981 in Hawaiian waters.

Testing is presently under way on the Seacoast Test Facility, a shore-based OTEC installation, on the island of Hawaii. Although it will require a large initial investment, Hawaii has high hopes that OTEC power generation will be a commercially successful means of power generation by the mid-1990s.

Another proposed use of the OTEC concept is producing ammonia. A floating factory that could produce 586,000 tn of ammonia per year has been described in a feasibility report from the Johns Hopkins Applied Physics Laboratory to the U.S. Maritime Administration. The proposed demonstration ship would weigh 68,000 tn, have a width of almost 60 m (197 ft), and have a length of over 144 m (472 ft). It would move at less than 2 km/h through tropical waters and use energy derived from the temperature difference between the warm surface waters and

cold deep waters to produce ammonia. Each such plant could produce about 1.4 percent of our nation's ammonia requirements—75 percent of which is used for fertilizer—and save 22.6 billion cubic feet of natural gas each year (figure 6–23).

SUMMARY

In ocean water of low biological productivity, the molecular-sized particles scatter the short wavelengths of visible light, producing a blue ocean. Greater amounts of dissolved organic matter in more-productive ocean water scatter more green light which, along with the chlorophyll pigmentation of the plants, produces a green ocean. With increasing depth in the ocean, the colors of the visible spectrum are absorbed by ocean water in a way that removes red, yellow, and violet at relatively shallow depths. The blue and green wavelengths of light are the last to be removed. High turbidity, the measure of the amount of suspended material in water, greatly reduces the depth to which light will penetrate.

Density of ocean water is increased by increased salinity and decreased temperature. Density change per unit of temperature change is greater in warmer water. Density, temperature, and salinity are conservative properties of water, properties that are affected only by mixing and diffusion once a surface water mass sinks. Zones of rapid change in the density, temperature, or salinity in the water column are called a pycnocline, thermocline, and halocline, respectively.

Velocity of sound transmission in the ocean increases with increases in temperature, salinity, and pressure. A low-velocity sound channel is caused by a thermocline, in which temperature decreases with increased depth.

Radiant energy reaching Earth from the sun is mostly in the ultraviolet and visible light range, while that radiated back to space from Earth is primarily in the infrared spectrum. Water vapor, carbon dioxide, and other trace gases absorb the infrared radiation and heat the atmosphere. This phenomenon is called the greenhouse effect. Because human activities are increasing the concentrations of trace gases that enhance the greenhouse effect, there is concern that Earth may be warming.

The atmosphere is unevenly heated and set in motion because more energy is received and radiated back into space at low latitudes and water vapor, which absorbs infrared radiation well, is unevenly distributed in the atmosphere. Belts of low pressure, where air rises, are generally found at the equator and at about 60° latitude. High-pressure regions, where dense air descends, are located at the poles and at about 30° latitude. The air at Earth's surface that is moving away from the subtropical highs produces the trade winds moving toward the equator and westerlies moving toward higher latitudes.

Because Earth's surface rotates at different velocities at different latitudes, increasing from 0 km/h at the poles to over

1600 km/h (994 mi/h) at the equator, objects in motion tend to veer to the right in the Northern Hemisphere and to the left in the Southern Hemisphere. This is called the Coriolis effect.

In the heat budget of the world ocean, the only source of heat gain is the sun, while heat may be lost by radiation, evaporation, and conduction into the atmosphere. The amount of heat lost by radiation, primarily of infrared wavelengths, is greatest near the equator, where the greatest amount of solar energy is absorbed. The greatest heat loss by evaporation is in the dry subtropics, while the heat lost through conduction reaches a maximum along the western margins of oceans in temperate latitudes where warm water currents carry water into regions where the atmosphere is much colder than the water.

The tilt of Earth's axis of rotation and the distribution of continents modify the idealized pressure belts discussed above. High-pressure cells form over continents in winter and are replaced by low-pressure cells in summer. There is a cyclonic movement of air around the low-pressure cells and an anticyclonic movement around high-pressure cells. Cold air masses of high latitudes meet warm air masses of lower latitudes and create cold and warm fronts that move from west to east across Earth's surface at midlatitudes.

Nonetheless, ocean climate patterns are closely related to the idealized pressure belts previously discussed.

Sea ice forms as seawater is frozen in high latitudes. The process usually involves the formation of a slush, which breaks into pancakes that ultimately grow into floes. Sea ice develops as pack ice that forms each winter and melts almost entirely each summer, polar ice that is a permanent accumulation in polar regions of the Arctic Ocean, and fast ice that forms frozen to shore during the winter. Icebergs form as large chunks of ice break away from the large continental glaciers that form on Ellesmere Island and Greenland in the Northern Hemisphere and Antarctica in the Southern Hemisphere.

From the motions and patterns of heat distribution in the atmosphere and oceans, renewable, nonpolluting sources of energy can be exploited. These include heat stored in the ocean and potential and kinetic energy stored in the winds, waves, tides, and currents. Ocean Thermal Energy Conversion (OTEC) is a process developed to use the difference in temperature between warm tropical surface water and the cold water below the thermocline to produce electricity and ammonia.

KEY TERMS

Adiabatic (p. 124)
Antarctic Circle (p. 130)
Arctic Circle (p. 130)
Cold front (p. 135)
Conservative property (p. 126)
Coriolis effect (p. 131)
Deep water (p. 127)
Density, In situ (sigma) (p. 124)
Density, Potential (sigma theta)
  (p. 124)
Doldrums (p. 135)
Electromagnetic spectrum (p. 124)
El Niño–Southern Oscillation (p. 123)
Fast ice (p. 141)
Greenhouse effect (p. 123)

Halocline (p. 127)
Heat budget (p. 133)
Ice floe (p. 139)
Iceberg (p. 139)
Intertropical convergence zone
  (p. 135)
Jet stream (p. 135)
Nonconservative property (p. 126)
Ocean Thermal Energy Conversion
  (OTEC) (p. 143)
Pack ice (p. 139)
Pancake ice (p. 139)
Polar (p. 138)
Polar easterly winds (p. 131)
Polynyas (p. 141)

Pycnocline (p. 126)
Sea ice (p. 138)
Shelf ice (p. 141)
Sofar channel (p. 128)
Subpolar (p. 135)
Subtropical (p. 135)
Temperate (p. 135)
Thermocline (p. 127)
Trade winds (p. 131)
Tropical (p. 135)
Tropic of Cancer (p. 130)
Tropic of Capricorn (p. 130)
Upper water (p. 127)
Warm front (p. 135)
Westerly winds (p. 131)

QUESTIONS AND EXERCISES

1. How is the color of the ocean surface water related to biological productivity? Why does everything in the ocean at depths below the shallowest surface water take on a blue-green appearance?
2. Describe the relative effect of temperature change on water density at high- versus low-temperature ranges.
3. The position of the pycnocline in the water column is determined by the combined effects of the thermocline and halocline. Describe these relationships in tropical waters (see figures 6–3 and 6–4).
4. How does the development of a thermocline produce a sofar, or sound channel, below the ocean's surface?
5. Explain why the atmosphere is heated primarily by back radiation from Earth rather than by direct radiation from the sun.
6. Discuss the "greenhouse gases" in terms of their relative concentrations and relative contributions to any increased greenhouse effect.
7. Describe the effect on Earth as a result of Earth's axis of rotation being angled 23.5° to the ecliptic.
8. Since there is a net annual heat loss at high latitudes and a net annual heat gain at low latitudes, why does the temperature difference between these regions not increase?
9. Why are there high-pressure caps at each pole and high-pressure belts at 30° latitudes compared to low-pressure belts in the equatorial region and at 60° latitudes?
10. Describe the Coriolis effect in the Northern and Southern hemispheres and include a discussion of why the effect increases with increased latitude.

11. In considering the heat budget of the world ocean over a long period, why should Q_c (rate of heat gain or loss through ocean currents) and Q_t (net rate of heat gain or loss) be considered to be zero?
12. Describe the ocean regions where the maximum rates of heat loss by radiation, evaporation, and conduction occur, and explain why.
13. Discuss why the idealized belts of high- and low-atmospheric pressure shown in figure 6–9 are modified (see figure 6–12).
14. Name the polar and tropical air masses that affect U.S. weather. Describe the pattern of movement across the continent and patterns of precipitation associated with warm and cold fronts.
15. How are the ocean's climatic belts (figure 6–16) related to the broad patterns of air circulation described in figure 6–9?
16. Describe the formation of sea ice from the initial freezing of the ocean surface water through the development of polar ice in the Arctic Ocean.
17. What is the difference in the average size and shape of icebergs in the Arctic and Antarctic? Why do these differences exist?
18. Construct your own diagram of how an Ocean Thermal Energy Conversion unit might generate electricity, or make a flow diagram presenting the steps of the process.

REFERENCES

Changing climate and the oceans. 1987. *Oceanus* 29:4, 1–93.

Charlson, R. J.; Schwartz, S. E.; Hales, J. M.; Cess, R. D.; Coakley, J. A., Jr.; Hansen, J. E.; and Hofmann, D. J. 1992. Climate forcing by anthropogenic aerosols. *Science* 255:5043, 423–30.

Miller, A. 1971. *Meteorology.* Columbus, Ohio: Merrill.

Ocean energy. 1979. *Oceanus* 22:4, 1–68.

Oceans and climate. 1978. *Oceanus* 21:4, 1–70.

The oceans and global warming. 1989. *Oceanus* 32:2, 1–75.

Pickard, G. L. 1975. *Descriptive physical oceanography.* 2nd ed. New York: Pergamon Press.

Rodhe, H. 1990. A comparison of the contribution of various gases to the greenhouse effect. *Science* 248:4960, 1217–19.

SUGGESTED READING

Sea Frontiers

Boling, G. R. 1971. Ice and the breakers. 17:6, 363–71. An interesting history of people in icy waters and the development and improvement of ice breakers.

Charlier, R. 1981. Ocean-fired power plants. 27:1, 36–43. The potential of ocean thermal-energy conversion is discussed.

Houghton, R. A., and Woodwell, G. M. 1989. Global climate change. 260:1, 36–47. The history of global climate change is discussed along with the potential for future climatic change.

Land, T. 1976. Europe to harness the power of the sea. 22:6, 346–49. An article emphasizing the clean, safe, permanent nature of ocean tides and waves as a source of power.

Mayor, A. 1988. Marine mirages. 34:1, 8–15. The nature of marine mirages and the research efforts that lead to our understanding them are considered.

Rush, B., and Lebelson, H. 1984. Hurricane! The enigma of a meteorological monster. 30:4, 233–39. An overview of the nature of hurricanes and the problem of predicting where they will go.

Scheina, R. L. 1987. The Titanic's legacy to safety. 33:3, 200–209. A brief summary of the *Titanic* sinking and an overview of the ice patrol and iceberg collision history subsequent to the sinking of the "unsinkable" luxury liner.

Smith, F. G. W. 1974. Planet's powerhouse. 20:4, 195–203. A description of how Earth's "heat engine" works, with an emphasis on the nature of tropical cyclonic storms.

———. 1974. Power from the oceans. 20:2, 87–99. A survey of the many tried and untried proposals for extracting energy from the oceans.

Sobey, E. 1979. Ocean ice. 25:2, 66–73. The formation of sea ice and icebergs as well as their climatic and economic effects are discussed.

———. 1980. The ocean-climate connection. 26:1, 25–30. The increasing amount of knowledge of the effect of the oceans on Earth's climate may be used to predict climatic trends of the future.

Scientific American

Gregg, M. 1973. Microstructure of the ocean. 228:2, 64–77. A discussion of the methods of studying the detailed movements of ocean water by observing temperature and salinity changes over distances of one centimeter and the motions they reveal.

Jones, P. D., and Wigley, T. M. 1990. Global warming trends. 263:2, 84–91. Data relating to evidence of global warming over the past 100 years is presented.

MacIntyre, F. 1974. The top millimeter of the ocean. 230:5, 62–77. Processes that are confined to a thin film at the surface of ocean water and their role in the overall nature of the oceans are discussed.

Penney, T. R., and Bharathan, D. 1987. Power from the sea. 256:1, 86–93. A prediction that generating electricity by ocean thermal energy conversion will be competitive with fossil fuel plants as the price of oil rises.

Revelle, R. 1982. Carbon dioxide and world climate. 247:2, 35–43. Some of the possible effects of increasing atmospheric temperature due to CO_2 accumulation are considered.

Stanley, S. M. 1984. Mass extinctions in the oceans. 250:6, 64–83. Geological evidence suggests most major periods of species extinction over the last 700 million years occurred during brief intervals of ocean cooling.

Stolarski, R. S. 1988. The Antarctic ozone hole. 258:1, 30–37. The discovery of the Antarctic ozone hole and its possible significance are discussed.

White, R. M. 1990. The great climate debate. 263:1, 36–45. The controversy over the degree of global warming we can expect in our future is discussed.

Chapter 7

Ocean Circulation

Currents, or water masses in motion, are driven ultimately by energy from the sun. The circulation of these masses can be categorized as being either **wind-driven** or **thermohaline.** The wind-driven currents are set in motion by moving air masses, and this motion is confined primarily to horizontal movement in the upper waters of the world ocean. The thermohaline circulation has a significant vertical component and accounts for the thorough mixing of the deep masses of ocean water. Thermohaline circulation is initiated at the ocean surface by temperature and salinity conditions that produce a high-density mass, which sinks and spreads slowly beneath the surface waters.

Even though we think of the ocean as having great depth, ocean basins are very shallow in comparison with their widths. If the ocean basins were scaled down so their widths equaled the width of this page, they would be no deeper than the page is thick. Yet within this thin shell of water, there is a rich dynamic structure in which the wind-driven surface current system is separated by the pycnocline from the deep thermohaline circulation. Nevertheless, these separate systems do communicate

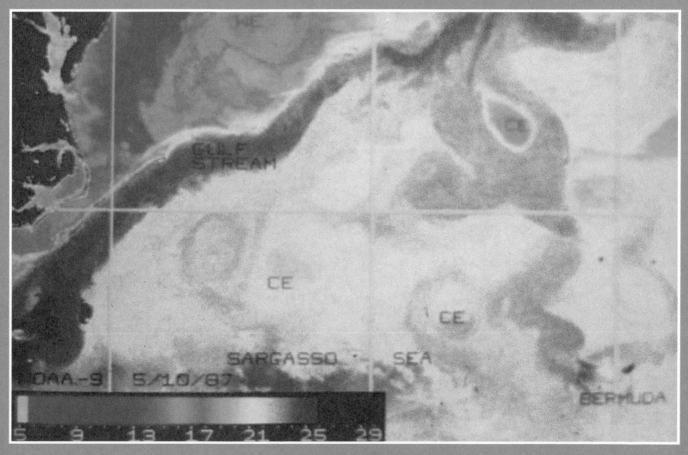

In this false color image, the Gulf Stream appears dark red, indicating ocean water temperature above 25°C. The sharp boundary north of the Gulf Stream, where blue indicates water temperatures between 5° and 15°C, is called the "North Wall." The meandering of the Gulf Stream results in pinchoffs where warm Sargasso Sea water is trapped in warm eddies (WE) north of the North Wall, and cold shelf water is trapped in cold eddies (CE) south of the Gulf Stream. (Photo courtesy of NOAA.)

across the pycnocline in a way that is only now being recognized.

HORIZONTAL CIRCULATION

Wind-driven horizontal circulation in the surface waters develops from stress at the interface between the ocean and the wind. The trade winds blowing out of the southeast in the Southern Hemisphere and out of the northeast in the Northern Hemisphere provide the backbone of the system of ocean surface currents. Setting the water masses between the tropics into motion, the trade winds develop the **equatorial currents** that can be found in all of the world's oceans.

These currents move west parallel to the equator, and owing to the Coriolis effect and the deflection along continental margins, they move away from the equator as warm **western boundary currents.** At the same time, the westerly winds blowing out of the northwest in the Southern Hemisphere and the southwest in the Northern Hemisphere drive surface water between 30° and 60° latitude in an easterly direction. The Coriolis effect and continental barriers turn this water toward the equator as a cold **eastern boundary current.** This process produces the dominant feature of ocean basin surface circulation, the **subtropical gyre.** It rotates in a clockwise direction in the Northern Hemisphere and in a counterclockwise direction in the Southern Hemisphere (figure 7–1).

Ekman Spiral

To explain why water masses move as they do relative to wind direction, we will recall the observations made by Fridtjof Nansen during the voyage of the *Fram.* Nansen determined that the Arctic sea ice moved 20° to 40° to the right of the wind blowing across its surface. He passed this information on to V. Walfrid Ekman, a physicist who developed the mathematical relationships that explain the observations.

Ekman developed a circulation model that has been called the **Ekman spiral** (figure 7–2). The model assumes that a homogeneous water column is being set in motion by wind blowing across its surface. Owing to the Coriolis effect, the surface current moves in a direction 45° to the right of the wind in the Northern Hemisphere. This surface mass of water, moving as a thin lamina, or sheet, sets another layer beneath it in motion. The surface layer moves with a velocity no more than 3 percent of the wind speed. The energy of the wind is passed through the water column from the surface down, with each successive layer of water being set in motion with a lower velocity than and in a direction to the right of the one that set it in motion. At some depth the momentum imparted by the wind to the moving water laminae will

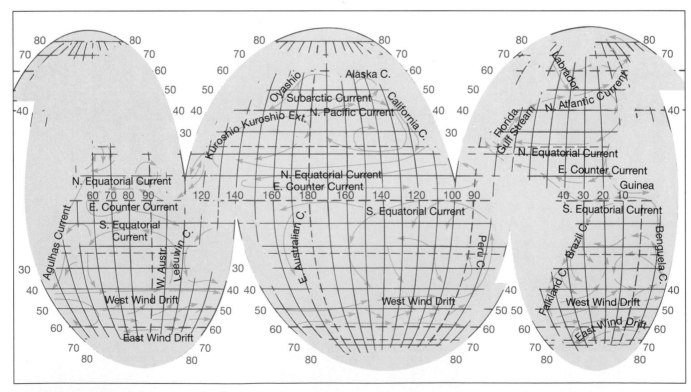

FIGURE 7–1
Wind-Driven Surface Currents in February and March. (After Sverdrup et al., 1942.)

FIGURE 7–2
Ekman Spiral. Wind drives surface water in a direction 45° to the right of the wind in the Northern Hemisphere. Deeper water continues to deflect to the right and move at a slower speed with increased depth. Ekman transport, the net water movement, is at right angles to the wind direction.

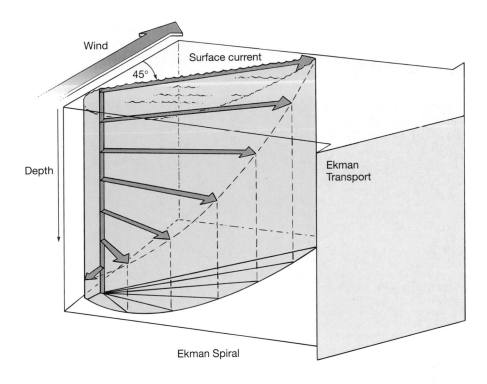

be lost, and there will be no motion as a result of wind stress at the surface. The depth at which motion ceases is called the depth of frictional influence. Although it depends on wind speed and latitude, this stillness normally occurs at a depth of about 100 m (328 ft). Figure 7–2 shows the spiral nature of movement with increasing depth from the ocean's surface. The length of each arrow in the figure is proportional to the velocity of the individual lamina, and the direction of each arrow indicates its direction of movement.

Under these theoretical conditions, the surface current should flow at an angle of 45° to the direction of the wind. From the surface to the depth of frictional influence, the net water movement, the **Ekman transport,** will be at right angles to the direction of the wind.

There are, however, no ideal conditions in existence in the ocean, and movements actually occurring as a result of wind stress on the ocean surface will deviate from this idealized picture. Generally, we can say that the surface current will move at an angle of less than 45° to the direction of the wind and that the Ekman transport will be at an angle less than 90° to the direction of the wind. This is particularly true in shallow coastal waters, where all of the movement may be in a direction very nearly that of the wind and the turning with increased depth is minor.

Geostrophic Currents

If we consider a gyre in the North Atlantic Ocean, for example, and remember the Ekman transport, it can be seen that a clockwise rotation in the Northern Hemisphere will tend to produce a **subtropical convergence** and a piling up of water in the center of the subtropical gyre. We find within all such ocean gyres hills of water that rise as much as 2 m (6.6 ft) above the water level at the margins of the gyres. As Ekman transport pushes water into the "hill" structure, the gravitational force acts to move particles down the surface slope. The Coriolis force deflects the water flowing down the slope to the right in a curved path (figure 7–3). The water piles up on these hills until the down-slope component of gravitational force acting on individual particles of water balances the Coriolis force. When these two forces finally balance, the net effect is a geostrophic current moving around the hill.

Again, this idealized flow would exist if there were no friction. Owing to the friction between water molecules, the water follows a path that moves it gradually down the slope of the hill.

Westward Intensification

The apex of the hills formed within the rotating gyres is not in the center of the gyre. The highest part of each hill is located closer to the western boundary of the gyre. The causes of **westward intensification** are complex, but a good part of the phenomenon can be explained in terms of the Coriolis effect. Remember that this effect increases in strength toward higher latitudes. Thus, the water that is driven eastward by the westerly winds tends more strongly to veer right toward the equator than the water that is driven westward near the equator tends to veer

FIGURE 7–3
Geostrophic Current. *A*, as Earth's wind systems set ocean water in motion, circular gyres are produced. Water piles up inside the subtropical gyre with the apex of the "hill" closer to its west margin because of Earth's rotation to the east. The theoretical geostrophic current flows parallel to the contour of the hill and represents an equilibrium between the Coriolis force pushing water toward the apex through Ekman transport and the downslope component of gravity acting to move the water down the slope. Because of friction between water molecules, the path of the current is gradually down the slope of the hill. *B*, because the Coriolis effect is stronger on water farther from the equator, the eastward-flowing high-latitude water tends to turn equatorward more strongly than the westward-flowing equatorial water tends to turn toward higher latitudes. This causes a broad, slow, equatorward flow of water across most of the subtropical gyre and forces the apex of the geostrophic hill to the west. This phenomenon is referred to as westward intensification, the main manifestation of which is a high-speed, warm, western boundary current flowing along a westward slope of the hill that is much steeper than that of the eastern slope associated with the slow drift of cold water toward the equator.

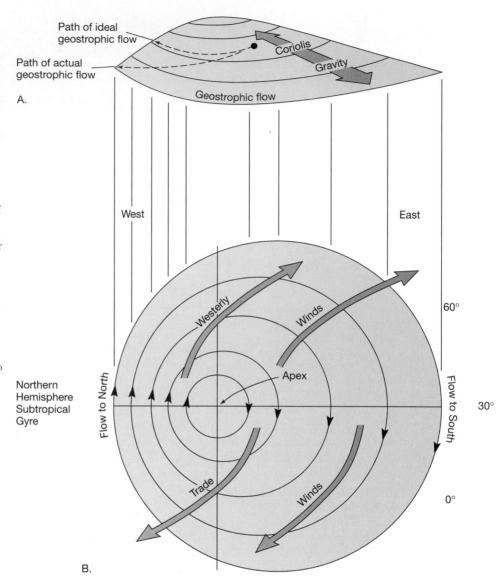

toward higher latitudes. This causes a broad equatorward flow of water across most of the subtropical gyres.

Assuming we have a steady state with a constant volume of water rotating around the apex of the hill, it can be seen in figure 7–3B that the velocity with which the water moves along the western margin will be much greater than that with which it will move around the eastern side of the hill. These relationships can be seen particularly well in the North Atlantic and North Pacific gyres, where the western boundary currents move with speeds in excess of 5 km/h (3.1 mi/h) in a northerly direction, while a flow along the eastern margin of each basin is better characterized as a drift moving at velocities well below 0.9 km/h (0.6 mi/h). Western boundary currents commonly flow 10 times faster and to greater depths than eastern boundary currents. The warm western boundary currents are usually less than one-twentieth the width of

the broad, cool drifts flowing toward the equator on the east side of the subtropical gyres.

Directly related to this difference in speed is the steepness of the hill's slope. The slope of the hill on the side of the slow-moving eastern boundary current is quite gentle; the western margin has a steep slope corresponding to the high velocity of the western boundary current.

An indirect method of determining current velocity in the upper-water mass is measuring the internal distribution of density and the pressure field it creates. This pattern of distribution is determined by collecting water samples along a number of vertical columns distributed throughout the area to be studied. At each station, temperature and salinity are measured. Average density of the water column is computed from the temperature-salinity characteristics measured above an arbitrary depth at which the surface current is assumed to have died out (possibly 100 m, or 328 ft). No current or horizontal pressure gradient is thought to exist at this depth. The pressures of the overlying water columns are assumed to be equal.

If our assumption of no horizontal pressure gradient existing at our reference level is correct, all measured water columns above this datum must contain equal mass. However, a water column with a low average density will, because of its greater volume, stand higher above the datum than a water column of high density. By computing the heights of columns necessary at all stations to produce the zero horizontal pressure gradient at the datum depth, a topography for the ocean surface can be determined (figure 7–4).

Maps showing this type of water motion can be prepared for any depth between the datum and the ocean surface; they represent the **dynamic topography** that is used to compute current velocity. The steeper the slope, the higher the velocity of the geostrophic current flowing in a direction generally parallel to the topography contours (lines of equal height above the equal pressure datum).

Rapid progress is being made in the use of radar altimeters mounted on satellites to map the dynamic topography of the oceans. Another technique that may replace the laborious procedures involved in the classic method of determining dynamic topography is ocean acoustic tomography, described later in this chapter.

Equatorial countercurrents are generated as a consequence of the large volume of water that is driven westward in the north and south equatorial currents. Since the Coriolis effect is minimal near the equator, much of this water is not turned toward higher latitudes but piles up at the western margin of an ocean basin. This is particularly true in the western Pacific Ocean (figure 7–1), where a dome of equatorial water is trapped in the island-filled embayment between Australia and Asia.

Continued influx of water carried by the equatorial currents forces an eastward counterflow of water that continues to South America.

ANTARCTIC CIRCULATION

Although the oceanic mass surrounding the continent of Antarctica is not officially recognized as an ocean, we will first consider the circulation that exists around Antarctica. We begin here because the mightiest of all ocean currents in terms of water volume transport, the Antarctic Circumpolar Current, dominates the movement of water masses in the southern Atlantic, Indian, and Pacific oceans south of 50°S latitude. This latitude may be considered the northern boundary of the Southern Ocean.

A surface flow called the **East Wind Drift** moves in a *westerly* direction around the margin of the Antarctic continent. This flow is driven by the Polar Easterlies moving out to sea from Antarctica. The East Wind Drift is most extensively developed to the east of the Antarctic Peninsula in the Weddell Sea region and in the area of the Ross Sea. If this terminology seems confusing, it is because this current is named after the winds that drive it, and winds are named on the basis of the direction they are coming from: East winds blow out of the east (figure 7–5).

The main circulation system in Antarctic waters is the **Antarctic Circumpolar Current,** extending northward to a position of approximately 40°S latitude. This mass is driven by the westerly winds that reach very great strength throughout much of the year. The surface portion of this flow is named the **West Wind Drift.** As a result of the Coriolis effect deflecting moving masses to the left in the Southern Hemisphere, there is a zone of divergence, the **Antarctic Divergence,** between the East Wind Drift and the West Wind Drift.

The Antarctic Circumpolar Current meets its greatest restriction as it passes through the 1000-km (620-mi)

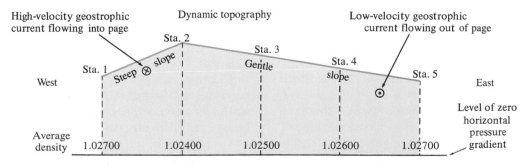

FIGURE 7–4
Dynamic Topography. Comparison of average densities for the water columns above the arbitrary no-current datum tells us the apex of the hill is near station 2 because the lower the density of the water, the greater the volume a given mass will occupy. Northern Hemisphere geostrophic flow is indicated by circles beneath slopes.

FIGURE 7–5
Antarctic Surface Circulation.
Near the Antarctic continent a westward-flowing current—the East Wind Drift—is driven by the polar easterly wind system. Ekman transport produces a zone of divergence—Antarctic Divergence—between the East Wind Drift and the West Wind Drift that circles Antarctica with an easterly flow. The cold West Wind Drift water converges with warmer water in each of the major ocean basins to the north at the Antarctic Convergence.

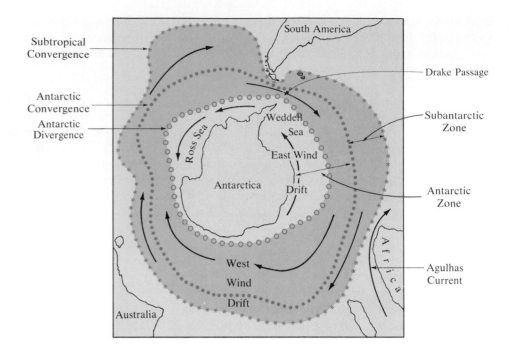

Drake Passage between the Antarctic Peninsula and the southern islands of South America. Although the current does not move at high velocity, reaching a maximum velocity at the surface of about 2.75 km/h (1.65 mi/h), it does transport more water than any other ocean current, an average of about 130 million m³/s.

As we continue to discuss ocean circulation, we will refer to the amount of water being transported by currents in the various oceans and will use the unit *sverdrup* (sv) in place of million m³/s (35 million ft³/s), as has been suggested by Dunbar in honor of Harald Sverdrup.

ATLANTIC OCEAN CIRCULATION

In the Atlantic Ocean, the basic surface circulation pattern is that of two large gyres. The North Atlantic gyre rotates in a clockwise direction, while the South Atlantic gyre rotates in a counterclockwise pattern. The driving force behind these rotations, which are partially separated by the **Atlantic Equatorial Countercurrent,** are the northeast and southeast trade winds.

The South Atlantic gyre is composed of the **South Equatorial Current,** which reaches its greatest strength just below the equator and is split in two by the topographic interference of the eastern extension of Brazil. Part of the South Equatorial Current moves off along the northeastern coast of South America toward the Caribbean Sea and the North Atlantic. The rest is turned southward as the **Brazil Current,** which ultimately merges with the West Wind Drift and moves in an easterly direction across the South Atlantic. The gyre is completed by a slow-drifting movement of cold water, the **Benguela**

Current, that flows up the western coast of Africa. There is also a significant flow of cold water moving in a northerly direction along the western margin of the South Atlantic. The **Falkland Current,** an important cold current, moves up the coast of Argentina as far north as 25°–30°S latitude, wedging its way between the continent and the Brazil Current. The Brazil Current has a much smaller volume than its Northern Hemisphere counterpart, the Gulf Stream. This results partly from the splitting of the South Equatorial Current by the configuration of South America (figure 7–6).

The **North Equatorial Current** moves parallel to the equator in the Northern Hemisphere, where it is joined by that portion of the South Equatorial Current that is shunted toward the north along the South American coast. This flow splits into two masses, the **Antilles Current** that flows along the Atlantic side of the West Indies and the **Caribbean Current** that passes through the Yucatan Channel into the Gulf of Mexico. These masses reconverge as the water that entered the Gulf of Mexico exits between Florida and Cuba as the **Florida Current.** The Florida Current flows close to shore over the continental shelf, carrying a volume that at times exceeds 35 sv. As it moves off Cape Hatteras and flows across the deep ocean in a northeasterly direction, it becomes the **Gulf Stream,** flowing at velocities up to 9 km/h (5.6 mi/h). The western margin of the Gulf Stream can frequently be defined as a rather abrupt boundary that moves periodically closer to and farther away from the shore. The eastern boundary becomes very difficult to identify, as it is usually masked by filamentous meandering masses that change their position continuously. Its character gradually merges with the water of the Sargasso

FIGURE 7–6
Atlantic Ocean Surface Currents.
(Base map courtesy of National
Ocean Survey.)

Cold →

Warm →

CONVERGENCES

ARC–Arctic
STC–Subtropical
ANC–Antarctic

CURRENTS

A–Antillean
Bg–Benguela
Br–Brazil
C–Canary
CC–Caribbean
EG–East Greenland
EW–East Wind Drift
EC–Equatorial Counter
Fa–Falkland
F–Florida

G–Guinea
GS–Gulf Stream
I–Irminger
L–Labrador
NE–North Equatorial
N–Norwegian
SE–South Equatorial
WW–West Wind Drift

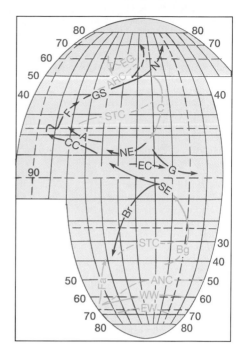

Sea. A volume transport of over 90 sv off Chesapeake Bay indicates a large volume of Sargasso Sea water has been added to the flow provided by the Florida Current. However, by the time the Gulf Stream reaches the Tail of the Banks, south of Newfoundland, the volume carried by the Gulf Stream has been reduced to 40 sv. This would indicate that most of the Sargasso Sea water that joined the Florida Current and made up the Gulf Stream has returned to the diffuse flow of the Sargasso Sea.

Just how this loss of water occurs is yet to be determined, but much of it may be achieved by meanders (sinuous curves in the course of a current) pinching off to form large eddies. As is shown in figure 7–7, meanders north of the Gulf Stream pinch off and trap warm Sargasso Sea water in eddies rotating in a clockwise direction. These eddies move southwest at speeds of 3–7 km/day (1.9–4.3 mi/day) toward Cape Hatteras, where they rejoin the Gulf Stream. South of the Gulf Stream and west of 50°W longitude, eddies with a core of cold slope water rotating counterclockwise move south and west. These large eddies, ranging in diameter from 100 to 500 km (62 to 310 mi) and extending to depths of 1 km (0.62 mi) in the warm rings and as deep as 3.5 km (2.2 mi) in the cold rings, could remove large volumes of water from the Gulf Stream.

At the Tail of the Banks, about 40°N latitude and 45°W longitude, the Gulf Stream continues in a more easterly direction across the North Atlantic. This current is composed of numerous branches into which the Gulf Stream flow is ultimately broken. Two branches that are composed of water produced through the mixing of the **Labrador Current** and the Gulf Stream are the **Irminger Current,** flowing up along the west coast of

Iceland, and the **Norwegian Current,** which moves north along the coast of Norway. The other major branch follows the 45°N latitude across the North Atlantic and turns south as the **Canary Current** passing between the Azores and Spain. This southward flow is very diffuse and spreads over a broad area as it moves southward and eventually joins the North Equatorial Current.

The effects of the westward intensification of current flow in the North Atlantic Ocean can be seen from the surface temperatures shown in figure 7–8. From 20° to 40°N latitude off the coast of North America, we can see a 20° temperature change from waters near the Dominican Republic to those near New York in February. By contrast, on the eastern side of the North Atlantic only a 5°–6° range in temperature can be observed between 20° and 40°N latitude from a point off the coast of Mauritania to waters off the coast of Spain.

PACIFIC OCEAN CIRCULATION

The surface circulation in the Pacific Ocean is generally similar to that which we described for the Atlantic Ocean, except that the *Equatorial Countercurrent* is much better developed in the Pacific Ocean than in the Atlantic (figure 7–9).

Embedded within the South Equatorial Current is a thin ribbonlike undercurrent, the **Cromwell Current.** This current flows east from the Samoa Islands in the western Pacific to the Galápagos. Only 0.2 km (0.12 mi) thick and approximately 300 km (186 mi) wide, the Cromwell Current at depths of 70–200 m (440–1256 ft) flows with velocities approaching 5 km/h (3 mi/h) in an

FIGURE 7–7

Sea-Surface Temperatures in Northwest Atlantic Ocean. *A*, sea-surface temperature data gathered by a NOAA satellite were processed to produce this false-color image of the northwest Atlantic Ocean. The warm Gulf Stream waters are shown as orange and red. The colder nearshore waters are blue and purple. Warm water from south of the Gulf Stream is transferred to the north as warm core rings (yellow) surrounded by cooler (blue and green) water. Cold nearshore water spins off to the south of the Gulf Stream as cold core rings (green) surrounded by warmer (yellow and red) water.

The rings form when meanders close to trap warm or cold water within them. The warm rings contain shallow, bowl-shaped masses of warm water about 1 km (0.62 mi) deep with diameters of about 100 km (62 mi). The cold rings have cones of cold water that extend to the ocean floor. They may be more than 500 km (310 mi) across at the surface. The diameter of the cone increases with depth. (Image courtesy of Dr. Charles McLain/Rosensteil School of Marine and Atmospheric Science, University of Miami.) *B*, this diagram is an overlay of the satellite image in *A*. It attempts to clarify the process of ring formation by showing the core of the Gulf Stream flow. It shows the warm ring and three cold rings as having formed by the pinching off of meanders. The cold rings have separated from the main Gulf Stream flow, but the warm ring has just formed as the meander that flowed around it pinches off from the main flow.

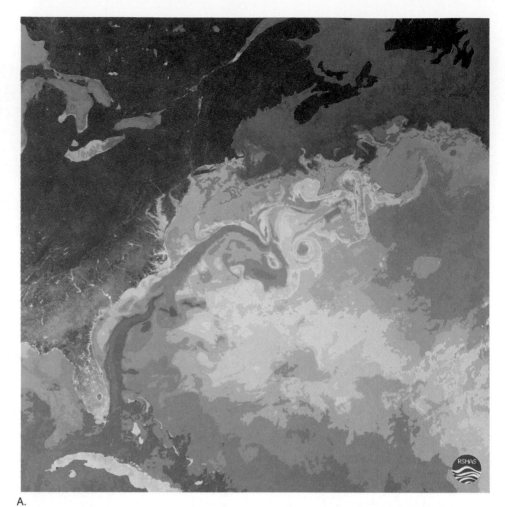

A.

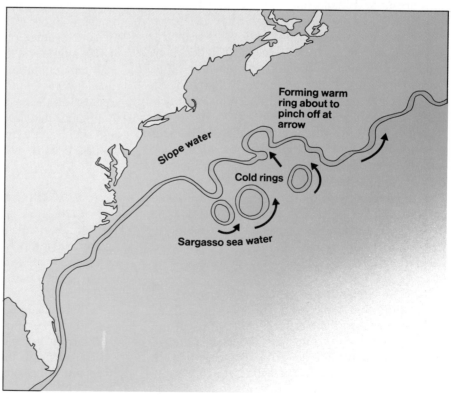

B.

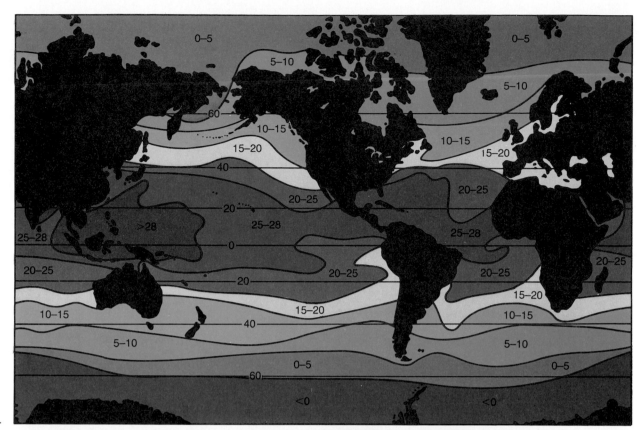

A.

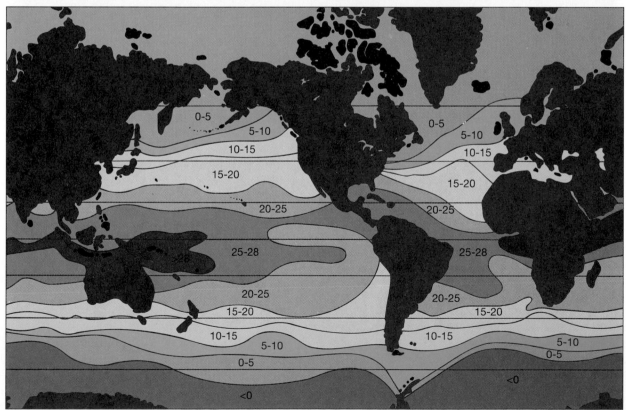

B.

FIGURE 7–8
Surface Temperature of the World Ocean (°C). Average surface temperature distribution for August *(A)* and February *(B)*. (After Sverdrup et al., 1942.)

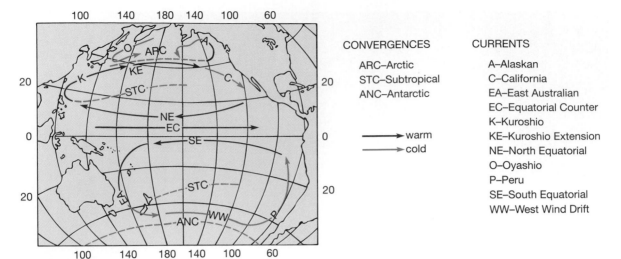

FIGURE 7–9
Pacific Ocean Surface Currents.

easterly direction. Its volume transport is approximately 40 sv.

El Niño–Southern Oscillation Events

A periodic phenomenon that is observed primarily in the Pacific Ocean is the El Niño–Southern Oscillation (ENSO). It is clearly related to other periodic climatic events observed throughout the world, and the relationships among these events are being studied intensely. Because knowledge of these phenomena is important to our understanding of the world's climate, we will present a brief overview of the present knowledge of ENSO and some of its associations with climatic events in other parts of the world.

During the 1920s, G. T. Walker identified what he called the Southern Oscillation (SO), associated with a condition in which the summer high-pressure system in the southeastern Pacific occurs in conjunction with a low-pressure system over the Indo-Australian region (*see* figures 7–10 and 6–12). This pressure difference lessens when trade winds diminish, the temperature of surface waters in the eastern Pacific increases, and the Equatorial Countercurrent flow increases. The average period of this oscillation is three years, but it ranges from two to ten years. When the oscillations are extreme—producing very widespread warm surface-water conditions and minimal pressure differences across the tropical Pacific—they are called El Niño–Southern Oscillations (ENSO). There were eight ENSO events between 1950 and 1990.

Normally the circulation cell called the **Walker Circulation** affects the southeast trade winds, which converge on the Indo-Australian low-pressure cell, rise, and produce high rates of precipitation in the low-pressure area. Dry air descends within the southeastern Pa-

cific high-pressure cell off the west coast of South America. This coast is characterized by high rates of evaporation.

One precursor of an ENSO event is the movement of the Indo-Australian low-pressure cell to the east beginning in October or November. In extreme cases such as the 1982–83 event, severe droughts can occur in Australia because the low-pressure cell moves so far east.

Concurrent with the eastward shift of the Indo-Australian low-pressure cell, the meteorological equator or **Intertropical Convergence Zone (ITCZ),** where the northeast trade winds and southeast trade winds meet and rise, moves south. Its normal seasonal migration is from 10°N latitude in August to 3°N in February, but during ENSO events it may move south of the equator in the eastern Pacific. Associated with this shift are weak trade winds, a decrease in coastal upwelling, and an unusually thick column of abnormally warm surface water in the eastern Pacific (figure 7–11A). These initial events are amplifications of normal seasonal fluctuations.

As the ENSO develops, the weakened trade winds and anomalous warmth of surface waters observed in the eastern Pacific spread toward the west. The coming of unusually warm surface waters to Kiritimati (Christmas Island; 2°N, 157°W) can be predicted by earlier observation of increased surface temperatures off the coast of Peru. The event is fully developed by January (figure 7–11B). Heavy rainfall from the southward shift of the ITCZ strikes the coast of Ecuador and Peru, which is usually arid, and spreads west across the tropical Pacific. The intense eastward flow of the Equatorial Countercurrent causes a rise in sea level along the western coast of the Americas that progresses poleward in both hemispheres. The event ends 12 to 18 months after it starts, with a gradual return to normal conditions that begins in

FIGURE 7–10
Walker Circulation. Normal oceanic and atmospheric conditions that are altered by El Niño–Southern Oscillation events.

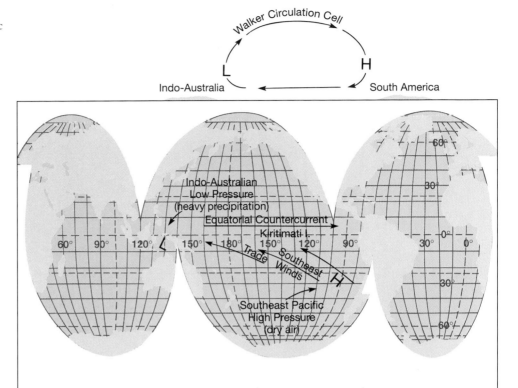

the southeastern tropical Pacific and spreads to the west (figure 7–11C).

Although the below-average temperatures in the eastern Pacific at the end of an El Niño, shown in figure 7–11C, are usual, such an event had not occurred between 1975 and 1988. This cooling has been given the name **La Niña** and seems to be associated with weather phenomena opposite to those of El Niño at other locations on Earth. For instance, Indian Ocean monsoons are drier than usual in El Niño years and wetter than usual in La Niña years.

A very intense 1982–83 ENSO event that caused a severe drought in Australia and Indonesia was anomalous in that it was initially confined to the central and western tropical Pacific and spread to the east late in its development.

Well before the overall pattern of development of the ENSO was recognized, a periodic movement of warm waters over the rich, cold fishing waters off the coast of Peru was recognized. This event, called El Niño, causes severe coastal rains and drives away the anchovy that are the basis of fishing there and serve as a food supply for a large bird population. El Niños have especially disastrous effects on the economy of Peru because of its heavy dependence on this fishery and the guano industry.

In November 1982, it was observed that virtually all of the 17 million adult birds that normally inhabited Kiritimati had abandoned their nestlings. This event indicated the severity of the ENSO in the central Pacific, as such an abandonment is not known to have occurred before. It is assumed that the spread of a thick layer of warm water over the surrounding ocean prevented the rise of nutrients into the surface waters, causing the fish to leave in search of better feeding grounds. The birds, in turn, were forced to leave in search of the fish. If the birds did not find their necessary supply of fish, a large percentage of the adults may have died in addition to the nestlings.

Figure 7–12 (top) shows the warming that occurred off southern California from January 1982 to January 1983 as a result of the 1982–83 ENSO. This warming was accompanied by a major reduction in biological productivity, as is shown by the Coastal Zone Color Scanner phytoplankton pigment images for April 1982 compared to those of March 1983 (figure 7–12, bottom).

A 1991–92 ENSO produced global weather modifications similar to those of the 1982–83 event. Although the causes of ENSO events are not fully understood, computer models were able to predict the 1991–92 ENSO.

INDIAN OCEAN CIRCULATION

Surface circulation in the Indian Ocean, which extends to only about 20°N latitude, varies considerably from that which we have discussed for the Atlantic and Pacific oceans. From November to March the equatorial circulation is similar to that which exists in the other oceans,

FIGURE 7–11
Sea Temperature Anomaly (°C) Resulting from the Averaging of ENSO Event Temperature Anomalies from 1950 through 1973. *A,* after onset: average of March, April, and May temperature anomalies. *B,* maximum development: average of December, January, and February anomalies. *C,* ending of event: average of May, June, and July anomalies. The typical progress of an ENSO event shows that the abnormally high surface-water temperatures first appear in the eastern Pacific off the coast of Ecuador and Peru *(A)*. This condition begins to be observable by December or January along the coast of South America and is well developed during the March–May period shown. In *B* it can be seen to develop in a westerly direction along the equator and reaches maximum development in the Central Pacific by the following February. By July *(C)*, conditions return to near normal, with the exception of a significant negative temperature anomaly in the eastern Pacific. (Data courtesy of NOAA.)

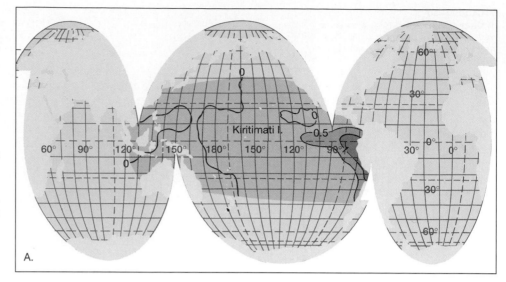

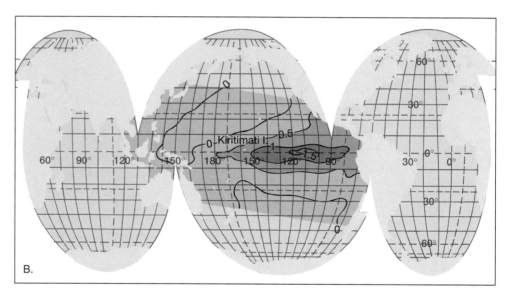

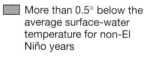 More than 0.5° below the average surface-water temperature for non-El Niño years

0 to 0.5°C below the average

0 to 0.5°C (0.9°F) above the average surface-water temperature for non-ENSO years

0.5° to 1.0°C (0.9° to 1.8°F) above average

1.0° to 1.5°C (1.8° to 2.7°F) above average

More than 1.5°C (2.7°F) above average

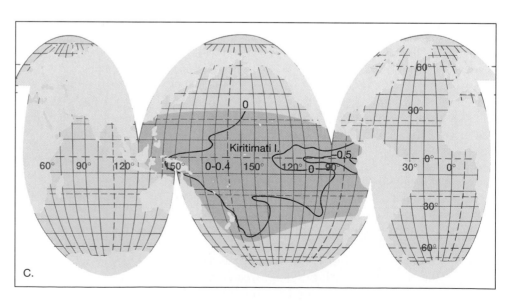

FIGURE 7–12
Water Temperatures off Southern California. The Advanced Very High Resolution Radiometer sea-surface temperature image (top) shows that waters off southern California in January 1983 are an average of 1.72°C warmer than a year before. The Coastal Zone Color Scanner phytoplankton pigment images (bottom) show that there was a reduction in plant productivity over the same time span. The strongest El Niño ever recorded occurred in 1982–1983 and pushed a layer of warm water north from the equator. Coinciding with this event was a reduction of upwelling of nutrient-rich, cold water along the coast. This reduction in the nutrient level of the surface water reduced the level of biological productivity. Spectacular physical phenomena such as this and their biological consequences help highlight the close relationship between the physical and biological phenomena of the oceans. (Courtesy Paul C. Fiedler/National Marine Fisheries Service.)

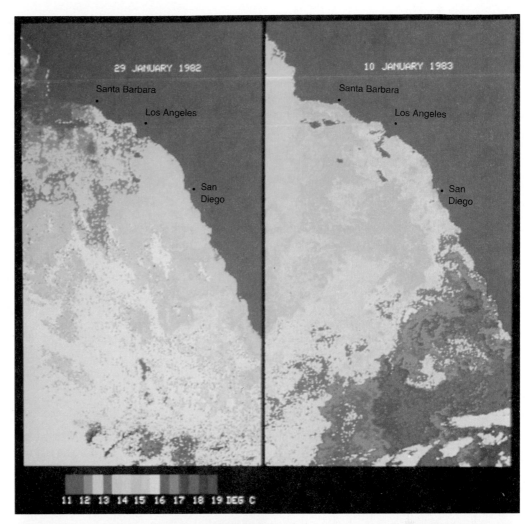

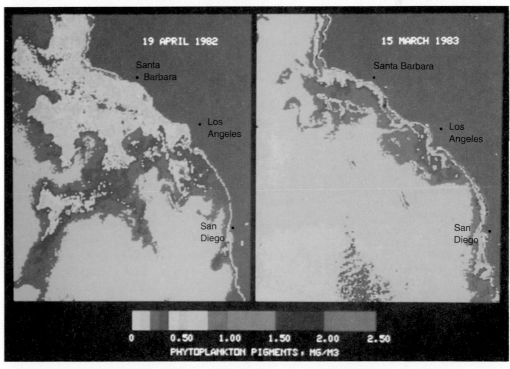

with two westward-flowing equatorial currents separated by an Equatorial Countercurrent. In contrast to the wind systems in the Atlantic and Pacific oceans, which are shifted to the north of the geographical equator, we find that in the Indian Ocean the meteorological equator is shifted to the south of the geographical equator. The Equatorial Countercurrent flows between 2° and 8°S latitude, bounded on the north by the North Equatorial Current, which extends as far as 10°N latitude, and on the south by the South Equatorial Current, which extends to 20°S latitude. During winter the typical northeast trade winds are developed and are referred to as the **northeast monsoon.** They are reinforced by the fact that rapid cooling of the air during the winter months over the Asian mainland creates a high-pressure cell that forces atmospheric masses off the continent out over the ocean, where the air pressure is lower (figure 7–13).

The Asian mainland warms up faster than the oceanic water due to the relatively lower heat capacity of continental crustal material compared to water. As a result, a low-pressure cell develops over the continent during summer, which "sucks" in the air masses overlying the ocean. This gives rise to the **southwest monsoon,** which may be thought of as a continuation of the south-

A.
WINTER: November–March, Northeast monsoon wind season

B.
SUMMER: May–September, Southwest monsoon wind season

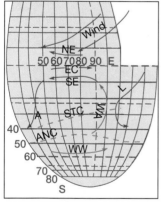
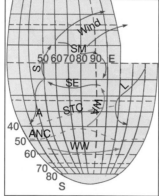

From November to March equatorial circulation in the Indian Ocean is similar to that of other oceans, except that the Equatorial Countercurrent is shifted south.

Low pressure over the mainland during summer draws the Southwest Monsoon winds over the North Indian Ocean. This wind produces the Southwest Monsoon Current which flows in an easterly direction and replaces the North Equatorial Current.

Warm → Cold →

CURRENTS A—Agulhas EC—Equatorial Counter L—Leeuwin
NE—North Equatorial S—Somali SE—South Equatorial
SM—Southwest Monsoon WA—West Australian
WW—West Wind Drift

CONVERGENCES STC—Subtropical ANC—Antarctic

FIGURE 7–13
Indian Ocean Surface Currents.

east trade winds across the equator. During this season the North Equatorial Current disappears and is replaced by the **Southwest Monsoon Current,** which flows from west to east across the North Indian Ocean. In September or October, the northeast trade winds are reestablished, and the North Equatorial Current reappears (figure 7–13).

The surface circulation in the southern Indian Ocean is similar to the counterclockwise circulation observed in other southern oceans. During the time the northeast trade winds blow, the South Equatorial Current provides water for the Equatorial Countercurrent and the **Agulhas Current,** which flows south along the eastern coast of Africa and joins the West Wind Drift. Turning north out of the West Wind Drift is the **West Australian Current,** which completes the gyre by merging with the South Equatorial Current.

During the southwest monsoon, a northward flow from the equator along the coast of Africa, the **Somali Current,** develops with velocities approaching 4 km/h (2.5 mi/h).

The eastern boundary current in the southern Indian Ocean is unique. Other eastern boundary currents of subtropical gyres are cold drifts toward the equator that produce arid coastal climates receiving less than 25 cm (10 in.) of rain per year. In the southern Indian Ocean, the West Australian Current is displaced offshore by a southward-flowing current called the **Leeuwin Current.** This current is driven south along the coast of Australia from the dome of warm water piled up in the East Indies by the equatorial currents in the Pacific Ocean. The Leeuwin Current produces a mild climate in southwestern Australia, which receives about 125 cm (50 in.) of rain per year. This current is weakened during ENSO events and contributes to the drought conditions that develop in Australia in association with these events.

VERTICAL CIRCULATION

Lateral movements of water masses may bring about vertical circulation within the upper-water mass. We refer to this shallow vertical circulation system as wind-induced circulation. Of greater oceanwide importance in producing the thorough mixing within the ocean is the vertical circulation resulting from density changes at the ocean surface, which cause the sinking of water masses. Since the changes that increase the density are normally changes in temperature and salinity, this circulation is referred to as thermohaline circulation.

Wind-Induced Circulation

There are regions throughout the world ocean where water moves vertically to the surface or down away from the surface as a result of wind-driven surface currents carrying water away from or toward these regions. **Up-welling** occurs in areas where the surface flow of water is away from the area. If volume is to be conserved and horizontal surface flows bring insufficient water into the area, water must come from beneath the surface to replace that which has been displaced. This condition may occur in the open ocean or along the margins of continents.

The North Equatorial Current of the Indian Ocean and South Equatorial currents in the Atlantic and Pacific oceans are driven by easterly winds in a westerly direction on either side of the equator. The Ekman transport of water on the north side of the equator will move it to the right, a higher latitude, while in the Southern Hemisphere it will move to the left to a higher latitude (figure 7–1). The net effect is a water deficiency at the surface between the two currents. Water from deeper within the upper-water mass comes to the surface to fill the void resulting from the deflection of equatorial water to the north and south (figure 7–14). This phenomenon is called **equatorial upwelling.**

Coastal upwelling is common along the margins of continents where the wind conditions are such that the surface waters adjacent to the continents are carried out to the open ocean via Ekman transport (figure 7–15). The replacement of this water comes from the lower portions of the upper water, as is the case with the Peru Current. Such areas are characterized by low surface temperatures and high concentrations of nutrients that make them

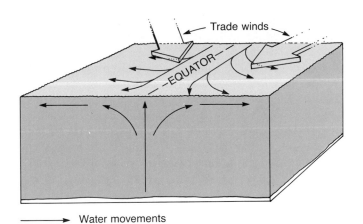

→ Water movements

FIGURE 7–14
Equatorial Upwelling. The Coriolis effect acting on trade-wind-driven westward-flowing equatorial currents pulls surface water away from the equatorial region. This water is replaced by subsurface water.

areas of high biological productivity. The reversal of the direction of coastal winds that are causing upwelling tends to push water toward shore, where it piles up and causes **coastal downwelling,** a sinking of surface water.

Near the center of the rotating gyres of the major ocean current systems the winds are relatively weak, and the water rotates at a very low rate. The winds that do blow in these regions may be steady in direction and are known to set up convection cells in the upper-water

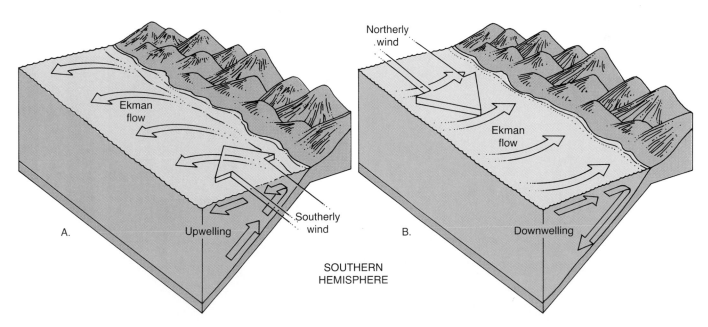

FIGURE 7–15
Coastal Upwelling and Downwelling. *A,* in areas where wind-driven coastal currents flow along the western margins of continents and toward the equator, the Ekman transport carries surface water away from the continent. An upwelling of deeper water replaces the surface water that has moved away from the coast. *B,* a reversal of the direction of the winds that cause upwelling will cause water to pile up against the shore and force downwelling.

mass. The phenomenon was first recognized by Irving Langmuir while crossing the Sargasso Sea in 1938. He observed streaks, or straight rows, of seaweed parallel to the direction of the wind and concluded that the plants were trapped in zones of convergence between cells. Not only is macroscopic plant material concentrated in these regions, but also microscopic plants and dissolved organic material. By contrast, in the regions of divergence, where water is surfacing as a result of convection, the concentrations of organic material are relatively low (figure 7–16). This phenomenon is called **Langmuir circulation.**

Thermohaline Circulation

The large-scale vertical circulation in the ocean results primarily from surface-density changes in oceanic water. Vertical mixing of ocean water is driven primarily through the sinking of water masses in high latitudes. It is only in the high-latitude areas that the water column has a gravitational stability sufficiently neutral to allow vertical movements of large masses of water. Surface masses may sink to the ocean bottom and deep-water masses may rise to the surface. The relatively strong density stratification, or pycnocline, that separates the upper- and deep-water masses throughout the lower-latitude regions does not exist in the high-latitude oceans.

Because the intensity of solar radiation is greater in the equatorial region, we might expect that heating of surface water there may cause the water to expand and move away from the equator toward the poles, spreading over the colder, more dense high-latitude waters (figure 7–8). We may further expect that at some depth a return flow from the high latitudes toward the equator would be set in motion to replace the surface water mov-

ing away from the equatorial region. This exchange does not seem to occur in the oceans, although it is theoretically proper to assume that such a pattern might develop. The energy imparted to the surface waters by winds greatly exceeds the energy that develops as a result of density changes due to temperature differences. Thus the effect of the wind overcomes any tendency that may exist for such a pattern of circulation to develop.

Another variable that affects the density of surface water is salinity (figure 6–19). Salinity appears to have a very minimal effect on the movement of water masses in the lower latitudes, and density changes resulting from salinity changes are of importance only in the very high latitudes, where low water temperature remains relatively constant. For example, the highest-salinity water in the open ocean is found in the subtropical regions, but there is no sinking water in these areas because water temperatures are high enough to maintain a low density for the surface-water mass and prevent it from sinking. In such areas, a strong halocline, or salinity gradient, may develop with a relatively thin surface layer of water having salinities in excess of 37‰. Salinity decreases rapidly with increasing depth to typical ocean-water salinities below 35‰.

An important sinking of cold surface waters that become deep-water masses occurs in the subpolar regions of the Atlantic Ocean. In the North Atlantic, the major sinking of surface water is thought to occur in the Norwegian Sea. From there it flows as a subsurface current into the North Atlantic. This flow becomes part of the **North Atlantic Deep Water.** Additional surface water may sink at the margins of the Irminger Sea off southeastern Greenland and the Labrador Sea. In the southern subpolar latitudes, the most significant area of deep-water-mass formation is the Weddell Sea, where rapid

FIGURE 7–16
Langmuir Circulation. Steady winds blowing across the ocean surface create convection cells with alternate right- and left- hand circulation. The axes of these cells run parallel to wind direction. Organic debris accumulates in downwelling zones of convergence and produces windrows that run parallel to the wind direction for great distances.

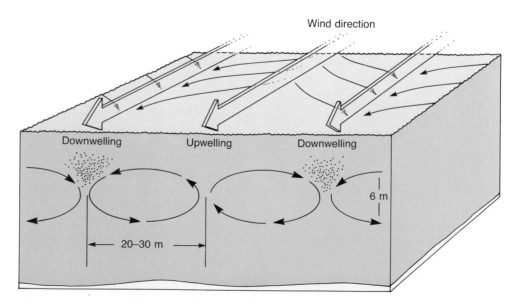

winter freezing produces high-density water that sinks down the continental slope of Antarctica and becomes **Antarctic Bottom Water,** the densest water in the open ocean (figure 7–17).

On a broad scale, there are latitudinal regions where surface-water masses converge and may cause sinking (figure 7–18). These regions of convergence are found within the subtropical gyres and in the Arctic and Antarctic. The subtropical convergences do not produce sinking because surface waters have relatively low densities as a result of high temperature. Major sinking does occur, however, along the Arctic and **Antarctic Convergence.** The major water mass formed from sinking at the Antarctic convergence is the **Antarctic Intermediate Water** mass (figure 7–18).

Figure 7–19 shows what is thought to be the general pattern of deep-water circulation in the world ocean. The most intense flow occurs as a western boundary current because of the rotation of Earth. For every unit of water that sinks from the surface into the deep ocean, a unit of deep water must return to the surface. It is much more difficult to identify specific areas where this vertical flow to the surface is occurring, and it probably occurs more as a gradual rather uniform return throughout the ocean basins. This return to the surface may be somewhat greater in the low-latitude regions, where surface temperatures are higher.

Figure 7–20 shows a probable pattern of integration between the deep thermohaline circulation and surface currents in which the deep ocean is provided

FIGURE 7–17
Atlantic Ocean Subsurface Circulation, as Evidenced by Physical Properties of Water. (Adapted from Sverdrup, Johnson & Fleming, *The oceans: their physics, chemistry, and general biology,* © 1942.)

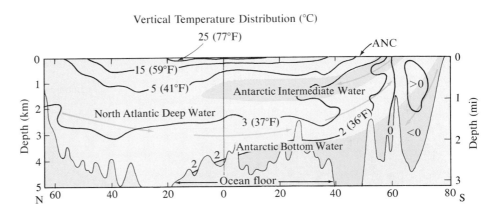

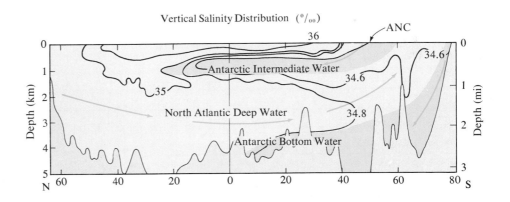

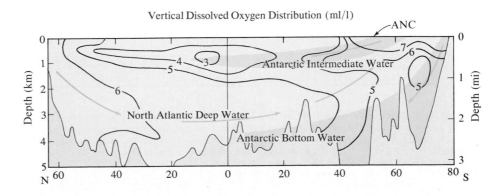

FIGURE 7–18
Regions Where Intermediate and Deep-Water Masses of the World Ocean Sink.

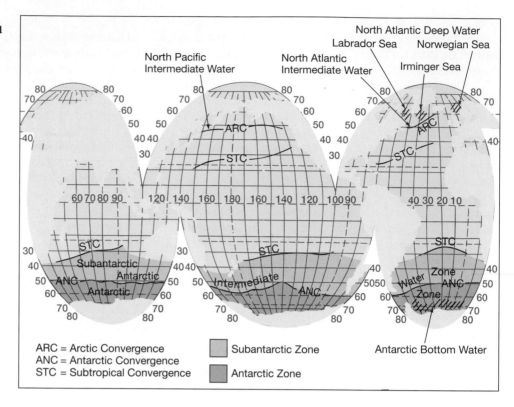

ARC = Arctic Convergence
ANC = Antarctic Convergence
STC = Subtropical Convergence

Subantarctic Zone
Antarctic Zone

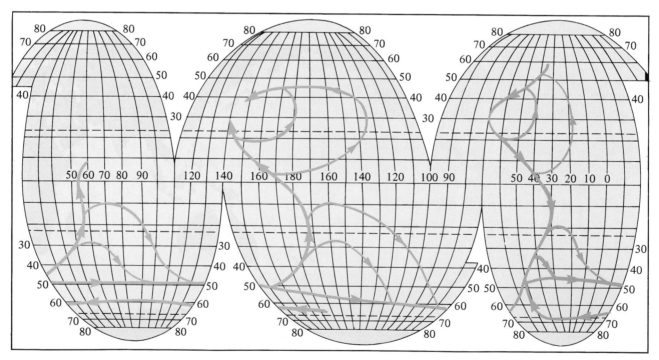

FIGURE 7–19
Stommel's Model of the Deep-Water Circulation of the World Ocean. Based on information obtained to date, this highly schematic model of deep-water circulation developed by Henry Stommel in 1958 appears to be reasonably correct. Heavy lines mark the major western boundary currents. They result from the same forces that produce the more intense western boundary currents in the surface circulation. (After H. Stommel.)

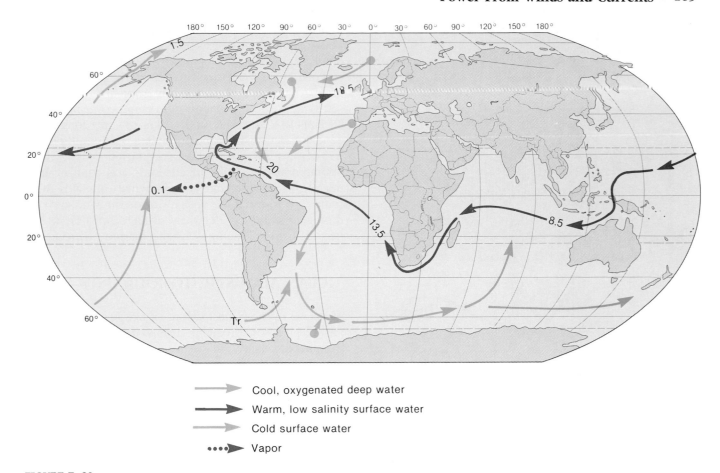

→ Cool, oxygenated deep water

→ Warm, low salinity surface water

→ Cold surface water

•••• Vapor

FIGURE 7–20
Global Cycle of Thermohaline Circulation. The deep ocean is replenished with oxygen by the sinking of the North Atlantic Deep Water (NADW). NADW is composed of water that sinks in the Norwegian Sea and is joined by water from Baffin Bay and the Mediterranean Sea. The total volume of this flow may be about 20 sv as it enters the South Atlantic. This deep flow is recooled in the Antarctic and flows on into the deep Indian Ocean and Pacific Ocean. A broad pattern of upwelling returns the bottom water to the surface in all oceans. A surface flow of warmer, low-salinity water of the North Pacific starts the return journey to the North Atlantic. A flow of 8.5 sv from the North Pacific increases to 13.5 sv after crossing the Indian Ocean and reaches a total of 18.5 sv by the time it is carried into the North Atlantic. A small amount of cold surface water enters the Atlantic from the Pacific through the Drake Passage and 1.5 sv flows into the Atlantic through the Bering Strait. At least 0.1 sv of water is transferred from the Atlantic Ocean to the Pacific Ocean as water vapor across the Isthmus of Panama. (After Gordon, 1986)

oxygen-rich water by the sinking of dense surface water primarily in the subpolar Atlantic Ocean. The return limb of this system begins with surface water flowing out of the northeastern Pacific Ocean. Water is added as the surface flow crosses the Indian Ocean and Atlantic Ocean on its return to the subpolar North Atlantic Ocean.

This pattern of return within the surface water may help to explain why the surface salinity of the North Pacific Ocean is the lowest and that of the North Atlantic is the highest in the world ocean. As the return flow begins in the north Pacific as low salinity water, its salinity increases with time by evaporation. Cooling of this high salinity water produces a density high enough to initiate sinking and formation of the North Atlantic Deep Water.

POWER FROM WINDS AND CURRENTS

The winds that transform solar energy into ocean currents have for centuries been used by society to drive its machines. Westerly winds off New England represent a near-shore wind supply that is reasonably sustained and contains a large amount of energy. There is some optimism that Offshore Windpower Systems (OWPS) such as that shown in figure 7–21 could provide electricity or hydrogen generated by electricity to meet the needs of large sections of the North Atlantic coast of the United States.

Many have considered the great amount of energy in the Florida–Gulf Stream Current System and dreamed

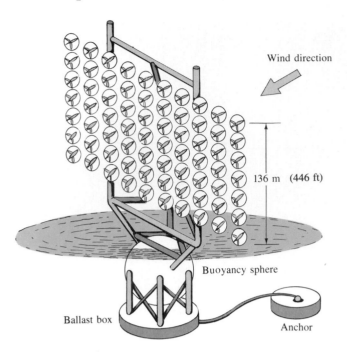

FIGURE 7–21
Offshore Windpower System. (Courtesy of Woods Hole Oceanographic Institution.)

of harnessing it to do the work of society. A group of scientists and engineers met in 1974 to consider this possibility and concluded that some 2000 megawatts (MW) of electricity could be recovered along the east coast of southern Florida. Devices proposed for extraction range from underwater "windmills" to a Water Low Velocity Energy Converter (WLVEC), which is operated by parachutes attached to a continuous belt. Again, calculations

indicate such systems can be economically competitive. The Coriolis Program was proposed in 1973 by William J. Moulton of Tulane University. Hydroturbines with diameters of 170 m (558 ft) are envisioned that could generate 43 MW of electricity from the movement of ocean currents (figure 7–22). An array of 242 units covering an area 30 km (18.6 mi) wide and 60 km (37 mi) long could generate 10,000 MW—the equivalent of 130 million barrels of oil. Tests are now being conducted to establish an optimum size, identify suitable sites, and confirm engineering, economic, and environmental estimates. What remains is to see if the designs for ocean wind and current generation of electricity actually function as expected in the sometimes hostile marine environment.

OCEAN ACOUSTIC TOMOGRAPHY (OAT)

When we observe the motions of ocean water, we find they range across broad spectrums of size and time. The identifiable patterns of ocean circulation range in size from eddies a few tens of kilometers across, created by water flowing over and around seamounts, to the slowly rotating subtropical gyres the size of ocean basins. Between these extremes are features such as the Gulf Stream rings, which are a few hundred kilometers in diameter, and coastal upwelling such as that responsible for the highly productive waters along the coast of Peru.

Fully understanding these motions and their effects on climate and biological productivity has long been the goal of oceanographers. However, it has not been possible to continuously gather data over sufficient area for the necessary length of time to achieve this goal because

FIGURE 7–22
Coriolis Program. Buoyant Coriolis hydroturbines anchored in ocean current. It is estimated that an array of 242 such units placed in the Gulf Stream could provide 10 percent of the present electricity needs of Florida. (Courtesy of U.S. Department of Energy.)

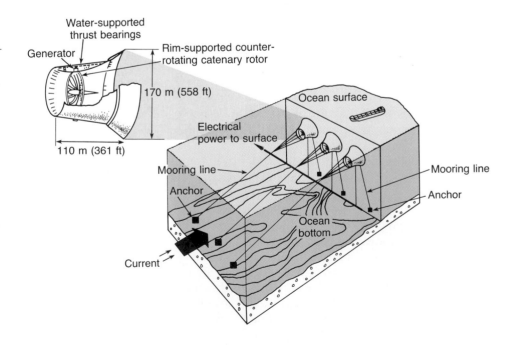

FIGURE 7–23
Munk and Wunsch. Walter H. Munk of Scripps Institution of Oceanography *(A)* and Carl Wunsch of Massachusetts Institute of Technology *(B)* are leaders in the development of ocean acoustic tomography techniques for the syntopic study of mid- to large-scale oceanic circulation patterns. (Walter Munk's photo courtesy of University of California at San Diego, Office of Learning Resources, Scripps Institution of Oceanography, Photo Lab.)

A.

B.

of the high cost of sending the required number of ships on long sea voyages.

Recently, Walter Munk of Scripps Institution of Oceanography (figure 7–23A) and Carl Wunsch (figure 7–23B) of Massachusetts Institute of Technology have led the way in developing a system that may make such data gathering possible—**ocean acoustic tomography** (OAT). "Tomo" is derived from the Greek word *tomos,* meaning a section or slice. A *tomogram* is a picture of a slice. Tomography has already found use in medicine through the use of x-rays. The CAT (computerized axial tomography) scan uses the x-rays to provide tomographic images of parts of the body, usually the brain.

The use of tomography to map the features of ocean circulation involves the transmission of low-frequency sound. Increases in water temperature, salinity, and pressure increase the speed of sound travel in ocean water. As sound travels through features such as cold or warm rings, the sound speed will be decreased or increased. Theoretically, if sound is transmitted across a slice of ocean from a series of transmitters on one side to a series of receivers on the other, features that cause a change in the speed of sound can be mapped.

A 1981 demonstration covering a 300-km (186-mi) square of ocean southwest of Bermuda involved the use of four transmitters and five receivers over a period of six months (figure 7–24). Two sensors of temperature and current flow were included and survey ships periodically entered the area to record data against which the tomography could be checked.

During the demonstration, a large cold eddy moved across the area and a frontal system separating cold and warm water moved into the area. The data gathered conventionally confirmed the existence of the features and proved the feasibility of such a system to provide instantaneous pictures of large slices of the world's oceans. Subsequent tests (1983) proved the technique could be used to measure current flow and deep-sea circulation at a depth of 760 m (2500 ft). The latter experiment, conducted in the North Pacific Ocean, involved three transceivers arranged in a triangle with sides 1000 km (620 mi) in length. A seven-transceiver system was set up off New England in 1987 to study the Gulf Stream, and a large, five-year study is under way to investigate the contribution of the Greenland Sea to the formation of North Atlantic Deep Water. In the near future it may well be possible to map such features within large ocean basins—something that was only a dream a few years ago.

In 1991, Walter Munk designed a test of OAT to see if it might aid in detecting global warming of the oceans that may be anticipated from the increase in greenhouse gas concentrations as a result of human activities. His group set off acoustical signals for 6 days at a depth of 150 m (490 ft) near Heard Island in the southern Indian Ocean. The signals were received after traveling as far as 18,000 km (11,000 mi). Theoretically, if these signals are sent out periodically over a period of years and if the ocean warms up, these signals will be observed to travel the same distances in less time.

FIGURE 7–24
Ocean Acoustic Tomography. *A*, during the 1981 demonstration of ocean acoustic tomography, transmitters (T1 to T4) emitted low-frequency sound signals that were received by the receivers (R1 to R5). Temperature and currents were measured by conventional means at moorings E1 and E2. *B*, results of the 1981 ocean acoustic tomography demonstration. Source of data: 1 and 3, ship surveys; 2, tomography. During the demonstration, data from a depth of 700 m showed a large cold eddy in the center of the survey area at the start of the survey. It was separated from warm water to the northeast by a well-defined temperature front. The cold water eddies observed moved off to the northwest, while the front progressed to the southwest during the demonstration. Cold eddies were indicated by the slowing of sound signals that passed through them. (Data courtesy of Woods Hole Oceanographic Institution.)

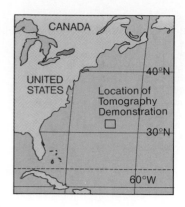

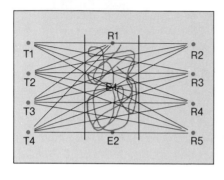

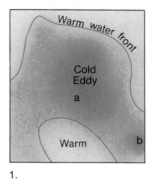

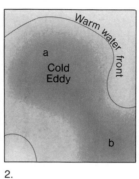

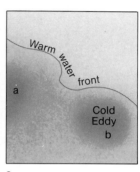

A.　　　　　　　　　B.　　　1.　　　　　　2.　　　　　　3.

SUMMARY

Horizontal currents set in motion by wind systems are characteristic of the surface waters of the world ocean. According to the Ekman spiral concept, winds will set the surface waters in motion in a direction 45° to the right of the wind in the Northern Hemisphere, and the net water movement will be at right angles to the wind direction. As a result of the Ekman spiral phenomenon, water is pushed in toward the center of clockwise gyres in the Northern Hemisphere and counterclockwise gyres in the Southern Hemisphere, forming "hills." As water in the Northern Hemisphere runs downslope on the hills of water that result, the Coriolis effect causes it to turn right and into the clockwise flow pattern. As a result, a geostrophic current flowing parallel to the contours of the hill is maintained. The apex of the hill is always located to the west of the geographical center of the gyre. Using dynamic topography, oceanographers can map the surface currents in ocean basins by determining the average density of water columns at locations throughout the basin and mapping this distribution. The speed of the current will increase with increased slope of the hills, which will normally be steepest on the west side.

The Circumpolar Current flows in a clockwise direction around the continent of Antarctica. It is the largest current in the world ocean, transporting over 130 sverdrups of water. The surface portion of this flow is often referred to as the West Wind Drift because it is driven by strong westerly winds. Two major deep-water masses form in Antarctic waters. The highest-velocity current in the oceans is the Gulf Stream as it flows along the southeast Atlantic coast. The best development of an equatorial countercurrent system is found in the Pacific, where large eastern movement of water is achieved by the Pacific Countercurrent and the Equatorial Undercurrent, a ribbonlike subsurface flow. The formation of deep-water masses is not pronounced in the Pacific Ocean, as there appears to be no deep sinking of surface water.

A periodic phenomenon observed in the Pacific Ocean that is also associated with climatic changes throughout the world is the El Niño–Southern Oscillation.

The Indian Ocean circulation is dominated by the monsoon wind systems that blow out of the northeast in the winter and the southwest in the summer.

The Antarctic Bottom Water, the densest water mass in the oceans, forms primarily in the Weddell Sea and sinks along the continental shelf into the Atlantic Ocean. Farther north, at the Antarctic Convergence, the low-salinity Antarctic Intermediate Water sinks to a depth of about 900 m (2952 ft). Sandwiched between these two masses is the North Atlantic Deep Water, rich in plant nutrients after hundreds of years in the deep ocean.

Continued investigation of the energy potential of ocean winds and currents may lead to development of a significant addition of renewable energy that can be exploited by society.

Ocean acoustic tomography is a complex acoustical system designed to image a slice of the ocean and provide a picture of temperature distribution over large areas. Some time in the future, it may provide the first subsurface images of in situ temperature conditions within whole ocean basins.

KEY TERMS

Agulhas Current (p. 160)
Antarctic Bottom Water (p. 163)
Antarctic Circumpolar Current (p. 151)
Antarctic Convergence (p. 163)
Antarctic Divergence (p. 151)
Antarctic Intermediate Water (p. 163)
Antilles Current (p. 152)
Benguela Current (p. 152)
Brazil Current (p. 152)
Canary Current (p. 153)
Caribbean Current (p. 152)
Coastal downwelling (p. 161)
Coastal upwelling (p. 161)
Cromwell Current (p. 153)
Downwelling (p. 161)
Dynamic topography (p. 151)

Eastern boundary current (p. 148)
East Wind Drift (p. 151)
Ekman spiral (p. 148)
Ekman transport (p. 149)
Equatorial Countercurrent (p. 151)
Equatorial Current (p. 148)
Falkland Current (p. 152)
Florida Current (p. 152)
Gulf Stream (p. 152)
Intertropical Convergence Zone (p. 156)
Irminger Current (p. 153)
Labrador Current (p. 153)
Langmuir Circulation (p. 162)
La Niña (p. 157)
Leeuwin Current (p. 160)

North Atlantic Deep Water (p. 162)
Northeast Monsoon (p. 160)
Norwegian Current (p. 153)
Somali Current (p. 160)
Southwest Monsoon (p. 160)
Southwest Monsoon Current (p. 160)
Subtropical convergence (p. 149)
Subtropical gyre (p. 148)
Thermohaline circulation (p. 147)
Upwelling (p. 161)
Walker Circulation (p. 156)
West Australian Current (p. 160)
Western boundary current (p. 148)
Westward intensification (p. 149)
West Wind Drift (p. 151)
Wind-driven circulation (p. 147)

QUESTIONS AND EXERCISES

1. Compare the forces directly responsible for creating horizontal and deep vertical circulation in the oceans. What is the ultimate source of energy for driving both circulation systems?
2. On a base map of the world, plot the major surface circulation gyres of the oceans, the meteorological equators of each ocean, and the Subtropical, Arctic, and Antarctic Convergences. Superimpose the major wind belts of the world on the gyres. Label the currents using the symbols used in figures 7–6, 7–9, and 7–13.
3. Diagram and discuss how the Ekman transport produces the "hill" of water within major ocean gyres that causes geostrophic current flow. As a starting place on the diagram, use the prevailing wind belts, the trade winds, and the westerlies.
4. What causes the apex of these geostrophic "hills" to be offset to the west of the center of the ocean gyre systems?
5. Explain how oceanographers determine the dynamic topography of the ocean waters.
6. The largest current in the world ocean in terms of volume transport is the Antarctic Circumpolar Current. Explain why its surface portion is referred to as the West Wind Drift. What is its maximum volume transport as compared to the maximum volume transport measure for the Gulf Stream?
7. Observing the flow of Atlantic Ocean currents in figure 7–6, compare the Brazil Current to the Gulf Stream and offer some possible explanations for the fact that the Brazil Current has a much lower velocity and volume transport than the Gulf Stream.

8. Explain why the Gulf Stream eddies that develop northeast of the Gulf Stream rotate clockwise and have warm-water cores while those that develop to the southwest rotate counterclockwise and have a core of cold water.
9. Describe the relationships between El Niño-Southern Oscillation events, Walker circulation, trade winds, equatorial countercurrent flow, and the Intertropical Convergence Zone.
10. Describe the relationships among the wind, the surface current it creates, and the development of equatorial and coastal upwellings.
11. Discuss why thermohaline vertical circulation is driven by sinking of surface water that occurs only in high latitudes.
12. Name the two major deep-water masses and give the locations of their formation at the ocean's surface.
13. The Antarctic Intermediate Water is identifiable throughout much of the South Atlantic on the basis of a temperature minimum, salinity minimum, and dissolved oxygen maximum. Why should it be colder, less salty, and contain more oxygen than the surface-water mass above it and the North Atlantic Deep Water below it?
14. The fact that surface salinities are high in the North Atlantic Ocean and low in the North Pacific Ocean may be explained by the pattern of return of deep-water masses to the surface where they will again sink. How?
15. Where is the potential for using ocean windpower systems and ocean current power systems greatest, and why?
16. On what basis may ocean acoustic tomography be used to check on whether or not global warming is occurring?

REFERENCES

Behringer, D.; Birdsall, T.; Brown, M.; Cornuelle, B.; Heinmiller, R.; Knox, R.; Metzger, K.; Munk, W.; Spiesberger, J.; Spindel, R.; Webb, D.; Worcester, P.; and Wunsch, C. 1982. A demonstration of ocean acoustic tomography. *Nature* 299, 121–25.

Gill, A. 1982. *Atmosphere-ocean dynamics.* Orlando, Florida: Academic Press.

Goldstein, R.; Barnett, T.; and Zebker, H. 1989. Remote sensing of ocean currents. *Science* 246:4935, 1282–86.

Gordon, A. L. 1986. Interocean exchange of thermocine water. *Journal of Geophysical Research* 91:C4, 5037–46.

Montgomery, R. 1940. The present evidence of the importance of lateral mixing processes in the ocean. *American Meteorological Society Bulletin* 21:87–94.

Pedlosky, J. 1990. The dynamics of the oceanic subtropical gyres. *Science* 248:4935, 316–22.

Philander, S. G. H. 1983. El Niño Southern Oscillation phenomena. *Nature* 302:5906, 295–301.

Pickard, G. L. 1975. *Descriptive physical oceanography.* 2nd ed. New York: Pergamon Press.

Stommel, H. 1987. *A view of the sea.* Princeton, N.J.: Princeton University Press.

———.1958. The abyssal circulation. Letters to the editor. *Deep Sea Research.* 5. New York: Pergamon Press.

Street-Perrott, A. F., and Perrott, R. 1990. Abrupt climate fluctuations in the tropics: The influence of Atlantic Ocean circulation. *Nature* 343:6259, 607–12.

Stuiver, M.; Quay, P. D.; and Ostlund, H. G. 1983. Abyssal water carbon-14 distribution and the age of the world oceans. *Science* 219:4586, 849–51.

Sverdrup, H. U.; Johnson, M. W.; and Fleming, R. H. 1942. Renewal 1970. *The oceans.* Englewood Cliffs, N.J.: Prentice Hall.

Wood, J. D. 1985. The world ocean circulation experiment. *Nature* 314, 501–11.

SUGGESTED READING

Sea Frontiers

Alper, J. 1991. Munk's hypothesis. 37:3, 38–43. Walter Munk's proposal to monitor global warming by transmitting sound signals across vast ocean reaches in the Sofar channel is discussed.

Frye, J. 1982. The ring story. 28:5, 258–67. A discussion of the Gulf Stream, its rings, and how the rings are studied.

Miller, J. 1975. Barbados and the island-mass effect. 21:5, 268–72. A discussion of the phenomenon by which waters around tropical islands are much more productive than the surface waters of the open ocean.

Smith, F. G. W. 1972. Measuring ocean movements. 18:3, 166–74. Discussed are some practical problems related to current flow and some methods used to determine current direction, speed, and volume.

Smith, F. G. W., and Charlier, R. 1981. Turbines in the ocean. 27:5, 300–305. A discussion of the potential of extracting energy from ocean currents is presented.

Sobey, E. 1982. What is sea level? 28:3, 136–42. The factors that cause sea level to change are discussed.

Scientific American

Baker, D. J., Jr. 1970. Models of ocean circulation. 222:1, 114–21. This article discusses observations of a model depicting a segment of the surface of Earth over which fluids move and helps explain geostrophic flow within ocean gyres. Some knowledge of basic physics is necessary for full comprehension of material presented.

Hollister, C. D., and Nowell, A. 1984. The dynamic abyss. 250:3, 42–53. Submarine "storms" are associated with deep, cold currents flowing away from polar regions toward the equator.

Munk, W. 1955. The circulation of the oceans. 191:96–108. A young physical oceanographer who has gone on to be highly honored by the scientific community presents his views of ocean circulation.

Spindel, R. C., and Worcester, P. F. 1990. Ocean acoustic tomography. 263:4, 94–99. The theoretical basis and practical application of OAT is clearly presented.

Stewart, R. W. 1969. The atmosphere and the ocean. 221:3, 76–105. The exchange of energy between the atmosphere and ocean and the resulting phenomena, currents, and waves are covered in this readable, comprehensive article.

Stommel, H. 1955. The anatomy of the Atlantic. 190:30–35. Another vintage article by a young oceanographer who went on to great achievements.

Webster, P. J. 1981. Monsoons. 245:5, 108–19. The mechanism of the monsoons and their role in bringing water to half Earth's population is discussed.

Weibe, P. H. 1982. Rings of the Gulf Stream. 246:3, 60–79. The biological implications of the large cold-water rings are considered.

Chapter 8

Waves

Waves are one of the most obvious phenomena of the ocean. Yet it was not until well into the nineteenth century that some understanding of what caused waves and how they behaved was developed. We will first look at the character of waves in general before proceeding to discuss oceanic waves. Wave phenomena, with the exception of electromagnetic waves, involve the transmission of energy and momentum by means of vibratory impulses through the various states of matter. Theoretically, the medium itself does not move in the direction the energy passes through. The particles that make up the medium simply oscillate in a back-and-forth or orbital pattern, transmitting energy from one particle to another.

Simple **progressive waves** (in which the wave form can be observed to move), shown in figure 8–1A, may be described as **longitudinal** or **transverse.** In longitudinal waves, such as sound waves, the particles that are in vibratory motion move back and forth in a direction parallel to the propagation of energy. Energy may be transmitted through all states of matter—gaseous, liquid, or solid—by longitudinal movement of particles.

A wave breaks spectacularly against a stack off Cape Kiwanda, Oregon. (Photo © Craig Tuttle/The Stock Market.)

LONGITUDINAL WAVE
Particles (color) move back and forth in direction of energy transmission. These waves transit energy through all states of matter.
A.

TRANSVERSE WAVE
Particles (color) move back and forth at right angles to direction of energy transmission. These waves transmit energy only through solids.

ORBITAL WAVE
Particles (color) move in orbital path. These waves transmit energy along interface between two fluids of different density (Liquids and/or gases).

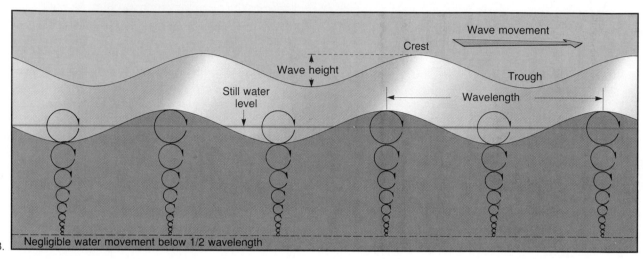

B.

FIGURE 8–1
Types of Progressive Waves. *A,* types of progressive waves. *B,* simple sea wave and its parts (*B* from the Tasa Collection: *Shorelines.* Published by Macmillan Publishing Co., New York. Copyright © 1986 by Tasa Graphic Arts, Inc. All rights reserved.)

Transverse wave phenomena involve the propagation of energy at right angles to the direction of particle vibration. An example of such a wave would be that created when one end of a rope is tied to a doorknob while the other end is moved up and down with the hand. A waveform is set up and progresses along the rope, and energy is transmitted from the motion of the hand to the doorknob. If one were to pick out a particular segment of the rope and watch it, the segment would be seen to move up and down at right angles to a line drawn from the doorknob to the hand that is putting energy into the rope. We generally consider that this type of wave can transmit energy only through solids, since it is only in a solid that particles are strongly attached to one another. Longitudinal and transverse waves are called **body waves,** as they transfer energy through a body of matter.

Waves on the ocean surface, as do all waves that transmit energy along an interface between any two fluids of different density, have particle movements that are neither longitudinal nor transverse. We may say more correctly that the movement of particles along such an interface involves components of both, since the particles

move in circular orbits at the interface between the atmosphere and ocean. They are **orbital waves,** or **interface waves.**

WAVE CHARACTERISTICS

When we observe the ocean surface, we see that there are waves of various sizes moving in various directions, resulting in a complex wave pattern that is constantly changing. We will, for the sake of introducing some characteristics that must be used to discuss waves, look at a much simpler form of an idealized wave representing the transmission of energy created by a single source and traveling along the ocean-atmosphere interface. Such a series of waves will have very uniform characteristics. We use such an idealized form in figure 8–1B.

As an idealized progressive wave passes a permanent marker, such as a pier piling, we will notice a succession of high parts of the waves, **crests,** separated by low parts, **troughs.** If we were to mark the water level on the piling when the troughs pass and then do the same for the crests, the vertical distance between the marks would be the **wave height, H.** The horizontal distance between corresponding points on successive waveforms, such as from crest to crest, is the **wavelength, L.** The ratio of **H/L** is **wave steepness.** The time that elapses during the passing of one wavelength is the **period, T.** Because the period is the time required for the passing of one wavelength, if either the wavelength or period of a wave is known, the other can be calculated, since $L(m) = 1.56 \ (m/s)T^2$ (figure 8–2):

$$\text{Speed } (S) = \frac{L}{T}$$

For example:

$$\text{Speed } (S) = \frac{L}{T} = \frac{156 \text{ m}}{10 \text{ s}} = 15.6 \text{ m/s}$$

Another characteristic related to wavelength and speed is **frequency, f.** Frequency is the number of wavelengths that pass a fixed point per unit of time and is equal to $1/T$. If 6 wavelengths pass a point in 1 minute, and we have the same wave system as in our previous example, then:

$$\text{Speed } (S) = Lf = 156 \text{ m} \times \frac{6}{\text{min}} = 936 \text{ m/min}$$

$$\frac{936 \text{ m}}{1 \text{ min}} \times \frac{1 \text{ min}}{60 \text{ s}} = 15.6 \text{ m/s}$$

Since the speed and wavelength of ocean waves are such that less than one wavelength passes a point per second, the preferred unit of time for scientific measurements, **period** (rather than frequency), is the more practical measurement to use when calculating speed.

Returning to the particle motion of ocean waves, the circular orbits followed by the water particles at the surface have a diameter equal to the wave height. While a particle is in the crest of a passing wave, it is moving in the direction of energy propagation. While it is in the

trough, it is moving in the opposite direction. The half of the orbit that is accomplished in the trough is at a lower speed than the crest half of the orbit. Therefore, there is a small net transport of water in the direction the waveform is moving. This condition results from the fact that particle speed decreases with increasing depth below the still-water line. Also, the diameters of particle orbits decrease with increased depth until particle motion associated with our idealized wave ceases at a depth of one-half wavelength, $L/2$.

Deep-Water Waves

The ocean waves with the characteristics just discussed belong to a category called **deep-water waves.** Such waves travel across the ocean where the water depth (d) is greater than one-half the wavelength (figure 8–3A). Included in deep-water waves are all wind-generated waves as they move across the open ocean. As can be seen by the speed calculations previously discussed, speed of deep-water waves is related to wavelength (L) [S (m/s) $= 1.25 \ \sqrt{L(m)}$] and period (T). The easier of these characteristics to measure is period. Thus the following equation is most commonly used for computing wave speed (g is acceleration due to gravity).

$$\begin{aligned}
\text{Speed (m/s)} &= \frac{gT}{2\pi} \\
&= \frac{9.8 \text{ m/s}^2 \times T \text{ (s)}}{2 \times 3.14} \\
&= 1.56T
\end{aligned}$$

Shallow-Water Waves

Waves in which $d < L/20$ are classified as **shallow-water waves** (long waves). Included in this category are wind-generated waves that have moved into shallow nearshore areas; **tsunami** (seismic sea waves), generated by disturbances in the ocean floor; and **tide waves,** generated by gravitational attraction of the sun and moon (figure

FIGURE 8–2
Speed—Deep-Water Waves. The theoretical relationships between wave speed, wavelength, and period for deep-water waves. Speed is equal to wavelength divided by period.
$L(m) = 1.56T^2$

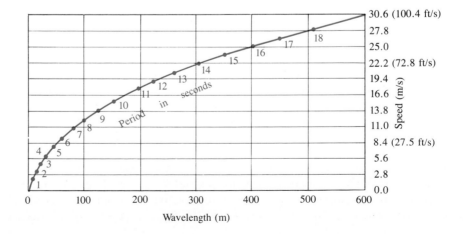

A. DEEP-WATER WAVE

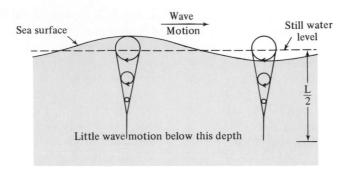

B. SHALLOW-WATER WAVE

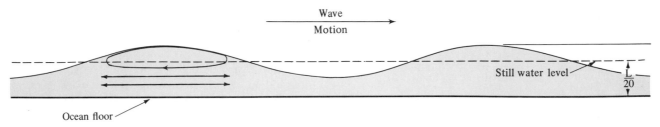

C. RELATIONSHIP OF WAVELENGTH TO WATER DEPTH

Shallow-water wave (d < L/20)	Transitional wave (d < L/2 but > L/20)	Deep-water wave (d > L/2)	> = greater than < = less than

FIGURE 8–3
Deep-Water and Shallow-Water Waves. *A,* wave profile and water-particle motions of a deep-water wave. Note the diminishing size of the orbits with increasing depth below the surface. Wave motion dies out at a depth of *L/*2. *B,* motions of water particles in shallow-water waves. Water motion extends to ocean floor because depth is less than *L/*20.

8–3B). In these waves, the wavelength is very great relative to water depth, and speed is determined by water depth, *d*:

$$\text{Speed (m/s)} = \sqrt{gd}$$
$$= 3.1 \sqrt{d(\text{m})}$$

Particle motion in shallow-water waves is in a very flat elliptical orbit approaching horizontal oscillation, and the vertical component of particle motion decreases with increasing depth. The presence of shallow-water waves can be detected to the ocean bottom and therefore can affect the bottom.

Transitional Waves

Transitional waves have wavelengths greater than twice, but less than 20 times the water depth. The speed of transitional waves is determined partially by wavelength and partially by water depth (figure 8–4). Deep-water waves generated by winds at the ocean surface usually have periods of 10–12 s. They maintain their periods after encountering shallow coastal water.

Since it is easier to measure the period of a wave than any of its other characteristics, Kinsman has used the system presented in figure 8–5 as a means of showing the concentration of energy in ocean waves with periods

ranging from less than 0.1 s to more than 1 day. The illustration also shows the principal causes of waves of different periods.

WIND-GENERATED WAVES

Sea

As the wind blows over the ocean surface, the pressure and stress deform the ocean surface into small rounded waves with wavelengths less than 1.74 cm (0.7 in.). These waves are called **capillary waves** (ripples) because the dominant **restoring force** that works to destroy them and smooth the ocean surface is surface tension (capillarity). Capillary waves characteristically have rounded crests and V-shaped troughs (figure 8–6).

As capillary wave development increases, the sea surface takes on a rougher character, which allows the wind and ocean surface to interact more efficiently. As more energy is transferred to the ocean, **gravity waves** with lengths over 1.74 cm (0.7 in.) develop with a shape more like that of a sine curve. Because they reach sufficient height at this stage, the force of gravity becomes the dominant restoring force. The length of "young" waves in a sea is generally 15–35 times their height. As additional

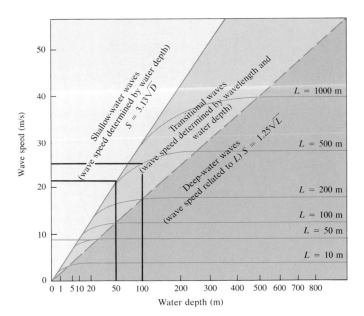

FIGURE 8-4
Wave Speed. To determine the wave speed of shallow-water waves, move up from the water depth to the line marked shallow-water waves, then over to the wave-speed scale [*see* red example for 50-m (164-ft) water depth]. To determine the wave speed of deep-water waves, continue the wavelength (*L*) line to the left until it intersects the wave-speed scale (*see* green example for 50-m wavelength). Combine these procedures for transitional waves [*see* purple example for 500-m (1640-ft) wavelength and 100-m (328-ft) water depth]. Note that the transitional waves have a lower wave speed than they would have had as deep-water waves.

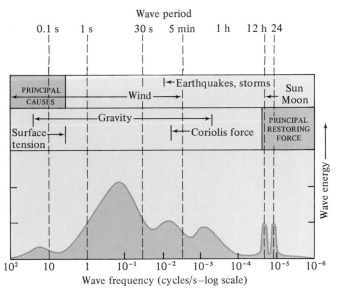

FIGURE 8-5
Distribution of Energy in Ocean Waves. This figure shows that most of the energy possessed by ocean waves is in wind-generated waves with a period of about 10 seconds. The long period peak above the 10^{-3} represents tsunami, while the two sharp peaks to the right represent the tides with semidaily and daily periods. (After Blair Kinsman, *Wind waves: Their generation and propagation on the ocean surface,* © 1965.)

energy is gained, the wave height increases more rapidly than wavelength, and the crests become pointed, while the troughs are rounded. When the steepness reaches 1/7, the speed with which the waves travel is 1.2 times that of typical deep-water waves, or $S = 1.56T \times 1.2 = 1.87T$.

Energy imparted by the wind increases the magnitude of height, length, and speed. When the wave speed reaches that of the wind, neither of these characteristics can change because there is no net energy exchange, and the wave is at its maximum size.

The area in which wind-driven waves are generated is called **sea** and is characterized by a choppy, short wave structure with waves moving in many directions and having many different periods and lengths. The variety of wave periods and wavelengths is caused by frequently changing wind speed and direction. The factors that are important in increasing the amount of energy that waves obtain are (1) wind speed, (2) the duration—the time during which the wind blows in one direction, and (3) the fetch—the distance over which the wind blows in one direction. Figure 8–7 shows the relationship among wind direction, fetch, and the sea.

Wave height is a characteristic that is directly related to the amount of energy possessed by the wave. Wave heights in a sea area are usually less than 2 m

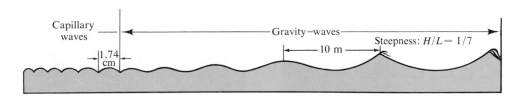

FIGURE 8-6
Capillary and Gravity Waves. As energy is put into the ocean surface by wind, small rounded waves with V-shaped troughs develop (capillary waves). As the water gains energy, the waves increase in height and length. When they exceed 1.74 cm (.07 in.) in length, they take on the shape of the sine curve and become gravity waves. Increased energy increases the steepness of the waves. The crests become pointed and the troughs rounded. As the steepness reaches 1/7, the waves become unstable, and whitecaps form as they "break." Not drawn to scale.

FIGURE 8–7
The Sea and Swell. As wind blows
across the "sea," wave size increases
with increased wind speed, fetch, and
duration. As waves advance beyond
the "sea," they continue to advance
across the ocean surface as "swell,"
free waves that are not driven by the
wind but sustained by the energy
they obtained in the "sea."

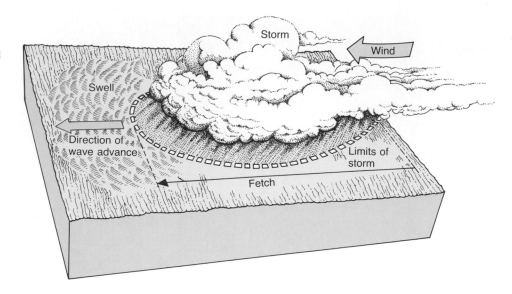

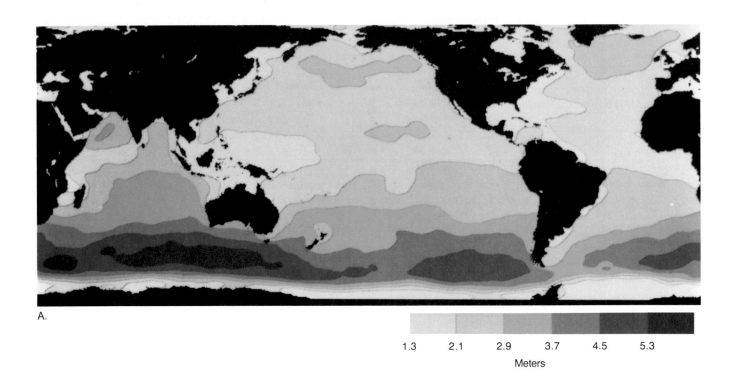

A.

1.3	2.1	2.9	3.7	4.5	5.3	

Meters

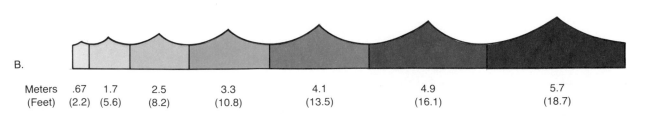

B.

Meters	.67	1.7	2.5	3.3	4.1	4.9	5.7
(Feet)	(2.2)	(5.6)	(8.2)	(10.8)	(13.5)	(16.1)	(18.7)

FIGURE 8–8
***SEASAT* Altimeter Wave Height, July 7–October 10, 1978.** Because the westerly wind belt in the
Southern Hemisphere reaches the highest average wind speed, this wind system produces the largest
ocean waves. The largest waves occur in the southern Indian Ocean. The wave heights represented
here are the averages for the largest one-third of the waves occurring at any location.

(6.6 ft), although it is not uncommon to observe waves with heights of 10 m (32.8 ft) and periods of 12 s. The largest wind-generated wave authentically measured was a 34-m (112-ft) wave with a period of 14.8 s seen in the North Pacific in February 1935 by the crew of the U.S. Navy tanker U.S.S. *Ramapo.* As sea waves gain energy, their steepness increases. When it reaches a critical value of 1/7, open ocean breakers called whitecaps form. Figure 8–8 is a satellite image of average wave height conditions throughout the world as they existed during the life of *Seasat* in 1978.

For a given wind speed there is a maximum fetch and duration of wind beyond which the waves will not grow. When both maximum fetch and duration are reached for a given wind velocity, the sea is said to be fully developed. The reason it can grow no further is the waves are losing as much energy through the breaking of whitecaps as they are receiving from the wind.

Table 8–1 shows the fetch and duration of wind required to produce a **fully developed sea** for given wind speeds. Table 8–2 describes the fully developed sea at given wind speeds in terms of average height, length, and period of waves. Also, the average height of the highest 10 percent of the waves is given.

TABLE 8–1
Fetch and Duration Required to Produce Fully Developed Sea for a Given Wind Speed.

Wind Speed km/h (mi/h)	Fetch km (mi)	Duration h
20 (12)	24 (15)	2.75
30 (19)	77 (48)	7
40 (25)	176 (109)	11.5
50 (31)	380 (236)	18.5
60 (37)	660 (409)	27.5
70 (43)	1093 (678)	37.5
80 (50)	1682 (1043)	50
90 (56)	2446 (1517)	65.25

Swell

As waves generated in a sea area move toward the margins of the generating area, where wind speeds are lower, they eventually are moving with a speed greater than that of the wind. When this occurs, the steepness of the waves decreases, and they become long-crested waves called **swell.** The swell moves with little loss of energy over large stretches of the ocean's surface, transmitting energy away from the input area of the sea to where most of it will eventually be released along the margins of continents. As a part of this transfer of energy away from the sea area, the waves with the longer length, traveling at higher speed, leave the sea area first. They are followed by the shorter-length, lower-speed **wave trains** (groups of waves). This progression illustrates the principle of **wave dispersion** (the sorting of waves by wavelength). In the generating area, waves of many wavelengths are present. In deep water, wave speed is a function of wavelength, so the longer waves "outrun" the shorter ones. To illustrate the distance that can be traveled by swell without having its energy depleted, swell that originated from storms in the Antarctic has been recorded breaking along the Alaskan coast after traveling a distance of more than 10,000 km (6210 mi).

As a set of waves leaves a sea and becomes a swell wave train, the group moves across the ocean surface at only half the velocity of an individual wave in the group. Progressively, the leading wave disappears. There will, however, always be the same number of waves in the group. As the front wave disappears, a new wave replaces it at the back of the group.

Interference Patterns. Since swells may be moving away from a number of storm areas in a given ocean, it is inevitable that the swell forms will run together and interfere with one another. This gives rise to one of the special features of wave motion—interference patterns. An interference pattern produced by the superposition of two or more wave systems will be the algebraic sum of the disturbance each wave would have produced individ-

TABLE 8–2
Description of Fully Developed Sea for a Given Wind Speed.

Wind Speed km/h (mi/h)	Average Height m (ft)	Average Length m (ft)	Average Period s	Highest 10% Waves m (ft)
20 (12)	.33 (1)	10.6 (34.8)	3.2	.75 (2.5)
30 (19)	.88 (2.9)	22.2 (72.8)	4.6	2.1 (6.9)
40 (25)	1.8 (5.9)	39.7 (130.2)	6.2	3.9 (12.8)
50 (31)	3.2 (10.5)	61.8 (202.7)	7.7	6.8 (22.3)
60 (37)	5.1 (16.7)	89.2 (292.6)	9.1	10.5 (34.4)
70 (43)	7.4 (24.3)	121.4 (398.2)	10.8	15.3 (50.2)
80 (50)	10.3 (33.8)	158.6 (520.2)	12.4	21.4 (70.2)
90 (56)	13.9 (45.6)	201.6 (661.2)	13.9	28.4 (93.2)

ually. The result may be a larger or smaller trough or crest, depending on whether the individual disturbances are of the same or opposite sign, in or out of phase (figure 8–9).

For instance, if swells are moving away from two storm areas and come together, the interference pattern may be constructive, destructive, or more likely mixed.

Constructive interference results if, in an ideal case, wave trains with the same wavelength come together in phase, crest to crest and trough to trough. Adding the displacements that would result from the waves individually, we will see that the interference pattern produced will be a wave with the same length as the two wave systems that are converging but with a wave height that is

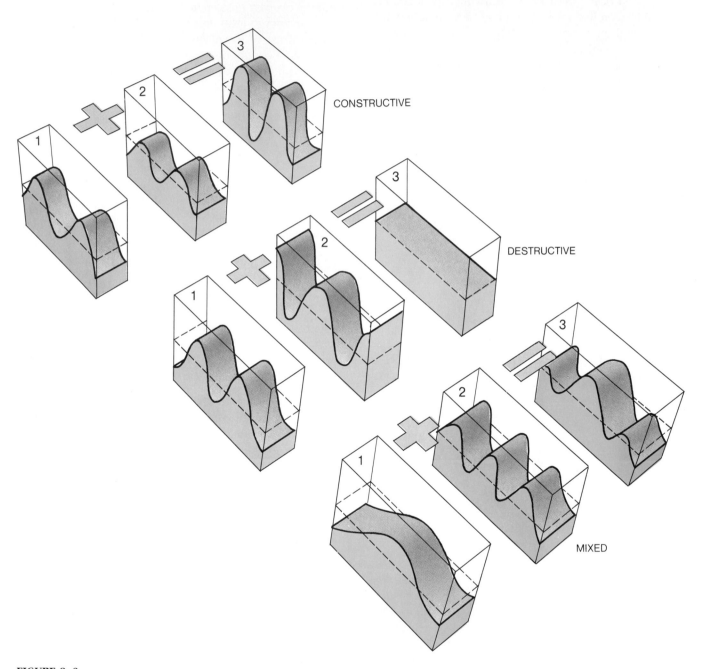

FIGURE 8–9
Wave Interference. As wave trains come together from different sea areas (1 and 2), three possible interference patterns may result (3). Very rarely, if the waves have the same length and come together in phase (crest to crest and trough to trough), totally constructive interference may occur. The algebraic sum of amplitudes produces a wave of the same length but of greater height. If two sets of waves have identical characteristics and come together 180° out of phase, the result will be no wave at all (destructive). More commonly, waves of different lengths and heights encounter one another and produce a mixed interference pattern.

equal to the sum of the wave heights of the individual wave systems.

Destructive interference results if the crests produced by the waves from one generating area coincide with the troughs produced by the waves from the second generating area. If the waves have identical characteristics, the algebraic sum of the crest plus the trough would be zero, and the waves would cancel each other out.

It is more likely that the two systems possess waves of different heights and lengths and would come together with both destructive and constructive interference. Thus, a more complex **mixed interference** pattern would develop. It is such interference that explains the occurrence of a sequence of high waves followed by a sequence of lower waves and other irregular wave distribution patterns observed as the swell approaches the margins of the continents.

Free and Forced Waves. In the swell we see what may be referred to as a **free wave,** which is moving with the momentum and energy imparted to it in the sea area but is not experiencing a maintaining force that keeps it in motion. In the wave-generating area, free and forced waves are present. A **forced wave** is one that is maintained by force that has a periodicity coinciding with the period of the wave. This force would be the wind; owing to the variability of the wind, many wave systems in the sea area alternate between being forced waves and free waves.

Surf

Most waves that are generated in the sea area by the force of storm-velocity winds move across the ocean as swell. They release their energy at the margins of the continents in the **surf zone,** where the swell forms breakers. As the deep-water waves making up the swell move toward the margin of the continent over gradually shoaling water, they eventually encounter water depths which are less than one-half wavelength.

These shoaling depths interfere with the particle movement at the base of the wave, and the wave slows down. As one wave is slowed down, the following waveform, which is still moving at an unaffected speed, tends to "catch up" with the wave that is being slowed, thus reducing the wavelength. Wave height increases, and the crests become narrow and pointed while the troughs become wide curves, a form that was previously described for high-energy waves in the sea. The increase in wave height accompanied by a decrease in wavelength increases the steepness (H/L) of the waves. As the wave steepness reaches 1/7, the waves break as surf.

If the surf is composed of swell that has traveled from distant storms, breakers will develop relatively near shore in shallow water, the shoaling of which is primarily responsible for their breaking. By the time waves break,

they have become shallow-water waves. The horizontal water motion associated with such waves translates water toward and away from the shore. The surf will be characterized by parallel lines of relatively uniform breakers. However, if the surf is composed of waves that have been generated by local wind, the waves may not have been sorted out into swell. The surf may be more nearly characterized by unstable, deep-water, high-energy waves with steepness already near 1/7. They will break shortly after feeling bottom some distance from shore and the surf will be rough and choppy with an irregular nature.

Ideally, when the water depth is about 1.3 times the breaker height, the crest of the wave breaks, producing surf. When the water depth becomes less than 1/20 the wavelength, the waves in the surf zone begin to behave as shallow-water waves. Particle motion is greatly interfered with by the bottom, and a very significant transport of water toward the shoreline occurs (figure 8–10).

The breaking experienced in the surf results from the fact that the particle motion near the bottom of the wave has been interfered with greatly by contact with the ocean bottom, and this tends to slow down the waveform. However, the individual particles that are orbiting near the ocean surface have not been slowed down as much as the entire waveform because they have experienced no contact with the bottom. This causes the top of the waveform to lean toward the shore as the bottom of the waveform is held back. The slowing down of the entire waveform compared to the increasing speed of motion in the circular orbits near the ocean surface because of increasing wave height results in the water particles at the interface between the ocean and the atmosphere moving faster toward shore than the waveform itself.

This motion can be seen particularly well in **plunging breakers** (figure 8–11), which have a curling crest that moves over an air pocket resulting from the fact that the curling particles have outrun the wave and there is nothing beneath them to support their motion. Plunging breakers form on moderately steep beach slopes. The more commonly observed breaker is the **spilling breaker** that results from a relatively gentle slope of the ocean bottom, which more gradually extracts the energy from the wave, producing a turbulent mass of air and water that runs down the front slope of the wave instead of a spectacular cresting curl. Because of the gradual extraction of energy of spilling breakers, they have a longer life span and give surfers a longer, if less exciting, ride than do the plunging breakers.

Recalling the particle motion of ocean waves in figure 8–1, we can see that in front of the crest the water particles are moving up into the crest. It is this force along with the buoyancy of the surfboard that helps maintain a surfer in front of a cresting wave. When this upward motion of water particles is interrupted by the wave passing over water that is too shallow to allow this

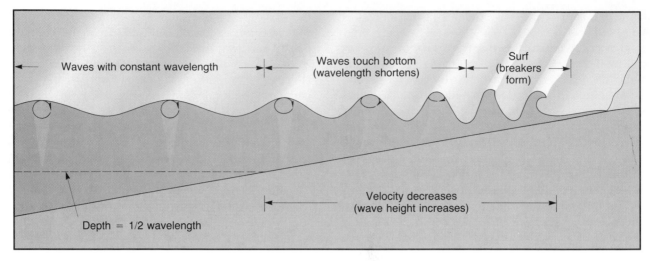

FIGURE 8–10
Surf Zone. As waves approach the shore and encounter water depths less than one-half a wavelength (*L*/2), friction removes energy and the waves slow down, wavelength decreases, and wave height increases. When the water depth is 1.3*H*, the wave reaches a steepness of ½ and breaks on the shore. (The Tasa Collection: *Shorelines*. Published by Macmillan Publishing, New York. Copyright © 1986 by Tasa Graphic Arts, Inc. All rights reserved.)

movement to continue, the ride is over. A skillful surfer, by positioning his board properly on the wave front, can regulate the degree to which the gravitational forces that propel him exceed the buoyancy forces, and high speeds can be obtained while moving along the face of the breaking wave.

Wave Refraction

We previously discussed that waves begin to bunch up and wavelengths become shorter as swell begins to "feel bottom" upon approaching the shore. It is seldom that the swell will approach the shore at right angles. Some

A.

B.

FIGURE 8–11
Breakers. *A*, plunging breaker, Oahu, Hawaii (photo © Tony Arruza/Bruce Coleman, Inc.). *B*, spilling breaker (photo © Peter Arnold, Inc.).

segment of the wave can be expected to feel bottom first and therefore be slowed down before the rest of the wave. This produces **refraction,** or bending, of the waves as they approach the shore. As shown in figure 8–12, an irregular shoreline might result from an irregular bottom topography that could slow down portions of a wave that was approaching the shore.

The effect of wave refraction is to unevenly distribute wave energy along the shoreline. **Orthogonal lines** are constructed perpendicular to the wave fronts and spaced in such a way that the amount of energy between lines is equal at all times. They are of great assistance in illustrating how energy is distributed along the shoreline by breaking waves. Orthogonals indicate the direction the wave is traveling and can be seen to converge on headlands jutting out into the ocean and diverge in bays. A concentration of energy is released against the headlands, while energy released in the bays is spread more thinly. This condition produces erosion on the headlands, while deposition may occur in the bays. The increased energy of the waves breaking on headlands is reflected in an increased wave height.

Wave Reflection

Swell can be reflected back into the ocean with little loss of energy from a vertical barrier such as a seawall. For this ideal reflection to occur without energy loss, the wave would have to strike the barrier at a right angle. Such a condition would be rare in nature. Less ideal reflections will nonetheless produce **standing waves,** which are the production of two waves of the same length moving in opposite directions.

The standing wave is a special interference pattern where no net momentum is carried because the waves are moving in opposite directions. The particles continue to move vertically and horizontally, but there is no more of the circular motion that we saw in the progressive wave. Standing waves are characterized by lines along which there is no vertical movement. There may be one or more such **nodes** or nodal lines. **Antinodes,** crests that alternately become troughs, represent the points of the greatest vertical movement within a standing wave (figure 8–13). There is no particle motion when an antinode is at its greatest vertical displacement, and the maximum particle movement occurs when the water surface is level. At this time, the maximum movement of the water is in a horizontal direction directly beneath the nodal lines. The movement of water particles beneath the antinodes is entirely vertical. We will consider standing waves further when we discuss the tides in the following chapter, as there are certain conditions under which the development of standing waves has a significant effect on the tidal character of coastal regions.

Reflection of wind-generated waves from coastal barriers will be at an angle equal to the angle at which the wave approached the barrier, as shown in figure 8–14. This type of reflection may produce a small-scale interference pattern similar to those we previously discussed.

An outstanding example of reflection phenomena is the Wedge that develops west of the 400-m (1312-ft) jetty at Newport Harbor, California. As the incoming waves strike the jetty and are reflected, a constructive

FIGURE 8–12

Refraction. As the waves "feel bottom" first in the shallow areas off the headlands, they are slowed. The segments of the waves that move through the deeper water leading into the bay are not slowed until they are well into the bay. As a result, the waves are refracted (bent) so the release of wave energy is concentrated on the headlands. Erosion is active on the headlands, while deposition occurs in the bay, where the energy level is low. Orthogonal lines, spaced so that equal amounts of energy are between any two adjacent lines, help to show the distribution of energy along the shore. (Adapted from Tarbuck, E. J., and Lutgens, F. K., *Earth science,* 6th ed. [New York: Macmillan, 1991].)

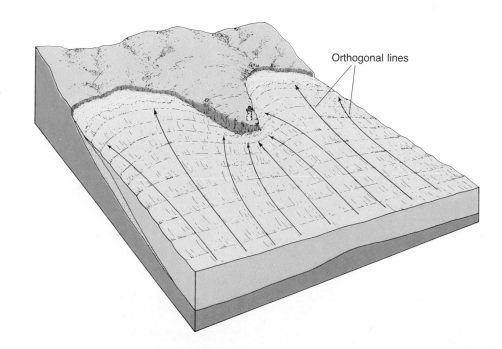

Orthogonal lines

FIGURE 8–13
Reflection—Standing Waves. An example of water motion at quarter-period intervals. Water is motionless when antinodes reach maximum displacement. Water movement is maximum when the water is level. Movement is totally vertical beneath the antinodes, and maximum horizontal movement occurs beneath the node. The circular motion of particles in progressive waves does not exist in standing waves.

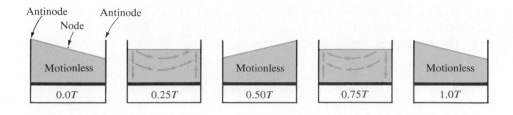

Antinode   Antinode
Node

| Motionless 0.0T | 0.25T | Motionless 0.50T | 0.75T | Motionless 1.0T |

interference pattern develops. When the crests of the incoming waves merge with the crests of the reflected waves, plunging wedge-shaped breakers that may reach heights in excess of 8 m (26 ft) form. These waves present a fierce challenge to the most experienced body surfers. Attracting some who were not up to the challenge, the Wedge has killed and crippled many who have come to try it.

Storm Surge

Large cyclonic storms that develop over the ocean will, owing to their low pressure, produce a hill of water beneath them that moves with the storm across the open ocean. Additionally, as the storm approaches shallow water near shore, the portion of the hill over which the wind is blowing shoreward will produce a **storm surge,** a mass of wind-driven water that produces an increase in sea level that may be extremely destructive to low-lying coastal areas. Storm surges can be particularly destructive when accompanied by high tide. The coincidence of a storm surge with high tide in areas that have particularly high tides frequently produces major catastrophes, with great loss of life and property damage (figure 8–15).

One of the most outstanding examples of storm surge occurred when the windows of a lighthouse atop a 90-m (295-ft) cliff at Dunnet Head, Scotland, were broken by stones tossed by waves breaking on the cliff. The lighthouse marks the western entrance to Pentland Firth, which connects the Atlantic Ocean with the North Sea between the Orkney Islands to the north and the Scottish mainland to the south.

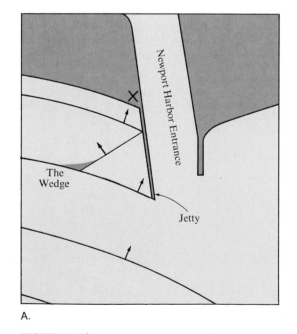
A.

B.

FIGURE 8–14
Reflection—The Wedge. *A,* the Wedge, a wedge-shaped crest that may reach heights in excess of 8 m (26 ft), develops as a result of interference between incoming waves and reflected waves near the jetty protecting the entrance to Newport Harbor, California. *B,* a view of a wedge crest taken from the landward end of the jetty. The three dots in the water in front of the wave are the heads of body surfers waiting to catch the wave.

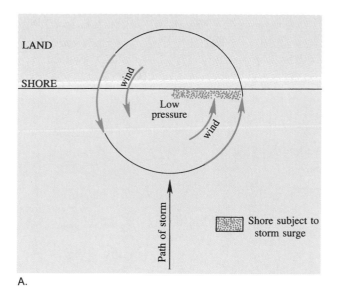

A.

B.

FIGURE 8–15
Storm Surge. *A,* as a cyclonic storm in the Northern Hemisphere moves ashore, the low-pressure cell around which the storm winds blow and the onshore winds associated with the storm will produce a high-water phenomenon called a *storm surge. B,* low pressure and high onshore winds of Hurricane Kate produced a storm surge at Key West, Florida, in November 1985. (Photo © M. Laca/ Weatherstock.)

In diagram A: LAND, SHORE, wind, Low pressure, wind, Path of storm, Shore subject to storm surge

TSUNAMI

The large waves referred to by the Japanese word tsunami originate as a result of disturbances within Earth's crust. We commonly hear them called tidal waves, which implies that they are related to the tides. They have no relationship to the tides, however. Tsunami are usually caused by fault movement, a displacement in Earth's crust along a fracture, that causes a sudden change in water level at the ocean surface. Secondary events, such as underwater avalanches, produced by the faulting may also produce tsunami (figure 8–16). The ocean that is most plagued by the tsunami is the Pacific. It is ringed by a series of trenches that are unstable margins of lithospheric plates along which large-magnitude earthquakes occur.

One of the most destructive tsunami ever generated came from the greatest release of energy from Earth's interior observed during historical times. On August 27, 1883, the island of Krakatoa in the Sunda Strait, between Sumatra and Java, exploded and essentially disappeared. The sound of the explosion was heard 4800 km (2981 mi) away at Rodriguez Island in the western Indian Ocean, and the dust that ascended into the atmosphere circled Earth and produced unusual and beautiful sunsets for nearly one year. A tsunami that rose to heights of more than 30 m (98 ft) devastated the coastal

region of the Sunda Strait, taking more than 36,000 lives. The energy carried by this wave was still detectable after crossing the Indian Ocean, passing the southern tip of Africa, and moving north in the Atlantic Ocean into the English Channel.

Since the wavelength of a typical tsunami will be in excess of 200 km (124 mi), it is obviously a shallow-water wave, the speed of which is determined by water depth. Moving at speeds well in excess of 700 km/h (435 mi/h) and with wave heights of approximately 0.5 m (1.6 ft) in the open ocean, tsunami are not readily observable until they reach shore. In shallow water, they are slowed down, and the water begins to pile up to form crests that may exceed heights of 30 m (98 ft).

Unlike the hurricane that represents a great hazard to ships at sea and may send them seeking the protection of a coastal harbor, a tsunami sends them from their coastal moorings into the open ocean. Until 1948, there was no possible warning of the coming of a tsunami that would be adequate to allow shippers to take the appropriate precautions to avoid the destructive waves. Since that time and as a result of a destructive wave that struck Hawaii in 1946, a tsunami warning system has been established throughout the Pacific Ocean. With every seismic disturbance that occurs beneath the ocean surface with the potential of producing a tsunami, observations are made at the closest tide-measuring stations to see if there

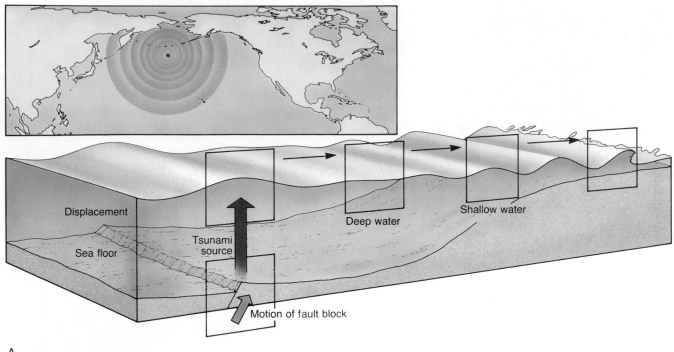

A.

B.

C.

FIGURE 8–16

Tsunami. *A,* movement along a fault in Earth's crust sends energy to the ocean surface. This energy then is distributed laterally along the atmosphere-ocean interface in the form of a tsunami. The energy is transmitted across the open ocean by undetectable waves over 200 km (124 mi) long and about 0.5 m (1.6 ft) high. They release their energy after reaching the shore and developing heights that may be in excess of 30 m (98 ft). Vertical uplift of ocean floor pushes up the ocean water column above the fault. The energy released into the water column by the earthquake travels away from the affected area as a tsunami. *B* shows Sandy Beach on the Island of Oahu moments before a tsunami generated by an earthquake in the Aleutian Trench area struck the beach. Note that the beach was exposed as the water moved out to sea. The arrow indicates the location of *C,* which shows the beach a few minutes later, after the tsunami arrived. A man who barely escaped the wave is circled at left; several automobiles swept off the highway are circled at right. (Photos by Y. Ishii. Reprinted by permission of *Honolulu Advertiser.*)

is any indication that such a wave has been created. Should one be detected, warnings are sent to all the coastal regions that might conceivably encounter a destructive wave along with the time at which the wave is to be expected. This warning allows for the removal of ships from harbors before the waves arrive if the disturbance has occurred at a great enough distance.

Prior to such a warning system, the first notice that most observers had of an impending tsunami would be the rapid seaward recession of the shoreline. The recession would then be followed in a few minutes by the destructive wave (figure 8–16B, C). Such a recession was observed in the port of Hilo, Hawaii, on April 1, 1946, as a result of an earthquake in the Aleutian Trench off the island of Unimak over 3000 km (1863 mi) away. The recession created by the formation of a trough preceding the first tsunami wave crest was followed by a wave that carried the waters to heights nearly 8 m (26 ft) above the normal high tide level. The tsunami also struck Scotch Cap, Alaska, on Unimak Island. The lighthouse on the island, the base of which stood 14 m (46 ft) above sea level, was totally destroyed by a wave that must have reached a height of 36 m (118 ft).

It was the wave produced by this disturbance in the Aleutian Trench that was recorded throughout the coastal regions of the Pacific Ocean at tide-recording stations and led to the development of what is now the International Tsunami Warning System (ITWS).

INTERNAL WAVES

We have to this point discussed waves that occur at an obvious density discontinuity existing between the atmosphere and the ocean. From our previous discussions we should be aware that there are also such discontinuities within the ocean water column itself. For instance, the pycnocline represents a density interface that is typically associated with the thermocline. Just as energy can be transmitted along the interface between the atmosphere and the ocean, it can also be transmitted as an interface wave along density interfaces beneath the ocean surface in what are called **internal waves.** Internal waves may have heights above 100 m (328 ft) (figure 8–17). The greater the difference in density between the two fluids, the faster the waves will move.

There is still much to be learned about internal waves, but their existence is well documented, and it is thought that many causes for them may be identified. Internal waves are known to have periods related to tidal forces, indicating that these forces may be a significant cause. Underwater avalanches in the form of turbidity currents, wind stress, and energy put into the water by moving vessels may also be causes of internal waves. It is thought that the parallel slicks seen on the surface waters may overlie the troughs of internal waves. The slicks are caused by a film of surface debris accumulating and dampening surface waves.

Internal waves reach greater heights from a smaller energy input than do the waves resulting from very large energy input observed at the ocean's surface. This is because they move along interfaces, such as the pycnocline, across which the density difference is considerably less than that which exists between the ocean surface and the atmosphere. They are thought to move as shallow-water waves at speeds considerably less than those of surface waves, with periods of 5–8 min and wavelengths of 0.6–0.9 km (0.37–0.56 mi).

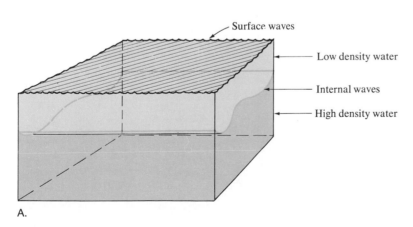

A.

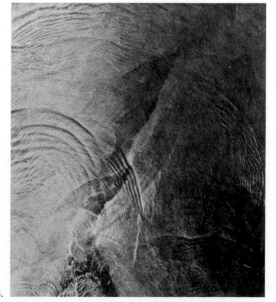

B.

FIGURE 8–17
Internal Wave. *A,* a simple internal wave moving along the density interface below the ocean surface. *B, Seasat* Synthetic Aperture Radar image of internal waves in the Gulf of California. They are thought to be associated with tidal currents in the shallow water around the island of Angel de La Guarda located at the bottom. The image is about 100 km (62 mi) on each side.

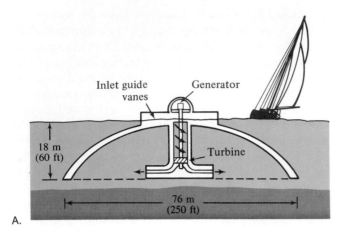

FIGURE 8–18
Dam-Atoll. Lockheed Corporation's Dam-Atoll designed to generate electricity from wave action. *A,* water entering at the surface spirals down an 18-m (60-ft) central cylinder to turn a turbine located at the bottom of the cylinder. Each unit is designed to produce from 1 to 2 MW. *B,* view from above. (Courtesy of Lockheed Corporation.)

POWER FROM WAVES

Extracting power from waves as they approach the shore is a possibility, although there are significant problems to be overcome. Owing to the refraction of waves, a large half-cone could be constructed to focus the energy of waves at the apex of the structure. Such a device might extract up to 10 MW of power per kilometer of shore, but could produce significant power only when large storm waves broke against it. Such a system would operate only as a power supplement, and a series of such structures along the shore would be required. Structures of this

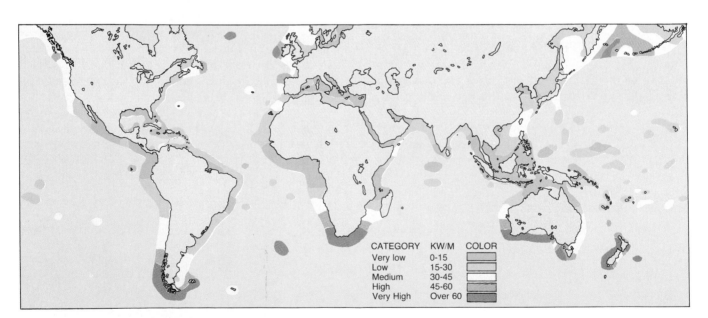

FIGURE 8–19
Global Coastal Wave Energy Resources. This map, constructed from data available in the U.S. Navy Summary of Synoptic Meteorological Observations (SSMO), shows the effect of west to east movement of storm systems in the temperature latitudes. The west coasts of continents are struck by larger waves than the east coasts. Also the larger waves appear to be associated with the westerly wind belts between 30° and 60° latitudes in both hemispheres. The significant wave height associated with the available wave power categories shown on the map are very low (1 m), low (2–2.5 m), medium (3 m), high (3.5 m), and very high (3.9 m). (Adapted from *Sea frontiers* (July–August 1987) "Sea Secrets," International Oceanographic Foundation (Vol. 33, No. 4) pp. 260–261; produced by The National Climatic Data Center with support from the U.S. Department of Energy.)

type could have a serious effect on longshore transport of sediment, which could lead to serious coastal erosion problems in areas deprived of sediment.

Along shores endowed with proper topography of the ocean floor, internal waves with their great heights may be effectively focused by refraction and induced to "break" against an energy-conversion device.

Two engineers at Lockheed Corporation have developed an interesting ocean wave energy device called Dam-Atoll, which is shown in figure 8–18. Waves enter the top of the unit, just at the ocean surface. Guide vanes at the opening cause the entering water to spiral into a whirlpool within an 18-m (60-ft) central core. The swirling column of water turns a turbine, the unit's only moving part. It can provide a continuous electrical power output of 1 to 2 MW according to the inventors, Leslie S. Wirt and Duane L. Morrow. In particularly good wave areas, such as the Pacific Northwest, they believe it would be possible to anchor 500 to 1000 units that would provide power in quantities comparable to those provided by Hoover Dam. These units, 76 m (250 ft) in diameter and made of concrete, have many potential uses other than the generation of electricity. They could be used in cleaning up oil spills, protecting beaches from wave erosion, creating calm harbors in the open sea, and desalinating seawater through reverse osmosis.

Economic conditions in the future may enhance the appeal of converting wave energy to electrical energy. The coastal regions with the greatest potential for exploitation are shown in figure 8–19.

SUMMARY

Wave phenomena transmit energy through various states of matter by setting up patterns of oscillatory motion in the particles that make up the matter. Progressive waves may be described as longitudinal, transverse, or orbital, depending on the pattern of particle oscillation. Particles related to ocean waves move primarily in orbital paths.

Characteristics used to describe waves are wavelength (L), wave height (H), period (T), and wave speed (S). If water depth is greater than 1/2 wavelength, a progressive wave travels as a deep-water wave with a speed that is directly proportional to wavelength. If water depth is less than 1/20 wavelength, it will move as a shallow-water wave, the speed of which increases with increased water depth.

As wind-generated waves form in the sea area, capillary waves with rounded crests and wavelengths less than 1.74 cm (0.7 in.) form first. As the energy of the waves increases, gravity waves with increased wave speed, wavelength, and wave height form. Energy is transmitted from the sea area across the ocean by low, rounded waves called swell. The swell releases its energy in the surf that forms as the waves break in the shoaling water near shore. If the waves break on a relatively flat surface, the result is usually a spilling breaker, while breakers forming on steep slopes have spectacular curling crests and are called plunging breakers.

When swell approaches the shore, the segments of the waves that encounter shallow water first will be slowed, and those parts of the wave that have not yet been interfered with by shallow water will move ahead, causing the wave to refract, or bend. Refraction causes a concentration of wave energy on the headlands, while low-energy breakers are characteristically found in bays. Reflection of waves off seawalls or other barriers can cause an interference pattern called a standing wave. In standing waves, crests do not move laterally as in progressive waves but form alternately with troughs at locations called antinodes. Separating the antinodes are locations where there is no vertical movement of the water—nodes.

During storms, the combination of low air pressure and onshore winds may produce storm surge that raises the water level at the shore many meters above normal sea level. Such surges are particularly destructive if they coincide with high tide.

Tsunami, or seismic sea waves, are generated by seismic disturbances beneath the ocean floor. Such waves have lengths in excess of 200 km (124 mi) and travel across the open ocean with undetectable heights of about 0.5 m (1.6 ft) at speeds in excess of 700 km/h (435 mi/h). On approaching shore, they may increase in height to over 30 m (98 ft). Single-wave tsunami have been known to cause millions of dollars worth of damage and take tens of thousands of lives. Internal waves are not well understood but are thought to form at density interfaces beneath the ocean surface, especially in connection with the pycnocline. They may be expected to have heights of up to 100 m (328 ft), with periods from 5 to 8 min.

Ocean waves can potentially be focused to produce hydroelectric power, but there are significant environmental considerations that will have to be incorporated into any such plan.

KEY TERMS

QUESTIONS AND EXERCISES

1. Discuss longitudinal, transverse, and orbital wave phenomena, including the states of matter in which they transmit energy.

2. Calculate the speed (S) for deep-water waves with the following characteristics:
   A. $L = 351$ m, $T = 15$ s
   B. $L = 351$ m, $f = 4$ waves/min
      Express speed (S) in meters per second (m/s)
   C. $T = 11$ s

3. Calculate the speed with which a shallow-water wave will travel across an ocean basin 4 km (2.5 mi) deep.

4. Describe the change in the shape of waves that occurs as they progress from capillary waves to increasingly larger gravity waves until they reach a steepness ratio of 1/7. A change in which variable will make gravity the dominant restoring force, H, L, S, T, or f?

5. What is the minimum fetch and duration of wind required to produce a fully developed sea with a wind speed of 45 km/h (27.9 mi/h)? What would be the average height, length, and period of waves in this fully developed sea?

6. Waves from separate sea areas move away as swell and produce an interference pattern when they come together. If the waves from Sea A and Sea B have wave heights of 1.5 m (5 ft) and 3.5 m (11.5 ft), respectively, what would be the height of waves resulting from constructive interference and destructive interference? Illustrate your answer with a diagram (refer to figure 8–9).

7. Describe changes, if any, in wave speed (S), length (L), height (H), and period (T), that occur as waves move across shoaling water to break on the shore.

8. Using orthogonal lines, illustrate how wave energy can be distributed along the shore. Identify areas of high and low energy release.

9. On the basis of the fundamental characteristics of standing waves shown in figure 8–13, construct a similar diagram of a standing wave in which two nodes and three antinodes exist.

10. List three factors that may affect the height of storm surge. Make a diagram of a hurricane coming ashore from the south along an east-west shore and indicate the segment of shore along which you think the storm surge will reach maximum height. Explain why.

11. Why is it more likely that a tsunami will be generated by faults beneath the ocean along which vertical rather than horizontal movement has occurred?

12. How long will it take for a tsunami to travel across 2000 km (1242 mi) of ocean if the average depth of ocean is 4500 m (14,760 ft)?

13. What ocean depth would be required for a tsunami with a wavelength of 220 km (136 mi) to travel as a deep-water wave? Is it possible that such a wave could become a deep-water wave any place in the world ocean?

14. Why is the development of internal waves likely within the thermocline?

15. Discuss some environmental problems that might result from the development of facilities for the conversion of wave energy to electrical energy.

REFERENCES

Bowditch, N. 1958. *American Practical Navigator*. Rev. ed. H. O. Pub. 9. Washington, D.C.: U.S. Naval Oceanographic Office.

Gill, A. E. 1982. *Atmosphere-ocean dynamics. International Geophysics Series,* Vol. 30. Orlando, Fla.: Academic Press, Inc.

Kinsman, B. 1965. *Wind waves: Their generation and propagation on the ocean surface.* Englewood Cliffs, N.J.: Prentice-Hall.

Melville, W., and Rapp, R. 1985. Momentum flux in breaking waves. *Nature* 317:6037, 514–16.

Pickard, G. L. 1975. *Descriptive physical oceanography: An introduction.* 2nd ed. New York: Pergamon Press.

Stewart, R. H. 1985. *Methods of satellite oceanography.* Berkeley, Calif.: University of California Press.

Sverdrup, H. U.; Johnson, M. W.; and Fleming, R. H. 1942. Renewal 1970. *The oceans: Their physics, chemistry, and general biology.* Englewood Cliffs, N.J.: Prentice-Hall.

van Arx, W. S. 1962. *An introduction to physical oceanography.* Reading, Mass.: Addison-Wesley.

SUGGESTED READING

Sea Frontiers

Barnes-Svarney, P. 1988. Tsunami: Following the deadly wave. 34:5, 256–63. The origin of tsunami, a history of major occurrences, and the methods used to detect them are covered.

Changery, M. J. 1987. Coastal wave energy. 33:4, 259–62. The wave energy resources of the world's shores are considered.

Ferrell, N. 1987. The tombstone twins: Lights at the top of the world. 33:5, 344–51. A short history of the two lighthouses on Unimak Island, Alaska.

Land, T. 1975. Freak killer waves. 21:3, 139–41. The British design a buoy that will gather data in areas where 30-m (98-ft) waves, which may be responsible for the loss of many ships, occur.

Mooney, M. J. 1975. Tragedy at Scotch Cap. 21:2, 84–90. A recounting of the events resulting from an earthquake off the Aleutians April 1, 1946. The resulting tsunami destroyed the lighthouse at Scotch Cap, Alaska.

Pararas-Carayannis, G. 1977. The International Tsunami Warning System. 23:1, 20–27. A discussion of the history and operations of the International Tsunami Warning System.

Robinson, J. P., Jr. 1976. Newfoundland's disaster of '29. 22:1, 44–51. A description of the destruction caused by a tsunami that struck Newfoundland on November 18, 1929.

———. 1976. Superwaves of southeast Africa. 22:2, 106–16. A discussion of the formation and destruction caused by large waves that strike ships off the southeast coast of South Africa.

Smail, J. 1982. Internal waves: The wake of sea monsters. 28:1, 16–22. An informative discussion of the causes of internal waves and their effect on surface ships and submarines.

Smail, J. R. 1986. The topsy-turvy world of capillary waves. 32:5, 331–37. Capillary waves are clearly described, and their role in transmitting wind energy to the motion of waves and currents is discussed.

Smith, F. G. W. 1970. The simple wave. 16:4, 234–45. This is a very readable explanation of the nature of ocean waves. It deals primarily with the characteristics of deep-water waves.

———. 1971. The real sea. 17:5, 298–311. A comprehensive and readable discussion of wind-generated waves.

———. 1985. Bermuda mystery waves. 31:3, 160–63. A discussion of the possible source of large waves that struck Bermuda on November 12, 1984.

Truby, J. D. 1971. Krakatoa: The killer wave. 17:3, 130–39. The events leading up to the 1883 eruption of Krakatoa and the tsunami that followed are described.

Scientific American

Bascom, W. 1959. Ocean waves. 201:2, 89–97. An informative discussion of the nature of wind-generated waves, tsunami, and tides.

Koehl, M. A. R. 1982. The interaction of moving water and sessile organisms. 274:6, 124–35. A discussion of the adaptations of benthic shore-dwelling animals to the stresses of strong currents and breaking waves.

Chapter 9

Tides

Humans have undoubtedly observed the rise and fall of the tides since they first inhabited the coastal regions of the continents. But until Herodotus (450 B.C.) observed tides on the Mediterranean, there was no known record of them. The first Greek observers concluded that the tides were related to the motion of the moon since both followed a similar cyclic pattern, but it was not until **Sir Isaac Newton** (1642–1727) developed his universal law of gravitation that the tides could be explained adequately.

The tides are the ultimate manifestation of shallow-water wave phenomena, possessing lengths measured in thousands of kilometers and heights ranging from zero to more than 15 m (49 ft). The ocean tides generated by the gravitational attraction of the sun and moon on the mass of the ocean affect every particle of water from the surface to the deepest part of the ocean basin. Thus, the tide undoubtedly has a much more far-reaching effect on ocean phenomena than we can observe at the ocean surface.

High tide (left) and low tide (right) in the Bay of Fundy, Nova Scotia. (Photos © Clyde H. Smith/Peter Arnold, Inc.)

TIDE-GENERATING FORCES

Law of Gravitation

Sir Isaac Newton published his *Philosophiae naturalis principia mathematica* in 1686 and stated in his preface, ". . . I derive from the celestial phenomena the forces of gravity with which bodies tend to the sun and several planets. Then from these forces, by other propositions which are also mathematical, I deduce the motions of the planets, the comets, the moon, and the sea." What followed is our first understanding of why tides behave as they do.

Newton's **law of gravitation** states that every particle of mass in the universe attracts every other particle of mass with a force that is proportional to the product of their masses and inversely proportional to the square of the distance between the masses. The greater the mass of the objects and the closer they are together, the greater will be the gravitational attraction. Mathematically this can be expressed as:

$$\text{Gravitational force} = G\,\frac{m_1 m_2}{r^2}$$

Here, G is the universal gravitational constant, m_1 and m_2 are the masses, and r is the distance between the masses. For spherical bodies, all of the masses can be considered to exist at the center of the sphere. Thus, r will always be the distance between the centers of bodies being considered.

Actually, tide-generating forces vary inversely as the cube of the distance from the center of Earth to the center of the tide-generating object, instead of varying inversely to the square of the distance as does the gravitational attraction. Therefore, the tide-generating force, although it is derived from the force of gravitational attraction, is not linearly proportional to it. Distance becomes a more highly weighted variable in the tide-generating forces than it is in gravitational attraction force:

$$\text{Tide-generating force} \propto \frac{m_1 m_2}{r^3}$$

Although the gravitational attraction between Earth and the sun is over 177 times that between Earth and the moon, the moon dominates the tides, as will be seen from the following comparison of the tide-generating force of the sun with that of the moon. Since the sun is 27 million times more massive than the moon, it should, solely on the basis of comparative masses, have a tide-generating force 27 million times greater than that of the moon. However, we must also consider the distance between Earth and the moon as compared to the distance between Earth and the sun. Since the sun is 390 times farther from Earth than the moon, its tide-generating force is reduced by 390^3, or about 59 million times compared to that of the moon. These conditions result in the sun's tide-generating force being $^{27}\!/_{59}$, or about 46 percent, that of the moon (figure 9–1 and table 9–1).

The full explanation is complex, but we will attempt a brief explanation of how the tide-generating force decreases in proportion to the cube of the increase in distance between the particle of Earth that is affected and the center of the tide-generating bodies (moon and sun). We will confine our consideration to the moon, but a similar explanation applies to the sun.

The tidal pattern we see on Earth is primarily the result of the rotation of the Earth-moon system around its center of mass. Figure 9–2A shows the path followed by the center of Earth as it rotates around the center of gravity for the Earth-moon system, a point about 4700 km (2918 mi) from the center of Earth. It shows also that the zenith (point on Earth's surface closest to the moon) and nadir (point on Earth's surface farthest from the moon) follow identical circular paths. Actually, all points on Earth must follow circular paths identical to that of the center of Earth. If we divide Earth into millions of tiny particles of equal mass, each must have an identical **centripetal** (center-seeking) **force** act on it to keep it in this circular path.

As an example of centripetal force, if you tie a string to a rock and swing the tethered rock around your head, the string pulls the rock toward your hand—a point in the center of the circular orbit the rock is following. The string is providing a centripetal force to the rock. If the

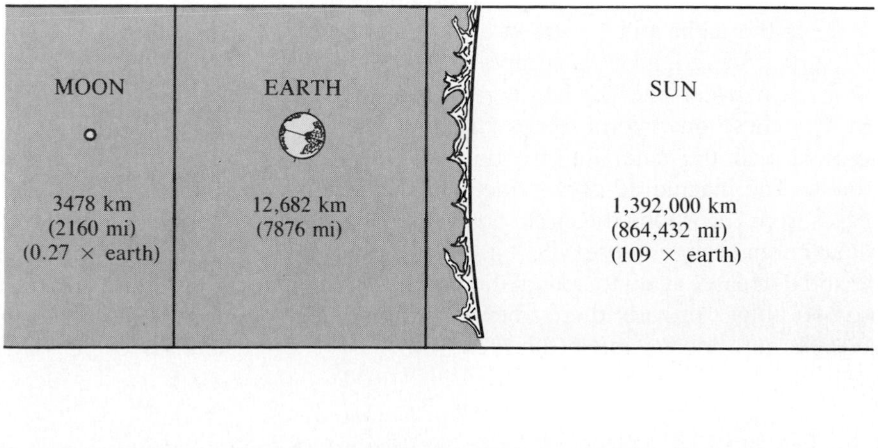

FIGURE 9–1
Relative Sizes of the Earth, Moon, and Sun. Earth has a diameter of 12,682 km (7876 mi). The diameter of the moon is 3478 km (2160 mi), which is 0.27 that of Earth. The diameter of the sun is 1,392,000 km (864,432 mi), which is 109 times the diameter of Earth. The relative sizes are shown to scale. See table 9–1 for a comparison of the masses of the sun and moon and their distances from Earth. These factors are important in determining their tide-generating effect on Earth.

MOON	EARTH	SUN
3478 km (2160 mi) (0.27 × earth)	12,682 km (7876 mi)	1,392,000 km (864,432 mi) (109 × earth)

TABLE 9–1
Relative Masses of Sun and Moon and Distances from Earth—Relative Tide-Generating Effect.

Tide-Generating Body	Distance from Earth (avg)	Mass (Metric tons)	Relative Tide-Generating Effect
Moon	384,835 km (238,483 mi)	7.3×10^{19}	Based on relative masses, the sun is 27 million times more massive than the moon and has 27 million times the tide-generating effect. However, since the sun is 390 times farther than the moon from the earth, its tide-generating effect is reduced by 390^3, or 59 million times.
Sun	149,758,000 km (93,016,845 mi)	2×10^{27}	

Moon

Earth   Distance of moon and sun from earth shown approximately to scale.   Sun

DETERMINATION OF TIDE-GENERATING FORCE OF SUN RELATIVE TO MOON

$$\text{Tide-generating force} \propto \frac{\text{Mass}}{(\text{Distance})^3} \propto \frac{\text{Sun—27 million times more mass}}{(\text{Sun—390 times farther away})^3}$$

$$(390)^3 = 59,000,000 \quad \text{Thus,} \quad \frac{27 \text{ million}}{59 \text{ million}} = 0.46 \text{ or } 46\%$$

The sun has 46% the tide-generating force of the moon.

string breaks, the rock can no longer maintain its circular orbit and will fly off in a straight line tangent to the circle (figure 9–2B).

Figure 9–3A shows that each particle of equal mass is located at a different distance and/or direction from the center of the moon. This causes the force of gravitational attraction between each particle and the moon, which must provide the required centripetal force to each particle, to be different. As a result of the differences between the required centripetal force and the gravitational force of attraction between each particle and the moon, small lateral (tangent to Earth's surface) or tide-generating forces are produced to push the ocean's water into bulges at the zenith and the nadir.

A plane passed through the center of Earth and perpendicular to a line through the centers of Earth and the moon intersects Earth's surface along a circle where tide-generating forces are zero. There is also no tide-generating force at the nadir and zenith. At all other points on Earth's surface there is a lateral component to the gravitational force between the particle it represents and the moon. It is these tiny lateral forces, the tide-generating forces, that push the water into the bulges at the zenith and nadir. The magnitude of the tide-generating forces increases from zero along the circle previously discussed to a maximum at an angle of 45° on either side of the circle and decreases again to zero at the zenith and nadir. These two bulges provide the framework within which the equilibrium tide operates (figure 9–3B).

EQUILIBRIUM THEORY OF TIDES

We will now consider the kinds of tide observations that would be predictable on an Earth with two tidal bulges, one toward the moon and one away from the moon as Earth rotates on its axis. In the following discussion, we assume an ideal ocean of uniform depth covers Earth and that there is no friction between the water in the ocean and the ocean floor. We will call this theoretical tide the **equilibrium tide.** Although this oversimplification will not give us an accurate means of predicting the types of tides that will occur throughout Earth's surface, it will give us the gross characteristics of tides in the world ocean. Only after we have established an understanding of the equilibrium theory will we attempt to consider the **dynamical theory** of tides that deals with changes in ocean depth, the existence of continents, and friction between the ocean water and ocean floor.

The Rotating Earth

This ideal ocean is modified only by the tide-producing forces that cause bulges on opposite sides of Earth, as shown in Figure 9–4. Let us assume that a stationary moon is directly above the equator so that the maximum bulge will occur on the equator on opposite sides of Earth. Since Earth requires 24 h for one complete rotation, an observer on the equator would experience two high tides per day. The time that would elapse between

FIGURE 9–2
Earth-Moon-System Rotation. *A*, the dashed line through the center of Earth is the path of Earth's center as it moves around the common center of gravity of the Earth-moon system. The circular paths followed by the nadir (point a) and zenith (point b) have the same radius as that followed by Earth's center. *B*, if you tie a string to a rock and swing it in a circle around your head, it stays in the circular orbit because the string exerts a centripetal (center-seeking) force on the rock. This force pulls the rock toward the center of the circle. If the string breaks with the rock at the position shown, the rock will fly off along a straight path tangent to the circle.

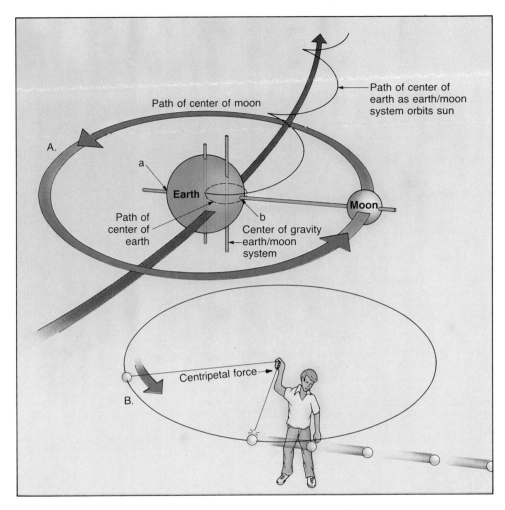

high tides, the tidal period, would be 12 h. An observer at any latitude north or south of the equator would experience a similar period, but the high tides would not be so high because the observer would be at the edge of the bulge rather than at the apex.

But high tides do not occur every 12 h on Earth's surface. This is due to the fact that the Earth-moon system is rotating about its center of mass while Earth is rotating on its axis.

The **lunar day,** the time that elapses between successive passages of the moon across the meridian (longitude line) of an observer, must be somewhat longer than the solar day of 24 h. It is actually 24 h 50 min. If one observes the time at which the moon rises on successive nights, it can be seen to rise 50 min later each night. Figure 9–5 shows that this results from the fact that as Earth is making its rotation on its axis in 24 h, the moon has moved 12.2° to the east. Earth must rotate another 50 min to have the moon again on the meridian of the observer. Therefore, if we wish to divide the lunar day of 24 h 50 min into 24 lunar hours, each **lunar hour** will be about 1 h 2 min of solar time in length. With the knowledge that the moon completes a 360° revolution in 29.53

days, this 12.2° of eastward revolution for the moon can be computed by

$$\frac{360°}{29.53 \text{ days}} = 12.2° \text{ per day}$$

We have so far considered the effects of Earth's rotation and the revolution of the Earth-moon system about its center of mass on the prediction of tides. We have ignored the effect of the sun. In the following discussion we will consider the combined effect of the moon and sun on Earth's tides.

Combined Effects of Sun and Moon

Figure 9–6 shows the path of Earth and the moon as the Earth-moon system revolves around the sun. It can be seen that approximately every 29½ days the moon is in the same phase. When the moon is between Earth and the sun it is said to be in **conjunction,** producing a **new moon.** The moon is in **opposition** when it is on the opposite side of Earth from the sun, causing a **full moon.** A **quarter moon** results when the moon is in **quadrature,** at right angles to the sun relative to Earth.

FIGURE 9–3
Tide-Generating Force. *A,* the length and direction of the arrows in this figure represent the magnitude and direction of the forces related to each of the eight identical masses located at each of the points shown on the figure. *C arrows*—The centripetal force required to keep each particle in the identical circular orbit it follows as a result of the Earth-moon rotation system. *G arrows*—The gravitational attraction between particles and the moon. This force provides the required centripetal force, but it is identical to it only for the particle at the center of Earth. It is in the required direction only for particles on a line connecting the centers of Earth and the moon. For all particles on the half of Earth facing the moon, the force is larger than required, and for the particles on the half of Earth facing away from the moon, it is smaller. *Blue arrows*—For all particles except at the center of Earth, there is a resulting residual force because the gravitational force varies from the required centripetal force. This force is small—averaging about 10^{-7} of the magnitude of Earth's gravity. Therefore, where forces act perpendicular to Earth's surface, as does gravity, they do not have any tide-generating effect. However, where they have a significant horizontal component—tangent to Earth's surface—they aid in producing tidal bulges on Earth. Since there are no other large horizontal forces on Earth with which they must compete, these small, ever-present residual forces can push water across Earth's surface. (Consider all vector arrows to extend from the point they represent.) *B,* along the intersection of Earth's surface and a plane running through Earth's center perpendicular to a line connecting the centers of Earth and the moon, the residual forces are directed toward the center of Earth and have no horizontal component—tide-generating forces are zero (dots). There is also no horizontal component at the zenith (*Z*) and nadir (*N*) because the residual forces point away from the center of Earth. Elsewhere on Earth's surface, there are horizontal tide-generating forces. The magnitude of the tide-generating force varies. It reaches maximum values along two circles that lie 45° on either side of the previously described circle of zero tide-generating force.

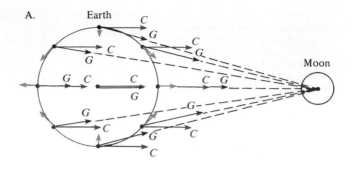

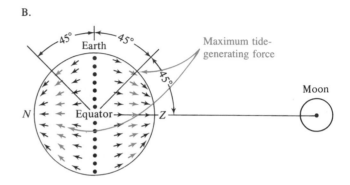

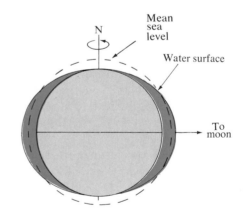

FIGURE 9–4
Equilibrium Tide. Assuming an ocean of uniform depth covering Earth, and the moon located above the equator, the tide-generating forces will produce two bulges on the ocean surface. One will extend in the direction of the moon and the other away from the moon. As Earth rotates, all points on Earth except the poles will experience two high tides each day as Earth rotates beneath the two bulges.

FIGURE 9–5
The Lunar Day. A lunar day is the time that elapses between successive appearances of the moon on the meridian directly over a stationary observer. As Earth rotates on its axis, the Earth-moon-system rotation moves the moon in the same direction (to the east). During one complete rotation of Earth on its axis (the 24-h solar day), the moon moves east 12.2°, and Earth must rotate another 50 min to put the observer in line with the moon.

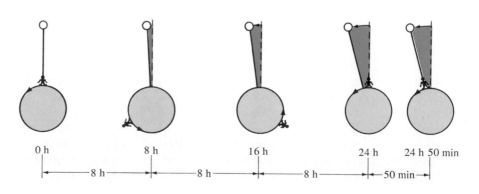

When the sun and moon are in opposition or conjunction, the tide-generating forces of the sun and moon are additive, and we experience maximum **tidal ranges,** the vertical difference between high and low tide. During quadrature, the tide-generating force of the sun is working at right angles to the tide-generating force of the moon, and we have minimum tidal range. The maximum tidal range condition that exists during the new and full moon phases is referred to as the **spring tide,** and during the quadrature phases we have **neap tide** (figure 9–6). The time that elapses between successive spring tides (full moon and new moon) or neap tides (first and third quarters) is a fortnight, about two weeks.

Effects of Declination

Up to this point we have considered that the moon and the sun remained at all times above the equator, or at least we did not describe specifically any deviation from such a position. We now must consider the fact that this is not the case. As Earth revolves around the sun, its axis of rotation is tilted 23.5° from vertical relative to the **ecliptic,** the plane of Earth's orbital path about the sun. It is owing to this tilt that we experience the seasons—spring, summer, fall, and winter. To further complicate our consideration, the plane of the moon's orbit is at an angle of 5° to the ecliptic. This angular distance of the sun or moon above or below the equatorial plane of Earth is called **declination.**

The tilt of Earth's axis relative to its plane of revolution around the sun is shown in figure 9–7A. It can be seen that the tilted axis assumes a constant direction in space throughout the yearly cycle that includes the equinoxes and solstices. At the **vernal equinox,** which occurs about March 21, the sun is directly above the equator and is moving from the Southern Hemisphere to the Northern Hemisphere. On about June 21 the **summer solstice** occurs. At this time the sun reaches its most northerly point in the sky, directly above the Tropic of Cancer, which is at 23.5°N latitude. Following this occurrence, the sun moves farther south in the sky each day, and on about September 23 it is directly above the equator again and produces the **autumnal equinox.** During the next three months the sun appears to be more southerly in the sky until the **winter solstice** on about December 22, when it is directly over the Tropic of Capricorn at 23.5°S latitude. Thus the sun may be found at declinations between 23.5° north and 23.5° south of the equator on a yearly cycle.

Since the plane of the moon's orbit intercepts the plane of the ecliptic at an angle of 5°, and since the plane of the moon's orbit **precesses,** or rotates, while maintaining its 5° angle with the ecliptic with a precessional cycle of 18.6 yr, we have a relatively complex consideration regarding the declination of the moon relative to the

plane of Earth's equator. In part 1 of figure 9–7B, the declination of the moon's orbit relative to Earth's equator is 28.5°. The declination will change from 28.5° south to 28.5° north and back to 28.5° south of the equator in a period of one month. Part 2 of the illustration shows the relationship of the ecliptic, the plane of the moon's orbit, and the plane of Earth's equator after one-fourth of the precession, or 4.65 yr later. The maximum declination of the moon's orbit relative to Earth's equator still approaches 28.5°. However, in part 3, when one-half precession is completed, which is 9.3 yr after the condition observed in part 1, the maximum declination of the moon relative to Earth's equator is 18.5°.

On the basis of these considerations we must alter our previous concept of the predicted equilibrium tide. We must now expect that tidal bulges will rarely be aligned with the equator and will occur for the most part north and south of the equator. Since the moon is the dominant force that creates tides in Earth's oceans, we would suspect that the bulges would follow the moon as it moves on its monthly journey across the equator, being found at a maximum of 28.5° north and south of the equator (figure 9–8).

Effects of Distance

Additional considerations that will affect the tide-generating force of the sun and moon on Earth are their changing distances from Earth. Earth ranges from **perihelion** with the sun, when the distance between the bodies is 148.5 million km (92.2 million mi) during the winter months in the Northern Hemisphere, to **aphelion,** when the distance between the bodies is 152.2 million km (94.5 million mi) during the summer months. The moon moves from **perigee,** its closest approach to Earth, of 375,200 km (233,000 mi) to **apogee** of 405,800 km (252,000 mi) and back to perigee during a period of 27½ days—the **anomalistic month.** (See figure 9–9.)

Because of these movements, spring tides have greater ranges during the Northern Hemisphere winter than in the summer. Also, as a result of changing distance between Earth and moon we will find that tidal ranges will become greater at perigee each anomalistic month. Keeping in mind that tide-generating forces vary inversely with the cube of the distance from the center of Earth to the center of the tide-generating body, it can be appreciated that these changes resulting from the elliptical nature of Earth's orbit around the sun and the moon's orbit around Earth can be readily observed in the tides.

Equilibrium Tide Prediction

To consider the tidal patterns we would predict for an idealized water-covered Earth, let us return our attention to the effect of declination. Let us assume that the declina-

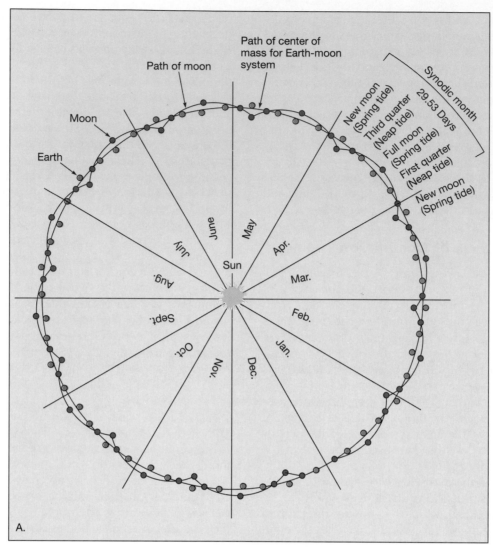

FIGURE 9–6
Sun-Moon-Earth Motion and Tides. As the Earth-moon system moves around the sun, the centers of each body follow a wavy path because they are also rotating about the center of mass of the Earth-moon system (*see* figure 9–2A). We can think of this system as a sledgehammer, with Earth being the heavy business end and the moon being a knob at the end of the handle. The "handle" that connects Earth and moon is gravity. If the hammer were thrown through space, the center of the knob (moon) at the end of the handle would follow an orbit around the center of mass with a much greater radius than that of the orbit of the center of the head of the hammer (Earth). Therefore, the wobble of the moon is more pronounced and is shown in exaggerated form, A. During a synodic month, the moon moves from a position between the sun and Earth (new moon) to a position that puts Earth between the moon and the sun (full moon) and back to its original position.

tion of the moon, which will determine the alignment of the tidal bulges, is 28° north of the equator. If we position an observer at this latitude on a permanent point, the observation of the tides that occur here will be different from observations at the equator.

If the observations begin when the moon is directly over the observer's head, high tide will be recorded. Six lunar hours (6 h 12½ min solar time) later a low tide will be recorded, followed by another high tide that will be much lower than the initial high tide (figure 9–10A–D). At the end of a 24-lunar-hour period, the observer will

have passed through a complete lunar-day cycle of two low tides and two high tides. A representative curve of the type of tide observed can be seen in figure 9–10E. Tide curves showing the heights of the same tides during one lunar day at the equator and at 28°S latitude are also provided. The vertical difference in successive high tides or successive low tides that occur as a result of the declination of the moon and the sun relative to Earth's equator is called the **diurnal inequality.**

To summarize the tides that we might predict as a result of the tide-generating forces of the moon and sun

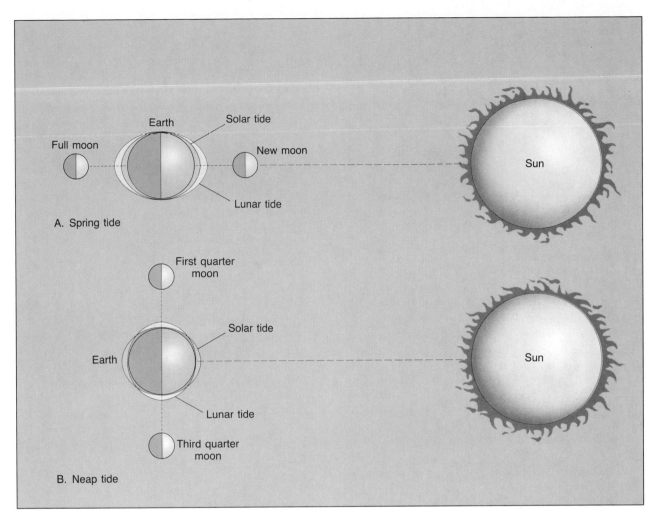

B, when the moon is in the new or full position, the tidal bulges created by the sun and moon are aligned, producing constructive interference and a larger bulge. When the moon is in the positions halfway between the new and full phases, the first and third quarters, the tidal bulge produced by the moon is at right angles to the bulge created by the sun. The bulges tend to "cancel" each other (destructive interference), and the resulting bulge is smaller. New moon and full moon phases produce spring tides with maximum tidal ranges, while the first and third quarter phases of the moon produce neap tides with minimal tidal ranges. (*B*, from The Tasa Collection: *Shorelines*. Published by Macmillan Publishing Co., New York. Copyright © 1986, by Tasa Graphic Arts, Inc. All rights reserved.)

on an Earth covered with a uniform depth of water for any point not on the equator, we would predict that the equilibrium tide would generally be observed at any location to have (1) two high tides and two low tides per lunar day. Except for the rare occasions when the sun and moon are simultaneously above the equator, we would expect that (2) neither the two high tides nor the two low tides would be of the same height because of the changing declination of the moon and the sun. We would expect (3) yearly and monthly cycles of tidal range related to the changing distances of Earth from the sun and moon. Lastly, we would expect that (4) each fortnight, half a lunar month, we would experience spring tides that would be separated by neap tides (figure 9–10F).

Considering the great number of variables that are involved in the prediction of tides, it is interesting to consider when the conditions might be right to produce the maximum tide-generating force. This will occur when the sun is at perihelion and in conjunction (new moon) or opposition (full moon) with the moon at perigee and when both the sun and moon have a zero declination. This condition occurs only once each 1600 years, and the next occurrence is predicted for A.D. 3300.

An indication of the significance of even a near coincidence of Earth's perihelion with the perigee of the moon during a spring tide was made woefully clear during the January 1983 storms in the North Pacific. The introduction of slow-moving low-pressure cells that developed over the Aleutians caused strong northwest winds

FIGURE 9–7
Orbital Planes of Earth and Moon. *A,* as Earth orbits the sun during one year, the axis of rotation is tilted 23.5° from perpendicular relative to the ecliptic. The sun shines down directly over the Tropic of Cancer (23.5°N) on the day of the summer solstice, June 21. Three months later the sun is directly above the equator (0°) during the autumnal equinox, September 23. The sun is directly over the Tropic of Capricorn (23.5°S) during the winter solstice, December 22, and returns to a position above the equator on the vernal equinox, March 21. Three months later, the yearly orbit is completed, and the sun is again directly over the Tropic of Cancer. *B,* the plane of the moon's orbit (gray plane) is tilted at an angle of 5° relative to the plane of the ecliptic (yellow plane) and rotates with a clockwise precession that has a period of 18.6 years. In *B1,* the declination of the moon's orbit is the sum of the angle of intersection of the plane of Earth's equatorial plane (blue plane) with the plane of the ecliptic (23.5°) plus the angle of intersection between the plane of the moon's orbit and the ecliptic (5°). This produces the maximum declination of the moon relative to Earth's equator of 28.5°. *B2* shows the positions of the planes 4.65 years later when the moon has achieved one-fourth of its precessional rotation. *B3* shows the relative positions of the planes after 9.3 years or one-half of the precession. The maximum declination of the moon relative to Earth's equator is now 18.5°, or 23.5° less 5°. (Adapted from C. Hauge, Tides, currents, and waves, *California Geology,* July 1972.)

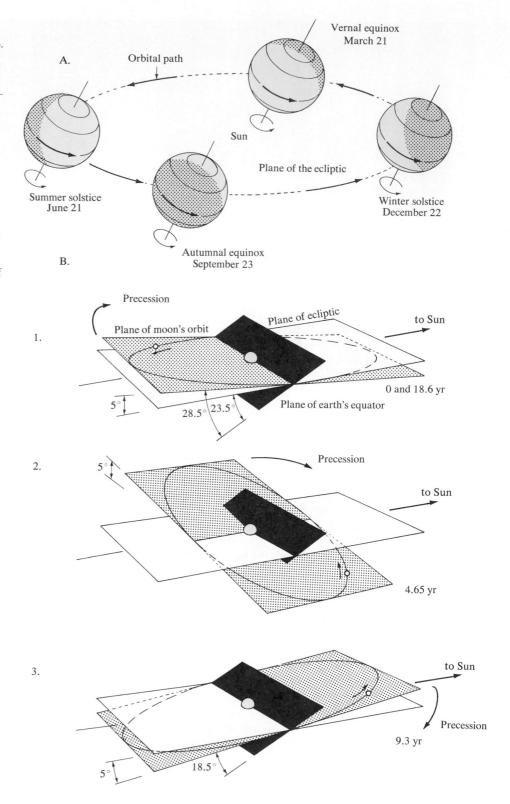

to blow across the ocean from Kamchatka to the U.S. coast. Averaging about 50 km/h (30 mi/h) the winds produced a near fully developed 3-m (10-ft) swell along the coast from Oregon to Baja California (see tables 8–1 and 8–2). The storm surge from this condition would have been trouble enough under average conditions, but the situation was made worse by the 2.25-m (7.4-ft) high spring tides.

The unusually high tides occurred because Earth was still near perihelion (January 2) when the moon

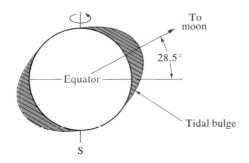

FIGURE 9–8
Maximum Declination of Tidal Bulges from Equator. The center of the tidal bulges may lie at any latitude from the equator to a maximum of 28.5° on either side of the equator.

reached perigee on January 28, 1983. Some of the largest waves came ashore January 26–28 when over $100 million in damage was done (figure 9–11). Twenty-five homes were destroyed, over 3500 homes were seriously damaged, and many commercial and municipal piers collapsed. At least a dozen lives were lost. With each such occurrence, we learn more of the cost of developing the shore.

DYNAMICAL THEORY OF TIDES

In our previous discussion of the equilibrium tide, we considered that the tidal bulges were directed at the moon and away from the moon on opposite sides of Earth. Maintaining this relationship with the moon as Earth rotates beneath the moon, the bulges (or wave crests), which are separated by a distance of half Earth's circumference (about 20,000 km or 12,420 mi), would be moving across Earth at a velocity of more than 1600 km/h (994 mi/h).

We previously stated that the tides were an extreme example of shallow-water waves whose speed is proportional to the square root of the water depth. In order

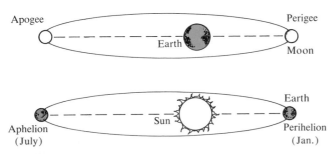

FIGURE 9–9
The Effects of Elliptical Orbits. The moon moves from perigee of 375,200 km (233,000 mi) to apogee of 405,800 km (252,000 mi) (top). Greater tidal ranges are experienced during perigean tides. Perihelion brings Earth within 148,500,000 km (92,200,000 mi) of the sun. Aphelion distance is 152,200,000 km (94,500,000 mi) (bottom). Greater tidal ranges are experienced during perihelion tides.

for the tidal wave to travel at 1600 km/h (994 mi/h), the depth of the idealized ocean would have to be 22 km (13.7 mi). Since the mean depth of the ocean is 3.9 km (2.4 mi), the tidal bulges move as forced waves whose velocity is determined by the ocean depth. Based on the mean ocean depth, the mean speed with which tidal waves can travel across the open oceans is about 700 km/h (435 mi/h). With this limitation on the speed of tidal waves, the theoretical equilibrium theory bulges that point toward and away from the tide-generating body cannot exist and break up into a number of cells.

In the open ocean, the crests and troughs of the tide wave actually rotate around an **amphidromic point** near the center of each cell. There is theoretically no tidal range at this point. Radiating from this point are **cotidal lines** that connect points along which high tide will simultaneously occur. Figure 9–12 shows the cotidal lines at 2-hour intervals for the world ocean. The cotidal lines are labeled with numbers indicating the time of high tide in hours after the moon crosses the Greenwich Meridian. They indicate that the rotation of the tide wave is counterclockwise in the Northern Hemisphere and clockwise in the Southern Hemisphere. The wave makes one complete rotation during the tidal period. The size of the cells is limited by the fact that the tide wave must make one complete rotation during the period of the tide (usually 12 lunar hours). Within an amphidromic cell, low tide is 6 hours behind high tide; for instance, if high tide is occurring along the cotidal line labeled 10, low tide is simultaneously occurring along the cotidal line labeled 4.

We must also consider the effect of the continents, which interrupt the free movement of the tidal bulges across the ideal unobstructed ocean surface we considered to exist in our discussion of the equilibrium tide. The ocean basins between continents have set up within them free-standing waves whose character modifies the forced astronomical tidal waves that develop within the basin. It is impossible for us to explain here the various tidal phenomena that occur throughout the world. For instance, high tide rarely occurs at the time the moon is at zenith, and the amount of time that elapses between the passing of the moon and the occurrence of high tides varies from place to place as a result of the many factors that determine the characteristics of the tide at any given location.

Types of Tides

Equilibrium tidal theory tells us that we should expect two high tides and two low tides of unequal heights during a lunar day. Due to modifications resulting from varying depths, sizes, and shapes of ocean basins, the tide predicted by the equilibrium theory is replaced in many parts of the world by either a **diurnal** (daily), **semidiurnal** (twice daily), or **mixed** tide (figure 9–13).

FIGURE 9–10
Predicted Equilibrium Tides. (*F*, from Anikouchine & Sternberg. *The world ocean: An introduction to oceanography.* © 1973. Reprinted by permission of Prentice Hall, Englewood Cliffs, N.J.)

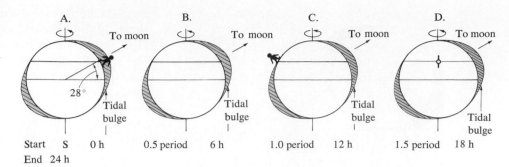

A. At start of lunar day the observer is at zenith in the center of the tidal bulge. He experiences a high tide. *B.* Six lunar hours later (0.5 period) the observer experiences low tide while located on the back side of diagram. *C.* After twelve lunar hours (1.0 period or 0.5 lunar day) the observer again experiences high tide. It is much lower than the first high tide, however, because the observer is now passing through the edge of the tidal bulge. *D.* Observer experiences low tide again eighteen lunar hours (1.5 period) after start. At the end of one lunar day (24 hours) the observer returns to *A* and experiences a high high tide.

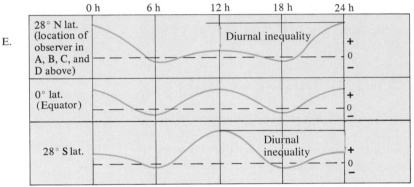

Tide curves for 28°N, 0°, and 28°S latitudes when the declination of the moon is 28° N (all curves for same longitude). Note that tide curves for 28°N and 28°S have identical highs and lows, but are out of phase by 12 hours. This results from the fact that the bulges in the two hemispheres occur on opposite sides of the earth.

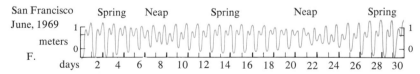

Along with the unequal heights for the two high and low tidal extremes occurring each lunar day, we expect to observe spring and neap tides. Thus the tide curve for the month of June 1969 demonstrates the general character of the predicted equilibrium tide.

The diurnal tide is characterized by a single high and low water each lunar day. These tides are common in the Gulf of Mexico and along the coast of Southeast Asia. Such tides have a tidal period of 24 h 50 min.

The semidiurnal tide has two high and two low waters each lunar day, and the heights of successive high waters and successive low waters are approximately the same. Since tides are always getting higher or lower at any location, due to the spring-neap tide sequence, successive high tides and successive low tides can never be exactly the same at that location. Semidiurnal tides are common along the Atlantic Coast of the United States. The tidal period is 12 h 25 min.

The mixed tide may have characteristics of both diurnal and semidiurnal tides. The diurnal inequality discussed earlier is a characteristic of this tide, as successive

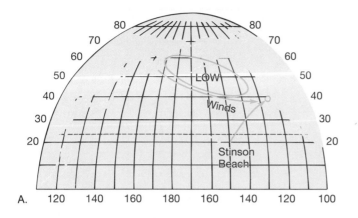

B.

FIGURE 9–11
High Tides of January 1983. *A,* January 1983 storm winds blow uninterrupted across the North Pacific Ocean from Kamchatka to the U.S. coast. *B,* homes threatened by storm waves and unusually high tides on January 27, 1983, at Stinson Beach north of San Francisco. (Wide World Photos.)

high tides and/or low tides will have significantly different heights. Mixed tides commonly have a tidal period of 12 h 25 min, which is a semidiurnal characteristic, but may also possess diurnal periods. This is the tide that is most common throughout the world and the type that is found along the Pacific Coast of the United States. Diurnal inequalities are greatest when the moon is at its maximum declination, and such tides are called **tropical tides** because the moon is over one of the tropic regions. When the moon is over the equator, the diurnal inequality is minimal, and tides with this characteristic are called **equatorial tides** (figure 9–14).

Tides in Narrow Bays

When tide waves enter coastal waters they are subject to reflection. In some cases the standing waves set up by reflections may have periods near that of the forced tidal

wave. Under such conditions, resonance (constructive interference) can produce significant increases in the tidal range.

The Bay of Fundy is such a place. With a length of 258 km (160 mi), it has a wide opening into the Atlantic Ocean. The Bay of Fundy splits into two narrow basins at its northern end, Chignecto Bay and Minas Basin (figure 9–15). The period of free oscillation in the Bay of Fundy is very nearly that of the tidal period. The resulting resonance, along with the narrowing of the bay toward the north end and the shoaling in that direction, produces maximum tidal ranges in the extreme northern end of Minas Basin. The maximum perigean tidal range of about 17 m (56 ft) occurs at the north end of the Minas Basin, and the minimum tidal range of about 2 m (6.6 ft) is found at the opening into the bay. The tidal range progressively increases from the mouth of the bay northward.

Coastal Tide Currents. The current that accompanies the slowly turning tide crest in a Northern Hemisphere basin will turn in a counterclockwise direction, producing a **rotary current** in the open portion of the basin. Because of increased effects of friction in shoaling, nearshore waters, the rotary current is changed to an alternating or **reversing current** that moves in and out rather than along the coast as would a rotary current. These reversing currents are of the greatest concern to navigators, since they are known to reach velocities of 44 km/h (27.6 mi/h) in restricted channels between islands of coastal British Columbia. Velocities of the rotating currents in the open ocean are usually well below 1 km/h (0.6 mi/h).

The reversing tidal current that develops throughout a lunar day for a mixed tide is depicted in figure 9–16. Beginning at high tide, the current velocity is zero as the water has just reached its highest stage and is momentarily to begin its outward flow. Following this high **slack water,** the lowering of the tide begins, and the ebb current velocity increases and reaches a maximum about 3 lunar hours after high slack water. The velocity decreases and eventually reaches zero again at the first low slack water. Following the change in current velocity associated with the tidal phase throughout the day, it can be seen that the maximum current velocity is reached midway through the ebb current that occurs between the higher high water and lower low water.

The illustration shows that the lower low water (LLW) is below the datum of the chart, the zero mark. How can the tide be less than zero? The datum that is commonly used for mixed tides is the mean lower low water, the average height of the lower low tides at the locality. Since the lower low tide that we are recording is below the average lower low tide for this locality, this tide will have a negative value. In areas where mixed

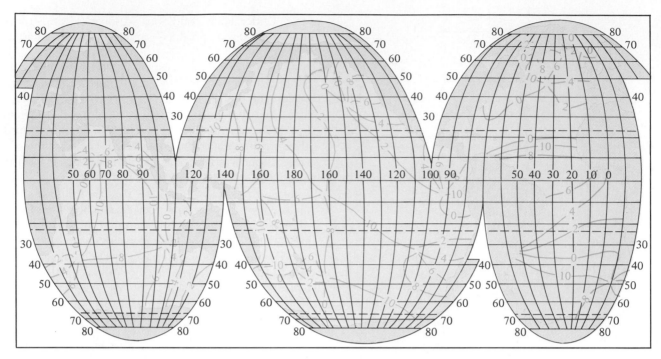

FIGURE 9–12
Cotidal Map of the World. Contour lines indicate times of M_2 high tide in lunar hours after the moon has crossed the Greenwich Meridian. Tidal ranges generally increase with increasing distance along cotidal lines away from the amphidromic points. Where cotidal lines terminate at both ends in amphidromic points, maximum tidal range will be near the midpoints of the lines. (Base map courtesy of National Ocean Survey. After von Arx, 1962; original by H. Poincaré, 1910, Leçons de Mécanique Céleste, Gauther-Crofts, Vol. 3.)

FIGURE 9–13
Types of Tides. *A*, types of tides. In a semidiurnal (twice daily) type of tide, there are two highs and lows during each lunar day, and the heights of each successive high and low are about the same. In the diurnal (daily) type of tide, there is only one high and one low each lunar day. In the mixed type of tide, both diurnal and semidiurnal effects are detectable, and the tide is characterized by a large difference in the high water heights, the low water heights, or both, during one lunar day. Even though a tide at a place can be identified as one of these types, it still may pass through stages of one or both of the other types. *B*, map showing the types that have been observed along portions of the coasts of North and South America. The numbers give the spring tide range in meters and are therefore near the maximum tidal range that can be expected. Storm waves, lower barometric pressure, ocean currents, and the concurrence of perigean and perihelion conditions could increase the range. (After C. Hauge, Tides, currents, and waves. *California Geology*, July 1972.)

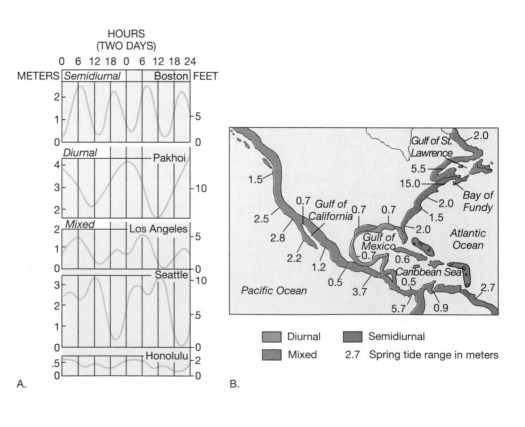

FIGURE 9–14
Tidal Curves for Three Locations.
The declinations of the moon and the sun, phase of the moon, and position of the moon in its orbit as noted across the top of the chart all contribute to the tidal variations that are depicted. Additional contributing factors include the position of Earth in its orbit around the sun and the configuration of the sea bottom and basin boundaries. Some names of water levels are listed along the right margin. Those that are used as the chart datum for a place are marked with an asterisk. MHWS, mean high water springs—the average height of the high water of the spring tides; MHW, mean high water—the average height of all the high tides at a place; MLW, mean low water—the average height of all the low tides at a place; MLWS, mean low water springs—the average height of all low waters of the spring tides; MHHW, mean higher high water—the average height of the higher high tides at a place where the tide is the mixed type and displays one inequality during a tidal day; MLLW, mean lower low water—the average height of the lower low tides at a place where the tide is of the mixed type. (After C. Hauge. Tides, currents, and waves, *California Geology,* July 1972.)

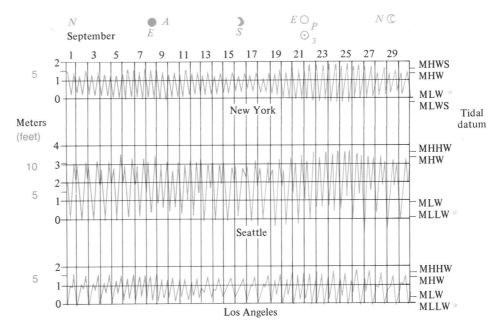

●, new moon; ☽, first quarter; ○, full moon; ☾, last quarter; *E*, moon on the equator; *N, S,* moon farthest north or south of the equator; *A, P,* moon in apogee or perigee; ₁☉₃, sun at autumnal equinox; ＊, chart datum.

tides are not observed, the tide datum is very commonly the mean low tide recorded at that place—the average low tide. Thus, most tide extremes recorded, even low tides, will have positive values. Only during spring tides are negative low tides observed (figure 9–14).

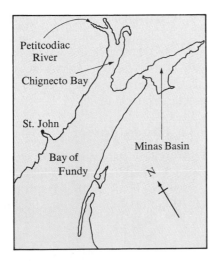

FIGURE 9–15
Bay of Fundy. The largest tidal range known in the world occurs at the north end of the Minas Basin in the Bay of Fundy. Because of its dimensions, this bay has a natural free-standing-wave period about equal to that of the forced tide wave. This combined with the fact that the bay narrows and becomes shallower toward its head causes a maximum tidal range at the northern end of 17 m (56 ft).

Tides in Rivers

The Amazon River probably possesses the longest estuarine stretch that is affected by oceanic tides. Tides can be measured as far as 800 km (497 mi) from the mouth of the Amazon, although the effects are quite small at this distance. Tidal waves that move up river mouths lose their energy due to the decreasing depth of water and the flow of the river water against the tide during the flood interval. As the wave moves up the river, it becomes more and more asymmetrical, developing a steep front (figure 9–17). This front produces a rapidly rising tide that falls slowly. An extreme development of this type produces a **tidal bore** in which a very steep wave front surges up the river. In the Amazon it is called pororoca and appears as a waterfall up to 5 m (16.4 ft) in height moving upstream at speeds up to 22 km/h (13.7 mi/h). Other rivers that experience bores are the Chientang in China, where they may reach 8 m (26.2 ft), the Petitcodiac in New Brunswick, Canada, the Seine in France, and the Trent in England.

TIDES AS A SOURCE OF POWER

The history of human efforts to harness the energy of the tides dates back at least to the middle ages. Interest in pursuing this renewable energy source waned during the time of cheap fossil fuel availability. But today, as we real-

FIGURE 9–16
Reversing Current. Note that tidal current velocity is zero at high and low tidal extremes. Maximum tidal current velocity occurs midway between tidal extremes.

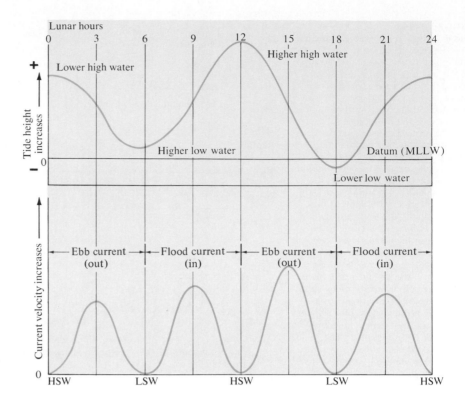

HSW—High slack water (velocity zero)   LSW—Low slack water (velocity zero)

ize that the end of cheap fuel could arrive at any time, there is increased interest in assessing the prospects of generating electricity using tidal energy.

Some Basic Considerations

The most obvious benefit of using the tides to generate electrical power would be in reduced operating costs as compared to the conventional thermal power plants that require radioactive isotopes or fossil fuels. Even though the initial cost of the tidal power-generating plant may be higher, there would be no ongoing fuel bill.

A negative consideration involves the periodicity of the tides. Power would be generated throughout only a portion of a 24-hour day unless special design features were included in the construction of the facility. Because the tides operate on a lunar period and our energy demand operates on a solar period, the energy available through generating power from the tides would only accidentally coincide with need.

Generators must run at constant speed, and we must use the flow of the tidal current in two directions. It is therefore obvious that we would benefit by using turbine blades whose angle could be varied and reversed. Adjusting the pitch of the turbine blades would set them flatter when the head of water was greatest and increase their angle as the head of water decreased, so that the generators could be maintained at a constant speed. The maximum head would usually not be great, so the flow channels constructed through dams would have to be of small diameter. This would require a series of small channels and small turbines, which would be less efficient than one large machine.

La Rance Power Plant

A successful tidal power plant was constructed in the estuary of La Rance River off the English Channel in France. The estuary, shown in figure 9–18, has a surface area of approximately 23 km² (8.9 mi²), and the tidal range at La

FIGURE 9–17
River Bores. As the tidal crest moves upriver it develops a steep forward slope through resistance to its advance by the river flowing to the ocean. Such crests (bores) may reach heights of 5 m (16.4 ft) and move at speeds up to 22 km/h (13.6 mi/h).

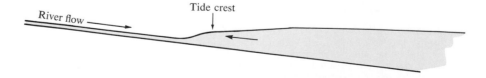

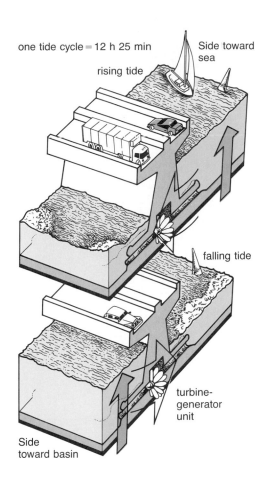

one tide cycle = 12 h 25 min

rising tide

Side toward sea

falling tide

turbine-generator unit

Side toward basin

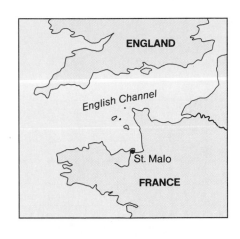

FIGURE 9–18
La Rance Tidal Power Plant at St. Malo, France. Block diagram shows the barrier between the open ocean to the right and the La Rance estuary to the left. The relative water levels during rising and falling oceanic tides are also shown. (Photo courtesy of Phototeque/ Electricite de France.)

Rance reaches a maximum of 13.4 m (44 ft). Usable tidal energy is proportional to the area of the basin and to the square of the amplitude of the tide. A barrier was built across the estuary a little over 3 km (1.9 mi) upstream, where it is 760 m (2493 ft) wide, to protect it from storm waves. The deepest water ranges from just over 12 m (39.4 ft) at low tide to more than 25 m (82 ft) at high tide. To allow water to flow through the barrier when the generating units are shut down, sluices (artificial channels) with adjustable vanes 10 m (33 ft) high and 15 m (49 ft) wide have been built into the barrier. Twenty-four generating units are located in conduits below the power plant. The bottoms of the conduits containing the turbine-generator units are 10 m (33 ft) below the surface of the

water at the lowest tide. The conduits containing the units are 53 m (174 ft) long, the cross-sectional area at each end is about 93 m² (995 ft²), and each unit can generate 10,000 kW at 3.5 kV.

The plant generates electricity only when sufficient head exists between the pool and the ocean—about one-half of the tidal period. Annual power production of about 540 million kWh without pumping can be increased to 670 million kWh by using the turbine-generators as pumps at the proper times to increase the head of water.

Within the Bay of Fundy, the province of Nova Scotia has constructed a tidal power plant that has generated up to 40 million kWh per year since completion in 1984.

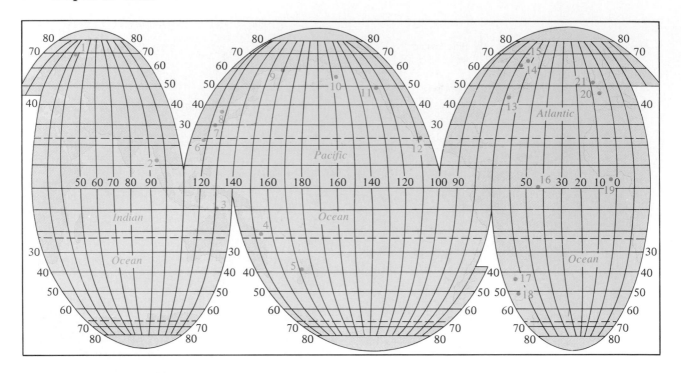

1. Mezan/Kislaya
2. Rangoon
3. Darwin
4. Broad Sound
5. Aukland
6. Amoy
7. Shanghai
8. Asan Bay
9. Sea of Okhotsk
10. Cook Inlet
11. Strait of Georgia
12. Gulf of California
13. Bay of Fundy/Passamaquoddy Bay
14. Ungava Bay
15. Frobisher Bay
16. Sao Luis
17. Golfo San Jorge
18. Straits of Magellan
19. Abidjan
20. Rance River/Chausey I.
21. Severn River

FIGURE 9–19
Sites with Major Potential for Tidal Power Generation.

It is constructed on the Annapolis River where the maximum tidal range is 8.7 m (26 ft).

Although some engineers think a tide-generating plant proposed for Passamaquoddy Bay near the United States–Canadian border at the south end of the Bay of Fundy could be made to generate electricity constantly, others who have studied the project are less optimistic. Potentially, the usable tidal energy seems great compared with La Rance because the volume of flow is about 117 times greater than that at La Rance. Whether or not tide-generating units are ever constructed on a large scale, this potential source of energy will receive increased attention as the cost of generating electricity by conventional means increases. Figure 9–19 shows the locations of some sites around the world that have potential for tidal generation of electrical power.

Any use of coastal waters for the tidal generation of electricity will have environmental costs resulting from the modification of current flow. It will interfere with many traditional uses of coastal waters such as transportation and fishing.

SUMMARY

The tides of Earth are derived from the gravitational attractions of the sun and moon. The moon has about twice the tide-generating effect of the sun. Small horizontal forces tend to push water into two bulges, one at Earth's zenith and one at the nadir relative to the tide-generating body—the sun or moon. Since the moon bulges are dominant, the tides we observe on Earth have periods dominated by lunar motions and modified by the changing position of the solar bulges.

If Earth were a uniform sphere covered with an ocean of uniform depth, and if we could ignore some of the considerations of physical motion, the tides on Earth would be those predicted by the equilibrium theory of tides. Such a tide would have a period of 12 h 25 min, or half a lunar day. A tide with maximum tidal range would occur each new moon and full moon, and tides with minimum range would occur with the first and third quarter phases of the moon. These would be the spring and neap tides, respectively. Since the moon may have a declination as much as 28.5° north or south of the equator, and the sun is directly over the equator only two times per year, the tidal bulges would usually be located so as to create two high tides of unequal height per lunar day. The same could be said for the low tides. Tidal ranges are greater when Earth is at perihelion in its orbit around the sun and when the moon is at perigee in its orbit around Earth because the tide-generating bodies are then closest to Earth.

Since Earth has an irregular surface with continents dividing the world ocean into irregularly shaped basins, the tides we actually observe on Earth are explained by the dynamic theory of tides. Open-ocean tides can be observed to rotate in a counterclockwise direction around an amphidromic point, a point of zero tidal range, in the Northern Hemisphere. The basic types of tides observed on Earth are a diurnal tide with a period of 1 lunar day, a semidiurnal tide with a period of half a lunar day (like that predicted for the equilibrium tide), and a mixed tide with characteristics of both. Mixed tides are usually dominated by semidiurnal periods and display a significant diurnal inequality. The diurnal inequalities are greatest when the moon is over the tropics and least when it is over the equator.

The effects of resonance and the shoaling and narrowing of coastal bays can be seen in the extreme tidal range at the north end of the Bay of Fundy. Tidal currents follow a rotary pattern in open-ocean basins but are converted to reversing currents at the margins of continents. The maximum velocity of reversing currents occurs during ebb and flood currents when the water is halfway between high and low standing waters. Tidal bores, tidal waves that force their way up rivers, are common in such rivers as the Amazon, Chientang, Seine, and Trent.

Since tides can be used to generate power without the requirement of fossil or nuclear fuel, the possibility of constructing such generating plants has always been attractive to engineers. One such plant is located in the estuary of La Rance River in France and is operating satisfactorily.

KEY TERMS

Amphidromic point (p. 199)
Anomalistic month (p. 195)
Aphelion (p. 195)
Apogee (p. 195)
Autumnal equinox (p. 195)
Centripetal force (p. 191)
Conjunction (p. 193)
Cotidal lines (p. 199)
Declination (p. 195)
Diurnal inequality (p. 196)
Diurnal tide (p. 199)
Dynamical tide theory (p. 192)
Ecliptic (p. 195)
Equatorial tide (p. 201)

Equilibrium tide theory (p. 192)
Full moon (p. 193)
Gravitational force (p. 191)
Law of gravitation (p. 191)
Lunar day (p. 193)
Lunar hour (p. 193)
Mixed tide (p. 199)
Neap tide (p. 195)
New moon (p. 193)
Newton, Isaac (p. 190)
Opposition (p. 193)
Perigee (p. 195)
Perihelion (p. 195)
Precession (p. 195)

Quadrature (p. 193)
Quarter moon (p. 193)
Reversing current (p. 201)
Rotary current (p. 201)
Semidiurnal tide (p. 199)
Slack water (p. 201)
Spring tide (p. 195)
Summer solstice (p. 195)
Tidal bore (p. 203)
Tidal range (p. 195)
Tide-generating force (p. 191)
Tropical tide (p. 201)
Vernal equinox (p. 195)
Winter solstice (p. 195)

QUESTIONS AND EXERCISES

1. Explain why the sun's influence on Earth's tides is only 46 percent that of the moon's, even though the sun exerts a gravitational force on Earth 177 times greater than that of the moon.
2. Discuss why the length of the lunar day is 24 h and 50 min of solar time.
3. Explain why the maximum tidal range (spring tide) occurs during new and full moon phases and the minimum tidal range (neap tide) with quadratures.
4. Discuss the length of cycle and degree of declination of the moon and sun relative to Earth's equator. Include a discussion of the effects of precession of the plane of the moon's orbit through the ecliptic.
5. Describe the effects of the declination of the moon and sun on the tides.
6. Diagram the moon's orbit around Earth and Earth's orbit about the sun. Label the positions on the orbits at which the moon and sun are closest to and farthest from Earth, stating the terms used to identify them. Discuss the effects of the moon's and Earth's positions on Earth's tides.
7. Define tropical and equatorial tides. Include the concept of diurnal inequality (see the days of September 1 and 21 in figure 9–14).
8. Describe the period and diurnal inequality of the following: diurnal tide, semidiurnal tide, and mixed tide.
9. Discuss the factors that help produce the world's greatest tidal range in the Bay of Fundy.
10. Describe the velocity of tidal currents in relationship to high and low tidal extremes.
11. Discuss at lease one positive and one negative factor related to tidal power generation.

REFERENCES

Clancy, E. P. 1969. *The tides: Pulse of the earth*. Garden City, N.Y.: Doubleday.

Defant, A. 1958. *Ebb and flow: The tides of earth, air, and water*. Ann Arbor: University of Michigan Press.

Gill, A. E. 1982. *Atmosphere and ocean dynamics. International Geophysics Series*, Vol. 30. Orlando, Florida: Academic Press.

Pond, S., and Pickard, G. L. 1978. *Introductory dynamic oceanography*. Oxford: Pergamon Press.

Sverdrup, H. U.; Johnson, M. W.; and Fleming, R. H. 1942. Renewal 1970. *The oceans: Their physics, chemistry, and biology*. Englewood Cliffs, N.J.: Prentice-Hall.

von Arx, W. S. 1962. *An introduction to physical oceanography*. Reading, Mass.: Addison-Wesley.

SUGGESTED READING

Sea Frontiers

Canove, P. 1989. The reclamation of Holland. 35:3, 154–64. A comprehensive history of how the Dutch have reclaimed and protected coastal lands from the threat of rising water.

Holloway, T. 1989. Eling tide mill. 35:2, 114–19. The tide mill at Eling Toll Bridge near Southampton, England, has been in existence since at least A.D. 1086. It is now a working museum, and the article describes its history and how it works.

Sobey, J. C. 1982. What is sea level? 28:3, 136–42. The role of tides and other factors in changing the level of the ocean surface is discussed.

Zerbe, W. B. 1973. Alexander and the bore. 19:4, 203–8. An account of Alexander the Great's encounter with a tidal bore on the Indus River.

Scientific American

Goldreich, P. 1972. Tides and the earth-moon system. 226:4, 42–57. The tide-generating force of the sun and moon on Earth and the effect of transfer of angular momentum from Earth to the moon as a result of tidal friction are discussed. Also considered are theories of lunar origin.

Greenberg, D. A. 1987. Modeling tidal power. 257:5, 128–31. The effects of using the large tidal ranges experienced in the Bay of Fundy to generate electricity are modeled on a computer.

Lynch, D. K. 1982. Tidal bores. 247:4, 146–57. Tidal bores can be spectacular walls of water rushing up rivers when the tide rises.

Chapter 10

Coastal Geology

Throughout our history, humans have been attracted to the coasts because of the moderate climates, food, and recreational opportunities they provide as well as the commercial benefits associated with nearness to the sea. The push toward the coasts has continued until 80 percent of the U.S. population lives within easy access of these regions. The coastal regions are undergoing increasing levels of stress that threaten to destroy this national resource to the degree that future generations may not have the opportunity to enjoy it. Before discussing the coastal region and its processes, I want to discuss a situation that has contributed to increasing the magnitude of this problem in hopes of stirring your interest in bringing about an improvement in the situation.

One of the saddest comments on government's ability to meet its responsibilities for protecting the health and safety of the citizenry derives from the federal government's development of the National Flood Insurance Program (NFIP) in 1968. Developed in response to Hurricane Betsy, which killed 75 people and did $1.4 million in property damage, the original goal of NFIP was

Waves break gently on Crescent Beach, Ecola State Park, Oregon. (Photo © Breck P. Kent.)

to encourage people to build beyond the flood-threatened areas in return for federally subsidized flood insurance in case some very unusual circumstance resulted in flood damage to their property. The goal was to reduce deaths and property damage from flooding. By the time the lobbyists for the banking, construction, and real estate industries had their fears alleviated by legislators, the program turned into a vast subsidy in support of overdevelopment of the U.S. coastal region.

Following passage of the bill, coastal development boomed. Compared to the ten years prior to passage of the flood insurance bill, when hurricanes killed 186 people and did $2.2 billion in damage, there were 411 deaths and $4.7 billion in property damage during the ten years following its passage. And the problem continues to get worse. Studies conducted by the General Accounting Office in 1983 into the actuarial soundness of the program showed that many premiums should be increased by 800 percent. As a result, only new properties were subject to higher rates.

In addition to the flood insurance paid out to help owners replace structures built in locations that are clearly in danger of damage by storms, the federal government subsidizes the development of these areas by paying most of the cost of infrastructure development. After the development gets into trouble, the federal government then assumes from 50 to 70 percent of the cost of erosion-control programs. It is clearly a case of the U.S. taxpayer subsidizing imprudent development decisions that benefit a small number of individuals.

Congress has passed the Coastal Barrier Rezoning Act, which excludes construction in undeveloped coastal areas from receiving federal assistance, and has encouraged states to develop setback regulations that realistically incorporate the threat of storm damage into all coastal building programs. Yet, if the National Academy of Sciences' warning of possible accelerated sea level rise in coming years is realized, the problem with existing structures could become immense in as few as 10 years.

What can be done to change the direction of such an unsound public policy? What general guidelines should we follow in developing a national program for conservative use of our coastal areas? Of course, politics is the major hurdle, but understanding the environment in question is an important first step. In the following pages we will discuss some of the features of the coast and shore, the processes that modify them, and some examples of the effects of human interference with these processes.

GENERAL DESCRIPTION OF THE COASTAL REGION

Moving from the oceans onto the continent, one encounters the **shore,** which is the zone that lies between low tide and the highest elevations on the continent that are affected by storm waves. The **coast** extends from the landward limit of the shore inland as far as features that seem to be related to marine processes can be found. The width of the coast may vary from less than 1 km (0.6 mi) to many tens of kilometers. As the waves beat against the shore they cause erosion, producing sediment that will be transported along the shore and deposited in the low-energy areas.

Erosional Shore Features

The landward limit of a shore that is dominated by erosion is commonly marked by a cliff. The **coastline,** which marks the boundary between the shore and the coast, is a line along the cliff, connecting points at which the highest effective wave action takes place.

The shore is divided into the **foreshore,** that portion exposed at low tide and submerged at high tide, and the **backshore,** which extends from the normal high tide to the coastline. The **shoreline** migrates back and forth with the tide and represents the water's edge. The **nearshore** zone is that region between the low tide shoreline and breakers. Beyond the low tide breakers is the **offshore** zone (figure 10–1.)

The **beach** is a deposit that consists of the wave-worked sediment that moves along the wave-cut bench of the shore area. It may continue from the coastline across the nearshore region to the line of breakers.

Wave Erosion. Due to refraction (bending of waves discussed in chapter 8), the wave energy is concentrated on **headlands** that jut out from the continent, while the amount of energy reaching the shore in bays is reduced. As the waves concentrate their energy on the headlands, erosion occurs and the shoreline retreats. The greatest concentration of wave energy, on a day-to-day basis, is in the foreshore region. However, during the rare periods when storm waves batter the shore, more erosion may occur across the entire shore in one day than may be achieved by average wave conditions over a period of years.

The cliff shown in figure 10–1 is referred to as a **wave-cut cliff,** which has been produced by wave action cutting away its base. The cliff develops as the upper portions collapse after being undermined by wave action. The undermining may be evident in the form of a notch at the base of the cliff that may be characterized by **sea caves.** Such caves are most commonly cut into hard sedimentary rock. Further wave action eroding the softer portions of the rock outcrops may develop the caves into openings running through the headlands, **sea arches.** Continued erosion and crumbling of arches will produce **stacks.** Such remnants rise from the relatively smooth **wave-cut bench** cut into the bedrock by wave erosion (figure 10–2). The rate of erosion by waves is determined by a number of variables:

FIGURE 10–1
Landforms and Terminology of Coastal Regions. The coastline marks the most landward evidence of direct erosion by ocean waves. It separates the shore from the coast. The shore extends from the coastline to the low-tide shoreline (water's edge). It is divided into the backshore, above the high-tide shoreline, which is covered with water only during storms, and the foreshore (intertidal or littoral zone). Never exposed but affected by waves that touch bottom is the nearshore that extends seaward to the low-tide breaker line. The greatest amount of sediment transport as beach deposit occurs within the shore and nearshore zones. Beyond the nearshore lies the offshore region, where depths are such that waves rarely affect the bottom.

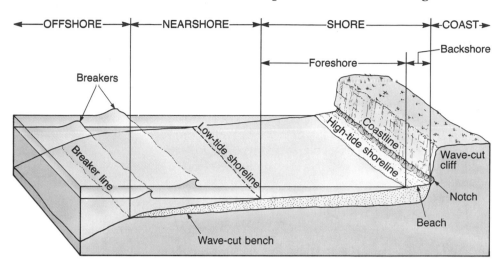

1. One of the most important variables is the *degree of exposure* of the coastal region to the open ocean. Coasts that are fully exposed receive higher-energy wave action and are likely to have rugged cliffs in areas of high topographic relief.

2. The *tidal range* is also an important variable affecting the amount of wave erosion. Given the same amount of wave energy, a region with a small tidal range will erode much more rapidly than one with a large tidal range that allows the wave energy to be spread over a

FIGURE 10–2
Sea Arch and Sea Stack along the Coast of Iceland. (Photo by Bruce F. Molnia, Terra-photographics/BPS.)

much broader shore. Although high-velocity tidal currents may develop in areas where a large tidal range exists, these currents are of limited importance in eroding the coastline.

3. The *composition of coastal bedrock* is very significant. Crystalline igneous rocks, such as granite, as well as metamorphic and hard sedimentary rocks, are relatively resistant and generally produce rugged shoreline topography. Weak sedimentary rocks, such as sandstone and shale, are more easily eroded, and a gentler topography associated with more extensive beach deposits is produced by their erosion.

Regardless of the rate of erosion, all coastal regions follow the same developmental path. As long as there is no change in the elevation of the landmass relative to the ocean surface, the cliffs will continue to retreat until the beaches widen sufficiently to prevent waves from reaching them, and the eroded material will be carried from the high-energy areas and deposited in the low-energy areas.

Depositional Shore Features

The coastal erosion we have just discussed, as well as erosion being carried on by running water inland, produces large amounts of sediment that must be distributed along the continental margin. As waves strike the shore at an angle, they set up a longshore movement of water— the **longshore current.** The velocity of the longshore current increases with increasing beach slope, angle of breakers with the beach, wave height, and decreasing wave period.

This current moves parallel to the shore between the shoreline and the breaker line, carrying with it the materials that make up the beach. At the landward margin of the surf zone the **swash,** a thin sheet of water, moves sediment onto the exposed beach at an angle, but the force of gravity causes the backwash to carry the sediment straight down the beach face. As a result, the pebbles and sand grains that are transported by the swash move in a zigzag pattern along the shore in the same direction as the longshore current within the surf zone (figure 10–3). The net annual longshore current is south along the Atlantic and Pacific shores.

Longshore drift is a term that has been applied to the movement of sediment by the processes just described. The amount of longshore drift in any coastal region is determined by an equilibrium between erosional and depositional forces. Any interference with the movement of sediment along the shore will destroy this equi-

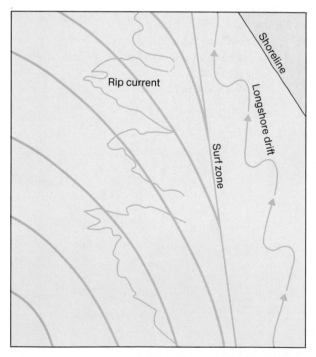

FIGURE 10–3
Longshore Currents and Rip Currents. As waves approach shore from a southerly direction, they produce a longshore current that flows north. Water in the current follows a zigzag path as waves push it up the beach slope from the direction of approach. The water runs back down the slope under the influence of gravity. Water of the longshore current finds its way offshore by passing through topographic lows as strong seaward flows. These rip currents are visible in the photograph as jets of water that appear light in color because of the turbidity resulting from sediment they have resuspended from the ocean floor. (Photo courtesy of Scripps Institute of Oceanography, University of California/San Diego.)

librium and result in a new erosional and depositional pattern determined by the nature of the interference.

Rip Currents. As longshore-current water moves onto the shore it must eventually run back into the ocean. This backwash of water finds its way into the open ocean as a thin sheet flow across the ocean bottom or in local **rip currents** that occur perpendicular to or at an angle to the coast where topographic lows or other conditions allow their formation. Rip currents may be less than 25 m (82 ft) wide and can attain velocities of 7 to 8 km/h (4 to 5 mi/h) but do not travel far from shore before they break up. If a light moderate swell is breaking, numerous rip currents moderate in size and velocity may develop. A heavy swell will usually produce fewer, more concentrated rips (figure 10–3).

Beach Composition. The material found in a beach deposit will depend on the source of the sediment that is locally available or transported by the longshore drift. In areas where the sediment is provided by coastal mountains, the beaches will be composed of minerals contained in the rocks of those mountains and may be of relatively coarse texture. If the sediment is provided primarily by rivers that drain lowland areas, sediment that reaches the coastal regions will normally be finer in texture; in many cases mud flats develop along the shore because only clay- and silt-sized particles are emptied into the ocean. In low-relief low-latitude areas such as southern Florida, where there are no mountains or other sources of rock-forming minerals nearby, most of the material of the beaches is derived from the remains of the organisms that live in the coastal waters. Beaches in these areas will be composed predominantly of shell fragments and the remains of microscopic animals, particularly foraminifers. Many volcanic island beaches in the open ocean will be composed of dark-colored fragments of the basaltic lava that makes up the islands or of coarse fragments of coral debris from the reefs that develop around the margins of islands in low latitudes.

Beach Slope. The slope of beaches is closely related to the size of the particles of which they are composed. Waves washing onto the beach carry sediment, thereby increasing the slope of the beach. If the backwash returns as much sediment as the waves carried in, the beach has reached equilibrium and will not steepen. A beach composed of fine-grained sand that is relatively angular will have a gently sloping, firm surface. Because these small grains interlock closely, little of the swash sinks down among the grains. Most of it runs back down the slope to the ocean, possessing enough energy to maintain equilibrium on a gentle slope. The backshores of such beaches are usually nearly horizontal.

Beaches composed of coarse sands or pebbles will usually contain more-rounded particles that are more loosely packed. The swash can quickly percolate into such deposits when it moves up the beach slope. Deposition of particles by the swash will continue until the beach slope becomes steep enough that the backwash running to the ocean has sufficient energy to maintain equilibrium. Such beaches are usually much less firm than beaches composed of finer material, and the backshores will slope significantly toward the coastline (table 10–1).

Special Geomorphic Features. Numerous depositional features that are partially or wholly separated from the shore are deposited by the longshore drift and other processes that are not well understood. A **spit** is a linear ridge of sediment attached at one end to land. The other end of the deposit points in the direction of longshore drift and ends in the open water. Spits are simply extensions of beaches into the deeper water near the mouth of a bay. The open-water end of the spit will normally curve into the bay as a result of current action.

If tidal currents or the currents initiated by river runoff are too weak to keep the mouth of the bay open, the spit may eventually extend across the bay and tie to the mainland to completely separate the bay from the open ocean. The spit then has become a **bay barrier.** A **tombolo** is a sand ridge that connects an island with another island or the mainland. Tombolos usually are aligned at a high angle to the main shore. This is because they are usually sand deposits laid down in the wave-energy shadow of the island; thus, their alignment is at right angles to the average direction of wave approach (figure 10–4).

Barrier islands are long offshore deposits of sand lying parallel to the coast. The origin of a barrier island is complex, and it may be that there are several explanations for its existence. It appears, however, that many such structures developed during the rise in sea level that

TABLE 10–1
The Relationship of Particle Size to Beach Slope.

Wentworth Particle Size	(mm)	Mean Slope of Beach
Cobble	256	24°
Pebble	64	17°
Granule	4	11°
Very coarse sand	2	9°
Coarse sand	1	7°
Medium sand	0.5	5°
Fine sand	0.25	3°
Very fine sand	0.125	1°
	0.063	

Source: After Table 9, "Average beach face slopes compared to sediment diameters" from *Submarine geology,* 3rd ed. by Francis P. Shepard, p. 127. Copyright © 1948, 1963, 1973 by Francis P. Shepard. Reprinted by permission of Harper & Row, Publishers, Inc.

FIGURE 10–4
Coastal Depositional Features. A *tombolo* is a deposit that connects an island to the mainland or another island. A *spit* is a deposit that extends from land into open water. A deposit that extends across a bay and closes the bay off from open water is a *bay barrier*. A *barrier island* is a long deposit separated from the mainland by a lagoon.

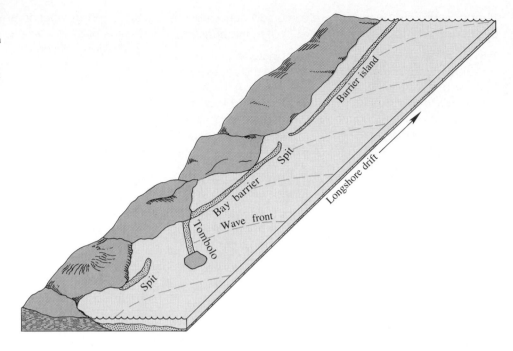

began with the last melting of the glaciers some 18,000 years ago.

Barrier islands are nearly continuous along the Atlantic Coast of the United States. They extend around Florida and along the Gulf of Mexico coast, where they may be found well past the Mexican border. Barrier islands may attain lengths in excess of 100 km (62 mi) and have widths of several kilometers. Examples of such features are Fire Island off the New York coast and Padre Island off the coast of Texas.

A typical barrier island has the following physiographic features from the ocean to the lagoon behind it: (1) ocean beach, (2) dunes, (3) barrier flat, and (4) salt marsh (figure 10–5A). The **ocean beach** is typical of the beach environment discussed earlier. During the summer, as gentle waves carry sand to the beach, it widens and becomes steeper. Higher-energy winter waves carry sand offshore and produce a narrow, gently sloping beach.

Winds blow sand inland during dry periods to produce **dunes,** which are stabilized by dune grasses that can withstand salt spray and burial by sand. Dunes are the estuary's primary protection against excessive flooding during storm-driven high tides. Numerous passes exist through the dunes, particularly along the southeast Atlantic Coast, where dunes are less well developed than to the north.

Behind the dunes the **barrier flat** forms as the result of deposition of sand driven through the passes during storms. These flats are quickly colonized by grasses. If for some reason the frequency of overwash by storms decreases, the grasses will successively be replaced by thickets, woodlands, and forests.

Salt marshes typically lie inland of the barrier flat. They are divided into the *low marsh,* extending from about mean sea level to the high neap-tide line, and the *high marsh,* extending to the highest spring-tide line. The low marsh is by far the most productive part of the salt marsh. New marsh is formed as overwash carries sediment into the lagoon, filling portions so they are intermittently exposed by the tides. Marshes may be poorly developed on parts of the island that are far from flood-tide inlets. Their development is greatly restricted behind barrier islands where artificial dune enhancement and inlet filling prevent overwashing and flooding.

Because of the gradual rise in sea level relative to the eastern North American coast, barrier islands are migrating landward, a fact clearly visible to those who build structures on these islands. Evidence for such migration can be seen in peat deposits cored beneath the barrier islands. Since the only barrier island environment in which peat forms is the salt marsh, the island must have moved inland over previous marsh development.

A possible cycle of salt marsh formation and destruction by landward migration of barrier islands is presented in figure 10–5B.

Deltas. Some rivers carry more sediment than can be distributed by the longshore current. Such rivers develop a **delta** deposit as sediment settles out at the river mouth. One of the largest such features is that produced by the Mississippi River. Deltas are fertile, flat areas that are subject to periodic flooding.

Delta formation begins when a river has filled its estuary with sediment. Once the delta forms, it grows through the distribution of sediment by **distributaries,**

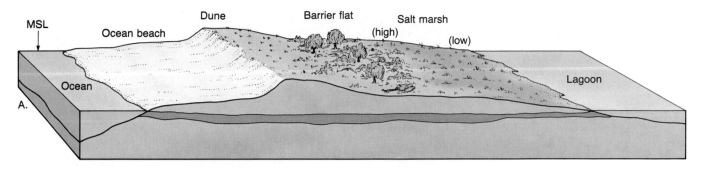

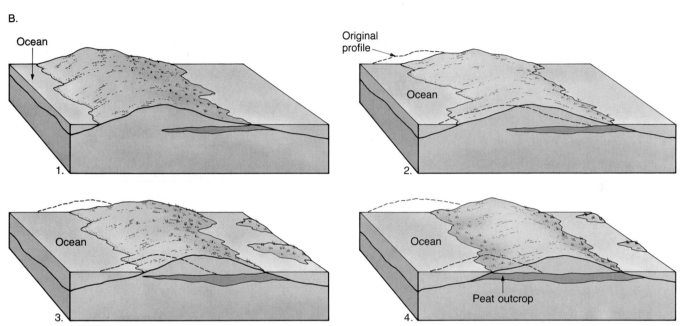

FIGURE 10–5
Barrier Islands. *A,* cross section through a barrier island. The major physiographic zones
of a barrier island are the *ocean beach, dunes, barrier flat,* and *high* and *low salt marsh.*
MSL is mean sea level. The peat bed represents ancient marsh environments that have
been covered by the island as it migrates toward the mainland with rising sea level. *B,*
(1) barrier island before overwash, with salt marsh development on mainland side; (2) overwash
erodes seaward margin of barrier island, cuts inlets, and carries sand into the lagoon, covering much
of the existing marsh; (3) new marsh forms in the lagoon that has retreated toward the mainland;
(4) with continued rise in sea level and migration of the barrier island toward the mainland, peat
formed from previous marsh deposits may be exposed by erosion of the foreshore of the ocean
beach.

branching channels that radiate out over the delta. Distributaries lengthen as they deposit sediment and produce fingerlike extensions to the delta. When the fingers get too long, they become choked with sediment. At this point, a flood may easily cause a shift in the distributary course and provide sediment to the low-lying areas between the fingers.

In areas where depositional processes are dominant over coastal erosion and transportation processes, the "bird foot" Mississippi-type delta results. Where erosion and transportation processes exert a significant influence, the shoreline of the delta will be smoothed to a gentle curve like that of the Nile Delta (figure 10–6A). The Nile Delta is presently eroding owing to the entrapment of sediment behind the Aswan High Dam, which was completed in 1964. Erosion is only one of the negative effects of the construction of this dam. This erosion, however, may have made possible the discovery of the ruins of ancient Alexandria beneath the Mediterranean waters at the edge of the delta. The problem of reduced availability of sediment may become very significant along the eastern end of the delta, where subsidence has occurred at a rate of 0.5 cm (0.2 in.)/yr for 7500 years.

FIGURE 10-6
Deltas. *A,* the relatively smooth, curved shoreline of the Nile Delta dominates this view. The Mediterranean Sea is to the left, and the Gulf of Suez is in the upper right. (Photo courtesy of NASA.) *B,* digitate structure of the Mississippi River Delta results from low-energy environment. Location of present main channel and older main channels (Atchafalaya River and Bayou Lafourche) show how river shifts position with time.

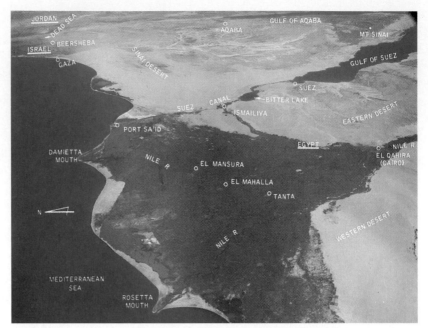

A.

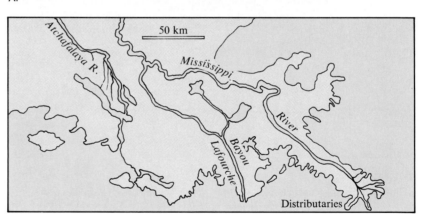

B.

CHANGING LEVELS OF THE SHORELINE

Along some coasts there are flat platforms, called **marine terraces,** backed by cliffs. **Stranded beach** deposits and other evidence of marine processes may exist many meters above the present shoreline. These features characterize **shorelines of emergence** that have reached their present positions relative to the existing shoreline by an uplift of the continent, by the lowering of sea level, or by a combination of the two.

In other areas one may find beneath the water overlying the continental shelf **drowned beaches** and **submerged dune topography.** These features along with the drowned river mouths along the present shoreline indicate **shorelines of submergence.** This submergence must have been caused by a subsidence of the continent, a rise in sea level, or a combination of the two.

Attempts to determine the causes of the change in relative level of the ocean and the continent have not met with great success. Whether the shoreline has become submerged because of a rising sea level or a subsiding continent in a particular region cannot be determined by examining the coastal features in that area, since both processes produce the same end results (figure 10–7).

Tectonic and Isostatic Movements

Changes in sea level relative to the continent may occur because of movement of the land—tectonic movement. Such movement includes large-scale uplift or subsidence of large portions of continents or ocean basins or more localized deformation of the continental crust involving folding, faulting, and tilting. Earth's crust also responds isostatically to the accumulation or removal of heavy loads of ice, sediment, or lava.

FIGURE 10–7
Evidence of Changing Levels of the Shoreline. Marine terraces resulting from ancient sea cliffs and wave-cut benches being exposed above present sea level mark ancient shorelines, as do drowned beaches that lie below sea level.

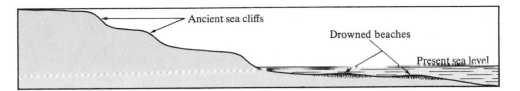

There is evidence that during the last 2.5 to 3 million years at least four major accumulations of glacial ice developed in high latitudes. Although Antarctica is still covered by a very large glacial accumulation, much of the ice cover that once existed in northern Asia, Europe, and North America has disappeared. The most recent period of melting began about 18,000 years ago. Accumulations of ice that were up to 3 km (2 mi) in thickness have disappeared from northern Canada and Scandinavia. While these areas were beneath the thick ice sheet, they were pushed down and are still in the process of recovery after the melting of the ice. Some geologists believe that by the time the isostatic rebound is finished, the floor of Hudson Bay, which is now about 150 m (492 ft) deep, will be above sea level. There is evidence in the Gulf of Bothnia, between Sweden and Finland, of 275 m (902 ft) of isostatic rebound during the last 18,000 years. Generally, tectonic and isostatic changes in the level of the shoreline are confined to a segment of the shoreline of a given continent.

Eustatic Movements

Changes in the level of the shoreline that can be measured on a worldwide basis because they are caused by the increase or decrease of water volume in the ocean or the capacity of the ocean basin are termed **eustatic.** This term refers to the highly idealized situation in which all of the continents remain static while the sea rises or falls. Small changes in sea level could be created by the formation or the destruction of large inland lakes. Probably more important is the locking into or the release from continental glaciers of Earth's water during glacial and interglacial stages.

Changes in sea-floor spreading rates can change sea level. Fast spreading produces larger rises like the East Pacific Rise that will displace more water than slow-spreading ridges like the Mid-Atlantic Ridge. Thus, fast spreading produces a rise in sea level.

During the Pleistocene Epoch, when the previously mentioned glacial advances were occurring, the amount of water in the ocean basin fluctuated considerably. Since the climate was colder during ice advances, we might account for some of the lowering of the shoreline by the contraction of the ocean volume as its temperature decreased. It has been calculated that for every 1°C (1.8°F) decrease in the mean temperature of the ocean water, sea level would drop 2 m (6.6 ft). Temperature indications derived from study of fossils from Pleistocene ocean sediments suggest the ocean surface temperature may have been as much as 5°C (9°F) lower than at present. Therefore, contraction of the ocean water may have lowered sea level by about 10 m (33 ft).

Although it is difficult to say with certainty what the range of shoreline fluctuation was during the Pleistocene, there is cause to believe that the shoreline was at least 120 m (394 ft) below the present shoreline. It is also estimated that if all the remaining glacial ice on Earth were to melt, sea level would rise by another 60 m (197 ft). This would give a minimum possible range of sea level during the Pleistocene of 180 m (590 ft), most of which must be explained through the capture and release of Earth's water by glaciers. Such changes in sea level are termed glacioeustatic oscillations.

During the last 18,000 years the ocean volume has been increasing owing to the expansion of the water resulting from warmer temperature and the melting of polar sea ice and glacial ice. Since the combination of tectonic and eustatic changes in sea level may be very complex and difficult to identify, it is hard to classify coastal regions as purely emergent or submergent. Most coastal areas show evidence of having experienced both submergence and emergence in the recent past. It is believed, however, that sea level has not risen significantly as a result of melting glacial ice during the last 3000 years.

There is at present much discussion resulting from the combined gleanings from historical research. There is clear evidence that atmospheric carbon dioxide has increased by 15 percent since 1958, when monitoring began. There has also been an apparent global warming of about 0.5°C (0.9°F) and eustatic rise in sea level of 10 cm (4 in.) since 1880 (figure 10–8). Are all of these observations tied together and the result of an increased greenhouse effect? Regardless of the answer to this question, which is at present not known, we have come to realize that we cannot dominate nature and must attempt to live within it. To do otherwise is futile. Some of the consequences of a rising sea level for U.S. coastal communities are discussed in the following paragraphs.

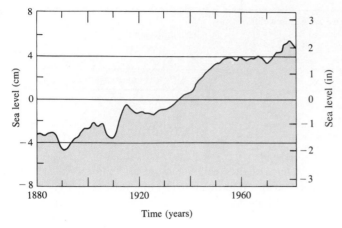

FIGURE 10–8
Sea-level Change from 1880 to 1980. Tide-gauge data averaged over 5-year periods show that global mean sea level has increased about 10 cm (4 in.) over the last 100 years. More recent data that remove the effect of isostatic recovery, however, indicate that for the last 50 years global sea level has risen 2.4 ± 0.90 mm/year. That rate is 2.4 times the rate shown in this figure. (After V. Gornits, S. Lebedeff, and J. Hausen. 1982. *Science* 215: 1611–14.)

PLATE TECTONICS AND COASTS

The most dramatic examples of sea level change during the last 3000 years have been due to tectonic processes. To understand the underlying tectonic cause of emergence or submergence, we must return to the concept of global plate tectonics.

Atlantic-type (Passive) Margins

Considering only the coasts of North America, we find the Atlantic Coast is subsiding and the Pacific Coast is emerging. Beginning with the breakup of Pangaea and the presently forming Atlantic Ocean, this emergence-submergence pattern can be readily understood. It was previously discussed that the lithosphere thickened, the ocean deepened, and heat flow decreased as the lithospheric plates cooled with age or with increasing distance from the spreading centers. As a result of this process, the ocean floor of this passive margin, the Atlantic-type margin, is thought to have subsided about 3 km (2 mi) over the last 150 million years.

Erosion of the continent produced sediment that has accumulated to a maximum thickness of 15 km (9.3 mi) along the Atlantic Coast. This thick deposit of sediment has been made possible by the plastic asthenosphere allowing the lithosphere to flex downward as the sediment burden is increased. It is this thick sedimentary wedge underlying the continental shelf, slope, and rise that has been exploited for its large petroleum reserves in the Gulf of Mexico.

These two processes account for the subsidence of the Atlantic and Gulf coasts. Although the rate of thermal subsidence decreases with time and the rate of subsidence due to sediment loading depends on sediment supply, it is not reversed unless the tectonic or isostatic conditions change.

Pacific-type (Active) Margins

In contrast to the passive and subsiding Atlantic-type margin, the Pacific-type margin, or active margin, is the scene of intense tectonic activity. Along such margins, the eruption of volcanoes and shock of earthquakes are common. The thick, broad sediment wedge characteristic of the Atlantic Coast is not well developed. Instead of being the product of the creation of a new ocean basin, the Pacific-type margin is the scene of lithospheric plate destruction.

The characteristic alignment of mountain ranges parallel to the continental margin can be seen. In all cases of continental margins associated with plate convergence, the compressive forces produce uplift. Along the California coast, the process has been further complicated with the development of the San Andreas Fault, a transform fault. This fault has made possible the local subsidence of the depositional basins in the Los Angeles area. But on the broader scale, emergence along Pacific-type margins is the characteristic condition.

UNITED STATES COASTAL CONDITIONS

Whether the dominant process along a coast is accretion or erosion depends on the combined effect of the variables we have been discussing—degree of exposure, tidal range, composition of coastal bedrock, tectonic or isostatic subsidence or emergence, and eustatic sea level change.

In its 1971 publication, "A Report on the National Shoreline Study," the U.S. Army Corps of Engineers found that of the nation's 135,870 km (84,240 mi) of coastline, 33,011 km (20,500 mi), or over 24 percent, was "seriously eroding." Subsequent studies supported by the U.S. Geological Survey along the shores of the contiguous coastal states and Alaska produced the rates of shoreline change presented in tables 10–2 and 10–3 and figure 10–9. These tables reflect conditions as they existed between 1979 and 1983. Table 10–2 shows the mean rate of change for each coastal state.

The Atlantic Coast

Most of the Atlantic coastline is exposed to storm waves from the open ocean, but barrier island development from Massachusetts south provides protection for the mainland from large storm waves. Tidal ranges generally increase from less than 1 m (3.3 ft) along the Florida coast to more than 2 m (6.5 ft) in Maine. Bedrock for most of Florida is a resistant limestone. From Florida

TABLE 10–2
Rates of Accretion (+) and Erosion (−) along Oceanic Coasts and in Bays of the United States (Contiguous and Alaska).

Region	Change (m/yr)
Atlantic Coast	−0.8
Maine	−0.4
New Hampshire	−0.5
Massachusetts	−0.9
Rhode Island	−0.5
New York	+0.1
New Jersey	−1.0
Delaware	+0.1
Maryland	−1.5
Virginia	−4.2
North Carolina	−0.6
South Carolina	−2.0
Georgia	+0.7
Florida	−0.1
Gulf of Mexico	−1.8
Florida	−0.4
Alabama	−1.1
Mississippi	−0.6
Louisiana	−4.2
Texas	−1.2
Pacific Coast	−0.0
California	−0.1
Oregon	−0.1
Washington	+0.5
Alaska	−2.4
Delaware Bay	−1.6
Chesapeake Bay	−0.7

TABLE 10–3
Rates of Accretion (+) and Erosion (−) for Coastal Landform Types.

Region	Change (m/yr)
Mud Flats	−2.0
Florida	−0.1
Louisiana–Texas	−2.1
Gulf of Mexico	−1.9
Sand Beaches	−0.8
Maine–Massachusetts	−0.7
Massachusetts–New Jersey	−1.3
Atlantic Coast	−1.0
Gulf Coast	−0.4
Pacific Coast	−0.3
Barrier Islands	−0.8
Louisiana–Texas	−0.8
Florida–Louisiana	−0.5
Gulf of Mexico	−0.6
Maine–New York	+0.3
New York–North Carolina	−1.5
North Carolina–Florida	−0.4
Atlantic Coast	−0.8
Rock Shorelines	+0.8
Atlantic Coast	+1.0
Pacific Coast	−0.5

north through New Jersey, the bedrock is composed primarily of poorly consolidated sedimentary rocks formed in the recent geologic past. These rocks do not provide great resistance to erosion and are sources of sand that is deposited as barrier islands and other depositional features common along the coast. From New York north there is considerable evidence of continental glaciers having affected the coastal region directly. Many coastal features, including Long Island and Cape Cod, are the products of glacial deposition. They are moraines that were deposited when the glaciers melted.

North of Cape Hatteras the coast is subject to very high energy conditions during fall and winter when the "nor'easters" blow out of the North Atlantic. The great energy of these storms is manifested in up to 6-m (20-ft) waves with a 1-m (3.3-ft) rise in sea level that follow the low pressure as it moves northward. Such high-energy conditions will significantly change coastlines that are predominantly depositional and may be expected to cause considerable erosion.

Most of the Atlantic coast appears to be experiencing a sea level rise of about 0.3 m (1 ft) per century.

This condition may be reversed in northern Maine because of isostatic rebound since removal of the continental ice sheet.

The Atlantic Coast has an average annual rate of erosion of 0.8 m (2.6 ft). Virginia leads the way with a 4.2-m (13.7-ft) loss per year, which is largely confined to barrier islands. Erosion rates for Chesapeake Bay are about average for the Atlantic coast, but Delaware Bay shows average erosion rates about twice (1.6 m/yr) the Atlantic coast average (figure 10–10). Seventy-nine percent of the observations made along the Atlantic coast showed some degree of erosion. In spite of the fact there are serious erosion problems in these states as well, Delaware, Georgia, and New York display accreting coasts.

The Gulf Coast

The Louisiana-Texas coastline is dominated by the Mississippi River Delta, which is being deposited in a microtidal and generally low-energy environment. The tidal range is normally less than 1 m (3.3 ft) and, with the exception of the hurricane season, wave energy is generally low. Tectonic subsidence is general throughout the Gulf Coast, and the average rate of sea level rise is similar to that of the southeast Atlantic coast, about 0.3 m (1 ft) per century.

The average rate of erosion is 1.8 m (6 ft) per year in the Gulf Coast. The Mississippi River Delta is experiencing the highest rate of erosion, which results in the

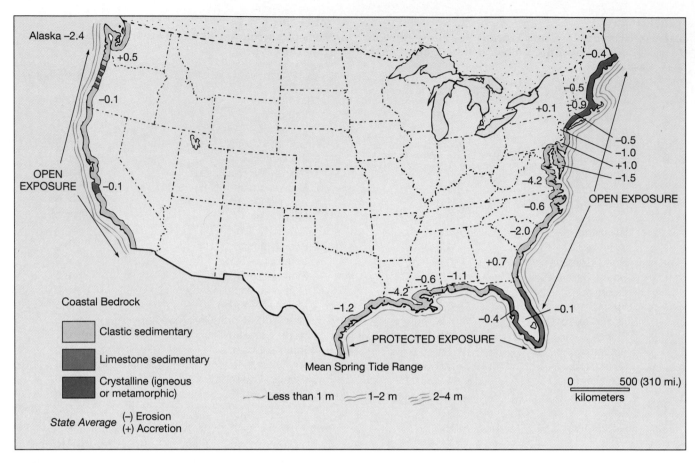

FIGURE 10–9
Distribution of Conditions Related to Coastal Erosion and Rates of Erosion vs. Accretion.

state of Louisiana losing an average of 4.2 m (13.7 ft) per year. Aided by dredging of barge channels through marshlands, Louisiana has lost more than one million acres of delta since 1900 and is losing it now at a rate in excess of 130 km² (50 mi²) per year. Although all Gulf states show a net loss of land, and the Gulf Coast has a higher rate of erosion than the Atlantic Coast, only 63 percent is receding because of erosion. The high average rate of erosion reflects the heavy erosion loss in the Mississippi River Delta area.

The Pacific Coast

Along the Pacific Coast, there is open exposure to large storm waves, with a tidal range mostly between 1 and 2 m. The bedrock varies from the predominant relatively weak marine deposits, which are geologically young, to local outcrops of granite, metamorphic, and volcanic rocks that are more resistant to erosion. Tectonically, the coast is rising (figure 10–11), but, with the exception of a segment along the coast of Oregon and the Alaskan coast, sea level still shows at least small rates of rise. High-energy waves may strike the coast in winter, with 1-m

waves being normal. Frequently, the wave height will increase to 2 m, and a few times per year 6-m (20-ft) waves hammer the shore.

During the winter, many beaches have the sand eroded from the foreshore area by high-energy storm waves. The exposed beaches, which are composed primarily of pebbles and boulders during the winter months, regain their sand as low-energy summer waves beat more gently against the shore. Damming of many of the rivers along the Pacific Coast for flood control and hydroelectric power generation has reduced the amount of sediment for the longshore transport. As a result, some areas experience severe threats of erosion.

With an average erosion rate of only 0.005 m (0.016 ft)/yr, and only 30 percent showing erosion loss, the Pacific Coast appears to be under much less of an erosion threat than the Atlantic and Gulf coasts. Yet, all along the coast there are areas that are experiencing high rates of erosion. Only Washington shows accretion, while Alaska is losing an average of 2.4 m (7.9 ft)/yr. The long, protected Washington shoreline within Puget Sound helps skew the Pacific Coast figures (figure 10–12). Although the average rate of erosion for California is only

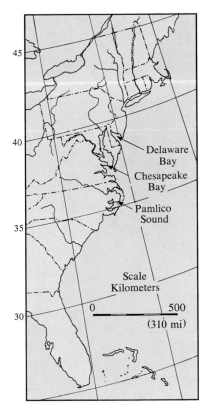

FIGURE 10–10
Locations of Major Bays along the Atlantic Coast.

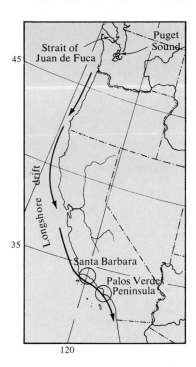

FIGURE 10–12
Features of Special Interest along the Pacific Coast.

0.1 m (0.33 ft)/yr, over 80 percent of the coast is experiencing erosion with local rates of up to 0.6 m (2 ft)/yr.

Erosion Rates and Landform Types

The variability we see in erosion rates along coasts is closely correlated to coastal landform types (table 10–3). Deltas and mud flats composed of fine-grained sediment erode most rapidly, with mean erosion rates of 2 m

(6.6 ft)/yr. This is not surprising, and it accounts for the fact the Gulf Coast is eroding at a greater rate than other coasts. Coarser deposits of sandy beaches and barrier islands erode more slowly (0.8 m/yr). The rock shore of Maine, which is experiencing isostatic rebound, shows great resistance to erosion, with an accretion rate of 1 m/yr. On the Pacific Coast, the rocks are not so durable and erode at an average rate of 0.5 m/yr.

Effects of Artificial Barriers

Jetties that have been constructed to protect harbor entrances from wave action are the most common barriers

FIGURE 10–11
Marine (Wave-cut) Terraces on San Clemente Island South of Los Angeles, California. Once at sea level, the highest terraces are now about 400 m (1320 ft) above it. (Photo by John S. Shelton.)

FIGURE 10–13
Santa Barbara Harbor. Construction of a breakwater to the west of Santa Barbara Harbor interfered with the eastward-moving longshore drift, creating a broad beach. Sand being deposited against the breakwater was no longer available to replace sand being removed by wave erosion to the east. As the beach extended around the break-water into the harbor, dredging operations were initiated to keep the harbor open and to put the sand back into the longshore drift. This helped reduce the coastal erosion east of the harbor. (U.S. Coast and Geodetic Survey Chart 5161.)

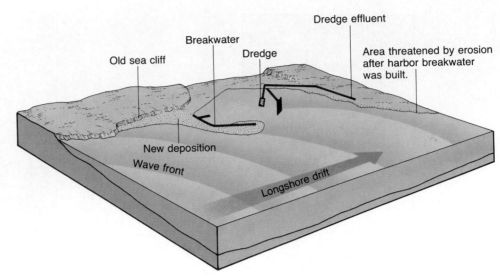

to the longshore transport of sediment. Many examples of the destruction resulting from the construction of such features along the shore can be cited. Along the southern California coast, excessive beach erosion related to the development of local harbor facilities is especially common. The longshore drift in California is predominantly from north to south, and the construction of a jetty results in the entrapment of sediment on the north side, causing increased erosion on the south side. A jetty constructed as a **breakwater** to protect harbor waters at Santa Barbara had a very significant effect upon the equilibrium established in this coastal region. The barrier created by

the breakwater on the west side of the harbor caused an accumulation of sand that was moving in an easterly direction along this portion of the California coast, which trends generally east-west. The beach to the west of the harbor continued to grow until finally the sand moved around the breakwater and began to fill in the harbor (figure 10–13).

While deposition was occurring at an abnormal rate to the west, erosion was proceeding at an alarming rate east of the harbor. The energy available east of the harbor was no greater than it had been previously, but the sand that had formerly moved down the coast to replace sand

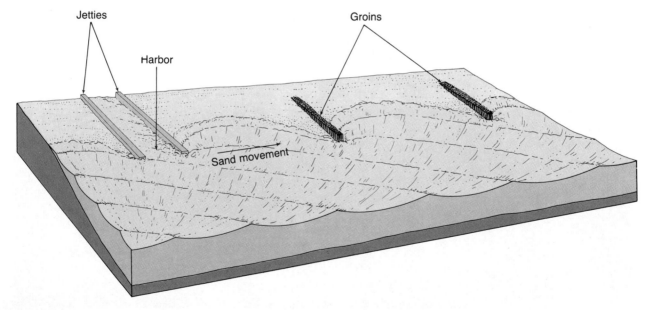

FIGURE 10–14
Jetties and Groins. Jetties and groins trap sand that would otherwise be moved along the shore by wave action. (From Tarbuck, E. J., and Lutgens, F. K., *Earth science.* 5th ed. (Columbus: Merrill, 1988). Reprinted by permission.)

removed was no longer available owing to its entrapment behind the breakwater. To compensate for this deficiency of sand downcurrent from the harbor, dredging had to be initiated within the harbor. The dredging keeps the harbor from filling in, and the sand is pumped down the coast so it can re-enter the longshore drift and replenish the eroded beach. The dredging operation has stabilized the situation, but at a considerable expense. It seems obvious that any time human beings interfere with natural processes in the coastal region, they will have to provide the energy needed to replace that which they have misdirected through modification of the shore environment.

Along many coasts, small jettylike structures called groins have been constructed at intervals along the beach (figure 10–14). They have helped beach development by causing sand to accumulate on the upcurrent sides of these structures. Being shorter than jetties built to protect harbors, the groins eventually allow the sand to migrate around their ends. An equilibrium may be reached that allows sufficient sand transport along the coast before excessive erosion occurs downcurrent from the last groin. Although some serious erosional problems in these regions have developed, construction of a series of groins usually results in much less beach erosion than the construction of one large jetty.

One of the most destructive of structures built to "stabilize" eroding shores is the **seawall.** Instead of extending out into the longshore current as do groins and jetties, they are built parallel to the shore to protect developments landward of them from the action of ocean waves. Once waves begin breaking against the seawalls, turbulence generated by the abrupt release of wave energy quickly erodes the sediment on their seaward sides, causing them to collapse into the surf (figure 10–15). Where they have been used to protect property on barrier islands, the seaward slope of the island beach is steepened and the rate of erosion increased.

Although most state laws say beaches belong to the public, government actions allow the owners of shore property to destroy beaches in an effort to provide short-term protection to their property. To add to the public insult, taxpayers subsidize property owners in many shore developments by helping to rebuild their storm-damaged properties at great public expense through the programs discussed earlier in the chapter.

An interesting possible alternative to the solid structures used to protect local shore areas from ocean waves is the **tethered-float breakwater** developed at Scripps Institution of Oceanography (figure 10–16). Tested successfully in San Diego Harbor, these inverted pendulums protect against coastal wave erosion by extracting energy from passing waves without stopping the transport of sediment. If they prove durable enough, they would certainly be a welcome substitute for traditional breakwaters.

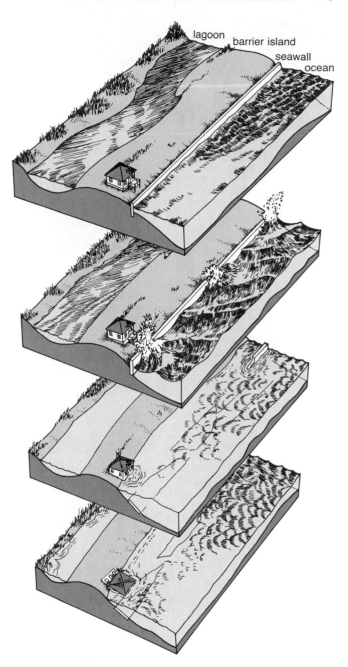

FIGURE 10–15
Seawalls. When a seawall is built on a beach to protect beachfront property, the first large storm will remove the beach from the seaward side of the wall and steepen its seaward slope. Eventually, the wall is undermined and falls into the sea, and the property is lost as the oversteepened beach slope advances landward in an effort to reestablish its natural slope angle.

PLASTICS: A MAN-MADE BEACH MATERIAL

During the last thirty years, the use of plastics has increased at a tremendous rate. The problem of disposing of this "throwaway" component of the western culture has already strained the capacity of our land-based dis-

FIGURE 10–16
Tethered-float Breakwaters.
A, deployment of tethered-float break-water at San Diego Harbor. (Photograph courtesy of Scripps Institute of Oceanography, University of California, San Diego.) *B*, artist's sketch of a tethered breakwater system incorporating 1.5-m (4.8-ft) spheres of steel placed 1.5 m apart. The spheres are anchored to the ocean floor and held just beneath the ocean surface. This system has the advantage over traditional breakwater structures in that it removes energy from waves but does not interfere with the transport of sediment along the shore.

A.

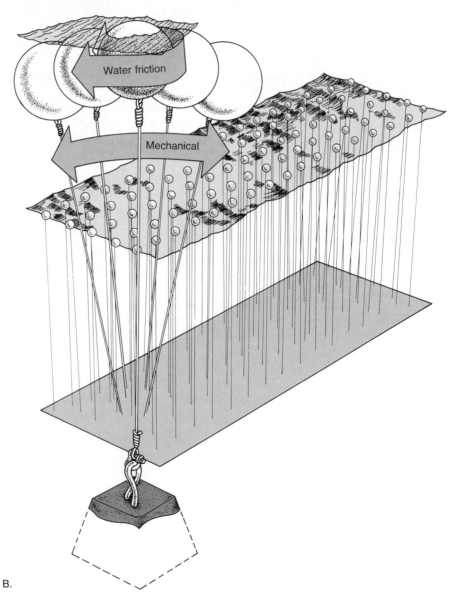

Water friction

Mechanical

B.

posal systems. Such waste is also an increasingly abundant component of oceanic flotsam.

Small pellets used in the production of essentially all plastic products are transported in bulk aboard commercial vessels. Found throughout the oceans, the pellets probably find their way into the oceans as a result of spillage at loading terminals. In coastal waters, plastic products used in fishing and thrown overboard by recreational and commercial vessels are common. They also find their way into the open ocean waters from careless dumping by commercial vessels.

The best documentation of negative effects of plastics in the ocean on marine organisms is found in the strangulation of seals and birds caught in plastic netting and packing straps (figure 10–17). Marine turtles are known to mistake plastic bags for jellyfish or other transparent plankton on which they typically feed.

It is believed that all plastics that enter the ocean may eventually be removed by shorelines of islands and continents that "filter" out the particles as ocean currents wash against their shores. Beaches throughout the world are probably increasing their plastic pellet content as a result of the filtering process. Some Bermuda beaches have up to 10,000 pellets/m². The beaches of Menemsha Harbor on Martha's Vineyard yielded 16,000 plastic spherules/m². Thin plastic film products will eventually be broken into smaller particles by photochemical degradation, and others may sink as they become denser because of photochemical degradation and encrustation by epifauna such as bryozoans and hydroids. These data are from a survey conducted between 1984 and 1987; the survey also found more than 10,000 plastic pieces and 1500 pellets/km² in the northern Sargasso Sea between

FIGURE 10–17
Elephant Seal with Plastic Packing Strap Tight Around Its Neck. This animal was found on a beach of San Clemente Island off the southern California coast. It was saved from strangulation by the removal of the plastic packing strap by scientists from the National Marine Fisheries Service. (Photo by Wayne Perryman, NMFS.)

28° and 40° north latitude. Compared to a survey conducted in 1972, the 1987 survey found that the concentration of plastic pellets had doubled.

The U.S. Congress is considering a bill to ban disposal of plastics in the 200-mile-wide Exclusive Economic Zone. Internationally, the Convention for the Prevention of Pollution from Ships prevents disposal of all plastics into the oceans. The acceptance of both of these regulations would aid greatly in turning around the increasing concentration of plastics in the world's oceans.

SUMMARY

Since passing the National Flood Insurance Program (NFIP) in 1968, the federal government has encouraged development of stretches of our shores subject to damage by storm waves. The direction of this and complementary programs must be reversed if the situation is be improved. One helpful piece of legislation in this direction was the Coastal Barrier Rezoning Act, which excludes construction in undeveloped coastal areas from receiving federal assistance.

The region of contact between the oceans and the continents is marked by the shore, lying between the lowest low tides and the highest elevation on the continents affected by storm waves. The coast extends inland from the shore as far as land features related to marine processes can be found. The shore is divided into the foreshore, extending from low tide to high tide, and the backshore, extending beyond the high-tide line to the coastline that separates the shore from the coast. Seaward of the low-tide are the nearshore zone, extending to the breaker line, and the offshore zone beyond.

Wave erosion of the shore produces a wave-cut cliff that constantly retreats, leaving behind such features as wave-cut benches, sea caves, sea arches, and stacks. The rate of wave erosion increases with increased exposure of the shore to the open ocean, decreasing tidal range, and decreasing strength of bedrock.

As waves break at an angle to the shore, a longshore current is set up that produces a longshore drift of sediment along the shore. Deposition of sediment transported by the longshore current may produce such features as beaches; spits, deposits attached at one end to land; tombolos, sand ridges connecting an island to another island or the mainland; and barrier islands, long offshore deposits lying parallel to the shore. From ocean side to lagoon side, barrier islands commonly exhibit the following divisions: ocean beach, dunes, barrier flat, and salt marsh. Deltas form at the mouths of rivers that carry more sediment to the ocean than can be distributed by the longshore current.

A drop in sea level may be indicated by ancient wave-cut cliffs and stranded beaches well above the present shoreline, and a rise in sea level may be indicated by submerged relict beaches and wave-cut cliffs, as well as drowned river valleys. Such changes in sea level may result from tectonic processes causing local movement of the land mass or from eustatic processes changing the amount of water in the oceans or the capacity of ocean basins, thus causing world-wide changes in sea level. Melting of continental ice caps during the last 18,000 years has caused a glacioeustatic rise in sea level of about 120 m (394 ft).

Along the Atlantic Coast, tidal range varies from less than 1 m (3.3 ft) along the Florida coast to over 2 m (6.5 ft) along the coast of Maine. Sea level is rising about 0.3 m (1 ft) per century along most of the coast except for Maine, where isostatic rebound keeps pace with the rising water level. The average annual rate of erosion is 0.8 m (2.6 ft), although accretion is dominant along the coasts of Delaware, Georgia, and New York. Along the Gulf Coast, wave energy levels are generally low and the tidal range is less than 1 m. Sea level is rising at a rate of 0.3 m (1 ft) per century, and the average rate of erosion is 1.8 m (6 ft) per year. The Mississippi River Delta is eroding at a rate of 4.2 m (13.7 ft) per year, resulting in an annual loss of wetlands set at 130 km^2 (50 mi^2). Along the Pacific Coast high energy winter waves and a 1 to 2 m (3.3 to 6.6 ft) tidal range are found. The average rate of erosion is only 0.005 m (0.016 ft) per year; because of the protected conditions of Puget Sound, Washington shows a net accretion. There are, however, many areas where the problem of erosion is significant. Based on coastal landform types, mud flats erode more rapidly than sand beaches, barrier islands, or rocky shores.

Structures constructed along the shore, such as jetties to protect harbors and groins used to widen beaches, will trap sediment on the upcurrent side, but erosion may then become a problem downcurrent. A possibly less damaging replacement for fixed breakwater structures is the tethered-float breakwater.

The amount of plastic accumulating in the oceans has increased dramatically. Certain forms of plastic are known to be lethal to marine mammals and turtles, and national and international legislation is being considered to ban the disposal of plastic in the oceans.

KEY TERMS

Backshore (p. 210)
Barrier flat (p. 214)
Barrier island (p. 213)
Bay barrier (p. 213)
Beach (p. 210)
Breakwater (p. 222)
Coast (p. 210)
Coastline (p. 210)
Delta (p. 214)
Distributary (p. 214)
Drowned beach (p. 216)
Dunes (p. 214)
Eustatic (p. 216)

Foreshore (p. 210)
Headland (p. 210)
Longshore current (p. 212)
Longshore drift (p. 212)
Marine terrace (p. 216)
Nearshore (p. 210)
Ocean beach (p. 214)
Offshore (p. 210)
Rip current (p. 213)
Sea arch (p. 210)
Sea cave (p. 210)
Seawall (p. 223)
Shore (p. 210)

Shoreline (p. 210)
Shoreline of emergence (p. 216)
Shoreline of submergence (p. 216)
Spit (p. 213)
Stack (p. 210)
Stranded beach (p. 216)
Submerged dune topography (p. 216)
Swash (p. 212)
Tethered-float breakwater (p. 223)
Tombolo (p. 213)
Wave-cut bench (p. 210)
Wave-cut cliff (p. 210)

QUESTIONS AND EXERCISES

1. Why has the National Flood Insurance Program had a negative impact on the nation's shores?
2. To help you reinforce your knowledge of the shore, construct and label your own diagram similar to that in figure 10–1.
3. Discuss the formation of such erosional features as sea cliffs, sea caves, sea arches, and stacks.
4. List and discuss three factors in the rate of wave erosion.
5. What variables affect the velocity of the longshore current?
6. What is the longshore drift, and how is it related to the longshore current?
7. Discuss the composition of beaches and the relationship of beach slope to particle size.
8. List and define the depositional features spit, tomobolo, bay barrier, and barrier island. Include discussion of how some barrier islands develop peat deposits running through them from ocean shore to marsh.
9. Discuss why some rivers have deltas and others do not. Also include the factors that determine whether a "bird foot" or a smoothly curved Nile-type delta will form.
10. Compare the causes and effects of tectonic vs. eustatic changes in sea level.
11. List the two basic processes by which coasts advance seaward, and list their counterparts that lead to coastal retreat.
12. Describe the tectonic and depositional processes causing subsidence along Atlantic-type margins.
13. How does the Pacific-type margin differ from the Atlantic-type margin?

14. Discuss the Atlantic Coast, Gulf Coast, and Pacific Coast by describing the conditions and features of emergence-submergence and erosion-deposition that are characteristic of each.

15. Describe the effect on erosion and deposition caused by putting a structure such as a breakwater or jetty across the longshore current and drift.

REFERENCES

Bird, E. C. F. 1985. *Coastline changes: A global review.* Chichester, U.K.: John Wiley & Sons.

Burk, K. 1979. The edges of the ocean: An introduction. *Oceanus* 22–3:2–9.

Coates, R., ed. 1973. *Coastal geomorphology. Publications in Geomorphology.* Binghamton, N.Y.: State University of New York.

Kuhn, G. G., and Shepard, F. P. 1984. *Sea cliffs, beaches, and coastal valleys of San Diego County: Some amazing histories and some horrifying implications.* Berkeley: University of California Press.

Leatherman, S. P. 1983. Barrier dynamics and landward migration with Holocene sea-level rise. *Nature* 301:5899, 415–17.

May, S. K.; Kimball, H.; Grandy, N.; and Dolan, R. 1982. The Coastal Erosion Information System. *Shore Beach* 50, 19–26.

Peltier, W. R., and Tushingham, A. M. 1989. Global sea level rise and the greenhouse effect: Might they be connected? *Science* 244:4906, 806–10.

Shepard, F. P. 1973. *Submarine geology,* 3rd ed. New York: Harper & Row.

———. 1977. *Geological oceanography.* New York: Crane, Russak & Company.

SUGGESTED READING

Sea Frontiers

Carr, A. P. 1974. The ever-changing sea level. 20:2, 77–83. A discussion of the causes of sea-level change is very well presented.

Emiliani, C. 1976. The great flood. 22:5, 256–70. An interesting discussion of the possible relationship of the rise in sea level resulting from the melting of glaciers 11,000–8,000 years ago and biblical and other ancient accounts of a great flood.

Feazel, C. 1987. The rise and fall of Neptune's kingdom. 33:2, 4–11. A discussion of factors that change the level of the sea, including atmospheric conditions, currents, climate, ocean topography, and sea-floor spreading.

Fulton, K. 1981. Coastal retreat. 27:2, 82–88. The problems of coastal erosion along the southern California coast are considered.

Grasso, A. 1974. Capitola Beach. 20:3, 146–51. The destruction of the beach of Capitola, California, shortly after the construction of a harbor by the U.S. Army Corps of Engineers at Santa Cruz to the north.

Mahoney, H. R. 1979. Imperiled sea frontier: Barrier beaches of the east coast. 25:6, 329–37. The natural alteration of barrier beaches is considered.

Pilkey, O. H. 1990. Barrier Islands. 36:6, 30–39. The origin, evolution, and types of barrier islands are discussed.

Schumberth, C. J. 1971. Long Island's ocean beaches. 17:6, 350–62. This is a very informative article on the nature of barrier islands. The specific problems observed on the Long Island barriers serve as examples.

Wanless, H. R. 1989. The inundation of our coastlines: Past, present and future with a focus on South Florida. 35:5, 264–7. This is an informative overview of the evidence that is useful in evaluating the past and future movements of the shorelines.

Wanless, H. R., and Tedesco, L. P. 1988. Sand biographies. 34:4, 224–32. Discusses how we can tell where beach sand came from and how it traveled to its present location.

Westgate, J. W. 1983. Beachfront roulette. 29:2, 104–9. The problems related to the development of barrier islands are discussed.

Scientific American

Bascom, W. 1960. Beaches. 203:2, 80–97. A comprehensive consideration of the relationship of beach processes, both large- and small-scale, to release of energy by waves.

Broecker, W. S., and Denton, G. H. 1990. What drives glacial cycles? 262:1, 48–107. The most up-to-date information is included in this consideration of the causes of glacial cycles.

Dolan, R., and Lins, H. 1987. Beaches and barrier islands. 257:1, 68–77. A discussion dealing with the ultimate futility of trying to develop and protect structures on beaches and barrier islands.

Fairbridge, R. W. 1960. The changing level of the sea. 202:5, 70–79. A discussion of what is known of the causes of changing level of the sea, which seems to be related mostly to the formation and melting of glaciers and changes in the ocean floor.

Chapter 11

The Coastal Ocean

In this chapter we will consider the increasingly stressed estuary and wetland environments, which are among the most biologically productive Earth environments. Acute and chronic marine pollution result from the accidental spilling of petroleum and the systematic accumulation of sewage, halogenated hydrocarbons, and mercury in coastal waters. The physical processes of the coastal ocean will be our initial focus, as it is these processes that account for its high biological productivity and ability to withstand the stresses of coastal contamination.

COASTAL WATERS

The primary difference between the coastal waters and the open ocean is depth. Generally, changes in the nature of water with respect to distance and time are much greater in the shallower coastal waters. Because of the shallowness of the coastal waters, river runoff and tidal currents have a very significant effect on the nature of coastal water.

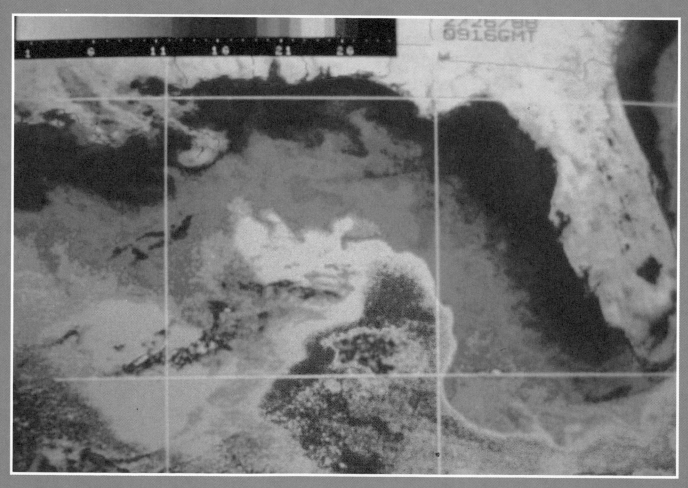

NOAA 9 satellite infrared image of sea-surface temperature in the Gulf of Mexico. The "Loop Current" flows clockwise around a dome of warm water, reaching temperatures of 26°C, then moves off through the Straits of Florida to feed the Gulf Stream. (Courtesy NOAA.)

Salinity

River runoff has the direct effect of reducing salinity of the surface layer in areas where mixing is not significant and throughout the water column where mixing does occur. In areas where the precipitation on the landmass is predominantly rain, the runoff of the rivers will be at the maximum during the season of maximum precipitation. However, if the runoff is fed to a great extent by the melting of snow and ice, the season of maximum runoff will always be the summer. In general, salinity will be lower in coastal regions than in the open ocean owing to the runoff of fresh water from the continents (figure 11–1A, C).

Counteracting the effect of runoff in some coastal regions will be the presence of prevailing offshore winds that will usually have lost most of their moisture over the continent. These winds evaporate considerable quantities of water as they move across the surface of the coastal waters. The increase in the evaporation rate in these areas tends to increase the surface salinity, as shown in figure 11–1B.

Temperature

In coastal regions of the ocean where the water is relatively shallow, very great ranges in temperature may occur on a yearly basis. Sea ice forms in many of the high-latitude coastal areas in which temperatures are determined by the freezing point of the water, which will generally be above −2°C (28.4°F). Maximum surface temperature in low-latitude coastal water may approach 45°C (113°F) in areas where the coastal water is somewhat restricted in its circulation with the open ocean and protected from strong mixing. The seasonal change in temperature can be most easily detected in the coastal regions of the midlatitudes, where surface temperatures are at a minimum in the winter and reach maximum values in the late summer.

Figure 11–1 shows how strong thermoclines may develop in areas where mixing does not occur. Very high-temperature surface water may form a relatively thin layer. Mixing reduces the surface temperature by distributing the heat through a greater vertical column of water, thus pushing the thermocline deeper and making it less pronounced. One major factor in mixing coastal water is the effect of tidal currents, which can have a considerable influence on the vertical mixing of shallow water near the coast. Prevailing winds can also have a significant effect on surface temperatures if they blow from the continent. These air masses we previously mentioned in regard to salinity will usually be of relatively high temperature during the summer and cause an increase in the surface temperature and rate of evaporation of ocean water. They will be of much lower temperature than the ocean surface during the winter and will cause a heat loss and cooling of the surface water near shore.

Coastal Geostrophic Currents

The geostrophic effect that was discussed in association with the current gyres in chapter 7 also develops in the coastal waters. The causes are basically the same. Wind blowing parallel to the coast in a direction that would cause water to pile up along the shore under the influence of the Coriolis effect produces a situation in which that water must eventually, under the influence of gravity, run back down the slope toward the open ocean. As the water runs down the slope away from the shore, the Coriolis effect causes it to veer to the north on the western coast and to the south on the eastern coast of continents in the Northern Hemisphere.

Another condition that produces geostrophic flow along the margins of a continent is the runoff of large quantities of fresh water that gradually mix with the oceanic water. This produces a surface slope of water away from the shore. The seaward slope is associated with salinity and density gradients, as both increase seaward (figure 11–2). These variable currents, which depend upon the wind and the amount of runoff for their strength, are bounded on the ocean side by the more steady boundary currents of the open-ocean gyres.

These local geostrophic currents frequently flow in the opposite direction of the boundary current, as is the case with the **Davidson Current** that develops along the coast of Washington and Oregon during the winter. A very great amount of precipitation occurs in the Pacific Northwest during the winter months, and the winds that generally blow out of the southwest are the strongest during these same months. Their combined effect produces a relatively strong northward-flowing geostrophic current between the southward-flowing California Current, which is part of the open-ocean circulation, and the continent of North America.

ESTUARIES

Estuaries (figure 11–3) are semienclosed coastal bodies of water in which the ocean water is significantly diluted by fresh water from land runoff. The mouths of large rivers throughout the world form the most economically significant estuaries, since many serve as ports and centers of ocean commerce. Many estuaries support important commercial fisheries as well, and the environmental changes that are occurring as a result of the commercial importance of the estuaries will be a major concern as the future development of these areas is planned. Many bays, inlets, gulfs, and sounds may be considered estuaries on the basis of their structure.

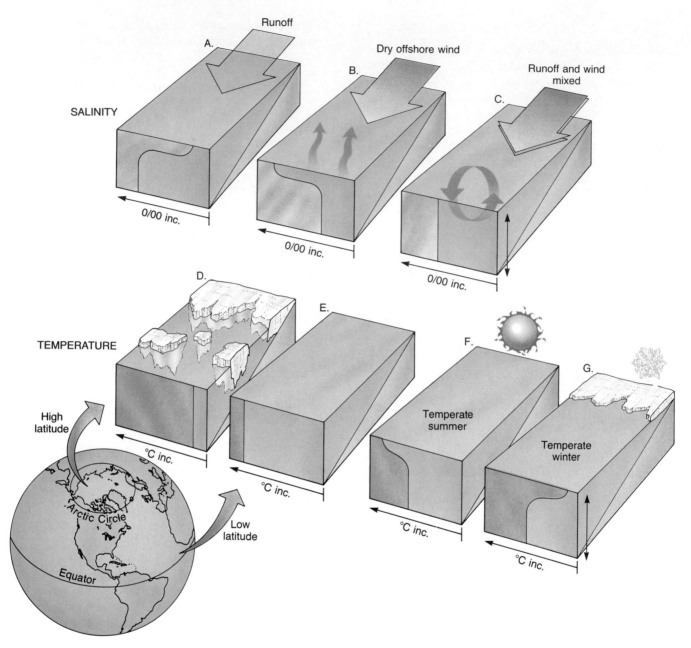

FIGURE 11–1
Temperature and Salinity Changes in the Coastal Ocean. *A,* freshwater runoff is not mixed into water column and forms surface layer of low salinity. A well-developed halocline occurs. *B,* warm, dry offshore winds cause a high rate of evaporation that may offset the effects of runoff, producing a halocline with a gradient that is the reverse of that seen in *A. C,* runoff is mixed with deeper water, producing an isohaline water column of generally lower salinity than the open ocean. *D,* in high latitudes where sea ice is forming or thawing throughout the year, the temperature of coastal water will remain uniformly near the freezing point. *E,* water in shallow, low-latitude coastal regions protected from free circulation with the open ocean may develop a high-temperature isothermal condition. *F,* in the midlatitudes, coastal water will undergo a significant warming during summer. A strong seasonal thermocline may develop. *G,* winter may produce a layer of low-temperature water at the surface. The thermocline shown may develop; however, cooling may cause the surface water to sink because of increased density. This would result in a well-mixed isothermal water column. Mixing due to strong winds may drive the thermoclines shown in *F* and *G* deeper and may even cause mixing of the entire water column, producing an isothermal condition.

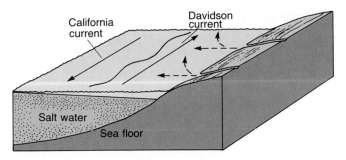

FIGURE 11–2
Davidson Current. During the winter rainy season, fresh runoff water produces a seaward slope away from low-salinity surface water near the shore. A surface flow of low-salinity water from the shore toward the open ocean occurs. As the flow of surface water is acted on by the Coriolis effect, it veers right. This creates a north-flowing current (Davidson Current) along the coast of Washington and Oregon between the shore and the southbound California Current.

Origin of Estuaries

Essentially all estuaries in existence today owe their origin to the fact that in the last 18,000 years sea level has risen approximately 120 m (394 ft) owing to the melting of much of the major continental glaciers that covered portions of North America, Europe, and Asia during the Pleistocene Epoch, more commonly referred to as the Ice Age. Four major classes of estuaries can be identified on the basis of their origin (figure 11–3):

1. **Coastal plain estuaries** were formed as the rising sea level caused the oceans to invade the existing river valleys. These estuaries are sometimes referred to as *drowned river valleys.* Chesapeake Bay is an example.

2. **Fjords** are glaciated valleys that are U-shaped with steep walls. They usually have a glacial deposit forming a sill near the ocean entrance. Fjords are common along Norwegian, Canadian, and New Zealand coasts.

3. **Bar-built estuaries** are shallow estuaries separated from the open ocean by bars composed of sand deposited parallel to the coast by wave action. Lagoons separating barrier islands from the mainland are bar-built estuaries.

4. **Tectonic estuaries** are produced by faulting or folding, which causes a restricted down-dropped area into which rivers flow. San Francisco Bay is in part a tectonic estuary.

Water Mixing in Estuaries

Generally, the freshwater runoff that flows into an estuary moves as an upper layer of low-density water across the estuary toward the open ocean. An inflow from the ocean takes place below the upper layer, and mixing takes place at the contact between these water masses. Stommel has classified estuaries on the basis of the degree of mixing as determined by the distribution of water properties into the following categories, which are shown in figure 11–4:

1. **Vertically mixed**—shallow, low-volume estuaries where the net flow always proceeds from the river at the head of the estuary toward the mouth. Salinity at any point in the estuary will be uniform from the surface to the bottom owing to the even mixing of the river water with ocean water by eddy diffusion at all depths. Salinity increases from the head to the mouth of the estuary.

FIGURE 11–3
Classification of Estuaries on the Basis of Origin. Four classes of estuaries are coastal plain (e.g., Chesapeake Bay, Maryland and Virginia), fjord (e.g., Strait of Juan de Fuca, Washington), bar-built (e.g., Laguna Madre, Texas), and tectonic (e.g., San Francisco Bay, California).

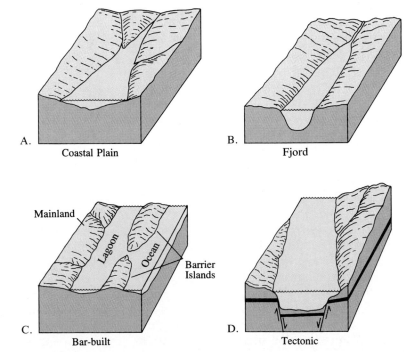

FIGURE 11–4
Classification of Estuaries on the Basis of Degree of Mixing. Note that due to the Coriolis effect surface freshwater flow toward the mouth of the estuary extends seaward much farther along the right-hand shore as one faces seaward. The opposite side of the estuary experiences a greater marine influence. In most estuaries, the marine flow of water into them is a subsurface flow.

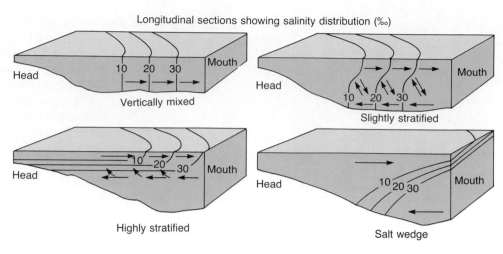

Longitudinal sections showing salinity distribution (‰)

Vertically mixed

Slightly stratified

Highly stratified

Salt wedge

2. **Slightly stratified**—a somewhat deeper estuary in which the salinity increases from the head to the mouth at any depth. Two basic water layers can be identified: the less saline upper water provided by the river and the deeper marine water separated by a zone of mixing. It is in this type of estuary that we begin to see the typical **estuarine circulation** pattern develop, as there is a net surface flow of low-salinity water into the ocean and a net subsurface flow of marine water toward the head of the estuary.

3. **Highly stratified**—typical of deep estuaries in which the salinity in the upper layer increases from the head of the estuary to the mouth, where it reaches a value close to that of open-ocean water. The deep-water layer, however, has a rather uniform marine salinity at any depth throughout the length of the estuary. The net flow of the two layers is similar to that described for the slightly stratified estuary, except that the mixing that occurs at the interface of the upper water and the lower water is such that the net movement is from the deep-water mass into the upper water. The less saline surface water does not seem to dilute the deep-water mass and simply moves from the head toward the mouth of the estuary, having its salinity increased as water from the deep mass joins it. Relatively strong haloclines develop in such estuaries at the contact between the upper and lower water masses. It is not unusual for these haloclines to reach magnitudes approaching 20‰ near the head of the estuary at times of maximum river flow.

4. **Salt wedge**—resulting where a saline wedge of water intrudes from the ocean below that of the river water; typical of the mouths of deep, large-volume-transport rivers. There is no horizontal salinity gradient at the surface in these deep estuaries, the water being essentially fresh throughout the length of, and even beyond, the estuary. There is, however, a horizontal salinity gradient at depth and a very pronounced vertical salinity gradient manifested as a strong halocline at any

station throughout the length of the estuary. This halocline will be shallower and more highly developed near the mouth of the estuary.

The mixing patterns described often cannot be applied to an estuary as a whole. Mixing within an estuary may change with longitudinal distance, season, or tidal conditions.

Estuaries and Human Activities

The overriding problem regarding the health of estuaries is not only that they are of great value in their contribution to natural ecosystems that have evolved within them over millennia, but that they also enhance the efficiency of economic activities that are not directly dependent upon the health of the estuary. These economic activities include shipping, logging, manufacturing, waste disposal, and others that can potentially damage the estuarine environment.

Even in areas where population pressure is still modest, estuaries can be severely damaged. The **Columbia River estuary,** which forms most of the border between Washington and Oregon, is flushed by tides that drive a salt wedge 42 km (26 mi) upstream and raise its water level by over 3.5 m (12 ft) (figure 11–5). When the tide falls, up to 28,000 m³/s (1,000,000 ft³/s) of fresh water push out hundreds of kilometers into the Pacific Ocean to aid returning salmon to find their "home" waters. In the late nineteenth century, farmers and dairymen moved onto the rich floodplains and eventually diked them to prevent the inevitable natural flooding. Now these lands are removed from the natural ecosystem and, ironically, neither are they suitable for agriculture. The river has been a great boon to the logging industry, which has dominated the economy of the region throughout most of its modern history. The river has survived some terrible insults inflicted by the logging industry, but the hydroelectric dams have dealt the system a serious blow. Many of these dams did not include salmon ladders, thereby closing off many of the previous spawning grounds. As is the case with most interactions between nature and humans, these structures were almost essential to the development of the region, but a high price was paid. At present, in an effort to diversify the local economy, the development of shipping facilities is occurring, with the accompanying dredging and increased potential for pollution. If such problems have developed at such sparsely populated areas as the Columbia River estuary, what must be the conditions at more highly stressed estuaries?

Chesapeake Bay is a classic example of a coastal plain estuary, produced by the drowning of the Susquehanna River and its tributaries. Most of the fresh water entering the bay enters along the western margin via rivers draining the slopes of the Appalachian Mountains. Also contributing to this pattern of salinity distribution is the Coriolis effect, acting on marine water entering the bay from the south, and the southward flow of river water (figure 11–6). The area has experienced intense pressure from human activities.

With maximum river flow in the spring, a strong halocline (pycnocline) develops, preventing mixing of the fresh surface water and saltier deep water. Beneath the pycnocline, which can be as shallow as 5 m (16 ft), waters may become **anoxic** (without oxygen) from May through August as dead organic matter decays in the deep water. Major kills of commercially important blue crab and oysters as well as other bottom-dwelling organisms occur during these events. For unknown reasons, the degree of stratification and kills of bottom-dwelling animals have increased over the last 40 years. It is, however, suggested that increased nutrients have been added to the bay as a result of human activities. The increased microscopic plant production resulting from this could cause the deposition of increased amounts of organic particulates on the bottom.

That much is yet to be learned about estuarine circulation is illustrated by the following observations made over an extended period in Chesapeake Bay. Just the opposite of the expected pattern, intervals of upstream surface flow accompanied by downstream deep flow have been recorded. Also, periods when all flow was upstream as well as periods of total down-stream flow have been observed. Even more complex patterns known to develop involve a surface and bottom flow in one direction separated by a middepth flow in the opposite direction or landward flows along the shores and seaward flows in the central portions of the estuaries.

Estuaries and the Continental Shelf Plankton Community

Studies of four estuarine environments and their influence on the continental shelf plankton community con-

ducted between Charleston, South Carolina, and Jacksonville, Florida, revealed similar conditions (figure 11–7A). As estuarine phytoplankton (microscopic photosynthetic cells) are carried seaward by the estuarine flow, the rate of photosynthetic production increased sixfold because of improved water clarity. Beyond a distance of 5 km (3 mi), production gradually dropped across the shallow shelf as the nutrient-poor Gulf Stream waters were approached (figure 11–7B). Seasonal changes in photosynthetic production were similar in the estuaries, nearshore shallow shelf waters (less than 20 m depth), and deeper offshore shelf waters. Photosynthetic productivity in all regions peaked in late summer, as did the abundance of zooplankton (microscopic animal populations), including fish eggs and larvae (figure 11–7C). Because these populations have different reproduction rates and would not be expected to peak together, it appears that the similar patterns of changes in photosynthetic production and the abundance of populations of animal plankton may be explained by the fact all were produced in the estuaries and flushed into the shelf waters by estuarine outflow. If this is true, these estuaries have a major influence on the plankton populations and the overall ecology of the continental shelf.

Understanding the dynamics of estuarine circulation is essential if desired advances are to be made in our ability to meaningfully describe (1) parameters of water quality such as dissolved oxygen and coliform bacteria, (2) suspended particulate matter and bottom sediment transport, and (3) biological activity, particularly in relation to microscopic plants and animals, fish eggs, and larvae.

WETLANDS

Bordering estuaries and other shore areas protected from the open ocean are **wetlands,** biologically productive strips of land delicately in tune with natural shore processes. They are of two types: **salt marshes** and **mangrove swamps.** Both are intermittently submerged by ocean water and are characterized by oxygen-poor mud and peat deposits. Marshes, characteristically inhabited by a variety of grasses, are known to occur from the equator to latitudes as high as 65°. Mangrove trees are restricted to latitudes below 30° (figure 11–8). Once mangroves colonize an area, they can normally outgrow and replace marsh grasses.

With our strong desire to live near the oceans, marsh and mangrove wetlands have suffered from systematic filling and development for housing, industry, and agriculture.

Research is showing that wetlands have a very high economic value when left alone. Salt marshes are believed to serve as nursery grounds for over half the spe-

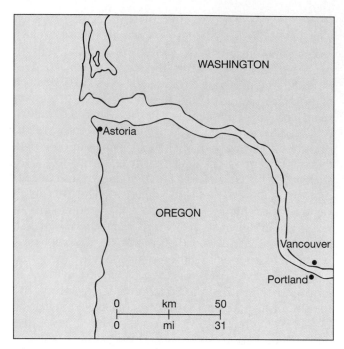

FIGURE 11–5
Columbia River Estuary. This large-volume salt wedge estuary has suffered more from interference with its flow by dams and diked floodplains than from chemical pollution. Like the Mississippi River, the large volume of its flow will aid it in fighting off the effects of chemical pollution.

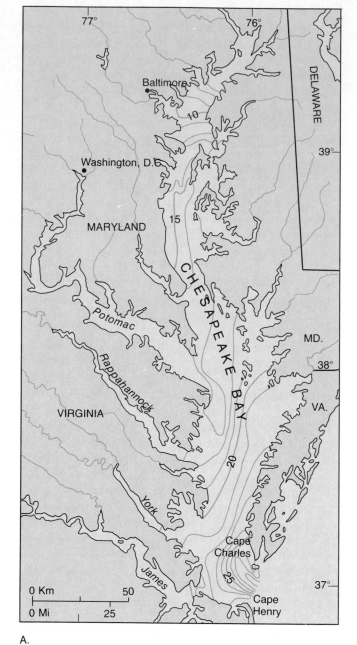

A.

FIGURE 11–6
Chesapeake Bay. *A*, location map and average surface salinity distribution. *B*, comparison of dissolved oxygen levels for the summers of 1950 and 1980. Deeper midbay waters are anoxic in 1980. (After Officer et al., 1984.)

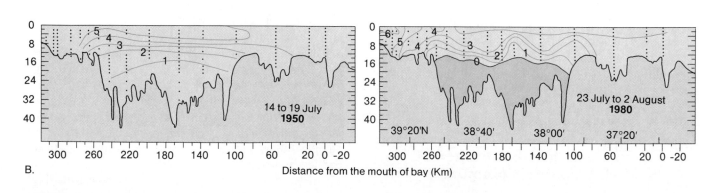

B.

Distance from the mouth of bay (Km)

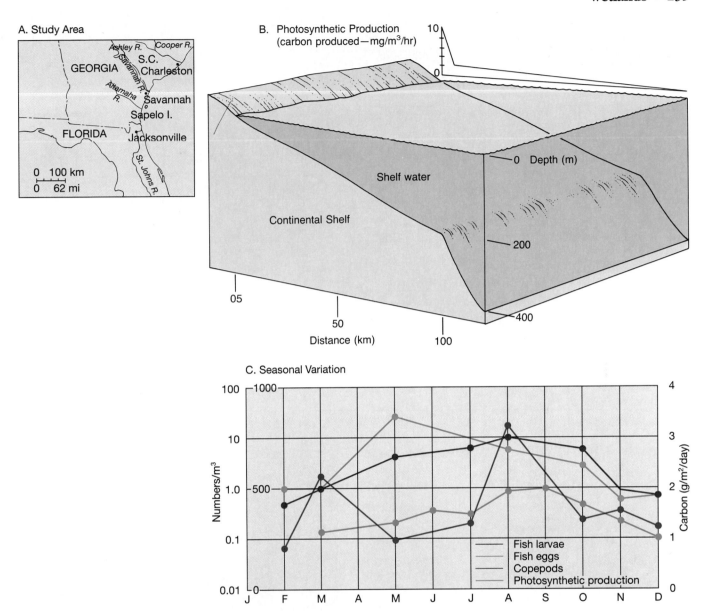

FIGURE 11–7
Estuaries and Shelf Waters. *A*, along the southeast U.S. coast, transects from estuaries across the continental shelf were conducted to study photosynthetic production and plankton abundance at the following locations: Charleston, S.C.; Savannah, Ga.; Sapelo Island, Ga.; and Jacksonville, Fla. Along this coast, relatively small-volume rivers flow onto a gently sloping continental shelf at least 100 km (62 mi) wide. *B*, this curve simulates average photosynthetic production pattern along the transects. Zero on the distance scale represents the nearshore shelf waters adjacent to the mouth of the estuary. Patterns of zooplankton abundance are similar. *C*, seasonal changes in photosynthetic production, copepods, an important population of small plankton that eat photosynthetic cells, fish eggs, and fish larvae. A similar pattern for all these quite different life forms indicates they may be produced in the estuary and flushed together into the shelf waters by estuarine outflow. (After Turner, R. E., et al., 1979.)

cies of commercially important fishes in the southeastern United States. Other fishes such as flounder and bluefish use them for feeding and overwintering, and fisheries of oysters, scallops, clams, and fishes such as eels and smelt are located directly in the marshes.

More than half of the nation's wetlands have been lost. Of the original 215 million acres of wetlands that once existed in the conterminous United States, only about 90 million acres remain. To help prevent the loss of remaining wetlands, the Environmental Protection Agency established an Office of Wetlands Protection (OWP) in 1986. At that time, wetlands were being lost to development at a rate of 300,000 acres per year. The OWP will actively pursue enforcement of regulations against wetlands pollution and identification of the most valuable wetlands that can be protected or restored.

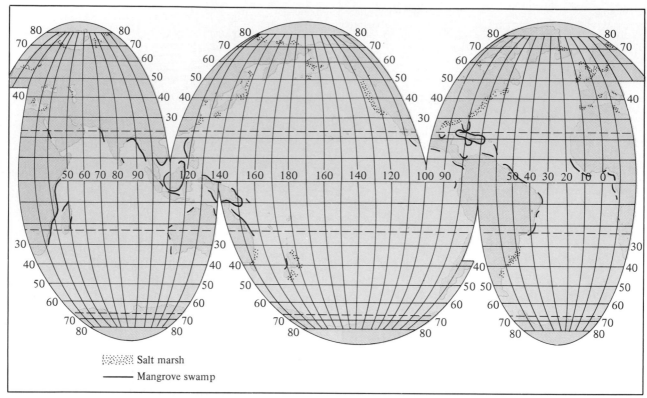

A.

B.

C.

FIGURE 11–8
World Distribution of Salt Marshes and Mangrove Swamps. *A*, note dominance of mangrove swamps in low latitudes and salt marshes in high latitudes. (Base map courtesy of National Ocean Survey.) *B*, salt marsh on San Francisco Bay at Shoreline Park, California. (Photo © Eda Rogers.) *C*, mangrove trees on Lizard Island, Great Barrier Reef, Australia. (Photo © R. N. Mariscal/Bruce Coleman, Inc.)

A very important characteristic of wetlands is their ability to remove inorganic nitrogen compounds and metals from groundwater polluted by land sources. Most of the removal is probably achieved through adsorption on clay-sized particles. Some of the nitrogen compounds trapped in the sediment are decomposed by denitrifying bacteria, releasing the nitrogen to the atmosphere as nitrogen gas. Much of the remaining compounds is used for plant production in this environment, which is one of the most productive in the world. With the death of the plants, the organic nitrogen compounds are either incorporated into the sediment and converted to peat or are broken up and become food for bacteria, fungi, or detritus-feeding shell- and fin fish.

LAGOONS

Landward of the barrier island salt marshes lie protected, shallow bodies of water called **lagoons.** Because of very restricted circulation between lagoons and the ocean, three distinct zones can usually be identified within a lagoon. There will usually be a *freshwater zone* near the mouths of rivers that flow into the lagoon, a *transitional zone* of brackish water (water with a salinity between that of fresh water and ocean water), and a *saltwater zone* close to the entrance, where the maximum tidal effects within the lagoon can be observed. Moving away from the salt-water zone near the mouth of the lagoon, the tidal effects diminish and are usually undetectable in the freshwater region that is well protected from tidal effects. Figure 11–9 shows the characteristics of a typical lagoon.

The salinity of lagoons is also determined by factors other than position relative to the mouth of a lagoon. In latitudes with seasonal variations in temperature and precipitation, ocean water will flow through the entrance during a warm, dry summer to compensate for the volume of water that is lost through evaporation. This flow results in an increased salinity within the lagoon. Lagoons may actually become hypersaline in arid regions where the inflow of seawater is not sufficient to keep pace with evaporation that is taking place at the surface of the lagoon. During the rainy season the lagoon will become much less saline as the amount of freshwater runoff increases.

Laguna Madre

A **hypersaline lagoon** along the coast of Texas between Corpus Christi and the mouth of the Rio Grande is **Laguna Madre,** which is a long, narrow body of water protected from the open ocean by Padre Island, a 160-km (100-mi)-long offshore island. Much of the lagoon, which probably formed about 6000 years ago as sea level was approaching its present height, is less than 1 m (3.3 ft) in depth. The tidal range of the Gulf of Mexico in this area is about 0.5 m (1.6 ft), and the inlets, one of which is shown in figure 11–10, at each end of the barrier island are quite small. There is, therefore, very little tidal interchange between the lagoon and the open ocean.

The shallowness of the water in the lagoon makes possible a very great seasonal range of temperature and

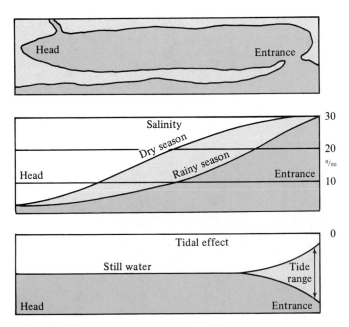

FIGURE 11–9
Lagoons. Typical configuration, salinity conditions, and tide conditions of a lagoon.

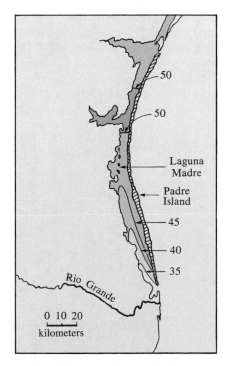

FIGURE 11–10
Laguna Madre. Typical summer surface salinity distribution (‰).

salinity in this semiarid region. The water temperatures are high in the summer and may fall below 5°C (41°F) in the winter. Salinities range from 2‰ to over 100‰ because of infrequent local storms that provide large volumes of fresh water in a short period of time and high evaporation that keeps the salinity generally well above 50‰. Because even the salt-tolerant marsh grasses cannot withstand such high salinities, the marsh is replaced by an open sand beach on Padre Island. At the inlets, the inflow from the ocean is as a surface wedge over the denser water of the lagoon. In turn, the outflow from the lagoon occurs as a subsurface flow, just the opposite of the circulation described for estuaries.

MEDITERRANEAN SEA

Although the Mediterranean Sea is not in its entirety coastal, we will discuss it here because the relationship between climate and the pattern of circulation between it and the Atlantic Ocean is classically represented here.

The Mediterranean Sea is an east-west-trending body of water that has a very irregular coastline dividing it into subseas with separate circulation patterns. Bounded by Europe and Asia Minor on the north and east and Africa to the south, it is surrounded by land except for a narrow connection with the Atlantic Ocean through the **Strait of Gibraltar** and the very narrow **Bosphorus** to the Black Sea. A man-made canal, the Suez Canal, connects the Mediterranean Sea to the Red Sea. The Mediterranean is divided into two major basins separated by a sill with a 400-m (1312-ft) depth, extending from Sicily to the coast of Tunisia. To the north of Sicily, strong currents run between Sicily and the Italian mainland through the Strait of Messina. The much broader connection between Sicily and Tunisia is the Strait of Sicily (figure 11–11).

Atlantic water enters the Mediterranean through the Strait of Gibraltar as a surface flow that is carried in to replace water evaporated at a high rate in the very arid eastern end of the Mediterranean Sea. The water level in the eastern Mediterranean is generally 15 cm (6 in.) lower than at the Strait of Gibraltar. This surface flow follows the northern coast of Africa throughout the length of the Mediterranean and spreads north across the sea (figure 11–12).

The remaining Atlantic water continues eastward to Cyprus where, during the winter, it sinks to form the Mediterranean Intermediate Water with a temperature of 15°C (59°F) and a salinity of 39.1‰. This water flows west along the North African coast at a depth of 200 to 600 m (650 to 2000 ft) and passes into the North Atlantic as a subsurface flow through the Strait of Gibraltar. By the time it passes through the strait, its temperature has dropped to 13°C (55°F) and the salinity is 37.3‰. It is still much denser than water at this depth in the Atlantic Ocean, so it moves down the continental slope until it reaches a depth of approximately 1000 m (3280 ft), where it encounters Atlantic water of the same density. It is at this level that it spreads out in all directions into the Atlantic water and becomes a detectable water mass over a broad area because of its high salinity.

The circulation between the Mediterranean Sea and the Atlantic Ocean is typical of closed restricted basins in areas where evaporation exceeds precipitation. In such situations the restricted basins will always lose water at a very high rate from the surface through evaporation, and this water will have to be replaced by surface inflow from the open ocean. The evaporation of the water that flows in from the open ocean increases its salinity to very high values, causing eventual sinking and the return to the open ocean as a subsurface flow. You will recall that during periods of hypersalinity, Laguna Madre has a pattern

FIGURE 11–11
General Bathymetry and Subseas of the Mediterranean Sea.
(Adapted from *Encyclopedia of oceanography,* edited by Rhodes Fairbridge, © 1966. Reprinted by permission of Dowden, Hutchinson, & Ross, Inc., Stroudsburg, Pa.)

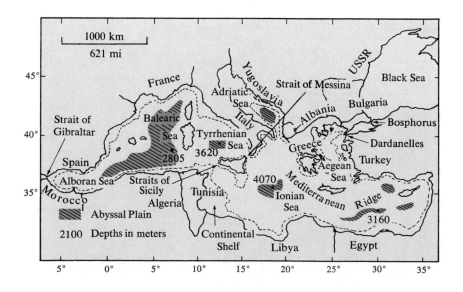

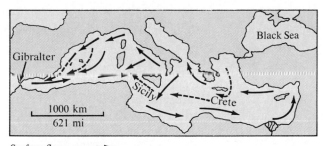

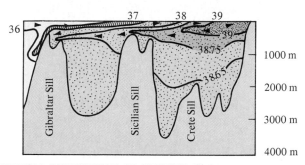

A. SURFACE AND INTERMEDIATE CIRCULATION

Surface flow ⟶
Intermediate flow (200–600 m; 650–2,000 ft) ----►

B. VERTICAL DISTRIBUTION OF SALINITY (°/₀₀)
Most of the mass of Mediterranean Water has salinity ranging between 38 °/₀₀ and 39 °/₀₀. Maximum salinities in excess of 39 °/₀₀ are found in the surface waters at the east end of the sea. After sinking and moving as Intermediate Water toward the Atlantic Ocean, salinity is reduced to about 37.3 °/₀₀ as it flows over the sill at Gibraltar.

FIGURE 11–12
Mediterranean Circulation.

of circulation between it and the Gulf of Mexico that represents Mediterranean circulation on a smaller scale. Since the Mediterranean Sea is the classic area for this type of circulation, it is given the name **Mediterranean circulation.** You will recognize it as being opposite the pattern we observed to be characteristic of estuaries, where the surface flow goes into the open ocean and subsurface flow of marine water is into the estuary. In areas where estuarine circulation develops between the marginal body of water and the ocean, the availability of fresh water from runoff and precipitation is greater than the rate of water loss to evaporation.

POLLUTION STRESSES THE COASTAL WATERS

Pollution of coastal waters has increased with increased use of coastal areas for residential, recreational, and commercial purposes. Some well-documented cases of the negative impact of chemical pollution of coastal waters by petroleum, sewage, halogenated hydrocarbons, and mercury are discussed in the following pages.

Petroleum

There have been a number of major oil spills since the *Torrey Canyon* ran aground at Seven Stones Rocks off the Cornwall coast on March 18, 1967. The spills have resulted from additional tanker accidents and the blowout of oil wells being drilled or produced in the coastal ocean. Although it is important to know how much oil enters the oceans and from what sources, our primary concern is the effect of hydrocarbon pollution on marine organisms and the marine environment in general.

Because hydrocarbons are organic substances biodegradable by microorganisms, they are considered by some to be among the least damaging pollutants of the ocean. Yet they are complex, containing molecules ranging in molecular mass from 16 to 20,000. Most crude oils will also contain, in addition to their hydrocarbon components, chemicals containing oxygen, nitrogen, sulfur, and trace metals. When this complex chemical mixture is combined with the ocean—another complex chemical mixture containing organisms—the result can indeed be negative to the organisms.

West Falmouth Harbor. Because of the complexity of petroleum, it is difficult to answer the question, How long does it take for a shore to recover from an oil spill? The oldest well-studied oil spill in the United States occurred on September 16, 1969, near West Falmouth Harbor in Buzzards Bay, Massachusetts. When the barge *Florida* came ashore and ruptured, currents carried the Number-2 fuel oil north into Wild Harbor, where the most severe damage occurred (figure 11–13). The initial kill was almost total for intertidal and subtidal animals in the most severely oiled area. A severe reduction in species diversity was accompanied by rapid increases in the

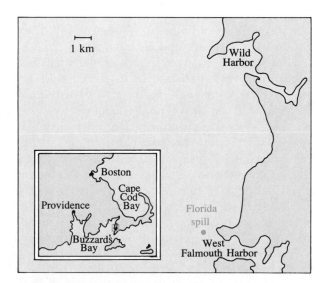

FIGURE 11–13
Florida **Oil Spill at West Falmouth Harbor.**

population of polychaete worms that were resistant to the oil. The species *Capitella capitata* accounted for up to 99.9 percent of the individuals taken in samples of the most severely oiled locations during the first year. Species diversity did not appreciably increase until well into the third year after the spill. Although conditions have returned to near normal, oil can still be found in the fine sediment of areas that underwent the most severe contamination.

Chedabucto Bay. A spill of heavy Bunker-C fuel oil in Chedabucto Bay, Nova Scotia, on February 4, 1970, by the vessel *Arrow* has provided some indication of the recovery time. Based on changes in visible contamination, about one-third of the initially oiled 200 km (124 mi) of shore was clear after three months. This cleaning was primarily the result of wave action on the rocky shore. By 1973, mechanical cleaning by waves, evaporation, photochemical decomposition, and bacterial decomposition had cleaned all but the restricted low-energy estuarine marshes. Fully 75 percent of the shore was clean. Visible contamination had been reduced to 15 percent by 1976 (figure 11–14).

Although the oil disappeared rapidly from the rocky shore, the low-energy sediments of the marshes will not be cleaned so quickly. The half-life for removal of the visible oiling in the bay as a whole is about two years, but it appears that the half-life for removal from the fine sediment will be up to 25 years.

Dr. John H. Vandermeulen of the Bedford Institute of Oceanography is at present gathering data on the long-term effects of the Chedabucto Bay spill. He says that sandy and muddy sediments still have detectable hydrocarbons and that boulder beach deposits are still cemented with heavy tar deposits. Oil released from local commercial and recreational fishing activities over the years has made it impossible to tie the present contamination to the *Arrow* spill in 1970. In terms of total biomass, the area seems to be fully recovered.

The biological consequences of the Chedabucto Bay spill have been indicated by the recovery of kelp, *Fucus vesiculosus;* the salt marsh grass, *Spartina alterniflora;* and the soft-shell clam, *Mya arenaria.* Over half of the *Fucus* was destroyed, and a recovery half-life of four years is indicated. *Spartina* suffered a similar reduction but was much slower to start recovery. *Fucus* made a gradual recovery that began immediately, but *Spartina* showed no signs of recovery until two years after the spill. Once recovery started, however, it was rapid, and an estimate of the recovery half-life is about 3 years. The clam, *Mya,* did not show such an encouraging pattern of recovery. Severely stressed by the contamination, *Mya* experienced a serious reduction in efficiency of food utilization. In some locations, *Mya* barely survived. Tissue and shell growth were greatly reduced, and the age-class

structure was altered because of the high degree of mortality that occurred with the spill. The recovery half-life of the *Mya* was in excess of 10 years.

Argo Merchant. When oil spills do not come ashore, the negative effects are not so obvious. The sinking of the *Argo Merchant* after it ran aground on Fishing Rip Shoals off Nantucket on December 15, 1976, provides the best-studied effects of such a spill. A full load of Number-6 fuel oil was dumped into the ocean 40 km (25 mi) southeast of Nantucket Island. Fortunately, the winds were such that no oil came ashore. The surface slick moved east out to sea and was gone by mid-January. Even though the oil was not visible for long, it did a significant amount of biological damage that was investigated by scientists from Woods Hole Oceanographic Institution, the National Marine Fisheries Service, and local institutions (figure 11–15).

The primary observable effect of the *Argo Merchant* spill on marine organisms was in pelagic fish eggs (pollock and cod) in plankton samples taken shortly after the spill. Of the 49 pollock eggs recovered, 94 percent were oil-fouled, and 60 percent of the 60 cod eggs were in that state. Twenty percent of the cod eggs and 46 percent of the pollock eggs were dead or dying (figure 11–16). Although little is known about the natural mortality rate of pelagic fish eggs, only 4 percent of a sample of cod eggs spawned in a laboratory were dead or dying at the same development stage.

Because most of the oil floated in a surface slick, contamination in subsurface water samples did not exceed more than 250 parts per billion. Other than the described damage to fish eggs and a significant amount of damage to other members of the plankton, little direct evidence of major biological damage was recovered. There were, however, numerous oiled birds that washed ashore at Nantucket and Martha's Vineyard.

Because pollock females spawn about 225,000 and cod about 1 million eggs each season over a range extending from New Jersey to Greenland, it is unlikely that a single occurrence would have a major effect on those fisheries. However, it is easy to see why the fishing communities of the northwest Atlantic are so happy about the marginal results of oil exploration on Georges Bank. It may be that a fishery already severely stressed by overexploitation could not survive even the lowest level of pollution brought about by the establishment of oil production on the bank.

Prince William Sound. The first major oil spill resulting from the development of the North Slope of Alaska petroleum reserves occurred March 23, 1989, when the California-bound *Exxon Valdez* went aground on rocks 40 km (25 mi) out of Valdez, Alaska (figure 11–17). The 32,000 MT of oil spilled makes this the largest spill to occur in U.S. waters. This spill in the pristine waters of

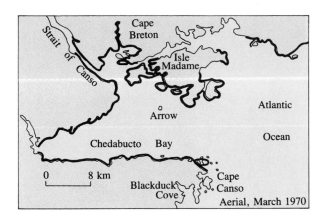

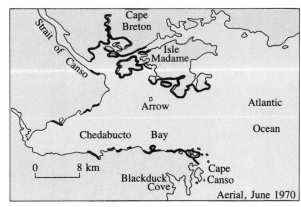

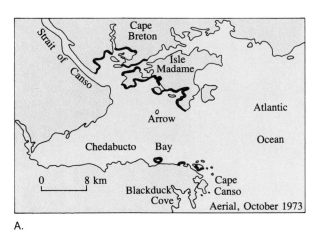

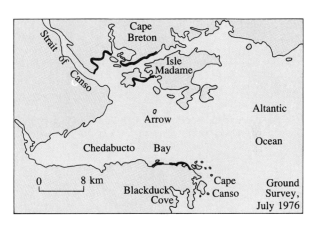

A.

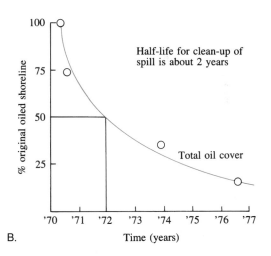

FIGURE 11–14
Oil Spill in Chedabucto Bay. *A*, distribution of oil residues in the shore zone of Chedabucto Bay, Nova Scotia, Canada, determined by aerial inspection in 1970 and 1973 and by ground survey in 1976. *B*, oil-removal pattern of stranded Bunker-C oil from shorelines from 1970 through 1976. No oil is visible at the surface now, but it can be found by removing the surface sediments in some marshes.

Prince William Sound spread to the Gulf of Alaska, and 1775 km (1000 mi) of shoreline were damaged. The U.S. Fish and Wildlife Service reported that at least 994 sea otters and 34,434 birds were killed by the spill. The actual kill could be ten times that amount by some official estimates. The total death toll will never be known. It was predicted that the Alaskan waters would have a long, slow recovery, but the fisheries closed in 1989 bounced back with record takes in 1990.

Encouraged by good results on small test plots, Exxon is spending $10 million to spread phosphorus- and nitrogen-rich fertilizers on Alaskan shorelines to boost the development of indigenous oil-eating bacteria. It is hoped that this will result in a cleanup rate that will produce in two to three years what would require up to seven years under natural conditions.

Oil spills are a problem that will be with us for many years, and they may become more common as the

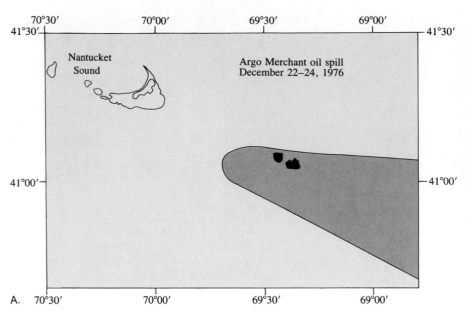

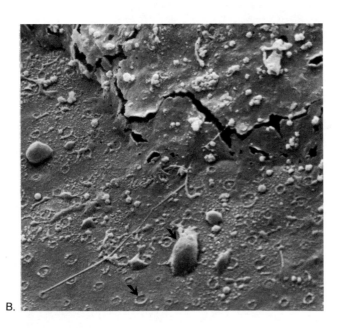

FIGURE 11–15
***Argo Merchant* Oil Spill, December 15, 1976.** *A,* location of the
grounding site and oiled area of the ocean. *B, Argo Merchant,* after
breaking up and spilling much of its cargo into the Atlantic Ocean
southeast of Nantucket. (*B,* photo courtesy of U.S. EPA/EMSL.)

B.

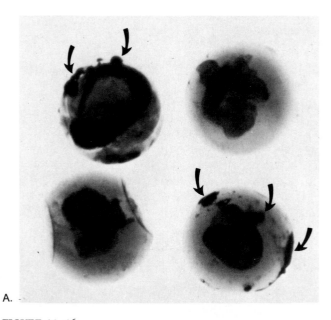

FIGURE 11–16
Pollock Eggs Affected by the *Argo Merchant* Spill. *A,* these eggs, in the tail-bud and tail-free
embryo stages, were taken from the edge of the oil slick. The outer membranes of the eggs at the
upper left and lower right are contaminated with a tarlike oil; arrows point to some of the oil masses.
The uncontaminated egg at the upper right has a malformed embryo; that at the lower left is collapsed
and also has an abnormal embryo. The actual size of pollock eggs is about 1 mm (0.04 in.); of cod
eggs, about 1.5 mm (0.06 in.). *B,* a portion of the surface of an oil-contaminated egg. The upper arrow
points to one of many oil droplets; the lower arrow to one of the membrane pores. (Scanning electron microscope: about 5000 ×. Courtesy of A. Crosby Longwell.)

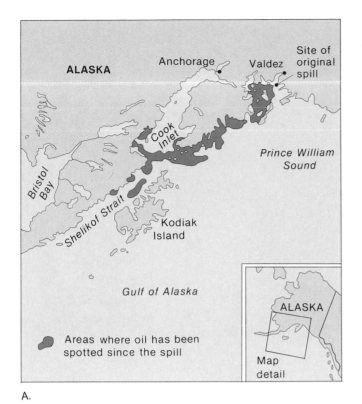

A.

B.

FIGURE 11–17
Exxon Valdez **Oil Spill.** *A*, location and extent of contamination resulting from the spill of heavy crude oil from the super-tanker *Exxon Valdez. B*, oil from the spill damaged shorelines and killed sea otters and birds. (Photo courtesy of U.S. Coast Guard.)

petroleum reserves underlying the continental shelves of the world are increasingly exploited. The largest oil spill on record occurred June 3, 1979, as a result of this exploitation. The Petroleus Mexicanos (PIMEX) oil-drilling station Ixtoc #1 in the Bay of Campeche off the Yucatán peninsula, Mexico, blew out and caught fire. Before it was capped on March 24, 1980, it spewed 468,000 MT of oil into the Gulf of Mexico (figure 11–18).

Oil can be broken down by microorganisms such as bacteria and fungi. Certain of these organisms are effective in breaking down only a particular variety of hydrocarbon, and none is effective against all forms. In 1980, Dr. A. M. Chakrabarty, now a microbiologist at the University of Illinois, worked at the General Electric Research and Development Center in Schenectady, New York. There he produced a microorganism capable of breaking down nearly two-thirds of the hydrocarbons in most crude oil spills. A similar strain was released into the Gulf of Mexico to test its effectiveness in cleaning up the crude oil spilled after a June 8, 1990, explosion disabled the *Mega Borg*. Preliminary results indicate that the bacteria are effective. Carl Oppenheimer of the University of Texas, who conducted the test, reports no indication of negative effect on the Gulf ecology.

Sewage

Each year over 500,000 MT of sewage sludge containing a mixture of human waste, oil, zinc, copper, lead, silver, mercury, PCBs, and pesticides have been dumped through sewage outfalls into the coastal waters of southern California. A more massive 8 million MT have been dumped into the New York Bight each year. Although the Clean Water Act of 1972 prohibited dumping of sewage into the ocean after 1981, the high cost of treating and disposing of it on land resulted in extended waivers being granted to these areas. Nonbiodegradable debris, probably carried into the ocean through storm drains by heavy rains, was washed up on Atlantic coast beaches during the summer of 1988. Ironically, this event, which was not related to sewage disposal, hurt the tourist business and was responsible for legislation again being passed to terminate disposal of sewage in the ocean.

It now appears that the political pressure to clean up the coastal water is sufficient to assure that land treatment and disposal will be the primary disposal procedure of the future. For the present, Los Angeles has stopped pumping sewage into the sea. If it is allowed to resume this practice, the sewage sludge pumped through

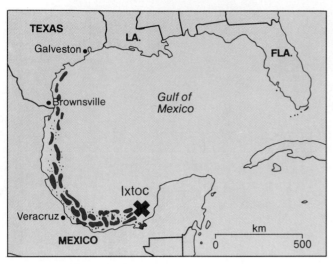

A.

B.

FIGURE 11–18
Ixtoc #1 Blowout, June 3, 1979. *A*, location of the Gulf of Campeche blowout and oil slick that threatened the Texas coast. *B*, fire-fighting equipment spraying water on the waters around the flaming oil directly above the wellhead. Oil can be seen spreading across the ocean surface. (Photo courtesy of NOAA.)

the lines will have had its toxic chemicals and pathogens removed.

Off the east coast over 8 million MT of sewage sludge was dumped by barge over dump sites totaling 150 km² (58 mi²) each year during the past few years. These dumps occurred at the New York Bight Sludge Site and the Philadelphia Sludge Site shown in figure 11–19. The New York Bight water depth is about 29 m (95 ft), and Philadelphia's site is 40 m (130 ft) deep. In such shallow water, the water column displays relatively uniform characteristics from top to bottom; even the smallest sludge particles reach the bottom without having undergone much horizontal transport, and the ecology of the dump site can be totally destroyed. At the very least, the concentration of organic and inorganic nutrients seriously disrupts the biogeochemical cycling. Greatly reduced species diversity results, and in some locations the environment becomes anoxic.

The shallow-water sites have been abandoned, and the sewage is now being transported to a deep-water site 106 miles (171 kilometers) out (figure 11–19). At the deep-water site, beyond the shelf break, there is usually a well-developed density gradient separating low-density, warmer surface water from high-density, colder deep water. Internal waves moving along this density gradient can retard the sinking rate of particles and may allow a horizontal transport rate 100 times greater than the sinking rate.

Local fishermen have been very concerned about the deep-water dumping and were reporting adverse ef-

fects on their fisheries soon after it began. As it stands now, this east coast dumping program for sewage is also doomed, and municipalities will have to find the money for land-based processing.

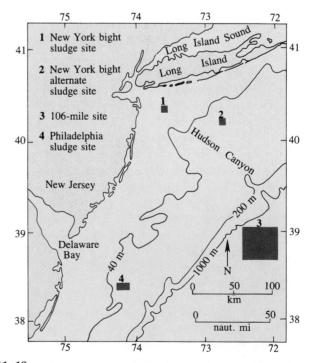

11–19
Previous and Present Atlantic Sewage Sludge Disposal Sites.

Halogenated Hydrocarbons (DDT and PCBs)

DDT (dichlorodiphenyltrichlorethane) and **PCBs** (polychlorinated biphenyls) are found throughout the marine environment. They are persistent, biologically active chemicals that have been put into the oceans entirely as a result of human activities.

At the time of a near-total ban on U.S. production of the pesticide DDT in 1971, 2.0×10^9 kg (4.4×10^9 lb) had been manufactured, most of it by the United States. Its use in the Northern Hemisphere since 1972 has virtually ceased. The danger of excessive use of DDT and similar pesticides was first manifested in the marine environment in the effects on marine bird populations. During the 1960s, there was a serious decline in the brown pelican population of Anacapa Island off the coast of California; high concentrations of DDT in the fish eaten by the birds had caused the production of excessively thin egg shells. A decline in the osprey population of Long Island Sound that began in the late 1950s and continued throughout the 1960s was also caused by thinning egg shells brought on by DDT contamination. Studies showed a 1 percent increase in brown pelican and osprey egg shell thickness from 1970 to 1976, while the concentration of pesticide residue decreased. Because there was an increased egg hatching rate associated with these changes, it was concluded that DDT was the cause of the decline in the bird populations during the 1960s (figure 11–20).

PCBs are industrial chemicals found in a variety of products from paint to plastics. They have been indicated

as causes of spontaneous abortions in sea lions and the death of shrimp in Escambia Bay, Florida.

The main route by which DDT and PCBs enter the ocean is the atmosphere. They are concentrated initially in the thin surface slick of organic chemicals at the ocean surface; then they gradually sink to the bottom attached to sinking particles. A study off the coast of Scotland indicated that open-ocean concentrations of DDT and PCBs are 10 and 12 times less, respectively, than in coastal waters. Long-term studies have shown that DDT residue content in mollusks along the U.S. coasts reached a peak in 1968. The pervasiveness of DDT and PCBs in the marine environment can best be demonstrated by the fact that Antarctic marine organisms contain measurable quantities of halogenated hydrocarbons. Obviously there has been no agriculture or industry on that continent to explain this presence; these substances have been transported from distant sources by winds and ocean currents.

Mercury

The stage was set for the first tragic occurrence of mercury poisoning with the establishment of a chemical factory on **Minamata Bay,** Japan, in 1938. One of the products of this plant was acetaldehyde, the production of which requires mercury as a catalyst. The first ecological changes in Minamata Bay were reported in 1950, human effects were noted in 1953, and the mercury poisoning known as Minamata disease became epidemic in 1956. It was not until 1968, however, that the Japanese government declared mercury the cause of the disease, which involves a breakdown of the nervous system. The plant was immediately shut down, but by 1969 over 100 people were known to suffer from the disease. Almost half of these victims died. A second occurrence of mercury poisoning resulted from pollution by an acetaldehyde factory, shut down in 1965, that was located in **Niigata,** Japan. Between 1965 and 1970, 47 fishing families contracted the disease.

During the 1960s and 1970s, much attention was given to the problem of mercury contamination in seafood. Studies done on the amount of seafood consumed by various populations of humans have led to the establishment of safe levels of mercury content in fish to be marketed. To establish these levels for a given population, three variables must be considered:

1. fish consumption rate of the human population under consideration,
2. mercury concentration of the fish being consumed by that population, and
3. minimum ingestion rate of mercury that induces symptoms of disease.

If dependable data on these variables are available and a safety factor of 10 is applied to the determination, a

11–20
Brown Pelican. Pelican populations in the Channel Islands off southern California are increasing, following a serious decline in the 1960s caused by DDT toxicity.

FIGURE 11–21
Mercury Concentrations in Fish vs. Consumption Rates for Various Populations. Plotted curves show the safe ingestion level of mercury until the first signs of Minamata disease can be expected (0.3 mg/day), and the safe ingestion level of mercury after the safety factor of 10 is included (0.03 mg/day). For the United States, it appears that tuna and swordfish are safe to consume, although some swordfish will be banned because their mercury concentrations are above the present 1.0 mg/kg (ppm) FDA limit.

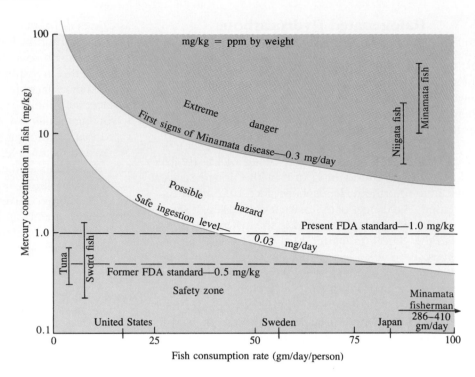

maximum concentration level allowable for the contaminant (in this case mercury) can be established with a high degree of confidence that it will protect the health of the human population.

Figure 11–21 shows how such data can be used to arrive at a safe level of mercury concentration in fish to be consumed by a population. For the general populations of Japan, Sweden, and the United States, the average individual fish consumption rates are 84, 56, and 17 g/day, respectively. By comparison, the members of the Minamata fishing community averaged 286 and 410 g/day during winter and summer seasons, respectively. Based on the minimum level of mercury consumption that is known to bring on symptoms of poisoning as determined by Swedish scientists using Japanese data—0.3 mg/day over a 200-day period—the three populations shown in

figure 11–21 would begin to show symptoms if they ate fish with the following mercury concentrations: Japan, 4 ppm; Sweden, 6 ppm; and United States, 20 ppm. If a safety factor of 10 is applied, the maximum concentrations of mercury in fish that could be safely consumed by these populations would be, for Japan, 0.4 ppm; for Sweden, 0.6 ppm; and for the United States, 2.0 ppm.

Although the U.S. Food and Drug Administration (FDA) initially established an extremely cautious limit of 0.5 mg/kg (ppm), the present limit of 1 mg/kg adequately protects the health of U.S. citizens. Essentially all tuna falls below this concentration, and most swordfish will be acceptable. The present limit amounts to the use of a safety factor of 20 instead of 10, so unless one eats an unusually large amount of tuna or swordfish, we should have little concern about consuming these fishes.

SUMMARY

Because of the shallow depth of the coastal ocean, river runoff, tidal currents, and seasonal changes in solar radiation, physical characteristics of the water such as temperature and salinity vary over a greater range than in the open ocean. Coastal geostrophic currents are produced as a result of runoff of fresh water and winds blowing in the proper direction along the coast.

Estuaries are semienclosed bodies of water where fresh water runoff from the land and ocean water mix. They are classified on the basis of origin as coastal plain, fjord, bar-built, and

tectonic. Mixing patterns range from that of a vertically mixed shallow estuary through slightly stratified and highly stratified to the salt wedge characteristic of large-volume river mouths. The typical pattern of estuarine circulation consists of a surface flow of low-salinity water toward the mouth and a subsurface flow of marine water toward the head.

Chesapeake Bay is a classic example of a coastal plain estuary with most of its stream inflow entering along its western margin. This fact—along with the Coriolis effect acting on the estuarine circulation—makes the eastern side of the bay saltier.

Possibly as a result of increased nutrient inflow resulting from human activities, greater amounts of the deep water in the estuary are undergoing greater periods of summer anoxia.

Wetlands are of two basic types: salt marsh and tropical mangrove swamp. These biologically productive regions are alternately covered and exposed by the tide. Marshes are also important in their ability to adsorb land derived pollutants before they reach the ocean. Despite their ecological importance, wetlands are fast disappearing from our coasts as they are encroached upon by human activities.

Long offshore deposits called barrier islands protect marshes and lagoons. Some lagoons have very restricted circulation with the ocean and may exhibit a great range of salinity and temperature conditions as a result of seasonal change.

The Mediterranean Sea has a classical circulation characteristic of restricted bodies of water in areas where evaporation greatly exceeds precipitation. It is the reverse of estuarine circulation and is called Mediterranean circulation.

Study of areas such as Wild Harbor, Massachusetts, and Chedabucto Bay, Nova Scotia, have shown that oil pollution reduces species diversity and persists longer on muddy bottoms than on sandy or rocky bottoms. Oil spills that do not come ashore, like that of the *Argo Merchant,* do considerably less environmental damage than those that do. The greatest damage resulting from the *Argo Merchant* spill was probably to plankton, especially fish eggs.

Although 1972 legislation required an end to dumping of sewage in the coastal ocean by 1981, exceptions continued to be made. Especially off the southern California coast and in the New York Bight, large amounts of sewage and sewage sludge continued to be dumped. Increased public concern resulted in new legislation that will prohibit sewage dumping in the ocean. Some scientists expected that the transfer of dump sites along the northeast Atlantic coast to deeper water farther offshore would make it safe to continue to dump sewage from Philadelphia and the New York area into the coastal ocean. However, fishermen have complained that the deep-water dumping has damaged fisheries less than one year after it began.

DDT pollution produced a decline in the Long Island osprey population in the 1950s and the brown pelican population of the California coast in the 1960s. Virtual cessation of DDT use in the Northern Hemisphere in 1972 allowed the recovery of both populations. The DDT had thinned the egg shells and reduced the number of successful hatchings.

The first major human disaster resulting from ocean pollution reached epidemic proportions in 1956 within Minamata Bay, Japan. Of more than 100 people affected by mercury poisoning, almost half died from its effects. A similar episode occurred during the late 1960s at Niigata, Japan. Mercury contamination levels have been set on the basis of fish consumption of populations throughout the world, and they appear to have been effective in preventing further poisonings.

KEY TERMS

Anoxic (p. 233)
Bar-built estuary (p. 231)
Bosphorus (p. 238)
Chesapeake Bay (p. 233)
Coastal plain estuary (p. 231)
Columbia River estuary (p. 232)
Davidson Current (p. 229)
DDT (p. 245)
Estuarine circulation (p. 231)

Estuary (p. 229)
Fjord (p. 231)
Highly stratified estuary (p. 232)
Hypersaline lagoon (p. 237)
Lagoon (p. 237)
Laguna Madre (p. 237)
Mangrove swamp (p. 233)
Mediterranean circulation (p. 239)
Minamata Bay (p. 245)

Niigata (p. 245)
PCBs (p. 245)
Salt marsh (p. 233)
Salt wedge estuary (p. 232)
Slightly stratified estuary (p. 232)
Strait of Gibraltar (p. 238)
Tectonic estuary (p. 231)
Vertically mixed estuary (p. 231)
Wetlands (p. 233)

QUESTIONS AND EXERCISES

1. For coastal oceans where deep mixing does not occur, discuss the effect offshore winds and freshwater runoff will have on salinity distribution. How will the winter and summer seasons affect the temperature distribution in the water column?

2. How does coastal runoff of low-salinity water produce a longshore geostrophic current?

3. Describe the difference between vertically mixed and salt wedge estuaries in terms of salinity distribution, depth, and volume of river flow. Which displays the more classical estuarine circulation pattern?

4. Discuss the factors that may explain why the surface salinity of Chesapeake Bay is greater on its east side, and periods of

summer anoxia in the deep-water mass are increasing in severity with time.

5. Name the two types of wetland environments and the latitude ranges where each will likely develop. How do wetlands contribute to the biology of the oceans and the cleansing of polluted river water?

6. What factors lead to a wide seasonal range of salinity in Laguna Madre?

7. Describe the circulation between the Atlantic Ocean and the Mediterranean Sea, and explain how and why it differs from estuarine circulation.

8. Describe the effect of oil spills on species diversity and recovery in the benthos of Wild Harbor and Chedabucto Bay.

9. Discuss whether oil spills that wash ashore or those that do not, such as that of the *Argo Merchant,* are more destructive to marine life.
10. How would dumping sewage in deeper water off the east coast help reduce the degradation of the ocean bottom?
11. Discuss the animal populations that clearly suffered from DDT pollution and the way in which this negative effect was manifested.
12. Refer to figure 11–21 and discuss the desirability of setting a mercury contamination level for fish at the 1.0 mg/kg level for the citizens of the United States, Sweden, and Japan.

REFERENCES

Borgese, E. M., and Ginsburg, N., eds. *1988 Ocean Yearbook,* Vol 7. Chicago: University of Chicago Press.

Cherfas, C. 1990. The fringe of the ocean: Under siege from land. *Science* 248:4952, 163–65.

Clark, W. C. 1989. Managing planet earth. *Scientific American* 261:3, 47–54.

Officer, C. B. 1976. Physical oceanography of estuaries. *Oceanus* 19:5, 3–9.

Officer, C. B.; Briggs, R. B.; Taft, J. L.; Tyler, M. A.; and Bovnton, W. R.

1984. Chesapeake Bay anoxia: Origin, development, and significance. *Science* 223:4631, 22–27.

Pickard, G. L. 1964. *Descriptive physical oceanography: An introduction.* New York: Macmillan.

Stommel, H. M. ed. 1950. *Proceedings of the colloquium on "The flushing of estuaries."* Woods Hole: Woods Hole Oceanographic Institution.

Valiela, I., and Vince, S. 1976. Green borders of the sea. *Oceanus* 19:5, 10–17.

SUGGESTED READING

Sea Frontiers

Baird, T. M. 1983. Life in the high marsh. 29:6, 335–41. The ecology of the salt marsh is discussed.

Baker, R. D. 1972. Dangerous shore currents. 18:3, 138–43. A discussion of the hazards to swimmers of various types of currents found near the shore.

Cardoza, Y., and Hirsch, B. 1991. Tidal creeks and scuba gear. 37:4, 32–37. The life forms inhabiting Florida tidal creeks are discussed.

de Castro, G. 1974. The Baltic: To be or not to be? 20:5, 269–73. A review of the pollution threat to life in the Baltic Sea and what has been done to solve the problem.

Edwards, L. 1982. Oyster reefs: Valuable to more than oysters. 28:1, 23–25. The value of oyster reefs to various estuarine life forms is detailed.

Heidorn, K. C. 1975. Land and sea breezes. 21:6, 340–43. A discussion of the causes of land and sea breezes.

Sefton, N. 1981. Middle world of the mangrove. 27:5, 267–73. The life forms associated with Cayman Island mangrove swamps are described.

Smith, F. 1982. When the Mediterranean went dry. 28:2, 66–73. The role of global plate tectonics in causing the Mediterranean Sea to dry up some 5 million years ago.

White, I. C. 1985. An organization to help combat oil spills. 31:1, 15–21. The efforts of the International Tanker Owners Pollution Federation Limited to increase the efficiency of cleanup operations and reduce the negative effects of oil spills have resulted in more-intelligent responses to oil spills.

Scientific American

Hsu, K. H. 1972. When the Mediterranean dried up. 227:6, 26–45. Evidence is presented that shows the Mediterranean Sea was a dry basin 6 million years ago.

Chapter 12

The Marine Habitat

In Chapter 2 we discussed some evidence for believing that life originated in the oceans. Primary among the considerations was the simple fact that the ocean constitutes the major part of the water environment. Considering the need organisms have for water, we can see that life would be totally impossible on the continents were it not for the hydrologic cycle that provides water that runs across and percolates through the surface and is temporarily stored in lakes and rivers before it returns to the oceans. Although availability of water is seldom a problem, the success of individual organisms living in the oceans is dependent upon their ability to adjust to the many physical variables that we will consider in this chapter.

GENERAL CONDITIONS

The fact that all the major divisions of the animal kingdom, phyla, have members that live in the marine environment attests to the ocean's being a fit biological realm. In fact, all phyla are thought to have originated in the ocean, and five are exclusively marine. Although it

The primary source of nutrition for these mussels, *Bathymodiolus* n. sp., is bacterial symbionts that live in their tissue. The bacteria depend primarily on methane, a hydrocarbon, for the energy and carbon needed to synthesize carbohydrates. The mussels were found on the edge of a nearly anoxic brine pool associated with a hydrocarbon seep at a depth of 650 m in the Gulf of Mexico. The brine layer (blue) has a salinity of more than 120‰. The mussels get methane from the brine layer and oxygen from the clear ambient water above the brine layer. (Reprinted by permission from I. R. MacDonald, Chemosynthetic mussels at a brine-filled pockmark in the northern Gulf of Mexico, *Science*: 248, 1096 ff., fig. 2. Copyright 1990 by the AAAS.)

may seem inconsistent at this point, the fact that only the simple one-celled plants inhabit the open oceans, whereas the more complex and highly developed plants are restricted to its margins and the continents, is further evidence pointing to the unique fitness of the oceans as a biological home.

The ocean environment is far more stable than the terrestrial environment, and as a result of this stability, the organisms that live in the oceans have generally not developed highly specialized regulatory systems to combat sudden changes that might occur within their environment. They are, therefore, affected to various degrees by quite small changes in concentrations of various salts in solution, temperature, turbidity, and other environmental variables.

Water constitutes over 80 percent of the mass of **protoplasm,** the substance of living matter. Over 65 percent of the weight of a human being and 95 percent of that of the jellyfish is accounted for by the presence of water (table 12–1). Water carries dissolved within it the gases and minerals needed by organisms and is itself one of the raw materials required by plants in the production of food through photosynthesis. Land plants and animals have developed very complex systems and devices to distribute water throughout their bodies and prevent it from being lost. The threat of atmospheric desiccation (drying out) does not exist for the inhabitants of the open ocean, as they are abundantly supplied with water.

Support

Another necessity of plants and animals is support. Land plants have vast root systems that fasten the plant securely to Earth. In the animal kingdom there are a number of support systems used, but each requires some combination of appendages that must totally support the land animal's weight.

In the ocean this is not the case. The water of the oceans, as it so lavishly bathes the organisms with their needed gases and nutrients, serves as their basic support as well. Organisms that live in the open ocean depend primarily upon buoyancy and frictional resistance to sinking to maintain them in their desired position. This is not to say that they do not have problems maintaining their position within the ocean; some of the special adaptations that are developed toward increasing their efficiency in this respect will be discussed in this and succeeding chapters.

Effects of Salinity

Marine animals are significantly affected by relatively small changes in their environmental conditions. One of these that is highly important is change of salinity. Some oysters that live in an estuarine environment at the mouths of rivers are capable of withstanding a considerable range of salinity. During floods the salinity will be extremely low. On a daily basis, as the tides force ocean water into the river mouth and draw it out again, the salinity changes considerably. The oysters, and most other organisms that inhabit coastal regions, have a tolerance for a wide range of salinity conditions. They are **euryhaline.** By contrast, other marine organisms, particularly those that inhabit the open ocean, can withstand only very small changes in salinity; these organisms are **stenohaline.**

Certain plants and animals cause important alterations in the amount of dissolved material in ocean water by extracting minerals to construct hard parts of their bodies that serve as protective coverings. The compounds primarily used for this purpose are silica (SiO_2) and calcium carbonate ($CaCO_3$). The silica is used by the most important plant population in the ocean, the diatoms, and the microscopic animals called radiolaria. Calcium carbonate is used by the foraminifera, most members of the phylum Mollusca, corals, and some algae that secrete a calcium carbonate skeletal structure.

Cell membranes are semipermeable and will allow the passage of molecules of some substances while screening out others. By **diffusion,** cells may take in the nutrient substances they need from the surrounding fluid medium that contains them in high concentration. The nutrient compounds, to which the cell wall is permeable, pass from the surrounding medium into the cell, where they are in lesser concentrations (figure 12–1).

The waste materials that a cell must dispose of after it has utilized the energy held in these nutrients are passed out of the cell by the same method. As the concentration of waste materials becomes greater within the cell than within the fluid medium surrounding the cell, these materials will pass out of this area of high concentration within the cell into the surrounding fluid. The waste products will then be carried away by the circulating fluid that services the cells in higher animals or by the surrounding water medium that bathes the simple one-celled organisms of the oceans.

Osmotic Pressure. When aqueous solutions of unequal salinity are separated by a water-permeable membrane, there will be diffusion of water molecules through the membrane into the more concentrated solution. This

TABLE 12–1
Water Content of Organisms.

Organism	Water Content (%)
Human	65
Herring	67
Lobster	79
Jellyfish	95

A.

DIFFUSION

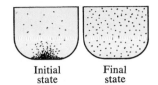

Initial state   Final state

If a substance soluble in water is placed in a pile on the bottom of a beaker of water, it will eventually become evenly distributed through random molecular motion. Diffusion means molecules of a substance move from areas of high concentration of the substance to areas of low concentration.

B.

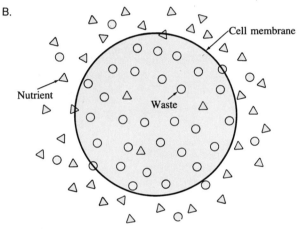

Nutrients are in high concentration outside the cell and diffuse into the cell through the cell membrane. Waste is in higher concentration inside the cell and diffuses out of the cell through the cell membrane.

FIGURE 12–1
Diffusion.

OSMOSIS

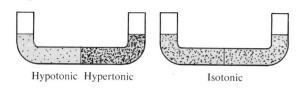

Hypotonic Hypertonic   Isotonic

Separating two water solutions of different salinities by a membrane that will allow the passage of water molecules but not the molecules of dissolved substances sets the stage for osmosis to occur. Water molecules will achieve a net motion by diffusion through the membrane into the more concentrated (hypertonic) solution from the less concentrated (hypotonic) solution. Osmosis will not produce a net movement of water through the membrane when the salinity of the two solutions is the same. They are then isotonic.

FIGURE 12–2
Osmosis.

process is called **osmosis** (figure 12–2). Such solutions can be compared in terms of their osmotic pressures. The **osmotic pressure** of a solution is that which must be applied to keep pure water that is separated from it by a water-permeable membrane from passing through the membrane and diluting the solution. The higher the salinity of the solution, the higher its osmotic pressure. We can compare the osmotic pressure of solutions by comparing their concentrations of dissolved solids—their salinity. This comparison is readily done by determining their freezing points, since the addition of dissolved solids to an aqueous solution lowers its freezing point (figure 12–3). This effect is due to the interference of the dissolved particles with the formation of ice crystals. The lower the freezing point, the higher the osmotic pressure of a solution.

If the salinity of body fluid and the external medium are equal, they are said to be **isotonic** and have

equal osmotic pressure. No net transfer of water will occur through the membrane. If the external fluid medium has a lower osmotic pressure and lower salinity than the body fluid within the cells of an organism, water will pass through the cell walls into the cells. This organism is **hypertonic** relative to the external medium. Should the salinity or osmotic pressure within the cells of an organism be less than that of the external medium, water from the cells will pass through the cell membranes into the external medium. This organism is described as **hypotonic** relative to the external medium.

In an oversimplification, we will consider osmosis as simply a diffusion process. If we consider the relative concentration of water molecules on either side of the semipermeable membrane, we will see that the net transfer of water molecules is from the side where the greatest concentration of water molecules exists through the membrane to the side that contains a lesser concentration of water molecules. The processes of water mole-

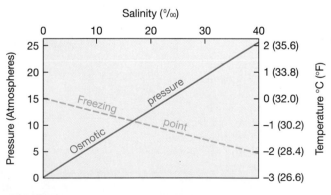

FIGURE 12–3
Osmotic Pressure and Freezing vs. Salinity. Osmotic pressure increases and the freezing point decreases with increased salinity.

cules moving through the semipermeable membrane (osmosis), nutrient molecules moving from the external medium where they are more concentrated into the cell where they will be used to maintain the cell, and the movement of molecules of waste materials from the cell into the surrounding medium that carries them away are all going on at the same time.

We must keep in mind that during this process molecules of all the substances in the system are passing through the membrane in both directions. Nevertheless, there will be a net transport of molecules of a given substance from the side on which they are most highly concentrated to the side where the concentration is less.

In the case of the marine invertebrates, the body fluids and the external medium in which they are immersed are in a nearly isotonic state. Since no significant difference exists, they do not have to develop special mechanisms to maintain their body fluids at a proper concentration. Thus, they have an advantage over their freshwater relatives, whose body fluids are hypertonic in relation to the low-salinity external medium (table 12–2).

Marine fish have body fluids slightly more than one-third as saline as ocean water. This level may be the result of their having evolved in low-salinity coastal waters. They are therefore, hypotonic in relation to the surrounding medium. This difference produces a problem in that the saltwater fish, without some means of regulation, would be dehydrated by loss of water from body fluids to the surrounding ocean environment. This loss is counteracted by the fact that marine fish drink ocean water and excrete the salts through "chloride cells" located in the gills. They also help maintain their body water by discharging a very small amount of very highly concentrated urine (figure 12–4). To summarize the process by which marine fish maintain low osmotic pressure

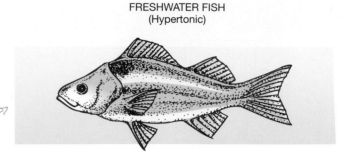

FRESHWATER FISH
(Hypertonic)

Do not drink
Cells absorb salt
Large volume of dilute urine

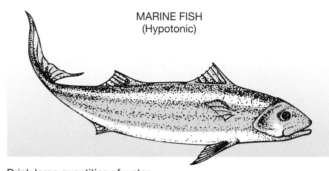

MARINE FISH
(Hypotonic)

Drink large quantities of water
Secrete salt through special cells

FIGURE 12–4
Freshwater and Marine Fish.

in their body fluids: They drink large quantities of water, extract the salts of that water, and dispose of the salts by excreting them. They conserve body water by excreting a highly concentrated urine solution.

TABLE 12–2
Osmotic Pressure of Body Fluids in Various Organisms.
A comparison of the freezing points of the body fluids of some marine and freshwater organisms and the waters in which they live. The lower the freezing point of the fluid, the higher the salinity and osmotic pressure. Marine invertebrates and sharks are essentially isotonic, while most marine vertebrates are hypotonic. Essentially all freshwater forms are hypertonic. The freezing points of ocean water corresponding to invertebrate specimens vary because the specimens were taken from coastal waters, where water salinity varied.

	Marine Animals			Freshwater Animals		
	Freezing Points (°C)				Freezing Points (°C)	
Animals	**Body Fluid**	**Water**		**Animals**	**Body Fluid**	**Water**
INVERTEBRATES						
Annelid worm	−1.70	−1.70		Mussel	−0.15	−0.02
Mussel	−2.75	−2.12		Water flea	−0.20	−0.02
Octopus	−2.15	−2.13		Crayfish	−0.80	−0.02
Lobster	−1.80	−1.80				
Crab	−1.87	−1.87				
VERTEBRATES						
Cod	−0.74	−1.90		Eel	−0.62	−0.02
Shark	−1.89	−1.90		Carp	−0.50	−0.02
Turtle	−0.50	−1.90		Water snake	−0.50	−0.02
Seal	−0.50	−1.90		Manatee	−0.50	−0.02

Freshwater fish are hypertonic in relation to the very dilute external medium in which they live. The osmotic pressure of the body fluids of such fish may be 20 to 30 times greater than that of the fresh water that surrounds them. Their problem is to avoid taking into their body cells large quantities of water that would eventually rupture the cell walls. In order to meet this problem, the freshwater fish does not drink water. Its cells have the capacity to absorb salt. The water that may be gained as a result of body fluids being greatly hypertonic is disposed of with large volumes of very dilute urine (figure 12–4). To summarize the means by which the freshwater fish avoid having their cells ruptured by the intake of water through osmosis: They do not drink, they have cells that can absorb salt to maintain their osmotic pressure, and they dispose of large quantities of water through copious flow of dilute urine.

We should all be aware that we cannot drink salt water when at sea because the osmotic pressure of our body fluids is about one-fourth that of ocean water. To drink ocean water would cause a drastic loss of water through the wall of the digestive tract by osmosis, causing eventual dehydration.

Availability of Nutrients

The distribution of life throughout the ocean's breadth and depth is basically dependent upon the availability of plant nutrients. In areas where the conditions are right for the supply of large quantities of nutrient materials, marine populations will reach their greatest concentration. Where are these areas in which we find the greatest concentration of biomass? To answer this question, we must consider the sources of nutrients.

Previously, we discussed the role played by water in eroding the continents, carrying the eroded material to the oceans, and depositing it as sediment at the margins of the continents. During this process, the water dissolves inorganic compounds in the form of nitrate and phosphate compounds that serve as the basic nutrient supply for plants. Through the process of photosynthesis, plants combine these with carbon dioxide and water to produce the carbohydrates, proteins, and fats that store the energy upon which the rest of the biological community depends.

If the continents are the major sources of these inorganic nutrients, then we might expect that the greatest concentrations of marine life will be found at their margins. This is the case. As we go to the open sea from the margins of the continents, we will travel through waters containing progressively lower concentrations of marine life. This situation is basically due to the vastness of the world's oceans and the great distance between the open ocean and the coastal regions where the nutrients are supplied.

Availability of Solar Radiation

To make use of these available nutrients, the plants must meet another requirement. Photosynthesis cannot proceed without the appropriate energy provided by solar radiation. The radiation from the sun penetrates the atmosphere quite readily, so plants on the continents would seem to have an advantage over those in the ocean, as there is hardly ever any shortage of solar radiation to drive the photosynthetic reactions for these land-based plants.

By contrast, the ocean water does represent a significant barrier to the penetration of solar radiation, although in the clearest water solar energy may be detected to depths in excess of about 1 km (0.62 mi). However, the amount of solar energy reaching these depths is extremely small and would not be sufficient to support the photosynthetic process. Photosynthesis in the ocean is restricted to a very thin layer, approximately the upper 100 m (328 ft), which is the depth to which sufficient light energy to carry on photosynthesis can penetrate. Near the coast this zone is much thinner as the water contains more suspended material that restricts the light penetration.

Considering these two factors necessary for photosynthesis, the supply of nutrients and the presence of solar radiation, we find that in the open ocean away from the continental margins the column of water having solar energy available extends deeper, but that this column contains only a small concentration of nutrients. Conversely, in the coastal regions where the water is more turbid, light penetrates to much shallower depths, but the nutrient supply is quite rich. Since the coastal zone is by far the most productive zone, we can see manifested in this comparison the overriding importance of nutrient availability to life in the oceans.

Margins of the Continents

We have mentioned that the stable condition of the ocean environment makes it ideal for the continuation of life processes, but we are now indicating that the richest concentration of marine organisms is found in the very margins of the oceans, where the conditions are the least stable. When we consider the physical conditions that would appear at first to be deterrents to the establishment of life, we will see that many are present in the coastal regions.

The water depths are shallow, allowing seasonal variations in temperature and salinity that are much greater than would be found in the open ocean. A varying thickness of water column exists in the nearshore region as a result of the tide movements that periodically cover and uncover a thin strip along the margins of the continents, where there is a constant beating of the surf. The surf condition represents a sudden release of energy that has been carried for great distances across the open ocean. During this transit, little energy was lost and little effect on the marine environment was observed as a consequence of its transfer.

The development of such a great concentration of biomass in the environment containing so many factors that would seem to inhibit the development of life serves to highlight the importance of the evolutionary process in developing new species by natural selection. Over the eras of geologic time, new life forms have developed to fit into all biological niches. Many of these forms have adapted to live under conditions that would seem deleterious to the conservation of life. This fact attests to the great range of physical conditions within which life can exist so long as those basic requirements for the production of the food supply are met.

We find that along continental margins there are those areas where life is more abundant than others. What are the characteristics of these areas that result in such an uneven distribution of life along the continental margins? Again, we need consider only those basic requirements for the production of food for the answer. If we measure the various properties of the water along the coast of the continents and compare those measurements from place to place, we will find that the areas containing the greatest concentrations of biomass are those where the water temperatures are lower. Due to the low temperature, it is possible for the water to maintain greater amounts of the important gases, oxygen and carbon dioxide, in solution than would be found in warmer waters. Of particular importance is the increased availability of carbon dioxide, and we see again an example of the greater availability of the basic requirements of plants affecting the distribution of life in the oceans.

Upwelling. In certain areas of the coastal margins we find an additional factor that enhances the conditions for life—upwelling. Upwelling occurs along the western margins of continents where surface currents are moving toward the equator. It consists of a flow of subsurface water to the surface that grows directly out of the relationship existing between the direction of the current flow and the continental margin.

In the case of northward-flowing currents along the western margin of the Southern Hemisphere continents or southward-flowing currents along the western margin of Northern Hemisphere continents, the Ekman transport tends to move the surface water away from the coast. As the surface water moves away from the continents, water surfaces from depths of 200 to 1000 m (656 to 3280 ft) to replace it. This water that surfaces from depths below those where photosynthesis occurs is rich in plant nutrients because there are no plants in these deeper waters to utilize them. This constant replenishing of nutrients at the surface enhances the conditions for life in these areas. Upwelling water is usually of low temperature, which produces the additional benefit of having a high capacity for dissolved gases (figure 12–5).

Water Color and Life in the Oceans

It is usually possible to visually determine areas of high and low organic production in the ocean from the color of the water. Coastal waters are almost always greenish in color because they contain more large particulate matter that disperses solar radiation in such a way that the wavelengths most scattered are those that represent the greenish or yellowish light. This condition is also partly the result of the presence of yellow-green microscopic marine plants in these coastal waters. In the open ocean, where particulate matter is relatively scarce and marine life exists in low concentration, the water appears blue due to the size of the water molecules and the scattering they effect on the solar radiation, which is similar to the process that produces the apparent blueness of the skies.

Green color in water usually indicates the presence of a lush biological population such as might correspond to the jungle environments on the continents. The deep indigo blue of the open oceans, particularly between the tropics, usually indicates an area that lacks abundant life and could be considered a biological desert.

Size

Earlier in this chapter we referred to the fact that marine plants are quite simple in comparison to the specialized forms we find throughout the continents. How can we explain the fact that so many marine plants have not followed the path of aggregation (becoming multicellular)?

Assuming the availability of nutrients, the major requirements of marine plants are that they maintain themselves within the upper layers of ocean water where solar radiation is available to carry on photosynthesis and that they take in nutrients from the surrounding waters and expel waste materials as efficiently as possible. These requirements lead us to a consideration of size and shape. The ease with which marine plants with no means of locomotion will maintain their position in the upper layers of the ocean is closely related to the ratio that exists between surface area and body mass. The greater the surface area per unit of body mass, the easier it will be for an organism to maintain its position because there is more surface per unit of body mass in frictional contact with the surrounding water. Greater surface area per unit of body mass amounts to a greater resistance to sinking per unit of body mass. An examination of figure 12–6 will help you understand that this ratio of surface area per unit of mass increases with decrease in size. It can readily be seen that it would benefit the one-celled plants making up the bulk of marine plant life to be as small as possible. They are, in fact, microscopic.

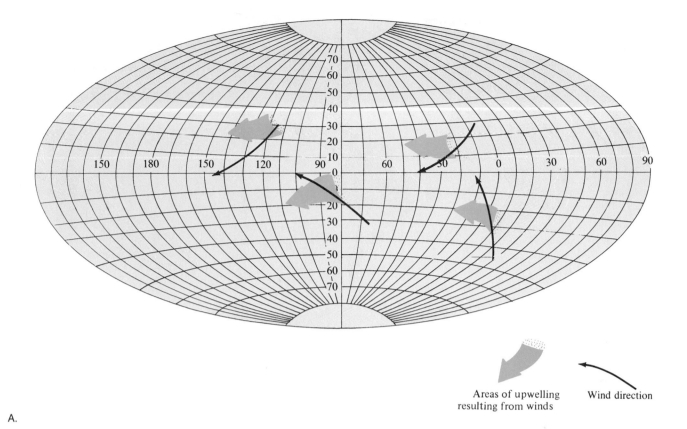

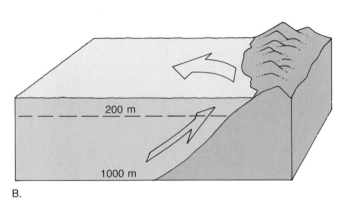

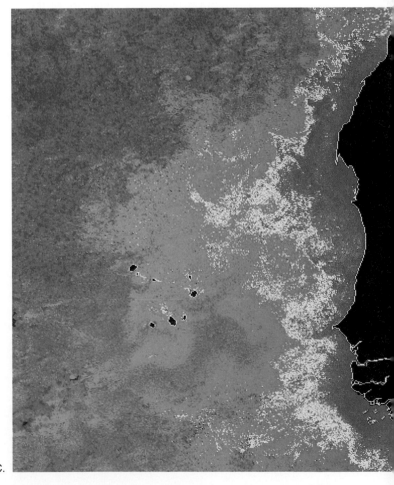

A.

B.

C.

Areas of upwelling
resulting from winds

Wind direction

FIGURE 12–5
Upwelling. *A,* the coastal winds usually drive currents south along
the western margins of continents in the Northern Hemisphere,
while they are usually driven north in the Southern Hemisphere.
The Ekman transport moves water away from the western margins
of the continents. (Base map courtesy of National Ocean Survey.)
B, as the surface water moves away from the continent, water from
depths of 200–1000 m (656–3280 ft) rises to replace it. Because this
water comes from below the photosynthetic zone, it is cold and rich
in nutrients. *C,* in this false-color CZCS image from the *Nimbus-7*
satellite, wind-driven upwelling along the coast of northwest Africa
results in high phytoplankton biomass and productivity. Highest productivity is indicated by colors of red and orange, decreasing
through yellow, green, and blue. (Image courtesy of NASA/Goddard
Space Flight Center.)

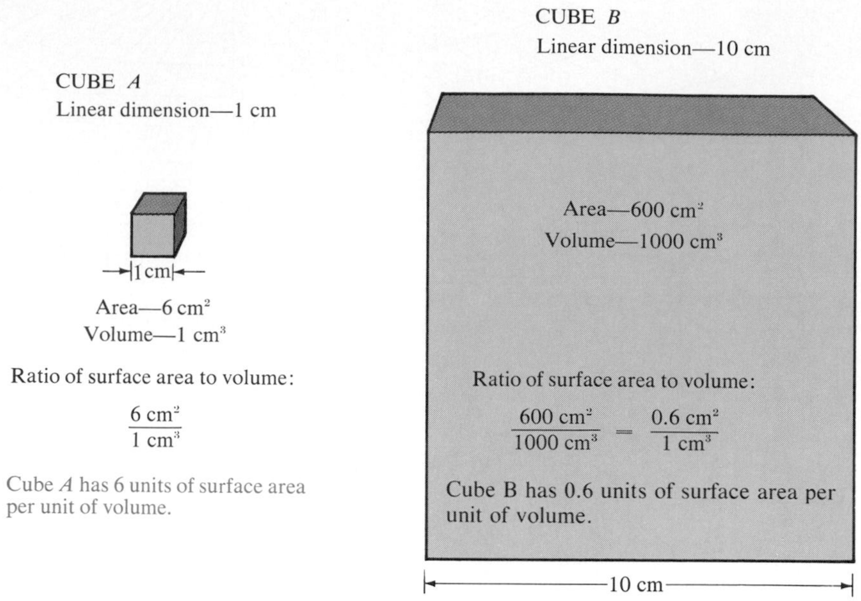

FIGURE 12–6
The Importance of Size. Cube *A* has ten times as much surface area per unit of volume as cube *B*. If both cubes are composed of the same substance, their masses are proportional to their volumes, since mass = volume × density. If cubes *A* and *B* were plankton, cube *A* would have ten times as much frictional resistance to sinking per unit of mass as cube *B*. It could stay afloat by exerting far less energy than cube *B*. Cube *A*, were it a planktonic alga, could also take in nutrients and dispose of waste through the cell wall ten times as efficiently as an alga with the dimensions of cube *B*.

CUBE *A*
Linear dimension—1 cm

←|1 cm|←

Area—6 cm²
Volume—1 cm³

Ratio of surface area to volume:

$$\frac{6 \text{ cm}^2}{1 \text{ cm}^3}$$

Cube *A* has 6 units of surface area per unit of volume.

CUBE *B*
Linear dimension—10 cm

Area—600 cm²
Volume—1000 cm³

Ratio of surface area to volume:

$$\frac{600 \text{ cm}^2}{1000 \text{ cm}^3} = \frac{0.6 \text{ cm}^2}{1 \text{ cm}^3}$$

Cube B has 0.6 units of surface area per unit of volume.

|←————————10 cm————————→|

The efficiency with which the plants take in the nutrients from the surrounding water and expel their waste through their cell membranes is also related to the amount of surface area per unit of mass. The greater the surface area per unit of mass, the easier it is for the organism to carry on these functions. Since both processes are dependent on diffusion, there would be a point at which increased size would reduce the surface area to body mass ratio to the point that the cell could not be serviced and would die. It is a direct result of this condition that we find cells in all plants and animals to be microscopic, regardless of the size of the organism. Each cell must take in nutrients and dispose of wastes by diffusion.

Examining diatoms carefully, we see that it is quite common for them, as well as other members of the microscopic marine community, to have long needlelike extensions from their siliceous tests, or protective coverings. These extensions serve to increase further the ratio of surface area to the mass of the organism.

Small plants and animals use another contrivance to increase their ability to stay in the upper layers of the ocean: They produce a drop of oil, which tends to decrease their overall density and increase their buoyancy.

With all these adaptations made in an attempt to increase the ability to float, the organisms still have a density in excess of the density of water and therefore tend to sink, if ever so slowly. This is not a serious handicap, since they are so small and are carried quite readily along with the water movements. In these surface layers, there is considerable turbulent action associated with the mixing that results from wind action on the surface. This turbulence tends to dominate the movement of these small organisms to the point that they are able to spend adequate time bathed in the solar radiation to carry on their role of producing the energy needed by the other members of the marine biological community.

Viscosity

We are familiar with the fact that liquids flow with varying degrees of ease, and we know that syrup flows less readily than water. This condition is indicative of the higher viscosity of syrup as compared to water. **Viscosity,** defined as internal resistance to flow, is a characteristic of all fluids.

The viscosity of ocean water is affected by two variables—temperature and salinity. Viscosity of ocean water increases with an increase in salinity but is affected even more by temperature, a lowering of which increases viscosity. As a result of this temperature relationship, we find that floating plants and animals in colder waters have less need for extensions to aid them in floating than those that occupy warmer waters. It has been observed that members of the same species of floating crustaceans will be very ornate with featherlike appendages where they occupy warmer waters, while these appendages are missing in the colder, more viscous environments. It appears then that high viscosity would benefit the floating members of the marine biocommunity because it would make it easier for them to maintain their position in the surface layers.

With increasing size of organisms and a change in the mode of life, viscosity ceases to enhance the ability of an organism to pursue its lifestyle and becomes an obstacle instead. This is particularly true of the large organisms that swim freely in the open ocean. They must pursue prey and displace water to move ahead. The more rapidly they swim, the greater the stress that is created on the organism. Not only must water be displaced ahead of the animal, but water must move in behind it to occupy the space that it has vacated. The latter is one of the more important considerations in **streamlining.**

The familiar shape of the free-swimming fish and the mammals manifests the type of adaptations that must be achieved in streamlining an organism to move with a minimum of effort through the water. We see that a common shape achieved to meet this need is a laterally flattened body presenting a small cross section at the anterior end with a gradually tapering posterior. This form, which is characteristic of the bony fishes, reduces stress that results from movement through the water and makes it possible for this movement to be achieved with the exertion of a minimum of energy (figure 12–7).

Temperature

Of all the conditions we can measure in the ocean, there are probably no physical characteristics we have discussed that are more important to the life in the oceans than temperature. It has been stated that the marine environment is much more stable than that of the land. A comparison of temperature ranges that exist in the sea with those that are found on land will serve to exemplify this condition.

The minimum temperature observed in the open sea is never much less than −2°C (28.4°F), and the maximum seldom exceeds 27°C (80.6°F). The continental temperatures range from a low of −88°C (−127°F) to a high of about 58°C (136°F). The temperature found on

A body with a low degree of streamlining. Its broad structure causes too much resistance area in front and too much wake region behind.

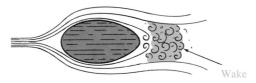

Wake

Teardrop shape (fusiform) front is well rounded and the body tapers gradually to the rear, producing a small wake region.

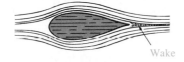

Wake

FIGURE 12–7
Streamlining. Due to the viscosity of water, any body moving rapidly through it must be designed properly to produce as little stress as possible as it displaces the water through which it moves. After the water has moved past the body, it must fill in behind the body with as little eddy action as possible.

the continent varies considerably on a seasonal and daily basis. Daily variations at the ocean surface rarely exceed more than 0.2°C (0.4°F) or 0.3°C (0.5°F), although they may be as great as 2° or 3°C (3.6 or 5.4°F) in shallower coastal waters. Annual temperature variations are also small. They range from 2°C (3.6°F) at the equator to 8°C (14.4°F) at 35° to 45° latitude and decrease again in the higher latitudes. Annual variations of temperature in shallower coastal areas may be as high as 15°C (27°F).

The temperature variations we are discussing as characteristic of surface waters are reduced in magnitude with an increase in water depth. In the deep oceans daily or seasonal variation in temperature becomes a consideration of little or no importance. Throughout the deeper parts of the ocean the temperature remains uniformly low. At the bottom of the ocean basin, where the depth is in excess of 1.5 km (0.9) mi), temperatures hover around the 3°C (37.4°F) mark regardless of the latitude.

We previously discussed the fact that a decrease in temperature will increase density, viscosity, and the capacity of water to hold gases in solutions. All these changes have significant effects upon the organisms inhabiting the ocean. A direct consequence of the increased capacity of water to contain dissolved gases is seen in the vast plant communities that develop in the high latitudes during the summer seasons when solar energy is available to carry on photosynthesis. One of the major factors in this phenomenon is the abundance of dissolved gases—carbon dioxide for the use of plants in carrying on photosynthesis and oxygen needed by the animals that feed upon the plants.

It has been observed that floating organisms are larger individually in colder waters than in warmer waters of the tropics, although the tropical populations appear to be characterized by a larger number of species. However, the total biomass of floating organisms in the colder high-latitude planktonic environments greatly exceeds that of the warmer tropics. The fact that organisms in the tropics are smaller than those observed in the higher latitudes could quite possibly be related to the lower viscosity and density found in the lower latitude waters. Being smaller, these tropical species can expose more surface area per unit of body mass, and they are also characterized by ornate plumage to further increase surface area per unit of body mass (figure 12–8). These plumose adaptations are strikingly absent in the larger cold-water species.

Increases in temperature increase the rate of biological activity, which more than doubles with an increase in the temperature of 10°C (18°F). Tropical organisms apparently grow faster, have a shorter life expectancy, and reproduce earlier and more frequently then their counterparts in the colder waters.

There are some species of animals that can successfully live only in cooler waters and others than can do so

FIGURE 12–8
Water Temperature and Appendages. *A,* copepod *(Oithona)* that displays the ornate plumage characteristic of warm water varieties. *B,* copepod *(Calanus)* that displays the less ornate appendages found on temperate and cold water forms. (After Sverdrup, Johnson, and Fleming.)

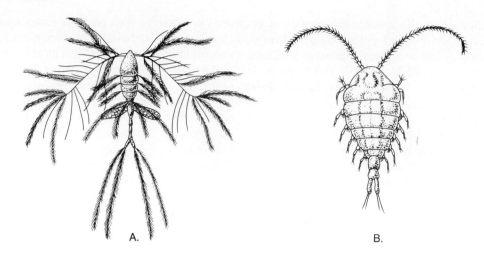

A.   B.

only in warmer waters. Many of these can withstand only a very small change in temperature and are called **stenothermal.** Other varieties apparently are little affected by temperature change and can withstand changes over a large range. These are classified as **eurythermal.** Stenothermal organisms are found predominantly in the open ocean and at depths where it is very unlikely that large ranges of temperature will occur. The eurythermal organisms are more characteristic of the shallow coastal waters, where the largest ranges of temperature are found, and the surface waters of the open ocean.

DIVISIONS OF THE MARINE ENVIRONMENT

We can readily divide the marine environment into two basic units: the ocean water itself is the **pelagic environment,** and the ocean bottom constitutes the **benthic environment** (figure 12–9A).

Pelagic Environment

The pelagic environment is further divided into two provinces. The **neritic province,** which extends from the shore seaward, includes all water overlying an ocean bottom less than 200 m (660 ft) in depth. Beyond this depth the **oceanic province** is found.

The oceanic province includes water with a very great range in depth from the surface to the bottom of the deepest ocean trenches. It is subdivided into the **epipelagic zone** from the surface to a depth of 200 m (660 ft), the **mesopelagic zone** from 200 to 1000 m (660 to 3280 ft), the **bathypelagic zone** from 1000 to 4000 m (3280 to 13,000 ft), and the **abyssopelagic zone,** which includes all the deepest parts of the ocean below 4000 m (13,000 ft) depth.

An important factor in determining the distribution of life in the oceanic province and boundaries of

biozones within it is the availability of light. The **euphotic zone** extends from the surface to a depth where there is enough light to support photosynthesis, rarely to more than a depth of 100 m (328 ft). Below this zone, there are small but measurable quantities of light within the **disphotic zone** to a depth of about 1000 m (3280 ft). Below this depth there is no light in the **aphotic zone.**

The upper epipelagic zone is the only oceanic biozone in which there is sufficient light to support photosynthesis. The boundary between it and the mesopelagic zone (200 m or 660 ft) is also the approximate depth at which the level of dissolved oxygen begins to decrease significantly. This decrease in oxygen results because no plants are found beneath about 150 m (490 ft), and the dead organic tissue (detritus) descending from the biologically productive upper waters is undergoing decomposition by bacterial oxidation (figure 12–9B). Nutrient content of the water also increases abruptly below 200 m (600 ft) (figure 12–9). This depth also serves as the approximate bottom of the mixed layer, seasonal thermocline, and surface water mass.

Within the mesopelagic zone, a dissolved oxygen minimum occurs at a depth of about 700 to 1000 m (2300 to 3280 ft). The intermediate-water masses that move horizontally in this depth range often possess the highest levels of plant nutrient content in the ocean. The base of the permanent thermocline and the boundary between the disphotic and aphotic zones characteristically occur at the 1000-m (3280-ft) boundary between the mesopelagic and bathypelagic zones.

Within the mesopelagic zone, we see evidence that animals sense the presence of light in the phenomenon known as the **deep scattering layer** (DSL). During sonar equipment tests by the U.S. Navy early in World War II, a sound-reflecting surface that changed depth on a daily basis was observed. The variation indicated a reflecting mass at a depth of 100 to 200 m (328 to 660 ft)

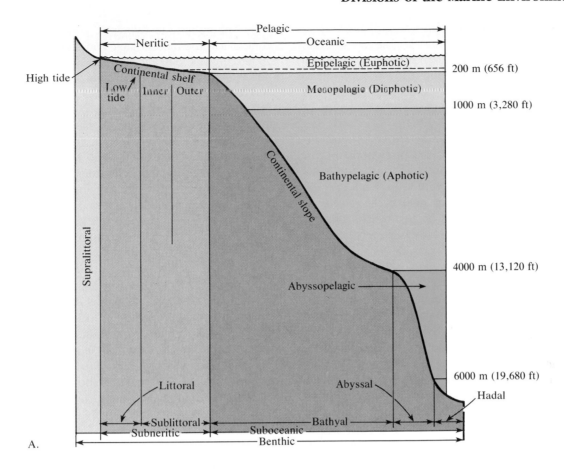

A.

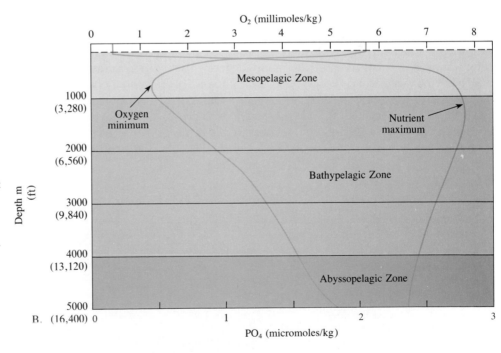

FIGURE 12–9
Oceanic Biozones. A, distribution of biozones of the pelagic and benthic environments. B, distribution of oxygen (O_2) and plant nutrient phosphate (PO_4) in the water column, mid- to low latitudes. Oxygen is abundant in surface water owing to mixing with the atmosphere and plant photosynthesis. Nutrient content is low in surface water owing to uptake by plants. Oxygen content decreases and nutrient content increases abruptly below the euphotic zone. An oxygen minimum and nutrient maximum are recorded at or near the base of the mesopelagic zone. Nutrient levels remain high to the bottom while oxygen content increases with depth as deep- and bottom-water masses carry oxygen into the deep ocean.

during the night that sank to depths as great as 900 m (2950 ft) during the day. After considerable investigation it was determined that this echo, the DSL, was produced by masses of migrating marine life that moved closer to the surface at night and then to a greater depth during the day. DSL appeared to be a response to the changing intensity of solar radiation (figure 12–10).

The use of plankton nets and submersibles has indicated that the DSL contains layers of siphonophores, copepods, euphausids, cephalopods, and small fish. Tests that have been done using fish with swim bladders suggest that they may be the primary cause of the reflection, since a very small concentration of such fish is sufficient to produce such a phenomenon. Fish, however, are predators and if their presence is responsible for the echoes, the organisms on which they prey are probably responsible for the cyclic nature of their movement. The smaller organisms responding to a changing intensity of light could trigger the vertical movements, while the presence of fish that prey upon them could be primarily responsible for the echoing phenomenon associated with the migration.

The mesopelagic zone is also inhabited by species of fish that have unusually large and sensitive eyes, capable of detecting light levels 100 times less than those which the human eye can detect. Another important inhabitant of this zone is the bioluminescent group, especially shrimp, squid, and fish. Approximately 80 percent of the inhabitants carry light-producing **photophores.** These are glandular cells containing luminous bacteria surrounded by dark pigments. Some contain lenses to amplify the radiation. This cold light is produced by a chemical process involving the compound luciferin. The molecules of luciferin are excited and emit photons of light in the presence of the enzyme luciferase and oxygen. Only a 1 percent loss of energy is required to produce the illumination. This system is similar to that of the firefly.

The aphotic bathypelagic zone and abyssopelagic zone below the mesopelagic zone represent in excess of 75 percent of the living space in the oceanic province. In this region of total darkness many totally blind fish exist. Very bizarre small predaceous species make up the total fish population. Many species of shrimp that normally feed on detritus become predators at these depths, where the food supply has been greatly reduced from what is available in the shallower waters. The animals that live in the aphotic bathypelagic and abyssopelagic zones feed mostly upon one another and have developed impressive warning devices and unusual apparatus to make them more efficient predators. They are characterized by small expandable bodies, extremely large mouths relative to body size, and very efficient sets of teeth (figure 12–11).

Aside from the depth of light penetration, the primary physical criteria used in defining the boundaries of the bathypelagic zone are the top and bottom of deep-water masses. Oxygen content increases with depth, as these water masses carry oxygen from the cold surface waters where they formed to the deep ocean. The abyssopelagic zone is the realm of the bottom-water masses that commonly move in the opposite direction of the overlying deep-water masses of the bathypelagic zone.

Benthic Environment

The benthic environment can be subdivided into two larger units, which are then subdivided on the basis of various criteria (figure 12–9). These are the **subneritic province,** which extends from the spring high tide shoreline to a depth of 200 m (660 ft) (approximately the continental shelf), and the **suboceanic province,** which includes all the benthic environment below the 200 m (660 ft) depth.

A transitional region above the high tide line is the **supralittoral.** The supralittoral zone, commonly called

FIGURE 12–10
Deep Scattering Layer. The deep scattering layer, which scatters and reflects sonar signals well above the bottom, may be caused by euphausids that grow to lengths greater than 2 cm (0.8 in.) and lantern fish (myctophid) that reach lengths of 7 cm (2.8 in.). They are predators that feed on smaller planktonic organisms migrating vertically in the water column on a daily cycle.

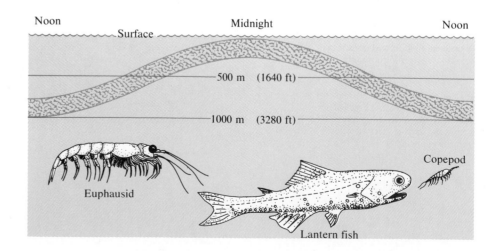

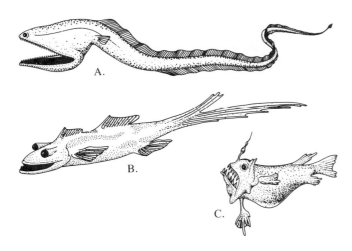

FIGURE 12–11
Deep-Sea Fish. A, *Eurypharynx pelecanoides.* Length, 50–60 cm
(20–24 in.) B, *Gigantura chuni.* Length, to 12 cm (5 in.). C,
Linophryne bicornis. Length, 5–8 cm (2–3 in.).

the **spray zone,** is covered with water only during periods of extremely high tides and when tsunamis or large
storm waves break on the shore.

The intertidal zone is classified as the **littoral
zone,** and from low tide to 200 m (660 ft), the subneritic
is called the **sublittoral** or shallow subtidal zone. The
inner sublittoral includes that part of the sublittoral to
a depth of approximately 50 m (160 ft). This seaward
limit will vary considerably because it is determined by
that depth at which we find no plants growing attached to
the ocean bottom. Its extent is determined primarily by
the amount of solar radiation that penetrates the surface
water. In areas of high turbidity the inner sublittoral will
have its seaward limit at shallower depths due to the decreased penetration of solar radiation, while in areas
where the water is unusually clear it may stand well
below the 50 m (160 ft) mark. The **outer sublittoral** includes that portion of the sublittoral zone from the seaward limit of the inner sublittoral to a depth of 200 m
(660 ft) or the shelf break, the seaward edge of the continental shelf.

With increased depth in the suboceanic system, we
find the **bathyal zone** extending from a depth of 200 to
4000 m (660 to 13,100 ft) and corresponding generally to
that geomorphic province found beyond the continental
shelf—the continental slope. From a depth of 4000 to
6000 m (13,100 to 19,700 ft) stretches the **abyssal zone,**
which includes in excess of 80 percent of the benthic
environment. A restricted environment, including all
depths below 6000 m (19,700 ft), found only in the
trenches along the margins of continents is the **hadal
zone.**

Once we have reached the outer sublittoral benthic
region, we will see no more attached plants as we go
from there into the deep-ocean basin. All the photosyn-

thesis seaward of the inner sublittoral zone is carried on
by the microscopic algae floating in the neritic province
above the outer sublittoral zone and in the epipelagic
zone of the oceanic province. The base, or seaward limit,
of the inner sublittoral can be said to coincide with the
base of the euphotic zone, which is that zone of water
near the ocean surface into which enough light penetrates to support photosynthesis. The seaward limit of the
outer sublittoral approximately coincides with the shelf
break. Currents are in general stronger across the outer
sublittoral bottom than along the upper continental slope
of the bathyal zone. Sediments are also coarser on the
continental shelf than on the continental slope.

Changes in the sediment type of the continental
slope to a depth of about 4000 m (13,100 ft) are subtle.
Throughout much of the ocean, the calcium carbonate
compensation depth occurs at about this depth, so sediments in the bathyal zone may contain considerable calcium carbonate, while those below contain little or none.
In general, the deposits of neritic sediment found around
the continents begin to be replaced by oceanic sediment
below 4000 m (13,100 ft).

The ocean floor representing the abyssal zone is
covered by soft oceanic sediment, primarily abyssal clay.
The tracks and burrows of animals that live in this sediment are frequently recorded in bottom photographs
(figure 12–12). The abyssal zone represents over 60 percent of the surface area of the benthic environment, or
almost 43 percent of Earth's surface.

The hadal zone below 6000 m (19,700 ft) is primarily ocean trenches. Isolation in these deep, linear depres-

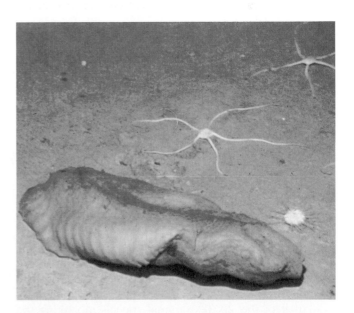

FIGURE 12–12
Benthic organisms. Deep-sea sea cucumber, brittle stars, and sea
urchin. (Photograph by Fred Grassle.)

sions allows the development of faunal assemblages found only in these trenches.

DISTRIBUTION OF LIFE IN THE OCEANS

It is difficult to describe the degree to which the sea is inhabited because of its immense space and the paucity of our knowledge of it. We do know, however, that some populations fluctuate greatly each season, and this fact increases the difficulty of describing with numbers the extent to which the marine environment is populated. Although it is perhaps meaningless to estimate the number of individual organisms in the ocean, we may derive some means of comparing the marine and terrestrial environment by comparing the number of marine species to land species.

Well over 1.2 million species of animals are known, and only about 17 percent of these live in the ocean. Many biologists believe there may be from 3 to 10 million unnamed and undescribed animals living on Earth. A large number of these may well inhabit the oceans. Yet, the following theory may explain why we can expect fewer species of marine animals to exist than terrestrial animals.

If the ocean is such a prime habitat for life and if life originated in this environment, why do we now see such a small percentage of the world's animal species living there? This lesser number may well result because the marine environment is more stable than the terrestrial environment. The relatively uniform conditions of the open ocean do not produce pressures for adaptation. Also, once we get below the surface layers of the ocean, the temperatures are not only stable but also relatively low. Chemical reactions are retarded by this lower temperature; this, in turn, may reduce the tendency for variation to occur.

If we are to consider briefly the great variety of species of organisms found on the continents, we can assume that this development was the product of an environment less stable than is found in the ocean, one possessing many opportunities for natural selection to produce new species to inhabit new niches within this terrestrial environment. At least 75 percent of all land animals are insects that have evolved species capable of inhabiting very restricted environmental niches. If we ignore the insects, the sea does possess in excess of 45 percent of the remaining animal species living in the marine and terrestrial environments.

In describing the distribution of life within the ocean, we can get a general idea of the relative concentrations in the various marine environments if we consider that of the 200,000 animal species inhabiting the marine environment, only 4000 (about 2 percent) live in the pelagic environment. The remainder inhabit the ocean floor (figure 12-13).

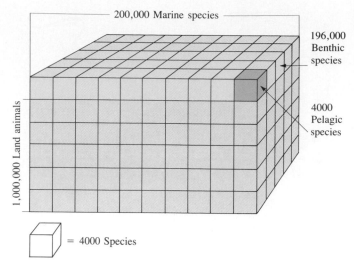

FIGURE 12–13
Distribution of Animal Species on Earth. The cube represents the 1,200,000 species of animals living on Earth. The blue-shaded portion indicates the proportion (17 percent) that lives in the oceans. Of the marine species, 98 percent live on or in the ocean floor.

Before further discussion of the organisms of the oceans, we need to define some terms to describe these organisms on the basis of the portion of the ocean they inhabit and the means by which they move.

Plankton

The **plankton** include all those organisms that drift with ocean currents. This does not mean that all are without the ability to move. Many plankters have this capacity but either are able to move only weakly or are so restricted to vertical movement that they cannot determine their horizontal position within the ocean. The plants that follow this life style in the upper layers of the ocean are called **phytoplankton,** and the animals are **zooplankton** (figure 12–14). It has recently been discovered that free-living bacteria are much more abundant in the plankton community than they were previously thought to be. Having an average dimension of only 0.5μm (0.00002 in.), the **bacterioplankton** were missed in earlier studies because of their small size.

In considering the plankton in more detail, we will try to develop some basic understanding of this portion of marine life that many of us have never seen or imagined. The fact is, most of the biomass of Earth is found adrift in the oceans as plankton. The volume of Earth's space inhabited by animals that either drift or swim greatly exceeds that occupied by all animals that live on land or the ocean's bottom.

The plankton range in size from larger animals and plants such as jellyfish, salps, and Sargassum (**macroplankton,** 2 to 20 cm or 0.8 to 8 in.) to bacteria that are too small to be filtered from the water by a silk net

PHYTOPLANKTON

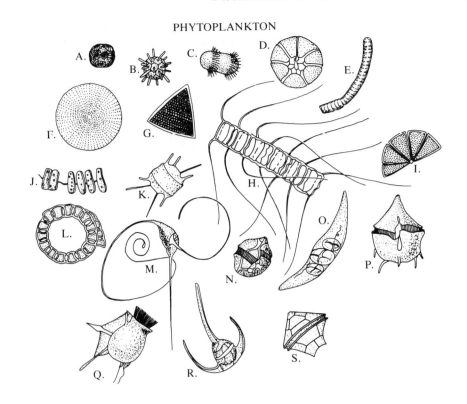

FIGURE 12–14
Phytoplankton and Zooplankton.
Maximum dimension in parentheses.
Phytoplankton *A* and *B,* coc-
colithophoridae (15 µm, or
0.0006 in.). *C–L,* diatoms (80 µm, or
0.0032 in.): *C, Corethron; D, Asterom-
phalus; E, Rhizosolenia; F, Coscinodis-
cus; G, Biddulphia favus; H,
Chaetoceras, I, Licmophora, J, Thalas-
siorsira; K, Biddulphia mobiliensis; L,
Eucampia. M–S,* dinoflagellates
(100 µm, or 0.004 in.): *M, Ceratium
recticulatum; N, Goniaulax scrippsae;
O, Gymnodinium; P, Goniaulax tri-
acantha; Q, Dynophysis; R, Ceratium
bucephalum; S, Peridinium.*

Zooplankton *A,* fish egg (1 mm, or
0.04 in.); *B,* fish larva (5 cm, or 2 in.);
C, radiolaria (0.5 mm, or 0.02 in.); *D,*
foraminifera (1 mm, or 0.04 in.); *E,*
jellyfish (30 m, or 98 ft); *F,* arrowworms (3 cm, or 1.2 in.); *G* and *H,*
copepods (5 mm, or 0.2 in.); *I,* salp
(10 cm, or 4 in.); *J,* doliolum; *K,* jellyfish (30 m, or 98 ft); *L,* worm larva
(1 mm, or 0.04 in.); *M,* fish larva
(5 cm, or 2 in.); *N,* tintinnid (50µm,
or 0.002 in.); *O,* foraminifera (1 mm,
or 0.04 in.); *P,* dinoflagellate *(Noc-
tiluca)*
(1 mm, or 0.04 in.).

ZOOPLANKTON

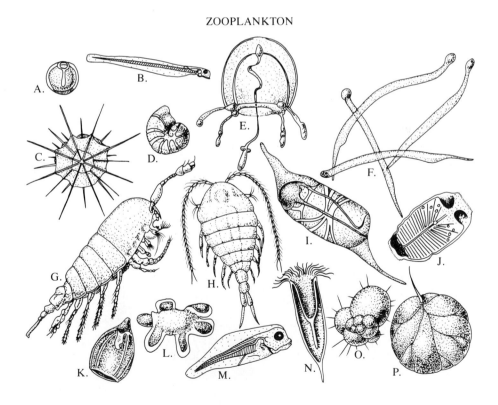

and must be removed by other types of microfilters
(**picoplankton,** 0.2 to 2µm or 0.000008 to 0.00008 in.).

An additional scheme for classifying plankton is
based on the portion of their life cycle spent within the
plankton community. Those organisms such as planktonic diatoms and copepods which spend their entire life

in the plankton are **holoplankton.** Many of the organisms we normally consider as nekton or benthos because
they spend their adult life in one or the other of these
living modes actually spend a portion of their life cycle as
plankton. Many of the nekton and very nearly all the benthos during their larval stages make the plankton com-

munity their home. These organisms, which as adults sink to the bottom to become members of the benthos or begin to swim freely as members of the nekton, are **meroplankton.**

Nekton

The **nekton** include all those animals capable of moving independently of the ocean currents. They are not only capable of determining their own positions within relatively small areas of the ocean but are also capable, in many cases, of long migrations. Included in the nekton are most adult fish and squid, marine mammals, and marine reptiles (figure 12–15).

The freely moving nekton, with some exceptions, are not able to move at will throughout the breadth of the ocean. By the presence of nonvisible barriers they are effectively limited in their lateral range. Gradual changes in temperature, salinity, viscosity, and availability of nutrients create impenetrable barriers. The death of large numbers of fish has frequently been caused by temporary lateral shifts of masses of ocean water. Vertical range is normally determined by pressure for all fish that contain air bladders and mammals that belong to the nekton.

Fish appear to be everywhere but are normally considered to be more abundant near continents and islands and in colder waters. Some fish, such as the salmon, ascend rivers to spawn. Many eels do just the reverse, growing to maturity in fresh water, then descending the streams to breed in the great depths of the ocean.

Benthos

The **benthos** live on or in the ocean bottom. There are those called **infauna** that live buried in the sand, shells, or mud. Those that live attached to rocks or move over the surface of the ocean bottom are the **epifauna.** Benthos such as some shrimp and demersal flounder that live in the bottom but move with relative ease through the water above the ocean floor are the **nektobenthos** (figure 12–16).

FIGURE 12–15
Nekton. Not drawn to scale; typical maximum dimension is in parentheses. *A,* bluefin tuna (2 m, or 6.6 ft); *B,* bottlenosed dolphin (4 m, or 13 ft); *C,* nurse shark (3 m, or 10 ft); *D,* barracuda (1 m, or 3.3 ft); *E,* striped bass (0.5 m, or 1.6 ft); *F,* sardine (15 cm, or 6 in.); *G,* deep-ocean fish (8 cm, or 3 in.); *H,* squid (1 m, or 3.3 ft); *I,* angler fish (5 cm, or 2 in.); *J,* lantern fish (8 cm, or 3 in.); *K* gulper (15 cm, or 6 in.).

FIGURE 12–16
Benthos—Some Intertidal and Shallow Subtidal Forms. Not drawn to scale; typical maximum dimension is in parentheses. *A,* sand dollar (8 cm, or 3 in.); *B,* clam (30 cm, or 12 in.); *C,* crab (30 cm, or 12 in.); *D,* abalone (30 cm, or 12 in.); *E,* sea urchins (15 cm, or 6 in.); *F,* sea anemones (30 cm, or 12 in.); *G,* brittle star (20 cm, or 8 in.); *H,* sponge (30 cm, or 12 in.); *I,* acorn barnacles (2.5 cm, or 1 in.); *J,* snail (2 cm, or 0.8 in.); *K,* mussels (25 cm, or 10 in.); *L,* gooseneck barnacles (8 cm, or 3 in.); *M,* sea star (30 cm, or 12 in.); *N,* brain coral (50 cm, or 20 in.); *O,* sea cucumber (30 cm, or 12 in.); *P,* lamp shell (10 cm, or 4 in.); *Q,* sea lily (10 cm, or 4 in.); *R,* sea squirt (10 cm, or 4 in.).

FIGURE 12–17

Seamounts and Benthos Zonation. *A,* Locations of Jasper Sea-mount and Volcano 7 in the eastern Pacific Ocean. *B,* a high density bed of spiral black corals, *Stichopathes* sp., near a peak on Jasper Seamount. A deep-sea sponge is visible on the right side of the photo. (Photo is courtesy of Paul K. Dayton, Scripps Institution of Oceanography, University of California, San Diego.) *C,* on the upper summit of Volcano 7 (746 m) rocks and foraminiferal sand are covered by protozoan mat and flocculent detritus. White dot is a small coelenterate. *D,* on the lower summit of Volcano 7 (780 m) rock and foraminiferal sand support two large and a number of smaller white sponges, brachyuran crabs (average length is 5 cm), shrimp, sea stars, and white serpulid worm tubes. The green, fluffy material is flocculent detritus. Rattail fish swim near the right margin of image. *E,* on the lower summit of Volcano 7 (819 m) rocky area supports crabs, shrimp, serpulid worm tubes, sea stars, two stalked barnacles in lower center, and many sea anemones. (Photos *C, D,* and *E* are courtesy of Karen Wishner, University of Rhode Island.)

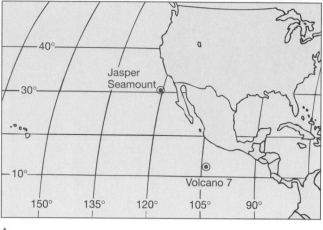

A.

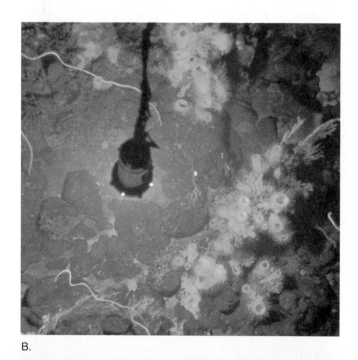

B.

C.

D.

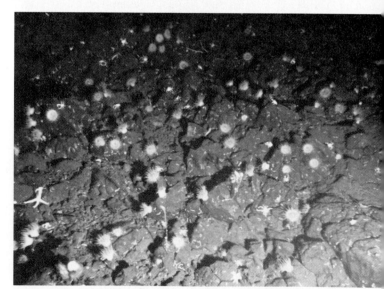

E.

The littoral and inner sublittoral are the only zones in which we find the macroscopic algae attached to the bottom because they are the only benthic zones to which sufficient light can penetrate. There is a great diversity of physical and nutritive conditions in these zones. Animal species have developed in great numbers within this nearshore benthic community as a result of the variations existing within the habitat. As one moves across the bottom from the littoral into the deeper benthic environments, one commonly observes that an inverse relationship exists between the distance from shore and the number of benthic individuals and biomass that can be found. The number of species per unit of surface area may remain relatively constant throughout the benthic environment.

Throughout most of the benthic environment, animals live in a region of perpetual darkness where photosynthetic production cannot occur. They must feed on each other or on whatever outside nutrients fall from the productive zone near the surface. The deep-sea bottom is an environment characterized by coldness, stillness, and darkness. Under these conditions, it would be expected that life would move at a relatively slow pace. For those animals that move around on the bottom, streamlining, which is very important to the nekton, is of little importance.

The organisms that live in the deep sea normally have quite a wide range of distribution because the physical conditions for life do not vary greatly, even over great distances, on the deep-ocean floor. A few species appear to be extremely tolerant of pressure changes in that members of the same species may be found in the littoral province and at depths of several kilometers.

The 1977 discovery of the first hydrothermal vent biocommunity in the Galápagos Rift has shown us that high concentrations of deep-ocean benthos are possible. It appears that the primary limiting factor for life on the deep-ocean floor is the availability of food. At these vents, food produced by chemosynthetic bacteria is abundant. Thus, the size of individuals and the total biomass in the hydrothermal communities far exceeds that previously known for the deep-ocean benthos.

A seamount is an interesting place to examine the effects of changing physical conditions on the benthos because significant depth changes occur over small horizontal distances. Jasper Seamount and Volcano 7 in the eastern Pacific have been investigated to determine the effects of current conditions and the oxygen minimum zone on benthic communities (figure 12–17A).

On Jasper Seamount there are eight peaks ranging in depth from less than 600 m (1925 ft) to 1200 m (3850 ft) as it rises from the 4200 m (13,480 ft) abyssal ocean floor. A Deep Tow photographic survey showed the benthos to include sponges, black corals, gorgonian corals, anemones, and tunicates. The survey indicated that black corals, which grew at depths above 1150 m, were up to three times more abundant near peaks than at the same depth on mid-slope (figure 12–17B). Gorgonian corals, which grew between the depths of 1170 and 1250 m, were five times more abundant near peaks than at the same depth on mid-slope.

Current-meter readings indicated that current speed was about twice as fast near peaks as at mid-slope (4.9 cm/s vs 2.7 cm/s). This acceleration of currents over the peaks is analogous to that which causes air to flow faster over the curved top of an airplane wing than along its flat bottom surface. The accelerated currents could increase the density of benthos by carrying in (1) more larvae ready to settle and (2) more particulate food per unit of time to areas near the peaks.

Volcano 7 rises from a depth of 3400 m (10,914 ft) to 730 m (2343 ft). In the overlying water, the base of the euphotic zone is at 85 m (273 ft). The oxygen minimum zone is well developed, and there is less than 0.2 ml O_2/litre of seawater from 288 to 1077 m depth. There is little plankton throughout this range of depth, so dead organic matter from the euphotic zone falls undegraded to the seamount surface where it produces a thick, fluffy, green flocculent rich in organic matter and protozoans above 750 m (figure 12–17C). There are no large benthos found near the peak above 750 m. The flocculent concentration decreases with depth and the concentration of large benthos increases to 19 individuals/m^2 at 810 m (figure 12–17D and E). Below that depth the concentration of large benthos decreases and is less than 0.5 individual/m^2 below 1000 m.

Considering the observations at Jasper Seamount, we would have expected that the large filter-feeding benthos should have been more abundant near the peak of Volcano 7. It appears that the effect of the peak penetrating the very well-developed oxygen minimum was to allow too much organic flocculent to settle onto the peak of Volcano 7 for the filter-feeding large benthos to cope with. At greater depths the concentration of flocculent decreased, and the sponges, tube worms, anemones, and barnacles were able to establish themselves while crabs, sea stars, and shrimps fed on the rocky surface and the near-bottom water.

SUMMARY

The relatively stable marine environment is thought to have given rise to all phyla of organisms. Those organisms that have established themselves in the terrestrial realm have had to develop complex systems for support and for acquiring and retaining water.

Although the relative proportions of the constituents of salinity in the ocean and in the body fluids of the organisms are often very nearly the same, the salinity of one may differ greatly from the other. If the body fluids of an organism and ocean water are separated by a membrane that allows water molecules to pass, problems requiring osmoregulation may develop. Marine invertebrates and sharks are essentially isotonic, having body fluids with a salinity similar to that of ocean water. They rarely face osmoregulatory problems. Most marine vertebrates are hypotonic, having body fluids with a salinity lower than that of ocean water, and tend to lose water through osmosis, the passing of water molecules from a region in which they are in higher concentration through a semipermeable membrane into a region where they are in lower concentration. Freshwater organisms are essentially all hypertonic, having body fluids much higher in salinity than the water in which they live, so they must compensate for a tendency to take water into their cells through osmosis.

For life to flourish in any environment, there must be a sufficient food supply. The basic producers of food are plants, so the requirements of plants must be met if food is to be plentiful. The availability of nutrients and solar radiation make plant life possible. Since solar radiation is available only in the surface water of the ocean, plant life is restricted to a thin layer of surface water usually no more than 100 m (330 ft) deep. Nutri-

ents derived ultimately from the continents are much more abundant near continental features. Although much is yet to be learned about the distribution of life in the oceans, it appears that the biomass concentration of the oceans decreases away from the continents and with increased depth. The color of the oceans ranges from green in highly productive regions to blue in areas of low productivity.

The plants that must stay in surface water to receive sunlight and the small animals that feed on plants do not have effective means of locomotion. They depend, therefore, on their small size and other adaptations to give them a high ratio of surface area per unit of body mass, which results in a greater frictional resistance to sinking. Large animals that swim freely face an altogether different problem and generally have streamlined bodies to reduce frictional resistance to motion.

Compared to life in colder regions, organisms living in warm water tend to be individually smaller, comprise a greater number of species, and constitute a much smaller total biomass. Warm water organisms also tend to live shorter lives and reproduce earlier and more frequently than their cold water counterparts.

The marine environment is divided into two basic units—the pelagic (water) and the benthic (bottom) environments. These regions, which are further divided primarily on the basis of depth, are inhabited by organisms we can classify into three categories on the basis of life style. These categories are the plankton, or free-floating forms with little power of locomotion; the nekton, or free swimmers; and the benthos, or bottom-dwellers.

KEY TERMS

Abyssal zone (p. 261)
Abyssopelagic zone (p. 258)
Aphotic zone (p. 258)
Bacterioplankton (p. 262)
Bathyal zone (p. 261)
Bathypelagic zone (p. 258)
Benthic environment (p. 258)
Benthos (p. 264)
Deep scattering layer (p. 258)
Diffusion (p. 250)
Disphotic zone (p. 258)
Epifauna (p. 264)
Epipelagic zone (p. 258)
Euphotic zone (p. 258)
Euryhaline (p. 250)
Eurythermal (p. 258)
Hadal zone (p. 261)

Holoplankton (p. 263)
Hypertonic (p. 251)
Hypotonic (p. 251)
Infauna (p. 264)
Inner sublittoral (p. 261)
Isotonic (p. 251)
Littoral zone (p. 261)
Macroplankton (p. 262)
Meroplankton (p. 264)
Mesopelagic zone (p. 258)
Nektobenthos (p. 264)
Nekton (p. 264)
Neritic province (p. 258)
Oceanic province (p. 258)
Osmosis (p. 251)
Osmotic pressure (p. 251)
Outer sublittoral (p. 261)

Pelagic environment (p. 258)
Photophores (p. 260)
Phytoplankton (p. 262)
Picoplankton (p. 263)
Plankton (p. 262)
Protoplasm (p. 250)
Spray zone (p. 261)
Stenohaline (p. 250)
Stenothermal (p. 258)
Streamlining (p. 257)
Sublittoral zone (p. 261)
Subneritic province (p. 260)
Suboceanic province (p. 260)
Supralittoral (p. 260)
Viscosity (p. 256)
Zooplankton (p. 262)

QUESTIONS AND EXERCISES

1. Discuss the major differences between marine plants and land plants, and explain the need for greater complexity of land plants.
2. Define the terms euryhaline, stenohaline, eurythermal, and stenothermal. Where in the marine environment will organisms displaying a well-developed degree of each characteristic be found?
3. Describe the relationships existing among osmotic pressure, salinity, and freezing point of a solution.
4. What is the problem requiring osmotic regulation that is faced by a hypotonic fish in the ocean? How have these animals adapted to meet this problem?
5. An important variable in determining the distribution of life in the oceans is the availability of nutrients. What are the relationships among the continents, nutrients, and the concentration of life in the oceans?
6. Another important determinant of plant productivity is the availability of solar radiation. Why is biological productivity relatively low in the tropical open ocean where the penetration of sunlight is greatest?
7. Discuss the characteristics of the coastal ocean where unusually high concentrations of marine life are found.
8. What factors create the color difference between coastal waters and the less productive open-ocean water?
9. Compare the ability to resist sinking of an organism with an average linear dimension of 1 cm (0.4 in.) to that of an organism with an average linear dimension of 5 cm (2 in.).

Discuss some adaptations other than size used by organisms to increase their resistance to sinking.
10. Changes in water temperature significantly affect the density, viscosity of water, and ability of water to hold gases in solution. Discuss how decreased water temperature changes these variables and may affect marine life.
11. Describe how higher water temperatures in the tropics may account for the greater number of species in these regions compared to low-temperature, high-latitude areas.
12. Construct a table listing the subdivisions of the benthic and pelagic environments and the physical factors used in assigning their boundaries.
13. Describe the vertical distribution of oxygen and nutrients in the oceanic province, and discuss the factors that are responsible for this distribution.
14. Discuss the probable cause and composition of the deep scattering layer.
15. List the relative number of species of animals found in the terrestrial, pelagic, and benthic environments, and discuss the factors that may account for this distribution.
16. Describe the lifestyles of plankton, nekton, and benthos. Why is it proper to consider that plankton account for a relatively larger percentage of the biomass of the oceans than the benthos and nekton?
17. List the subdivisions of plankton and benthos and the criteria used for assigning individual species to each.

REFERENCES

Borgese, E. M., Ginsburg, N., and Morgan, J. R., eds. 1991. *Ocean yearbook 9*. Chicago: The University of Chicago Press.

Coker, R. E. 1962. *This great and wide sea: An introduction to oceanography and marine biology*. New York: Harper and Row.

Genin, A.; Dayton, P. K.; Lonsdale, P. F.; and Spiess, F. N. 1986. Corals on seamount peaks provide evidence of current acceleration over deep-sea topography. *Nature* 322:6074, 59–61.

Hedpeth, J., Hinton, S. 1961. *Common seashore life of southern California*. Healdsburg, Calif.: Naturegraph.

Isaacs, J. D. 1969. The nature of oceanic life. *Scientific American* 221:65–79.

May, R. M. 1988. How many species are there on Earth? *Science* 241:4872, 1441–48.

Sieburth, J. M. N. 1979. *Sea microbes*. New York: Oxford University Press.

Sumich, J.L. 1976. *An introduction to the biology of marine life*. Dubuque, Iowa: Wm. C. Brown.

Sverdrup, H.; Johnson, M.; and Fleming R. 1942. Renewal 1970. *The oceans*. Englewood Cliffs, N.J.: Prentice-Hall.

Thorson, G. 1971. *Life in the sea*. New York: McGraw-Hill.

Wishner, K.; Levin, L.; Gowing, M.; and Mullineaux, L. 1990. Involvement of the oxygen minimum in benthic zonation on a deep seamount. *Nature* 346:6279, 57–59.

SUGGESTED READING

Sea Frontiers

Burton, R. 1977. Antarctica: Rich around the edges. 23:5, 287–95. The high level of biological productivity around the continent of Antarctica is the topic.

Gruber, M. 1970. Patterns of marine life. 16:4, 194–205. Many varieties of life in the ocean are discussed in terms of how their form and size fit them for life in a particular environmental niche.

Hammer, R. M. 1974. Pelagic adaptations. 16:1 2–12. A comprehensive discussion of the adaptations of pelagic organisms to reduce the energy required to maintain their position in the open ocean.

Patterson, S. 1975. To be seen or not to be seen. 21:1, 14–20. A discussion of the possible role of color in the protection and behavior of tropical fishes.

Perrine, D. 1987. The strange case of the freshwater marine fishes. 33:2, 114–19. Explains how marine crevalle jacks are able to inhabit the fresh waters of Crystal River, Florida.

Schellenger, K. 1974. Marine life of the Galápagos. 20:6, 322–32. A discussion of the unique life forms of the Galápagos Islands, 950 km (590 mi) from South America.

Thresher, R. 1975. A place to live. 21:5, 258–67. An interesting discussion of how bottom-dwelling animals compete for space on the ocean floor.

Williams, L. B., and Williams, E. H., Jr. 1988. Coral reef bleaching: Current crisis, future warning. 34:2, 80–87. Corals and related reef animals underwent "bleaching" along the Central American coast in 1983 and in the Caribbean Sea in 1987.

Wu, N. 1990. Fangtooth, viperfish, and black swallower. 36:5, 32–39. Strange adaptations help fish survive in the food-scarce and dark waters below 1,000 meters' depth.

Scientific American

Denton, E. 1960. The buoyancy of marine animals. 203:1, 118–29. The means by which some marine animals reduce the energy expenditure required to live in the ocean water far above the ocean floor are discussed.

Eastman, J. T., and DeVries, A. L. 1986. Antarctic fishes. 255:5, 106–14. Explains how one group of fish survived when the Antarctic turned cold.

Horn, M. H., and Gibson, R. N. 1988. Intertidal fishes. 258:1, 64–71. Intertidal fishes have undergone remarkable adaptation to survive this physically harsh environment.

Isaacs, J. D. 1969. The nature of oceanic life. 221:3, 146–65. A well-developed survey of the conditions for life in the ocean as they relate to the variety and distribution of marine life forms.

Isaacs, J. D., and Schwartzlose, R. A. 1975. Active animals of the deep-sea floor. 233:4, 84–91. A surprisingly large population of large fishes on the deep-sea floor is suggested by automatic cameras dropped to the ocean bottom.

Palmer, J. D. 1975. Biological clock and the tidal zone. 232:2, 70–79. This article investigates the mechanism of biological clocks set to the rhythm of the tides, which are found in organisms from diatoms to crabs.

Partridge, B. L. 1982. The structure and function of fish schools. 246:6, 114–23. Schooling benefits and the means by which fish maintain contact with the school are considered.

Vogel, S. 1978. Organisms that capture currents. 239:2, 128–39. The manner in which sponges use ocean currents is an important part of this discussion.

Chapter 13

Biological Productivity—
Energy Transfer

The major primary producers of the oceans, marine algae, capture most of the energy used to support the marine biological community. When we think of marine plants, most of us undoubtedly first consider the large macroscopic plants we see growing near shore in many coastal areas. These large plants, however, play only a minor part in the production of energy for the ocean population as a whole. Instead, marine organisms depend primarily on the small planktonic varieties of marine plant life that inhabit the upper sunlit water of the world's oceans. They are not so obvious. All are microscopic, but they are scattered throughout the breadth of the ocean surface layer and represent the largest biomass community in the marine environment—phytoplankton.

In this chapter we first consider the classification of organisms. We then discuss the composition of the marine algae, the ecological factors related to plant and bacterial productivity, and the passing of stored chemical energy on to other marine organisms within food webs.

A CZCS image of the northeast Atlantic shows the concentration of photosynthetic pigments (mg/m^3) ranging from 0.2 (darker blue) to more than 1.0 (darkest red). The highest levels are in shallow nearshore waters; the lowest are in the warm waters of the Gulf Stream (bottom). (Courtesy of NOAA.)

TAXONOMIC CLASSIFICATION

All living things belong to one of the five kingdoms shown in figure 13–1 (see appendix VI). The simplest of all organisms are classified as **Monera.** These organisms are single-celled and have their nuclear material spread throughout the cell. Included in this kingdom are the cyanobacteria (blue-green algae) and heterotrophic bacteria.

Representing a higher stage of evolutionary development is the kingdom **Protista.** It includes those organisms that are single-celled but have their nuclear material contained within a nuclear sheath. Here we find the algae (simple plants) and the single-celled animals, **Protozoa.**

Kingdom **Mycota,** the fungi, appears to be poorly represented in the oceans. Less than 1 percent of the known 50,000 species are found there. Fungi are found throughout the marine environment, but they are much more common in the intertidal zone in a mutualistic relationship with cyanobacteria or green algae. In this relationship, called lichen, the fungus provides a protective covering that retains water during periods of exposure, while the algae provide food for the fungus through photosynthesis. Other fungi function primarily as decomposers in the marine ecosystem: They remineralize organic matter.

Although there is one more kingdom of marine plants—**Metaphyta,** the multicelled plants—they are mostly restricted to the shallow coastal margins of the ocean and are not a major component of the marine productive community. They do play a very important role as producers within restricted communities, such as mangrove swamps and salt marshes.

The **Metazoa** consists of multicelled animals. Metazoans range in complexity from the simple sponges to the vertebrates (animals with backbones).

The kingdoms are divided into increasingly specific groupings (taxa). The system of taxonomic classification introduced by Carl von Linné (Linnaeus) in 1758 includes the following major categories:

Kingdom

Phylum

Class

Order

Family

Genus ⎫
⎬ Species name
Species ⎭

Every organism's scientific name includes its genus and species. For example, the toothed whale called the common dolphin is the species *Delphinus delphis.*

MACROSCOPIC PLANTS

As we discuss the marine plants, we will first consider those groups of plants with which we are most familiar. These are the attached forms of macroscopic algae and metaphyta found in shallow waters along the margins of the ocean. In classifying algae, we will use criteria that are based partly on the pigment content of the plants within the group (figure 13–2).

Phaeophyta (Brown Algae)

The phylum Phaeophyta contains the largest members of the marine plant community that are attached to the bottom wherever suitable substrate is available in the littoral and inner sublittoral zones. The dominant pigment in the brown algae is **fucoxanthin,** and the color may range from a very light brown to black. The brown algae occur primarily in temperate and cold water areas. Their sizes range from the small black encrusting patch of the *Ralfsia,* found in the upper and middle intertidal zones, where it is exposed to an extensive threat of desiccation and may become crisp and dry in the sun without dying, to the *Pelagophycus* (bull kelp), which may grow from small holdfasts in water depths in excess of 30 m (98 ft).

Chlorophyta (Green Algae)

The predominant forms of algae found in freshwater environments belong to the phylum Chlorophyta, green algae, and are not well represented in the ocean. Most species are intertidal, or grow in shallow waters of bays. Because of the pigment chlorophyll they range in color from yellow-green to a very dark green, but most are grass green in color. They grow only to moderate size, seldom exceeding 30 cm (12 in.) in the largest dimension. Forms range from that of finely branched filaments to flat thin sheets. The various species of the genus *Ulva* (sea lettuce), a thin membranous sheet two cell layers thick, may be found widely scattered throughout cold water areas, whereas the genus *Codium* (sponge weed), a dichotomously branched form more commonly found in warm waters, can exceed 6 m (20 ft) in length.

Rhodophyta (Red Algae)

Red algae belong to the phylum Rhodophyta and are the most abundant and widespread of the marine macroscopic algae. Over 4000 species are found, many of them attached, from the very highest intertidal levels to the outer edge of the inner sublittoral zone. They are very rare in fresh water. The red algae range in size from just visible to the naked eye to lengths of up to 3 m (10 ft). While Rhodophyta are found in both warm and cold water areas, the warm water varieties are relatively small. The characteristic pigment of the red algae is **phycoery-**

FIGURE 13–1
The Five Kingdoms of Organisms.
Monera—single cells without a nucleus; Protista—single cells with a nucleus; Mycota—the fungi; Metaphyta—complex, many-celled plants; Metazoa—complex, many-celled animals.

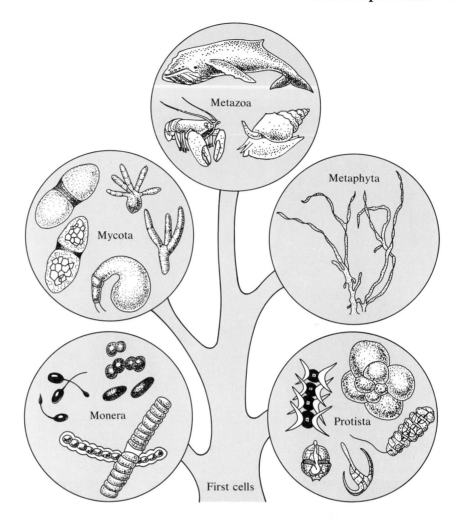

FIGURE 13–1
The Five Kingdoms of Organisms.
Monera—single cells without a nucleus; Protista—single cells with a nucleus; Mycota—the fungi; Metaphyta—complex, many-celled plants; Metazoa—complex, many-celled animals.

thrin. The color of the red algae will vary considerably depending on their depth in the intertidal or inner sublittoral zones. In the upper, well-lighted areas it may be green to black or purplish in color, and it changes through a brown to a pinkish red in the deeper water zones, where the light concentrations are lower. Although the bulk of marine plant productivity is believed to occur above water depths where the amount of light is reduced to 1 percent of that available at the surface (approximately 100 m, or 330 ft), a red alga has been observed growing at a depth of 268 m (880 ft) on a seamount near San Salvadore, Bahamas. Available light at this sighting was thought to be only 0.05 percent of the light available at the ocean's surface.

Spermatophyta (Seed-Bearing Plants)

The only metaphyta observed in the marine environment belong to the highest group of plants, the seed-bearing Spermatophyta. Two seed-bearing plants found in the marine environment are *Zostera* (eelgrass) and *Phyl-*

lospadix (surf grass). *Zostera,* a grasslike plant with true roots, is found primarily in quiet waters of bays and estuaries from the low tide zone down to a depth of some 6 m (20 ft). *Phyllospadix* prefers the high-energy environment of an exposed rocky coast and can be found from the intertidal zones down to a depth of 15 m (50 ft). Both of these plants are considered to be important sources of the detrital food for the marine animals that inhabit their environment. Found in salt marshes are grasses belonging mostly to the genus *Spartina,* while mangrove swamps contain primarily the mangrove genera *Rhizophora* and *Avicennia.*

MICROSCOPIC PLANTS

The phyla discussed next include all the members of the important phytoplankton that produce in excess of 99 percent of the food supply for marine animals. They are primarily floating forms, although some live on the bottom in the nearshore environment.

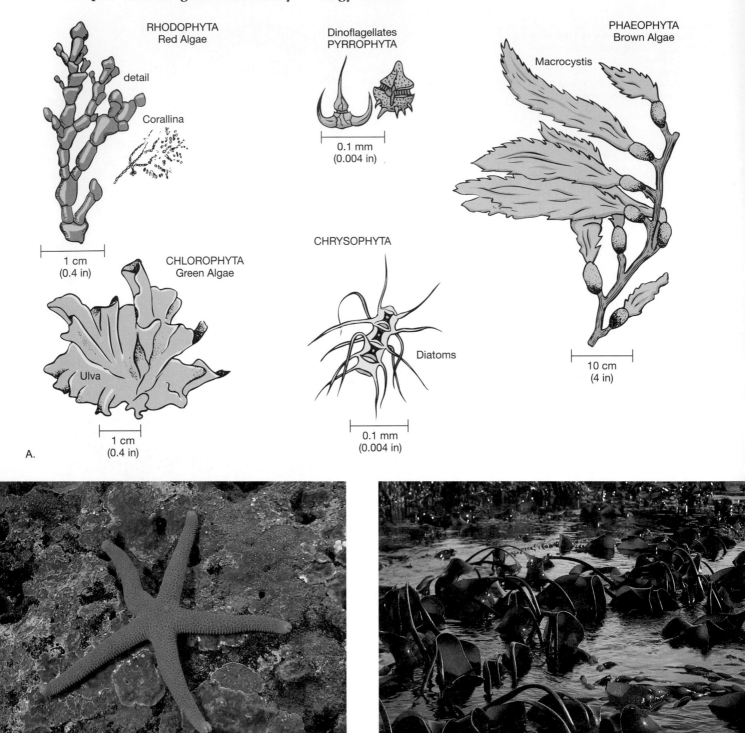

FIGURE 13–2
Algae. *A,* detailed drawing of several algae. (Line drawing by Phil David Weatherly.) *B,* encrusting red algae, *Lithothamnion,* with a sea star, *Henricia.* (Photo by Shane Anderson.) *C,* brown algae, *Laminaria;* oarweed at low tide. (Photo by D.J. Wrobel, Monterey Bay Aquarium/BPS.) *D,* green algae, *Codium fragile.* The prominent fingerlike plant is called *sponge weed. E,* photomicrograph of *Gonyaulax polyedra* magnified 1100 times. *Gonyaulax* is a large genus of phosphorescent marine dinoflagellates which, in great abundance, cause what is known along the shoreline as red tide. (Photo from Scripps Institution of Oceanography, University of California, San Diego.) *F,* coccoliths, disc-shaped $CaCO_3$ plates that cover coccolithophore cells. Diameter is 0.06 mm. (Courtesy of Deep Sea Drilling Project, Scripps Institution of Oceanography, University of California, San Diego.)

D.

E.

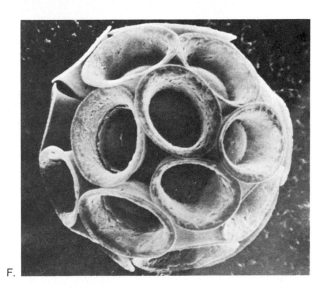

F.

Chrysophyta (Golden Algae)

Containing predominantly the yellow pigment **carotin,** these microscopic plants belong to the phylum Chrysophyta and store food in the form of leucosin (a carbohydrate) and oils.

Diatoms. The diatoms are a class of plants contained in a shell, or **frustule,** composed of opaline silica ($SiO_2 \cdot nH_2O$). These silica housings are important geologically because they accumulate on the ocean bottom and produce a siliceous sediment, **diatomite.** Some raised deposits of diatomite on land are mined and used primarily in the manufacture of filtering devices.

The frustule of the diatom is similar in structure to a microscopic pill box. The top and bottom of the box are called valves; the larger of the valves is the **epitheca** and the smaller valve is the **hypotheca.** The protoplasm of the plant is contained within this housing and exchanges nutrients and waste products with the surrounding water through slits or pores in the valves.

The reproduction of diatoms is by simple cell division. To achieve this division, the valves of the frustule must separate. Each valve then serves as one-half the housing for each newly formed cell. A new valve must form to complete the enclosure of each daughter cell, and each of these newly formed halves forms as a hypotheca.

As can be seen in figure 13–3, this process leads to the formation of smaller and smaller organisms. Eventually, the newly formed daughter cell approaches a size so small that if it were decreased further, abnormalities would result. At this point, a change in the process oc-

FIGURE 13–3
Diatom Reproduction. Diatom cell division. As the epitheca and hypotheca of a diatom separate, each becomes the epitheca of a new cell. The new frustule is completed by the generation of a new hypotheca by each new diatom. Following the arrow through the formation of three generations of diatoms, it can be seen that some cells will be crowded into ever smaller frustules. When the size of the frustule becomes critically small, an auxospore forms to allow growth of a larger cell.

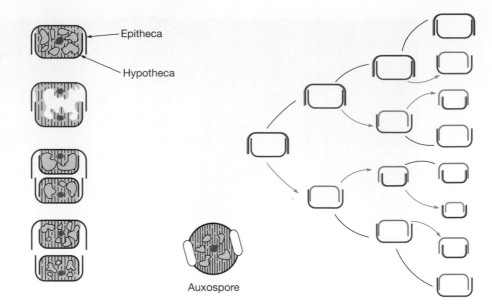

curs, namely, the formation of an auxospore. The auxospore forms between the two separating valves within a membrane that allows the auxospore to grow into a full-sized diatom so that the process of reproduction by the splitting of the frustule can commence again.

Coccolithophores. **Coccolithophores** are for the most part flagellated organisms covered with small calcareous plates, **coccoliths,** made of calcium carbonate ($CaCO_3$). The name of the group means "bearers of coccoliths." The individual plates are about the size of a bacterium, and the entire organism is too small to be captured in plankton nets. Coccolithophores are included in the nanoplankton, with dimensions of less than 0.06 mm (0.002 in.). The coccolithophores contribute significantly to calcareous deposits in all the temperate and warmer oceans.

Pyrrophyta (Dinoflagellate Algae)

A group second in importance to the diatoms in marine productivity is the **dinoflagellates,** which belong to the phylum Pyrrophyta. Possessing **flagella** (whiplike structures) for locomotion, they have a slight capacity to move into more favorable areas for plant productivity. The dinoflagellates are not important geologically because many have no protective covering, and those that do, have coverings made of cellulose that is easily decomposed by bacterial action once the organism dies. Many of the 1100 species undergo structural changes as a result of changes that occur in the environment. Many are luminescent.

Red tides result from conditions where up to 2 million dinoflagellates may be found in 1 liter (1 qt) of water. Mainly responsible for this red tide phenomenon are the genera *Ptychodiscus* and *Gonyaulax,* which pro-

duce water-soluble toxins. *Gonyaulax* toxin is not poisonous to shellfish but will concentrate in their tissue and is poisonous to humans who eat the shellfish. *Ptychodiscus* toxin kills fish and shellfish. April through September are particularly dangerous months in the Northern Hemisphere. In most areas there is a quarantine against the taking of shellfish that feed on these microscopic organisms and concentrate the poisons that they secrete to levels that are dangerous to humans. *Ptychodiscus breve* is an important contributor to Gulf of Mexico red tides. *Gonyaulax* is an important participant in the cooler waters of the New England and West Coast areas.

A potentially tragic epidemic of paralytic shellfish poisoning (PSP) from a *Gonyaulax tamarensis* red tide occurred in the coastal waters of Massachusetts in the fall of 1972. Fortunately, no deaths occurred, although 30 cases of poisoning were reported. Symptoms of PSP are similar to those of drunkenness, including incoherent speech, uncoordinated movement, dizziness, and nausea. Documented cases throughout the world include 300 deaths and 1750 nonfatal cases. There is no known antidote for the toxin, which attacks the human nervous system, but the critical period usually passes in 24 h.

PRIMARY PRODUCTIVITY

Primary productivity is defined as the amount of carbon fixed by autotrophic organisms through the synthesis of organic matter from inorganic compounds such as CO_2 and H_2O using energy derived from solar radiation or chemical reactions. The major process through which primary productivity occurs is thought to be photosynthesis. Although new knowledge of the role of chemosynthesis in supporting hydrothermal vent communities on oceanic spreading centers has emerged, chemosynthesis is much less significant in overall marine primary production than is photosynthesis. Because of the far greater importance of photosynthesis in the overall productivity of the oceans and our greater knowledge of the factors that affect marine photosynthesis, the following considerations will relate primarily to this process.

Photosynthetic Productivity

The total amount of organic matter produced by photosynthesis per unit of time represents the **gross primary production** of the oceans. Plants use some of this organic matter for their own maintenance through respiration. That which remains is the **net primary production,** which is manifested as growth and reproduction products. It is the net primary production that supports the heterotrophic marine populations—animals and bacteria.

New production is that part of the gross primary production that is supported by nutrients brought in from outside the local ecosystem by processes such as upwelling. The higher the ratio of new production to gross primary production in an ecosystem, the greater its ability to support heterotrophic populations that we depend on for fisheries such as pelagic fishes and benthic scallops. **Regenerated production** results from nutrients being recycled within the ecosystem.

In the 1920s, the **Gran method** of measuring net primary productivity was developed based on the fact that oxygen is produced by photosynthesis in proportion to the amount of organic carbon synthesized. The method involves putting equal quantities of phytoplankton into a series of bottles, all of which contain the same amount of dissolved oxygen. The bottles are then arranged in pairs, one being transparent and the other totally opaque. These pairs are suspended on a hydrographic line through the euphotic zone, where they are left for a specific period of time. After the bottles are brought to the surface, the oxygen concentration is determined for each bottle.

Photosynthesis, which is confined to the transparent bottles, will add oxygen to the water, whereas respiration will reduce the oxygen content in both the transparent and opaque bottles. Increased oxygen concentration in the transparent bottles is proportional to the amount of photosynthesis that has occurred minus the oxygen consumed by plant respiration and thus represents the net production of the plants within the transparent bottle. Decreased oxygen content within the opaque bottles corresponds to the respiration rate. For any depth, an assessment of gross production can be made by adding the oxygen gain in the clear bottles to the oxygen loss in the opaque bottles.

The depth at which the oxygen production and the oxygen consumption are equal is called the **oxygen compensation depth,** and it represents that light intensity below which plants do not survive. Since respiration goes on at all times, during the daylight hours plants must produce, through photosynthesis, biomass in excess of that which is consumed by respiration in any 24-h period if the total biomass of the community is to increase (figure 13–4). An analogy can be made with the common paycheck: Gross photosynthesis (gross pay earned) = oxygen change in clear bottle (take home pay) + oxygen loss in dark bottle (income tax and other withholding.)

During the 1950s, a method involving the use of radioactive carbon (^{14}C) was developed. It has been refined and is currently the most often used method for determining marine primary productivity. The procedure is similar to that described above, because it involves the use of a series of paired clear and opaque bottles. Each

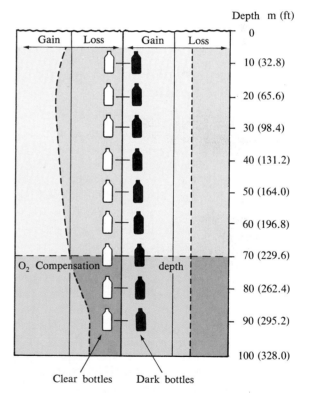

FIGURE 13–4
Oxygen Compensation Depth. Biological assay of phytoplankton production and consumption of oxygen (O_2) shows to what depth light penetrates with sufficient intensity to sustain photosynthesis at a level sufficient to increase phytoplankton mass. This occurs only above the oxygen compensation depth, the depth at which oxygen production by photosynthesis is equal to consumption through respiration. The change in oxygen content in the clear bottles represents a net photosynthesis resulting from gross photosynthesis less the consumption of oxygen by respiration. The loss of oxygen in the dark bottles gives a value for the amount of respiration that occurred in the clear bottles. This amount added to the net photosynthesis that occurred in the clear bottles gives a value for the gross photosynthesis in the clear bottles.

bottle contains identical phytoplankton samples, equal amounts of CO_2 containing carbon-14, and equal amounts of CO_2 containing stable carbon.

The phytoplankton sample is filtered from each bottle after the system has been suspended in the ocean for a sufficient period of time. The amount of beta radiation emitted by each sample is measured by a radiation counting device, and the rate of assimilation of carbon-14 is computed from these measurements. High levels of radioactive emission indicate high levels of productivity.

A third method used in studies of marine primary productivity is the measurement of chlorophyll *a* content of living phytoplankton samples taken from ocean surface waters. Although this method is not as precise as the measurement of net primary productivity using carbon-14, there is a direct relationship between the amount of chlorophyll *a* in a phytoplankton sample from a given volume of ocean water and gross primary productivity. The satellite sensor, Coastal Zone Color Scanner (CZCS), which failed in June 1986, was used to sense the effect of phytoplankton pigment on the color of ocean water. (Data from this system were used to prepare the map of pigment concentration used in figure 13–8.

Distribution of Photosynthetic Productivity

The production of organic compounds through the process of photosynthesis will vary considerably throughout the areal extent of the ocean as well as in time. Photosynthetic productivity in the oceans depends on (1) the availability of solar radiation and (2) the availability of nutrients. Both variables are related to the seasonal pattern that grows out of Earth's revolution around the sun while rotating on an axis that is tilted relative to the ecliptic.

Primary photosynthetic productivity of the oceans varies from about 0.1 g of carbon per square meter per day ($gC/m^2/d$) in the open ocean to over 10 $gC/m^2/d$ in highly productive coastal areas. This variability is primarily the result of the uneven distribution of nutrients throughout the photosynthetic zone. Some of the causes of this variability are discussed below.

Temperature Stratification and Nutrient Supply

Solar radiation penetrates only the surface layer of the ocean, and infrared wavelengths that can be converted directly to heat are essentially 100 percent absorbed in the upper 2 m (6.5 ft) of the ocean. It is not surprising that the surface waters are warmed. This condition produces a thermocline that separates the surface waters, which are relatively warmer, from the deep-water masses that are colder and more dense.

Thermoclines are relatively permanent in low latitude areas, while a seasonal thermocline may be present during the summer months in the midlatitudes. They are generally nonexistent in high latitude areas. In the following discussion of productivity, we will see that the presence or absence of the thermocline will have an important effect on nutrient availability.

In considering the availability of solar radiation and nutrients, we can rather conveniently divide the oceans into polar, temperate, and tropical regions. We will concern ourselves primarily with the nature of the open ocean in all three regions. The polar region will be discussed first as the productivity considerations are simplest there.

Polar Region. As an example of productivity on a seasonal basis in a polar sea, we will consider the Barents Sea off the northern coast of Europe, where there is peak diatom productivity during the month of May that tapers off through July (figure 13–5A). This brings about a peak period of zooplankton development, consisting primarily of small crustaceans of the genus *Calanus*. The zooplankton biomass reaches a peak in June and continues at a relatively high level until winter darkness begins in October. In this region above 70°N latitude, there is continuous darkness for about 3 months of winter and continuous illumination for a period of 3 months during summer.

In the Antarctic region, particularly at the southern end of the Atlantic Ocean, productivity patterns are similar to those found in the Barents Sea, except that the seasons are reversed and productivity is somewhat greater. The most likely explanation for this greater productivity in Antarctic waters is the continual upwelling of water that has sunk in the North Atlantic. Moving southward as a deep-water mass, the North Atlantic Deep Water surfaces hundreds of years later carrying high concentrations of nutrients (figure 13–5B).

To illustrate the very great productivity that occurs during the short summer season in polar oceans, we can consider the growth rate of baby whales. The largest of all whales, the blue whale, migrates through the temperate and polar oceans at the times of maximum zooplankton productivity. As a result of this excellent timing, they are able to develop and support calves that during a gestation of 11 months reach lengths in excess of 7 m (23 ft) at birth. The mother suckles the calf for 6 months with a teat that actually pumps the youngster full of rich milk. By the time the calf is weaned, it is over 16 m (53 ft) in length and will in a period of 2 years reach a length approaching 23 m (75 ft). In 3½ years a 60-ton blue whale has developed. This phenomenal growth rate gives some indication of the enormous biomass of copepods and krill that these large mammals feed upon.

In polar waters there is little density stratification to prevent the mixing of deeper water with shallow water. Polar waters are relatively isothermal (figure 13–5C). In

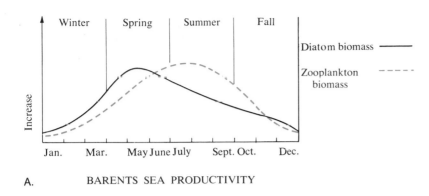

A. BARENTS SEA PRODUCTIVITY

Winter | Spring | Summer | Fall

Increase

Jan. Mar. May June July Sept. Oct. Dec.

Diatom biomass ——
Zooplankton ─ ─ ─ ─
biomass

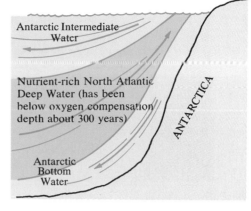

B. ANTARCTIC UPWELLING

Antarctic Intermediate Water

Nutrient-rich North Atlantic Deep Water (has been below oxygen compensation depth about 300 years)

ANTARCTICA

Antarctic Bottom Water

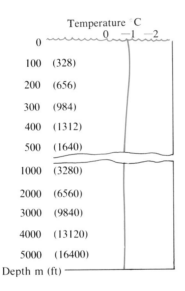

Temperature °C
0   −1   −2

Depth m (ft)	
0	
100	(328)
200	(656)
300	(984)
400	(1312)
500	(1640)
1000	(3280)
2000	(6560)
3000	(9840)
4000	(13120)
5000	(16400)

C.

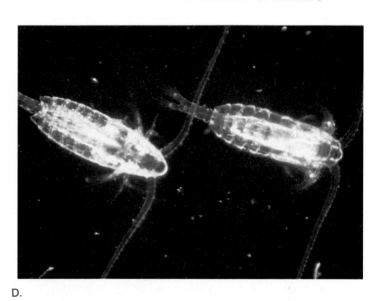

D.

FIGURE 13–5
Productivity in Polar Oceans. *A,* diatom mass increases rapidly in the spring when the sun rises high enough in the sky to cause deep penetration of sunlight. As soon as this diatom food supply develops, the zooplankton population begins feeding on it, and the zooplankton biomass peaks early in the summer. *B,* the continuous upwelling of North Atlantic Deep Water keeps the Antarctic waters rich in nutrients. When the summer sun provides sufficient radiation, there is a huge period of biological productivity. *C,* polar water shows little temperature stratification. Nearly isothermal conditions from the surface to the bottom are common. *D,* copepods, *Calanus,* each about 8 mm in length. (Photo courtesy of Scripps Institution of Oceanography, University of California, San Diego.)

most polar areas the surface waters freely mix with the deeper nutrient-rich water. There can be, however, some density stratification of water masses due to the summer melting of ice, which lays down a thin low-salinity layer that does not readily mix with the deeper waters.

There are usually high concentrations of phosphates and nitrates in the surface waters. Thus, plant productivity in the high latitudes is more commonly limited by the availability of solar energy than by the availability of nutrients. The productive season in these waters is relatively short but is characterized by an outstandingly high rate of production.

Tropical Region. In direct contrast with high productivity associated with the summer season in the polar seas, low productivity is the rule in the tropical regions of the open ocean. Light penetrates much deeper into the open tropical ocean than into the temperate and polar waters. This produces a very deep compensation depth. In the tropical ocean, however, a permanent thermocline produces a stratification of water masses and prevents mixing between the surface waters and the nutrient-rich deeper waters (figure 13–6A).

At about 20° latitude the concentrations of phosphate and nitrate are commonly less than 1/100 of the con-

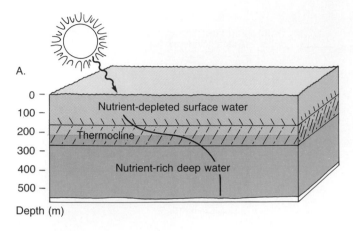

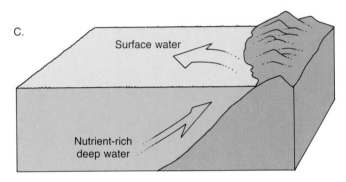

FIGURE 13–6
Productivity in Tropical Oceans. In normal tropical regions (A) deep penetration of sunlight produces a deep compensation depth with a good supply of solar radiation available for photosynthesis. A permanent thermocline serves as an effective barrier to mixing of surface and deep water. As plants use nutrients in the surface layer, productivity is retarded because the thermocline prevents replenishment of nutrients from deeper water. Tropical regions of high productivity (local areas where nutrients are brought to the surface) include areas of equatorial upwelling (B), where nutrient-enriched water from 200 to 300 m (656 to 984 ft) replaces diverging equatorial surface water, and areas of coastal upwelling (C), where surface water moved away from the shore is replaced by deep water that is rich in nutrients because it comes from below the euphotic layer.

centrations of these nutrients in temperate oceans during winter. Nutrient-rich waters within the tropics lie for the most part below 150 m (492 ft), and the highest concentration of nutrients occurs between 500 and 1000 m (1640 and 3280 ft) depth.

There is generally a steady, low rate of primary productivity in tropical oceans. Although the rate of tropical productivity is low, when we compare the total annual productivity of tropical oceans with that of the more productive temperate oceans, we find that the tropical productivity is generally at least half of that found in the temperate regions on an annual basis.

Within tropical regions there are three environments where productivity is unusually high—regions of equatorial upwelling, coastal upwelling, and coral reefs (figure 13–6B). In areas where trade winds drive westerly equatorial currents on either side of the equator, surface water diverges as a result of the Ekman spiral. This surface water that moves off toward higher latitudes is replaced by nutrient-rich water that surfaces from depths of up to 200 m (660 ft). This condition of equatorial upwelling is probably best developed in the eastern Pacific Ocean. Where the prevailing winds blow toward the equator and along the western margin of continents, the surface waters are driven away from the coast (figure 13–6C). They are replaced by nutrient-rich waters from depths of 200 to 900 m (660 to 2950 ft). As a result of this upwelling of nutrient-rich water, there is a high rate of

primary productivity in these areas, which supports large fisheries. Such conditions exist along the southern coast of California and the southwest coast of Peru in the Pacific Ocean. Upwelling also occurs along the northwest coast of Morocco and the southwest coast of Africa in the Atlantic Ocean. The relatively high productivity of coral reef environments is not related to the upwelling process and is discussed in Chapter 15.

Temperate Region. We have discussed the general productivity picture in the polar regions, where productivity is limited primarily by the availability of solar radiation, and in the tropical low-latitude areas, where the limiting factor is the availability of nutrients. We will next consider the temperate regions, where an alternation of these factors controls productivity in a pattern that is somewhat more complex.

Productivity in temperate oceans is at a very low level during the winter months, although high concentrations of nutrients are available in the surface layers. In fact, the nutrient concentration is higher during the winter season than at any time throughout the year. The limiting factor on productivity during the winter season in the temperate ocean is the availability of solar energy. Since the sun is at its lowest elevation above the horizon during this season, a higher percentage of solar energy is reflected and a smaller percentage absorbed into the surface waters. The compensation depth for basic producers such as diatoms is so shallow that it does not allow growth of the diatom population (figure 13–7).

As the sun rises higher in the sky during the spring season, the compensation depth deepens as the amount of solar radiation being absorbed by the surface water increases. Eventually, there is sufficient water volume included within the water column above the compensation depth to allow for the exponential growth of the diatom population. This expanding population puts a tremendous demand on the nutrient supply in the euphotic zone. In most Northern Hemisphere areas decreases in the diatom population will occur by May as a result of insufficient nutrient supply.

As the sun rises higher in the sky during the summer months, the surface waters in the temperate parts of the ocean are warmed, and the water becomes separated from the deeper water masses by a seasonal thermocline that may develop at depths of approximately 15 m (50 ft). As a result of this thermocline development, there is little or no exchange of water across this discontinuity, and the nutrients that are depleted from surface waters cannot be replaced by those available in the deep waters. Throughout the summer months the plant population will remain at a relatively low level but will again increase in some temperate areas during the autumn months.

The autumn increase is much less spectacular than that of the spring because of the decreasing availability of solar radiation resulting as the sun drops lower in the sky. This causes a decrease in surface temperature and the breakdown of the summer thermocline. A return of nutrients to the surface layer occurs as increased wind strength mixes it with the deeper water mass in which the nutrients have been trapped throughout the summer months. This bloom is very short-lived. The phytoplankton population begins to decrease rapidly. The limiting factor in this case is the opposite of that which reduced the population of the spring phytoplankton. In the case of the spring bloom, solar radiation was readily available, and the decrease in nutrient supply was the limiting factor.

Additional Considerations.   Coastal waters with high nutrient levels are highly productive or **eutrophic.** Most of the open ocean has a low level of nutrients and productivity. This low productivity is referred to as an **oligotrophic** condition. Figure 13–8 shows the general pat-

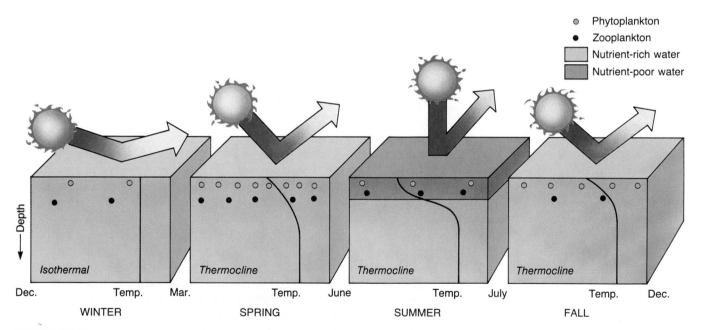

FIGURE 13–7
Productivity in Temperate Oceans. In winter, the sun is low in the sky. Much solar energy is reflected with little being absorbed into the ocean. The water column is *isothermal.* *Nutrients* are present throughout the water column. The plant and animal populations are at low levels due to lack of solar energy for photosynthesis. In spring, the sun is higher in the sky. More solar energy is absorbed by the ocean. A *thermocline* begins to develop. A "spring bloom" of plants occurs as a result of the availability of both solar energy and *nutrients.* In summer, the sun is high in the sky. Surface water warms and strong thermal stratification prevents mixing of surface and deep water. The *thermocline* reaches maximum development. When *nutrients* in the surface water are used up, the supply cannot be replenished from deep water. Plants become scarce in late summer. The increase in animal population that followed "spring bloom" is followed by a decrease. In fall, the sun is lower in the sky. Surface water cools. Fall winds aid in mixing surface and deep water. The *thermocline* begins to disappear. A small "fall bloom" of phytoplankton is terminated as too much solar energy is lost by reflection with the approach of winter.

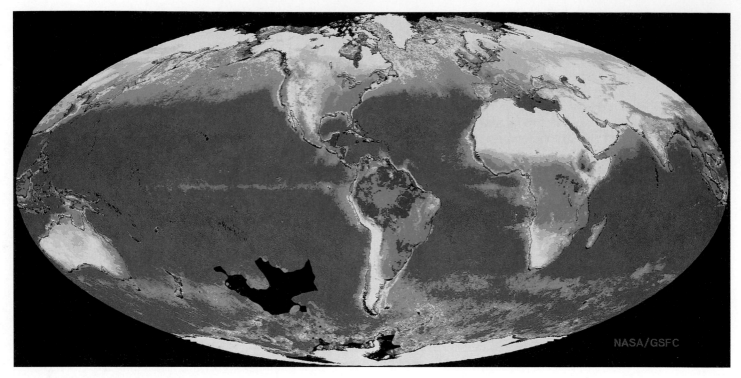

NASA/GSFC

FIGURE 13–8
Photosynthetic Production in the World Ocean. Data obtained by the Coastal Zone Color Scanner (CZCS) aboard the satellite *Nimbus* 7 between November 1978 and June 1981 were used to produce this false-color image. The CZCS senses changes in the color of ocean water caused by changing concentrations of photosynthetic pigment. Increasing photosynthetic productivity is correlated with increasing concentration of photosynthetic pigment and shows values below 0.1 mg/m³ (purple) in the oligotrophic open ocean and high values in excess of 10.0 mg/m³ (red) along eutrophic continental margins. Productivity of 1 mg/m³ is represented by the boundary between the yellow and green colors. The accuracy of these observations is 50 percent. Black represents areas of insufficient data. (Image is courtesy of Jane A. Elrod and Gene Feldman, NASA/Goddard Space Flight Center.)

terns of ocean productivity based on the concentration of phytoplankton pigment throughout the world ocean.

New measurements of photosynthetic productivity of oligotrophic waters in the North Atlantic and North Pacific oceans indicate they may be from 2 to 7 (or even more) times as productive as ^{14}C data indicate. Instead of using the small bottles used in the ^{14}C measurements and suspending them for a short time in the ocean, some physical oceanographers have used a much larger "bottle." In their studies of the circulation within the subtropical gyres, physical oceanographers use bottles that are "capped" by the pycnoclines produced by the thermoclines beneath the warm layers of surface water within the subtropical gyres of the open oceans. These bottles contain (depending on the design of the study) from tens to thousands of cubic kilometers of water that record the average results of photosynthetic activity over periods of from a few months to decades.

The studies analyze the effect of photosynthesis on the oxygen concentrations in (1) the euphotic zone and (2) beneath the euphotic zone. In the North Pacific Ocean, oxygen saturations in **subsurface oxygen maximums (SOM)** at depths between 50 and 100 m (165 and 330 ft) were from 110 percent to 120 percent at latitudes of from 30° to 40°N during summer months based on data obtained from 1962 through 1979. The excess oxygen (anything over 100 percent) may be due to trapped photosynthetically produced oxygen. The North Atlantic study determined the rate at which oxygen is used up beneath the euphotic zone (100 m, or 330 ft) by decomposing organic matter falling toward the ocean floor. **Oxygen utilization rates (OUR)** of twice the accepted mean oligotrophic rate indicate there may be much more organic matter synthesized in the oligotrophic euphotic zone than is indicated by ^{14}C data. The problem with the ^{14}C method may be that by using such a small sample and exposing it for such a short time, periods of unusually high productivity may be missed and not averaged into the results.

Some interesting findings may help explain how some of the biological productivity of oligotrophic water may have been missed. Mats of diatoms belonging to the genus *Rhizosolenia* have been found floating in the North Pacific gyre and the Sargasso Sea. Averaging 7 cm

(2.8 in.) in length, these mats possess symbiotic bacteria and cyanobacteria that fix molecular nitrogen (N_2) into nitrate (NO_3) usable as a nutrient by the diatoms (figure 13–9). Also, bundles of nitrogen-fixing cyanobacteria *Oscillatoria* spp. and *Trichodesmium* spp. have been verified in the Sargasso Sea as well as the tropical Indian and Pacific oceans. Such aggregations are readily broken up by net tows and missed by standard bottle casts.

Studies in the Pacific and Atlantic oceans indicate that a large percentage (60 percent in one study) of oligotrophic photosynthesis is achieved by picoplankton (0.2–2.0 μm; 0.000008–0.00008 in.) that pass through many filters used in productivity studies. To add to the confusion, recent studies in the tropical North Atlantic indicate nitrogen-fixing cyanobacteria account for more than 60 percent of the production, and picoplankton account for less than 10 percent. Continued investigation of this problem may reveal the overall productivity of the oligotrophic ocean waters has been greatly underestimated.

Chemosynthetic Productivity

Another source of potentially significant biological productivity in the oceans is occurring in the rift valleys of the oceanic spreading centers at depths of over 2500 m (8200 ft), where there is no light for photosynthesis. In regions where ocean water seeps down fractures in the ocean crust to depths where it is heated by underlying magma chambers, hydrothermal springs exist. Once the water is heated, it rises to the ocean floor dissolving min-

erals from the crustal rocks as it does so. Potentially significant deposits of minerals are associated with biotic communities. A significant component of these communities is the vestimentiferan worm population, a sample of which can be seen in figure 13–10. These 1-m (3-ft) tube worms and other chemosynthetic symbionts such as clams and mussels reach unusually large size in association with autotrophic bacteria that derive energy to produce their own food from the hydrogen sulfide (H_2S) gas dissolved in the water of the hydrothermal springs. By oxidizing the gas to form free sulfur (S) and sulfate (SO_4^{2-}), the bacteria release chemical energy that they use much as plants use solar energy to carry on photosynthesis. Similar communities have been discovered near cold water seeps at the base of the Florida Escarpment in the Gulf of Mexico and at other locations that will be discussed in Chapter 15. Since the bacterial synthesis of inorganic nutrients into organic molecules depends on the release of chemical energy, it is called chemosynthesis. The true significance of bacterial productivity on the deep-ocean floor will not be fully understood until much more research on the phenomenon is conducted. It has the potential, however, of increasing our estimates of the biological productivity of the ocean.

Biochemists have recently discovered that bacteria can obtain the chemical energy needed for the synthesis of organic molecules through the oxidation of a large variety of compounds containing the metals iron, manganese, copper, nickel, and cobalt. These microorganisms may well be an important factor in the deposition of ore-

A.

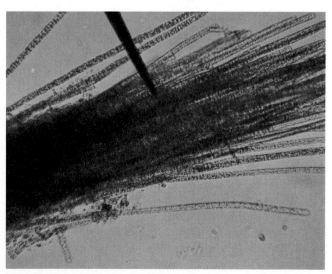

B.

FIGURE 13–9
Nitrogen Fixation by Aggregates of Diatoms and Cyanobacteria in Oligotrophic Ocean Waters. *A,* typical mat of *Rhizosolenia* about 5 cm long. It is composed of intertwining chains of *R. castracanei* (wider cells) and *R. imbricata* (narrower cells). Within these cells, symbiotic bacteria fix nitrogen for uptake by the plant cells. (Photo by James M. King.) *B,* an aggregation of the nitrogen-fixing cyanobacteria *Oscillatoria* spp. with an O_2 probe inserted into it. The probe has a maximum diameter of about 5 μm. (Photo courtesy of Hans W. Paerl, University of North Carolina.)

A.

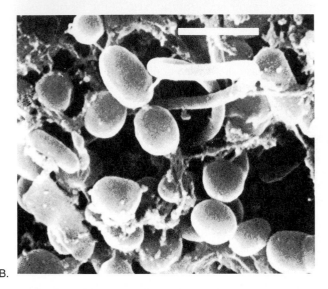

B.

FIGURE 13–10
Chemosynthetic Life of the Galápagos Rift. *A,* this photograph taken at the Galápagos Rift shows the large tube worms found there and at other deep-sea hydrothermal vents. These worms possess symbiont sulfur-oxidizing bacteria and chemosynthetically produce their own food by combining inorganic nutrients dissolved in the deep-ocean water. *B,* sample of sulfur-oxidizing bacteria from the Galápagos Rift with small particles of what is probably free sulfur attached to them. These and similar bacteria live symbiotically within the tissue of the tube worms, clams, and mussels found at the vents. They chemosynthetically produce food to support these organisms. Magnification 20,000×. Bar is 1μm. (Courtesy of Woods Hole Oceanographic Institution; *A* by Fred Grassle.)

quality deposits of the oxides of these metals on the ocean floor in the form of manganese nodules.

ENERGY TRANSFER

We have been discussing the general considerations related to availability of nutrients. Our attention will now be turned to the cycling of specific important classes of nutrients and the flow of energy.

Marine Ecosystem

The term **biotic community** refers to the assemblage of organisms that live together within some definable area. An **ecosystem** includes the biotic community and abiotic environment with which it interacts in the exchange of energy and chemical substances. Within an ecosystem there are generally three basic categories of organisms—**producers, consumers,** and **decomposers.** Plants and some bacteria are the autotrophic producers and have the capacity to nourish themselves through chemosynthetic and photosynthetic processes. The consumers and the decomposers are heterotrophic organisms that depend on the organic compounds produced by the autotrophs for their food supply.

Animals may be divided into three categories: **herbivores,** which feed directly on the plants; **carnivores,** which feed only on other animals; and **omnivores,** which feed on both. As the role of bacteria in the marine ecosystem becomes better understood, a fourth category of animals, the **bacteriovores,** which feed on bacteria, may be identified as an important component of the marine ecosystem. The decomposers, such as bacteria, break down the organic compounds of dead plants and animals and animals' excretions while taking some of these decomposition products for their own energy requirements. They characteristically release simple inorganic salts that are used by the plants as nutrients.

Symbosis is a relationship in which two or more organisms are closely associated in a way that benefits at least one of the participants. In the simplest terms, these relationships may be classified as commensalism, mutualism, or parasitism (figure 13–11). In **commensalism,** a smaller or less dominant participant, or symbiont, benefits without harming the host, the participant that affords subsistence or lodgment to the other. An example is the *Remora,* a fish that attaches to a shark or other fish to obtain food and transportation without harming the host. In **mutualism,** both symbionts benefit. Such a relationship exists between the large reef fish and the cleaner fish that keep them free of parasites. **Parasitism** describes a relationship where one participant, the parasite, benefits at the expense of the host. Many fish are hosts to isopods that attach to the fish and derive their nutrition from the body fluids of the fish, thereby robbing the host of its energy supply.

Energy Flow

Before considering the biogeochemical cycles that involve the transfer of organic and inorganic matter, we will consider the flow of energy in general. Most energy is put into a biotic community through plants. From the plants, it follows a unidirectional path (although cycles of energy exchange occur at many intermediate biological

A.

B.

C.

FIGURE 13–11
Symbiotic Relationships. *A, commensalism:* A remora swims below a Caribbean jewfish in hopes of sharing the food of its host. The remora uses a sucking organ on the top of its head to attach itself to larger fish. *B, mutualism:* Juvenile bluehead wrasses clean parasites from a stoplight parrot fish, *Sparisoms viride,* at a reef at Bonaire Island, Netherlands Antilles. *C, parasitism:* A parasitic isopod on the head of a blackbar soldierfish. (Photos *A* and *C* © Marty Snyderman; *B* © Fred Bavendam/Peter Arnold, Inc.)

levels) that leads to a continual degradation of energy culminating in **entropy,** or energy converted to a form where it is no longer available to do work. As can be seen in figure 13–12, which depicts the flow of energy through a plant-supported biotic community, energy enters the system as the high-grade radiant form, solar energy, which is absorbed by the plants. Photosynthesis converts it to a medium-grade, chemical energy, which is used for plant respiration and passes on to the animals to be used for growth and carrying on their various life-sustaining functions. Energy is expended by the animals as mechanical and heat energy, which are progressively lower forms of energy. Finally, it becomes biologically useless as entropy increases.

Composition of Organic Matter

Having discussed the noncyclic nature of the flow of energy through the biotic community and observed that it is a unidirectional flow, let us now consider the biogeochemical cycles involving matter that is not lost to the biotic community but is cycled by being converted from one chemical form to another by the various members of the community.

Biological mass is made up of compounds involving a number of elements that are abundant on Earth. All the elements that are naturally occurring can be assumed to be present in the oceans, at least in extremely small concentrations. However, some of these elements are in concentrations so small that we cannot detect their presence. Fewer than 70 have been identified by chemical analysis as being present in the oceans.

In view of the fact that living organisms are composed of carbohydrates, fats, and proteins, let us first concern ourselves with the elements that go into the makeup of these substances. About 20 elements in these basic organic compounds are considered to be essential to their production:

1. The major constituents of organic materials, those that individually make up at least 1 percent each of organic material on a dry weight basis, are, in the order of their abundance, carbon, oxygen, hydrogen, nitrogen, and phosphorus.
2. Occurring in concentrations from 500 ppm to 10,000 ppm by weight are the elements sulfur, chlorine, potassium, sodium, calcium, magnesium, iron, and copper.
3. A third group occurring in concentrations of less than 500 ppm of dry weight is composed of boron, manganese, zinc, silicon, cobalt, iodine, and fluorine.

Considering the availability of these elements as solutes in ocean water, we find that the second group, which we may refer to as the *secondary constituents* of organic composition, is in sufficient concentration to

FIGURE 13–12
Energy Flow through Photosynthetic Ecosystem. Energy enters the system as radiant solar energy. It is converted to chemical energy through photosynthesis. Metabolism releases the chemical energy for conversion to mechanical energy and energy is lost from the biotic community as heat, which increases the entropy of the ecosystem.

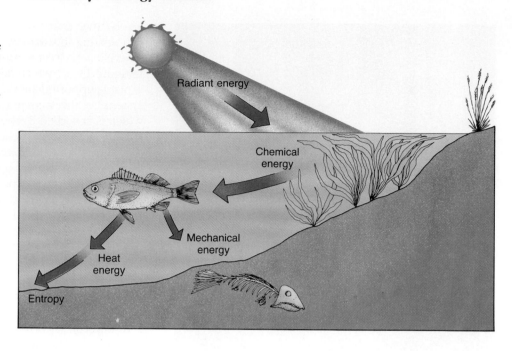

make it unlikely that members of this group would ever be limiting factors in plant productivity. However, the third group, the *tertiary constituents,* which occur in very low concentrations in organic compounds, are also found to be in very low concentrations in the marine environment. It seems probable that the tertiary constituents might create by their absence conditions that would limit productivity.

Returning to a consideration of the major constituents we find that carbon dioxide and water should certainly, on the basis of their concentrations, provide enough carbon, oxygen, and hydrogen to assure that these elements would never limit productivity. Nitrates and phosphates are the nutrients containing nitrogen and phosphorus that we need to consider in more detail since they do, under many conditions, limit marine productivity.

Biogeochemical Cycling

In biogeochemical cycles, elements will follow a pattern where an inorganic form is taken in by an autotrophic organism that synthesizes it into organic molecules. The food is passed through a food web that usually ends in bacterial decomposition of the organically produced compounds into the inorganic forms that may again be used in plant production.

A New Role for Bacteria. Although it has been widely accepted that zooplankton are the primary grazers of phytoplankton, new evidence indicates that free-living bacteria may consume up to 50 percent of the production of phytoplankton. The bacteria are thought to consume

the dissolved organic matter that is lost from the conventional food web by three processes:

1. Phytoplankton *exudate.* As phytoplankton age they lose some of their cytoplasm directly into the ocean.
2. Phytoplankton *"munchates."* As phytoplankton are eaten by zooplankton, cytoplasm is "spilled" into the ocean.
3. Zooplankton *excretions.* The liquid excretions of zooplankton are dissolved into ocean water.

Free-living heterotrophic bacteria absorb this dissolved organic matter and reenter the conventional food web primarily through the grazing of microscopic flagellates. Other larger zooplankton may be important in this process locally.

The details of the process by which free-living bacteria participate in the marine food web are still under investigation, but microbial ecologists studying the problem are convinced the evidence indicates bacteria have an important role in the transfer of energy through the marine ecosystem. Thus, bacteria may have more varied roles in the biogeochemical cycling of matter in the oceans than had previously been known.

Carbon, Nitrogen, and Phosphorus. Although there is no concern that carbon concentrations in the marine environment are a limiting factor in marine productivity, we will examine the carbon cycle, since this element is the basic component of organic compounds. To appreciate the excess of carbon dioxide in the world's oceans as it is related to the requirements for plant productivity, we should know that only about 1 percent of the total carbon present in the oceans is involved in plant productivity.

Comparatively, the soluble nitrogen compounds involved in plant productivity may be 10 times the total nitrogen compound concentration that can be measured as a yearly average. This level implies that the soluble nitrogen compounds must be cycled completely up to 10 times per year. Available phosphates may be turned over up to 4 times per year.

Comparing the ratio of carbon to nitrogen to phosphorus in dry weights of diatoms, we find that proportions are $41:7:1$. This ratio is also observed in the zooplankton that feed on the diatoms as well as in ocean water samples taken throughout the world. Thus, phytoplankton take up nutrients in the ratio in which they are available in the ocean water and pass them on to zooplankton in the same ratio. When these planktonic plants and animals die, carbon, nitrogen, and phosphorus are restored to the water in this ratio.

The Carbon Cycle. This cycle involves the uptake of carbon dioxide by plants that use it in the photosynthetic process. Carbon dioxide is returned to the ocean water primarily through respiration of plants, animals, and bacteria and secondarily by autolytic breakdown of dissolved organic materials. **Autolytic decomposition** results from the action of enzymes present in the cells of organic tissue and does not require bacterial action (figure 13–13A).

With the threat of an increased greenhouse effect resulting in part from the increased CO_2 concentration in the atmosphere, the marine carbon cycle has become an important focus of research. The oceans are believed to remove about 25 to 50 billion tonnes of carbon from the atmosphere each year. The "biological pump" that is believed to incorporate this atmospheric CO_2 into organic matter and transport it to the ocean sediments where it may be held for millions of years is poorly understood. To fully understand the role of the oceans in controlling the CO_2 concentration of the atmosphere will require much more study of this process throughout the breadth of the oceans.

The Nitrogen Cycle. Nitrogen is primarily important in the production of amino acids, the building blocks of proteins that are synthesized by plants. These photosynthetic products are consumed by animals and free-living bacteria. They are then passed on to the saprobic (decomposing) bacteria, along with dead plant tissue, as dead animal tissue and excrement.

The saprobic bacteria gain energy from breaking down these compounds. This breakdown leads to the liberation of inorganic compounds, such as nitrates, that are the basic nutrient salts used by plants. Most nitrogen in the ocean is found as molecular nitrogen (N_2). The most common combined forms are nitrite (NO_2) and nitrate (NO_3). These are highly oxidized states of nitrogen. The most abundant reduced form is ammonia (NH_3).

Different bacteria involved in the nitrogen cycle make it somewhat complicated. Although most of the bacteria play the heterotrophic roles of consuming dissolved organic matter or converting organic compounds into inorganic salts, there are some that are able to fix molecular nitrogen into combined forms. These are **nitrogen-fixing bacteria. Denitrifying bacteria** make up another special group whose metabolism depends upon the breakdown of nitrates and the liberation of molecular nitrogen. The nitrogen cycle is graphically presented in figure 13–13B.

Autotrophic organisms of various types can use ammonium nitrogen and nitrites. However, the most important nutrient form of inorganic nitrogen, nitrate, can be utilized more efficiently by plants. Studies of nutrient cycles in various portions of the world's oceans indicate that the availability of nitrogen is clearly a limiting factor in productivity during summer months. This condition arises from the fact that the processes involved in converting particulate organic substance to nitrate salts through bacterial action may require up to 3 months. The conversion begins in the lower portions of the photosynthetic zone as the particulate matter is sinking toward the ocean bottom.

There are three basic stages in the conversion of particulate organic matter to nitrate salts that involve three distinct bacterial types. The first reduces the particulate organic nitrogen into ammonium nitrogen, which is then acted upon by the second bacterial community that converts it to nitrite. A third action is required to convert the nitrite to the prime organic salt species, nitrate. By the time this conversion is completed, the nitrogen compounds are usually below the euphotic zone and thus unavailable for photosynthesis. They cannot readily be returned to the euphotic zone during summer due to the strong thermostratification that exists throughout much of the ocean surface. If nitrates are to again become available to plants in these regions, their return must wait until the thermostratification disappears, allowing upwelling and mixing during the winter to effect the movement of the nutrients into the surface waters.

The Phosphorus Cycle. The phosphorus cycle is simpler than the nitrogen cycle primarily because the bacterial action involved in breaking down the organic phosphorus compounds is simpler. This difference can be studied by comparing figure 13–13B and C. The rate at which the organic compounds can be decomposed into inorganic orthophosphates, which are the phosphorus compounds primarily used by plants, is much more rapid than that of nitrogen breakdown. This quicker rate is due to autolytic breakdown of phosphatic organic material by enzymes and the single step required in the bacterial breakdown. As a result of the greater rate of breakdown for organic phosphorus, much of it can be completed above the oxygen compensation depth. It is therefore

FIGURE 13–13
Biogeochemical Cycling. The chemical components of organic matter enter the biological system through plant photosynthesis (or bacterial chemosynthesis). They are passed on to animal populations through feeding patterns. When the plants and animals die, their organic remains are converted to inorganic form by bacterial or other decomposition processes. In this form, they are again available for uptake by plants. A, carbon cycle. There is a large supply of inorganic carbon in the oceans, and it is not a limiting factor in biological productivity. Only about 1 percent of the carbon dissolved in the ocean is involved in plant productivity.

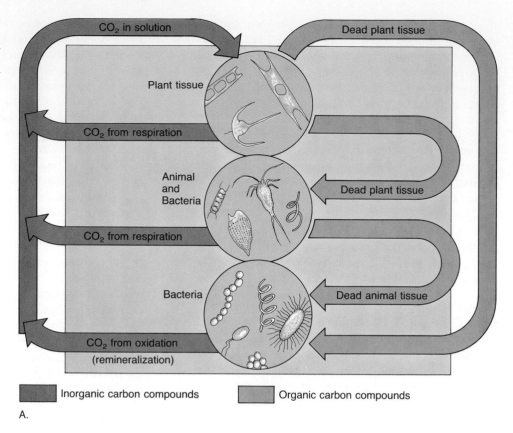

Inorganic carbon compounds   Organic carbon compounds

A.

made available to plants within the photosynthetic zone. Although the concentration of phosphorus in the oceans is only about one-seventh that of the concentration of nitrogen, the fact that the recycling can take place within the photosynthetic zone usually allows adequate quantities of inorganic phosphorus to be available for plant productivity. The lack of phosphorus is rarely a limiting factor of plant productivity.

The Silicon Cycle. As previously discussed, frustules of diatoms are composed of silica (SiO_2). Although the availability of silica can be a limiting factor in the productivity of diatoms, it is rarely a limiting factor of total primary productivity, since not all phytoplankton require silica as a protective covering. There is very little likelihood that silicon will ever be a limiting nutrient in total productivity because it is very abundant in the rocks that make up Earth's crust, and the small fraction of Earth's crustal silica that will eventually find its way into the sea is very great in relation to the silicon needs of the diatom population. Silicon concentrations range from unmeasurable up to 400 mg/m³. Fluctuations in concentration roughly coincide with those observed in nitrogen and phosphorus. However, the fluctuation in silica concentrations displays a much greater amplitude than fluctuations for nitrogen and phosphorus. This condition is probably due to the fact that silica does not undergo bacterial

decay and is taken directly into solution after undergoing autolytic breakdown.

TROPHIC LEVELS AND BIOMASS PYRAMIDS

Trophic Levels

The transfer of chemical energy stored in the mass of the ocean's plant population to the animal community occurs, in part, through the feeding process. Zooplankton feed as herbivores on the diatoms and other microscopic marine plants, while larger animals feed on the macroscopic algae and "grasses" found growing attached to the ocean bottom near shore. The herbivores will be fed upon by larger animals, carnivores, who in turn will be fed upon by another population of larger carnivores, and so on. Each of these feeding levels is a **trophic level.**

In general, the individual members of a feeding population will be larger but not too much larger than the individual members of the population on which they feed. Although we can consider this to be generally true, there are outstanding exceptions to this condition. The blue whale, which is the largest animal known to have existed on Earth, feeds upon the krill that grow to maximum lengths of 6 cm (2.4 in).

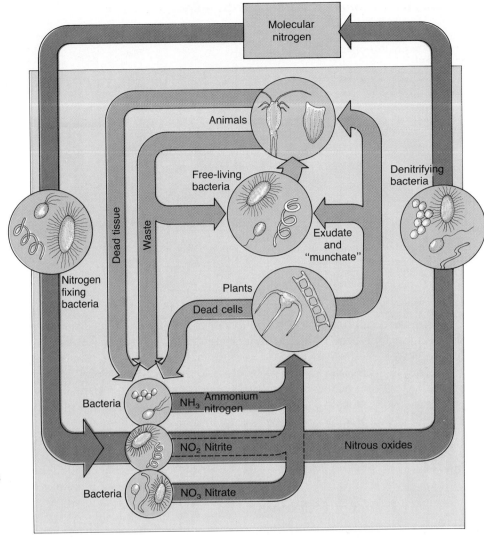

FIGURE 13–13 continued.

B, nitrogen cycle. The total amount of nitrogen fixed into organic molecules at any given time may be 10 times as great as the yearly average of soluble nitrogen compounds dissolved in ocean water. Therefore, each nitrogen atom must be recycled biogeochemically about 10 times per year. Also, the decomposition of organic nitrogen compounds back into the preferred inorganic form of nitrogen, nitrate (NO_3), requires three steps of bacterial decomposition. As a result, some of this process may not be completed until the molecules sink beneath the euphotic zone, where they are unavailable to plants. Nitrogen is considered to be the nutrient that is most likely to limit biological productivity as a result of its depletion. C, phosphorus cycle. Although each phosphorus atom may need to be recycled up to four times per year to maintain biological productivity in the oceans, it is seldom depleted to the point of limiting biological productivity. Two factors that help get organic phosphorus back into a form usable as nutrients by plants rather quickly are autolytic breakdown and the single step of bacterial decomposition required.

Inorganic nitrogen compounds

Organic nitrogen compounds

B.

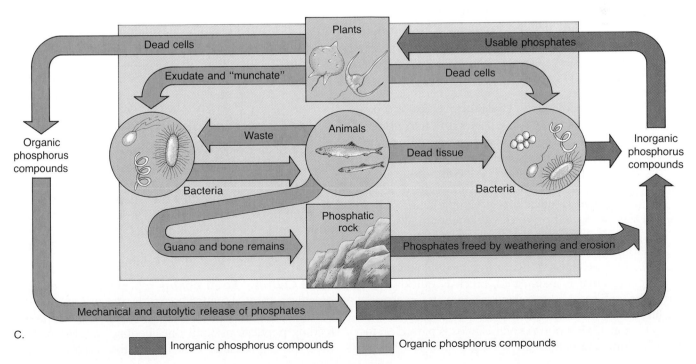

C.

Inorganic phosphorus compounds   Organic phosphorus compounds

289

FIGURE 13–14
Passage of Energy through a Trophic Level.

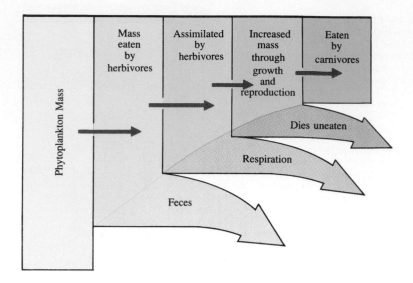

We should recall that all consideration of energy transfer must be approached with the understanding that the transfer of energy from one population to another represents a continuous flow of energy. Small-scale recycling and conservation of this energy occurs and slows the process of conversion of potential (chemical) energy to kinetic energy, to heat energy, to be lost finally to entropy. However, despite the cycling of energy, all energy that enters the organic community is inevitably lost to entropy in the end.

Transfer Efficiency

In the transfer of energy between feeding, or trophic, levels we are greatly concerned with efficiency. There is a relatively high degree of variability in the efficiency of various plant species under various laboratory conditions. As an average, we consider the percentage of light energy absorbed by plants and ultimately synthesized into organic substances made available to herbivores to be about 2 percent.

The **gross ecological efficiency** at any trophic level is the ratio of energy passed on to the next higher trophic level divided by the energy received from the trophic level below. When we examine the transfer of energy from feeding population to feeding population within the ocean, we can readily see that of the biomass representing the food intake of a given population only a portion will be passed on to the next feeding level.

Figure 13–14 shows that some of the energy taken in as food by herbivores is passed by the animal as feces and the rest is assimilated by the animal. Of that portion of the energy that is assimilated, much is quickly converted through respiration to kinetic energy for maintaining life, while what remains is available for growth and reproduction. Only a portion of this mass is passed on to the next trophic level through feeding. Figure 13–15 rep-

resents diagrammatically the passage of energy between trophic levels through an entire ecosystem from the solar energy assimilated by plants to the mass of the ultimate carnivore.

Many investigations have been conducted into the efficiency of energy transfer between trophic levels. There are many variables, such as the age of the animals involved, to be considered. Young animals display a higher growth efficiency than older animals. The availability of food can also alter efficiency. When food is plentiful, animals are observed to expend more energy in digestion and assimilation than when food is not readily available. Most efficiencies that have been determined range between 6 percent and 15 percent. It is well accepted that ecological efficiencies in natural ecosystems average approximately 10 percent. There is some evidence that in populations important to our present fisheries this efficiency may run as high as 20 percent. The true value of this efficiency is of practical importance to us because it determines the fish harvest we can anticipate from the oceans.

Biomass Pyramid

It is obvious from considering the energy losses that occur in each feeding population that there is some limit to the number of feeding populations in a food chain and that each feeding population must necessarily be smaller than the population upon which it feeds. It was previously mentioned that the individual members of a feeding population would be larger than their prey, but it seems they should also be less numerous.

Food Chains. **Food chains** represent a sequence of organisms through which energy is transferred from the primary producers through the herbivore and carnivore feeding populations until one feeding population does not have any predators, which marks the end of the chain.

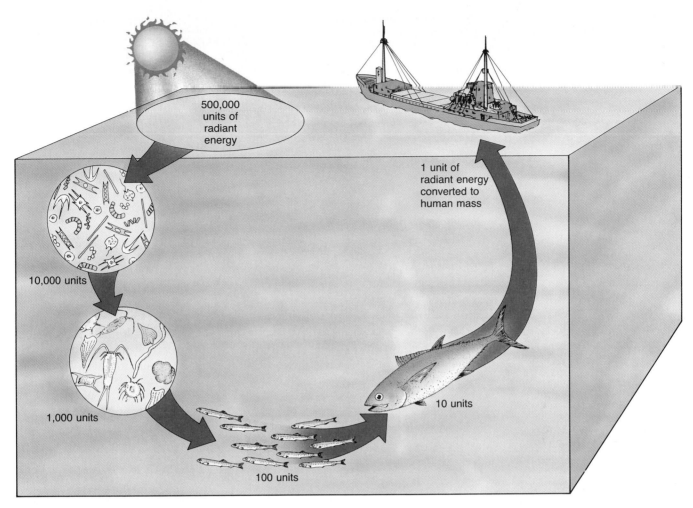

FIGURE 13–15
Ecosystem Energy Flow—Efficiency. Based on probable efficiencies of energy transfer within the ecosystem, one unit of mass equivalent is added to the fifth trophic level (the tertiary carnivores) for every 500,000 units of radiant energy available to the producers (plants). This value is based on a 2 percent efficiency of transfer by plants and 10 percent efficiency at all other levels.

In nature it is rare to find food chains comprising more than five trophic levels. Based on our previous considerations of the efficiency of energy transfer between trophic levels, we see that for a population within a food chain it may be beneficial to feed as close to the primary producing population as possible. This arrangement increases the biomass available for food and the number of individuals in the population to be fed upon.

An example of an animal population that is an important fishery and usually represents the third trophic level in a food chain is the herring off the coast of Newfoundland. Although some herring populations are involved in longer food chains, the Newfoundland herring feed primarily on a population of small crustaceans, copepods, that in turn feed upon diatoms (figure 13–16).

Food Webs. It is, however, uncommon to see simple feeding relationships of the type described above in na-

ture. More commonly the animals representing the last step in a linear food chain feed on a number of animals that have simple or complex feeding relationships, constituting a **food web.** The overall importance of food webs is not well understood, but one consequence for animals that feed through a web rather than a linear chain is their greater likelihood of survival if population extinctions or sharp decreases occur within the web at or below their feeding level. Those animals involved in food webs, such as the North Sea herring in figure 13–16, are less likely to suffer from the extinction of one of the populations upon which they depend for food than the Newfoundland herring that feed only on copepods. The extinction of the copepods in the latter case certainly would be expected to have a catastrophic effect on the herring population.

The Newfoundland herring population does, however, have an advantage over its relatives who feed

FIGURE 13–16
Food Chains—Food Webs.

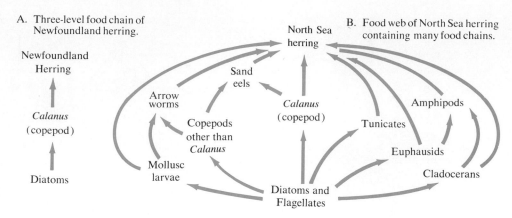

A. Three-level food chain of Newfoundland herring.

Newfoundland Herring

Calanus (copepod)

Diatoms

B. Food web of North Sea herring containing many food chains.

North Sea herring

Sand eels

Arrow worms

Copepods other than *Calanus*

Calanus (copepod)

Amphipods

Tunicates

Mollusc larvae

Euphausids

Cladocerans

Diatoms and Flagellates

through the broader-based food web. The Newfoundland herring more likely have a larger biomass to feed on since they are only two steps removed from the producers, while the North Sea herring represent the fourth level in some of the food chains within its web.

The ultimate effect of the transfer of energy between trophic levels can be seen in figure 13–17, which depicts the progressive decrease in numbers of individu-

als and total biomass at successive trophic levels resulting from decreased amounts of available energy.

A study of food webs in 55 marine ecosystems indicated an average of 4.45 trophic levels. Only two trophic levels were encountered at rocky shores in New England and Washington and at a Georgia salt marsh. Seven links were found in the longest food chains in Antarctic seas, while two tropical plankton communities in the Pacific

FIGURE 13–17
Biomass Pyramid. The higher on the food chain an organism is found, the larger it is as an individual. However, the total biomass represented by a population high on the food chain is less than that for a population at a lower level.

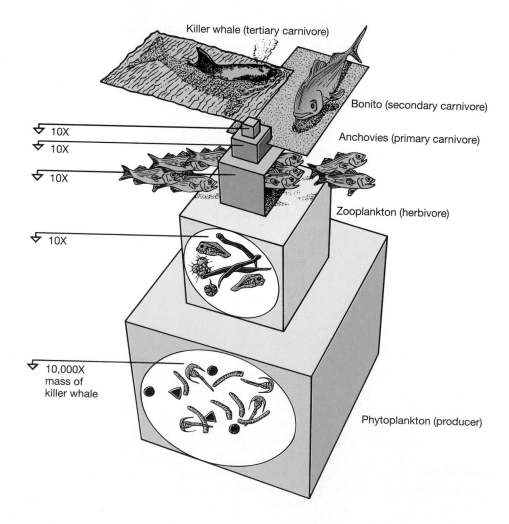

Killer whale (tertiary carnivore)

Bonito (secondary carnivore)

10X

10X

Anchovies (primary carnivore)

10X

Zooplankton (herbivore)

10X

10,000X mass of killer whale

Phytoplankton (producer)

Ocean were observed to have seven and ten trophic levels. Tropical epipelagic seas and Pacific Ocean upwelling zones had eight trophic levels. The study found that the amount of primary productivity and the degree of variability of the environment have little influence on the length of food chains within an ecosystem. The only variable that showed a clear relationship to the length of food chains was whether or not the environment was two- or three-dimensional. Three-dimensional (pelagic) marine ecosystems averaged 5.6 trophic levels compared to an average of 3.5 for two-dimensional (benthic) ecosystems.

SUMMARY

Organisms are divided into five kingdoms: Monera, single-celled organisms without a nucleus; Protista, single-celled organisms with a nucleus; Mycota, fungi; Metaphyta, many-celled plants; and Metazoa, many-celled animals. Taxonomic classification of organisms involves dividing the kingdoms into the increasingly specific groupings: phylum, class, order, family, genus, and species.

Protistan plants include the macroscopic algae Phaeophyta (kelp and *Sargassum*), Chlorophyta (green algae), and Rhodophyta (red algae). The microscopic algae include Chrysophyta (diatoms and coccolithophores) and Pyrrophyta (dinoflagellates). The more complex Spermatophyta are represented by a few genera of nearshore plants such as *Zostera* (eelgrass), *Phyllospadix* (surf grass), *Spartina* (marsh grass), and mangrove trees.

Photosynthetic productivity of the oceans is limited by the availability of solar radiation and of nutrients. The depth to which sufficient light penetrates to allow plants to produce only that amount of oxygen required for their respiration is the oxygen compensation depth. Plants cannot live successfully below this depth, which may occur at less than 20 m (65 ft) in turbid coastal waters or at a probable maximum of 150 m (492 ft) in the open ocean. Nutrients are most abundant in coastal areas due to runoff and upwelling. In high-latitude areas thermoclines are generally absent, so upwelling can readily occur, and productivity is commonly limited more by the availability of solar radiation than lack of nutrients. In low-latitude regions, where a strong thermocline may exist year-round, productivity is limited except in areas of upwelling. The lack of nutrients is generally the limiting factor. In temperate regions, where distinct seasonal patterns are developed, productivity peaks in the spring and fall and is limited by lack of solar radiation in the winter and lack of nutrients in the summer.

In addition to the primary productivity through plant photosynthesis, organic biomass is produced through bacterial chemosynthesis. Chemosynthesis, observed on oceanic spreading centers in association with hydrothermal springs, is based on the release of chemical energy by the oxidation of hydrogen sulfide.

Radiant energy captured by plants is converted to chemical energy and passed through the biotic community. It is expended as mechanical and heat energy and ultimately reaches a state of entropy, where it is biologically useless. There is, however, no loss of mass. The mass used as nutrients by plants is converted to biomass. Upon the death of organisms, the mass is decomposed to an inorganic form ready again for use as nutrients for plants. Of the nutrients required by plants, compounds of nitrogen are most likely to be depleted and restrict plant productivity. Since the total decomposition of organic nitrogen compounds to inorganic nutrients requires three stages of bacterial decomposition, these compounds may have sunk beneath the photosynthetic zone before decomposition was complete and therefore are unavailable to plants.

As energy is transferred from plants to herbivore and the various carnivore feeding levels, only about 10 percent of the mass taken in at one feeding level is passed on to the next. The ultimate effect of this decreased amount of energy that is passed between trophic levels higher in the food chain is a decrease in the number of individuals and total biomass of populations higher in the food chain.

KEY TERMS

Autolytic decomposition (p. 287)
Bacteriovore (p. 284)
Biotic community (p. 284)
Carnivore (p. 284)
Carotin (p. 275)
Coccolith (p. 276)
Commensalism (p. 284)
Decomposer (p. 284)
Denitrifying bacteria (p. 287)
Diatomite (p. 275)
Dinoflagellates (p. 276)
Ecosystem (p. 284)

Entropy (p. 285)
Epitheca (p. 275)
Eutrophic (p. 281)
Flagella (p. 276)
Food chain (p. 290)
Food web (p. 291)
Frustule (p. 275)
Fucoxanthin (p. 272)
Gran method (p. 277)
Gross ecological efficiency (p. 290)
Gross primary production (p. 277)
Herbivore (p. 284)

Hypotheca (p. 275)
Metaphyta (p. 272)
Metazoa (p. 272)
Monera (p. 272)
Mutualism (p. 284)
Mycota (p. 272)
Net primary production (p. 277)
New production (p. 277)
Nitrogen-fixing bacteria (p. 287)
Oligotrophic (p. 281)
Omnivore (p. 284)
Oxygen compensation depth (p. 277)

Oxygen utilization rate (OUR) (p. 282)
Parasitism (p. 284)
Phycoerythrin (p. 272)
Primary productivity (p. 276)
Producer (p. 284)

Protista (p. 272)
Protozoa (p. 272)
Red tide (p. 276)
Regenerated production (p. 277)

Subsurface oxygen maximum (SOM) (p. 282)
Symbiosis (p. 284)
Trophic level (p. 288)

QUESTIONS AND EXERCISES

1. List the five kingdoms of organisms and the fundamental criteria used in assigning organisms to them.
2. Compare the macroscopic algae in terms of pigment, maximum depth in which they grow, common species, and size.
3. The Chrysophyta contains two classes of important phytoplankton. Compare their composition and the structure of their hard parts as well as their geologic significance.
4. Discuss and compare the contributions of the Pyrrophyta genera *Ptychodiscus* and *Gonyaulax* to red tide development.
5. Define oxygen compensation depth, and explain the use of the dark and transparent bottle technique for its determination. Discuss how the quantity of oxygen produced by photosynthesis in each clear bottle (gross photosynthesis) is determined.
6. Compare the biological productivity of polar, temperate, and tropical regions of the oceans. Include a discussion of seasonal variables, thermal stratification of the water column, and the availability of nutrients and solar radiation.
7. Discuss chemosynthesis as a method of primary productivity. How does it differ from photosynthesis?
8. Describe the components of the marine ecosystem.

9. Describe the flow of energy through the biotic community; include the forms to which solar radiation is converted. How does this flow differ from the manner in which mass is moved through the ecosystem?
10. What are the proportions by weight of carbon, nitrogen, and phosphorus in ocean water, phytoplankton, and zooplankton? Suggest how these amounts may support or refute the idea that life originated in the oceans.
11. Explain why nitrogen is much more likely than phosphorus to be a limiting factor in marine productivity.
12. How is the energy taken in by a feeding population lost so that only a small percentage is made available to the next feeding level? What is the average efficiency of energy transfer between trophic levels?
13. If a killer whale is a third-level carnivore, how much phytoplankton mass is required to add each gram of new mass to the whale? Assume 10 percent efficiency of energy transfer between trophic levels. Include a diagram.
14. Describe the probable advantage to the ultimate carnivore of the food web over a single food chain as a feeding strategy.

REFERENCES

Ducklow, H. W. 1983. Production and fate of bacteria in the oceans. *Bioscience* 33:8, 494–501.

Carpenter, E. J., and Romans, K. 1991. Major role of the cyanobacterium, *Trichodesmium,* in nutrient cycling in the North Atlantic Ocean. *Science* 254:5036, 1356–58.

George D., and George, J. 1979. *Marine life: An illustrated encyclopedia of invertebrates in the sea.* New York: Wiley-Interscience.

Grassle, J. F., et al. 1979. Galápagos '79: Initial findings of a deep-sea biological quest. *Oceanus* 22:2, 2–10.

Jenkins, W. J. 1982. Oxygen utilization rates in North Atlantic subtropical gyre and primary production in oligotrophic systems. *Nature* 300, 246–48.

Littler, M. M.; Littler, D. S.; Blair, S. M.; and Norris, J. N. 1985. Deepest known plant life discovered on an uncharted seamount. *Science* 227:4683, 57–59.

Macdonald, R. W., and Carmack, E. C. 1991. Age of Canada Basin deep waters: A way to estimate primary production for the Arctic Ocean. *Science* 254:5036, 1348–50.

Martinez, L.; Silver, M.; King, J.; and Alldredge, A. 1983. Nitrogen fixation by floating diatom mats: A source of new nitrogen to oligotrophic ocean waters. *Science* 221:4606, 152–54.

Paerl, H. W., and Bebout, B. M. 1988. Direct measurement of O_2-depleted microzones in marine *Oscillatoria:* Relation to N_2 fixation. *Science* 241:4864, 442–45.

Parsons, T. R.; Takahashi, M.; and Hargrave, B. 1984. *Biological oceanographic processes,* 3rd ed. New York: Pergamon Press.

Pimm, S. L.; Lawton, J. H.; and Cohen, J. E. 1991. Food web patterns and their consequences. *Nature* 350:6320, 669–74.

Platt, T., and Sathyendranath, S. 1988. Oceanic primary production: Estimation by remote sensing at local and regional scales. *Science* 241:4873, 1613–19.

Platt, T.; Subba Rao, D. V.; and Irwin, B. 1983. Photosynthesis of picoplankton in the oligotrophic ocean. *Nature* 301:5902, 702–4.

Russell-Hunter, W. D. 1970. *Aquatic productivity.* New York: Macmillan.

Sathyendranath, S.; Platt, T.; Horne, E. P. W.; Harrison, W. G.; Ulloa, O.; Outerbridge, R.; and Hoepffner, N. 1991. Estimation of new production in the ocean by compound remote sensing. *Nature* 353:6340, 129–33.

Sherr, B. F.; Sherr, E. B.; and Hopkinson, C. S. 1988. Trophic interactions within pelagic microbial communities: Indications of feedback regulation of carbon flow. *Hydrobiologia* 159:1, 19–26.

Shulenberger, E., and Reid, J. L. 1981. The Pacific shallow oxygen maximum, deep chlorophyll maximum, and primary productivity reconsidered. *Deep Sea Research* 28A:9, 901–19.

SUGGESTED READING

Sea Frontiers

Arehart, J. L. 1972. Diatoms and silicon. 18:2, 89–94. A very readable description of the important role of silicon and other elements in the ecology of diatoms, including microphotographs showing the varied forms of diatoms.

Coleman, B. A.; Doetsch, R. N.; and Sjblad, R. D. 1986. Red tide: A recurrent marine phenomenon. 32:3, 184–191. The problem of periodic red tides along the Florida, New England, and California coasts is discussed.

Idyll, C. P. 1971. The harvest of plankton. 17:5, 258–67. An interesting discussion of the potential of zooplankton as a major fishery.

Jensen, A. C. 1973. Warning—red tide. 19:3, 164–75. An informative discussion of what is known of the cause, nature, and effect of red tides.

Johnson, S. 1981. Crustacean symbiosis. 27:6, 351–60. A description of various symbiotic relationships entered into by tropical shrimps and crabs.

McFadden, G. 1987. Not-so-naked ancestors. 33:1, 46–51. The nature of the coverings of marine phytoplankton cells is revealed by the electron microscope.

Oremland, R. S. 1976. Microorganisms and marine ecology. 22:5, 305–10. The role of such microorganisms as phytoplankton and bacteria in cycling matter in the oceans is discussed.

Philips, E. 1982. Biological sources of energy from the sea. 28:1, 36–46. The potential for converting marine biomass to energy sources useful to society is discussed.

Scientific American

Benson, A. A. 1975. Role of wax in oceanic food chains. 232:3, 76–89. A report on the findings from observations made of the content of wax in the bodies of many marine animals from copepods to small deep-water fishes and their implications.

Childress, J. J.; Feldback, H.; and Somero, G. N. 1987. Symbiosis in the deep sea. 256:5, 114–21. Deep-sea hydrothermal vent animals have a symbiotic relationship with sulfur-oxidizing bacteria that allows them to live in the darkness of the deep ocean.

Govindjee, and Coleman, W. J. 1990. How plants make oxygen. 262:2, 50–67. The process of oxygen production by plants is explained.

Levine, R. P. 1969. The mechanism of photosynthesis. 221:6, 58–71. Reveals what is known of the process by which energy is captured by plants and converted to useful forms of chemical energy while freeing oxygen to the atmosphere.

Pettit, J.; Ducker, S.; and Knox, B. 1981. Submarine pollination. 244:3, 134–44. Discusses the pollination of sea grasses by wave action.

Chapter 14

Animals of the Pelagic Environment

In the sunlit waters of the pelagic environment live the phytoplankton that provide the basis for nearly all of the marine biomass through their photosynthetic activities. This important marine population was discussed in chapter 13, so our emphasis here will be on the animal populations. The discussion will deal primarily with the adaptations that allow them to live successfully in the pelagic environment.

STAYING ABOVE THE OCEAN FLOOR

Since the tissues of the zooplankton and nekton (muscle, cartilage, scales, bone, and shell) are more dense than ocean water, pelagic animals can remain off the bottom only through the application of buoyancy or energy. Chapter 12 discussed how microplankton's decreased size aided in this process by an increased frictional resistance to sinking. However, the larger animals we will discuss in this chapter need additional means to remain off the ocean floor.

A deep-sea anglerfish, *Lasiognathus saccostoma,* attracts two deep-water shrimp to its bioluminescent "lure." The fish is about 10 cm (4 in.) long. (Photo © Peter Arnold, Inc.)

Gas Containers

Since air is approximately 0.001 the density of water at sea level, a small amount of it inside an organism significantly increases its buoyancy. Some cephalopods have rigid gas containers (figure 14–1). The genus *Nautilus* has an external shell; the cuttlefish *Sepia* and deep-water squid *Spirula* have an internal chambered structure. These animals become neutrally buoyant and can maintain their positions in the water easily. Since the air pressure in their air chambers is always 1 atmosphere, they are limited in the depth to which they may venture. The *Nautilus* must stay above a depth of approximately 500 m (1640 ft), or its chambered shell will collapse as the external pressure approaches 50 atmospheres. The *Nautilus* is rarely observed below a depth of 250 m (820 ft).

Neutral buoyancy is achieved by some slow-moving bony fish through filling a gas bladder, or **swim bladder,** with gases. The swim bladder is normally not present in very active swimmers, such as the tuna, or in fish that live on the bottom. Some fish have a **pneumatic duct** that connects the swim bladder to the esophagus (figure 14–2). These fish can add or remove air through this duct. In other fish, the gases of the swim bladder must be added or removed more slowly by an interchange with the blood. Since change in depth will cause the gas in the swim bladder to expand or contract, the fish removes or adds gas to the bladder to maintain a constant volume. Those fish without the pneumatic duct are limited in the rate at which they can make these adjustments and, therefore, cannot withstand rapid changes in depth.

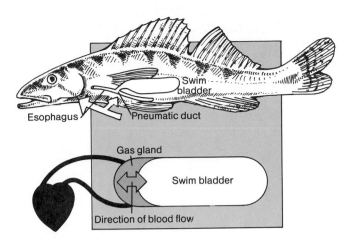

FIGURE 14–2
Swim Bladder of Some Bony Fishes. Example of a swim bladder connected to the esophagus by the pneumatic duct, allowing air to be added or removed rapidly. In fish with no pneumatic duct, all gas must be added or removed through the blood. This requires more time and is achieved by a network of capillaries associated with the gas gland.

The composition of gases in the swim bladders of shallow-water fishes is similar to that of the atmosphere. With increasing depth (fish with swim bladders have been taken from a depth of 7000 m, or 22,960 ft) the concentration of oxygen in the swim bladder gas increases from the 20 percent common near the surface to more than 90 percent. At the 700-atmosphere pressure that exists at a depth of 7000 m, gas is compressed to a density of 0.7 gm/cm³. This is about the density of fat, so many deep-water fish have air bladders filled with fat instead of compressed gas.

Floating Forms

Floating at the surface are some relatively large members of the plankton, the evolutionarily primitive siphonophore and scyphozoan **coelenterates,** as well as the pelagic tunicates. Somewhat smaller are the ctenophores and arrowworms. They all have soft gelatinous bodies with little if any hard tissue. A major strategy of this body type is the replacement in body fluids of heavy sulfate ions with chloride ions to maintain osmotic equilibrium.

The **siphonophore** coelenterates are represented in all oceans by the genera *Physalia* (Portuguese man-of-war) and *Velella* (by-the-wind sailor). Their gas floats, pneumatophores, serve as floats and sails that allow the wind to push these colonial forms across the ocean surface (figure 14–3). A colony of tiny sea-anemonelike **polyps** and jellyfishlike **medusae** (figure 14–3B) are suspended beneath the float. The long tentacles of the polyps primarily capture food, and the medusae carry on the sexual reproduction required to create a new colony. The tentacles of the *Physalia* may be many meters long and possess **nematocysts** long enough to penetrate the

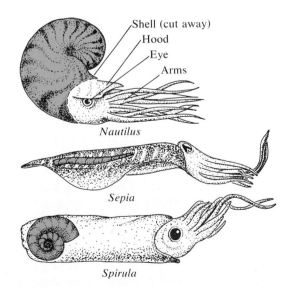

FIGURE 14–1
Gas Containers of the Cephalopods *Nautilus*, *Sepia*, and *Spirula*. The *Nautilus* has an external chambered shell, and the *Sepia* and *Spirula* have internal chambered structures that can be filled with gas to provide buoyancy.

FIGURE 14–3
Planktonic Coelenterates. *A,*
Physalia and Jellyfish. *B,* medusa.
(Photo by Larry Ford.)

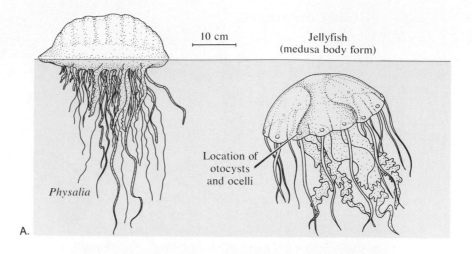

Jellyfish
(medusa body form)

10 cm

Location of
otocysts
and ocelli

Physalia

A.

B.

skin of humans; they have been known to inflict a painful and occasionally dangerous neurotoxin poisoning. The colonies grow from the initial polyp through asexual budding.

The **scyphozoans,** or jellyfish, have a medusoid, bell-shaped body with a fringe of tentacles and a mouth at the end of a clapperlike extension hanging down under the bell-shaped float. Ranging in size from near microscopic to 2 m (6.6 ft) in diameter with 60-m (197 ft) tentacles, most jellyfish have bells with a diameter of less than 0.5 m (1.6 ft). Jellyfish are capable of movement through muscular contractions (see figure 14–3A). Water enters the cavity under the bell and is forced out by contractions of muscles that circle the bell, jetting the animal ahead in short spurts. To allow the animal to swim generally in an upward direction, there are sensory organs spaced around the outer edge of the bell. These may be light-sensitive **ocelli** or gravity sensitive **otocysts.** This orientation ability is important because the jellyfish feed by swimming to the surface and sinking slowly through the life-rich surface waters.

Pelagic **tunicates** are generally transparent and barrel-shaped, with an anterior incurrent and posterior excurrent opening (figure 14–4). They move by a feeble form of jet propulsion effected by the contraction of bands of muscles that force water out the excurrent opening. Reproduction of solitary forms is complicated, involving alternate sexual and asexual reproduction. The genus *Salpa* includes solitary forms reaching a length of 20 cm (8 in.) and smaller aggregate forms that produce new individuals by budding. Individual members of an aggregate chain may be 7 cm (2.8 in.) long, and the chain of newly budded members may reach great lengths (see figure 14–4B). The genus *Pyrosoma* is luminescent and colonial. Individual members have their incurrent openings facing the outside surface of a tube-shaped colony that may be a few meters long. One end of the tube is closed, and the excurrent openings of the thousands of individuals all empty into the tube. Muscular contraction forces water out the open end to provide propulsion.

Ctenophores are comb-bearing animals closely related to the coelenterates in that their radially symmet-

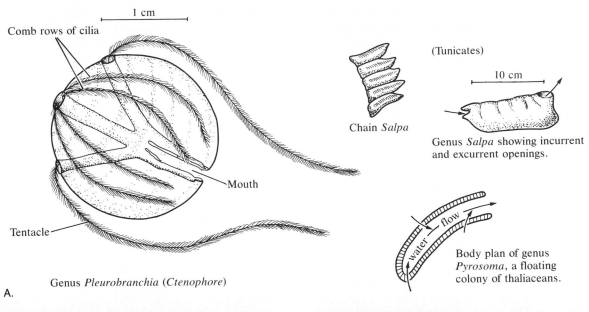

Genus *Pleurobranchia* (*Ctenophore*)

A.

Chain *Salpa*

(Tunicates)

Genus *Salpa* showing incurrent and excurrent openings.

Body plan of genus *Pyrosoma*, a floating colony of thaliaceans.

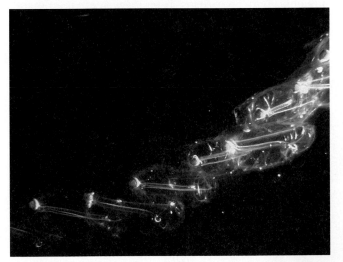

B.

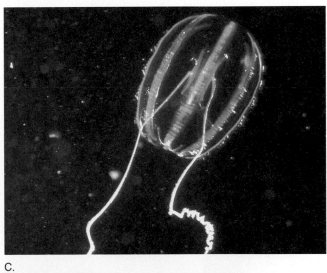

C.

FIGURE 14–4
Pelagic Tunicates and Cteno-phores. *A,* body structure. *B,* chain of salps. *C,* ctenophore, *Pleurobranchia. D,* ctenophore, *Beroe.* (Photos *B, C,* and *D* by James M. King, Graphic Impressions.)

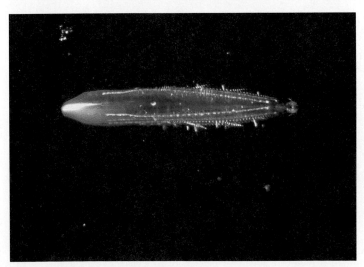

D.

rical bodies develop around their mouths. (Some forms, however, also possess an anal opening from the digestive cavity. Often referred to as comb jellies, this group of animals is entirely pelagic and confined to the marine environment.) The body form of most members of the phylum Ctenophora is basically spherical, with eight rows of cilia spaced evenly around the sphere. It is from these structures that the name comb-bearing is derived; when magnified, they look like miniature combs. If a ctenophore possesses tentacles, there will be two that contain adhesive organs instead of stinging cells to capture prey. The genera *Pleurobranchia* (sea gooseberries) (figure 14–4C) and the pink, elongated *Beroe* (figure 14–4D) range in size from the gooseberry dimension of the former to over 15 cm (6 in.) in the latter.

The **arrowworms** (chaetognaths—bristle jawed) are transparent and difficult to see, although they may grow to more than 2.5 cm (1 in.) in length (figure 14–5). The name chaetognath refers to the hairlike attachments around the mouth, which are used to grasp prey while they devour it. They are voracious feeders, eating primar-

ily the small zooplankton; in turn, they are eaten by fishes and larger planktonic animals such as jellyfish. These exclusively marine, hermaphroditic animals are usually more abundant in the surface waters some distance from shore.

Swimming Forms

This section examines larger pelagic animals, the *nekton,* with substantial powers of locomotion. They include rapid-swimming invertebrate squids, as well as fish and marine mammals.

Swimming squid include the genera *Loligo* (common squid), *Ommastrephes* (flying squid), and the *Architeuthis* (giant squid). Active predators of small fish, the smaller varieties of squid have long slender bodies with paired fins (figure 14–6). Unlike the less-active *Sepia* and *Spirula,* they have no hollow chambers in their bodies and therefore require more energy to remain in the upper water of the oceans. Trapping water in their mantle cavities and forcing it out through a siphon, these inverte-

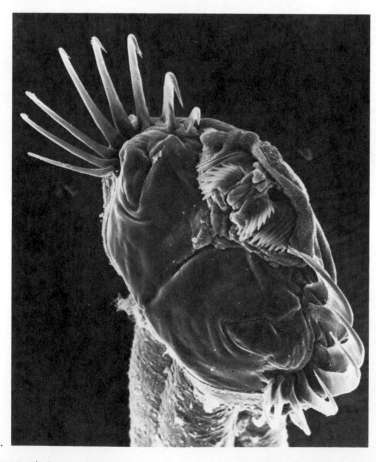

A.

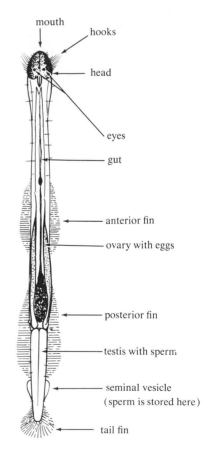

B. General Anatomy of an Arrowworm

FIGURE 14–5

Arrowworm. *A,* head of chaetognath *Sagitta tenuis* from Gulf of Mexico, magnified 161 times. (Photo courtesy of Howard J. Spero.) *B,* adult chaetognaths reach lengths ranging from 1 to 5 cm (0.4 to 2 in.).

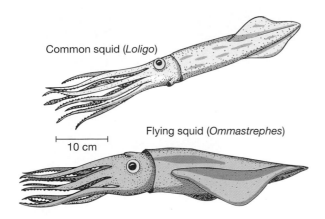

Common squid (*Loligo*)

Flying squid (*Ommastrephes*)

10 cm

FIGURE 14–6
Squids.

brates can swim about as fast as any fish their size. Two long arms with pads containing suction cups at the ends are used to capture prey, and eight shorter arms containing suckers take the prey to the mouth, where it is crushed by a beaklike mouthpiece.

The consideration of locomotion in fish is more complex than that of the squid because of body motion and the role of fins. The basic movement of swimming fish is the passage of a wave of lateral body curvature from the front to the back of the fish. This is achieved by the alternate contraction and relaxation of muscle segments, **myomeres,** along the sides of the body. The backward pressure of the fish's body and fins produced by the movement of this wave provides the forward thrust (figure 14–7).

Most active swimming fish have two sets of paired fins, pelvic and pectoral, used in maneuvers such as turning, braking, or balancing. When they are not in use, they can be folded against the body. Vertically oriented fins, dorsal and anal, serve primarily as stabilizers. The fin most important in propelling the high-speed fish is the tail, or caudal, fin. Caudal fins flare dorsally and ventrally to increase the surface area available to develop thrust. Increased surface area also increases frictional drag. The efficiency of the design of a caudal fin depends on its shape and can be expressed mathematically as an **aspect ratio,** calculated as follows:

$$\frac{(\text{fin height})^2}{\text{fin area}}$$

There are five basic shapes of caudal fins: rounded (as on sculpin), truncate (bass), forked (yellowtail), lunate (tuna), and heterocercal (shark) (figure 14–8). The rounded fin, with an aspect ratio of approximately 1, is flexible and useful in accelerating and maneuvering at slow speeds. The somewhat flexible truncate (ratio 3) and forked (ratio 5) tails will be found on faster fish and may still be used for maneuvering. The lunate caudal fin, with an aspect ratio of up to 10, is found on the fast-cruising fishes such as tuna, marlin, and swordfish; it is very rigid and useless in maneuverability but very efficient in propelling. The heterocercal fin is asymmetrical, with most of its mass and surface area in the dorsal lobe.

The heterocercal fin produces a significant lift to sharks as it is moved from side to side. This lift is important because sharks have no swim bladder and tend to sink when they stop moving. To aid this lifting, the pectoral fins are large and flat. Located on the anterior of the shark's body like the wings of an airplane, they serve as hydrofoils that lift the front of the shark's body to balance the posterior lift applied by the caudal fin. The shark loses a lot in maneuverability as a result of this modification of the pectoral fins.

FIGURE 14–7
Fish—Swimming Motions and Fins.

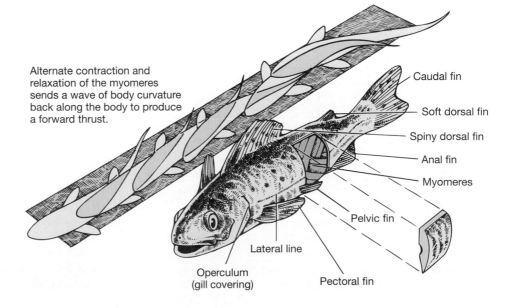

Alternate contraction and relaxation of the myomeres sends a wave of body curvature back along the body to produce a forward thrust.

Caudal fin

Soft dorsal fin

Spiny dorsal fin

Anal fin

Myomeres

Pelvic fin

Lateral line

Operculum (gill covering)

Pectoral fin

A. Rounded

B. Truncate

C. Forked

D. Lunate

E. Heterocercal

FIGURE 14–8
Caudal Fin Shapes. *A,* rounded (sculpin, flounder, angel). Queen angel. (Photo © Marty Snyderman.) *B,* truncate (salmon, bass, angelfish). Gray angelfish. (Photo © Wayne and Karen Brown.) *C,* forked (herring, yellowtail, goatfish). Goatfish, Red Sea. (Photo © Pechter Photo/The Stock Market.) *D,* lunate (bluefish, tuna, marlin). Blue marlin. (Photo © Bob Gomel/The Stock Market.) *E,* heterocercal (sharks). Silvertip shark. (Photo © Valerie Taylor/Peter Arnold, Inc.)

Pectoral fins in many fish do more than just help maneuvering (figure 14–9). Wrass and sculpins use them to "row" themselves through the water in jerky motions while the rest of the body remains motionless. Skates and rays swim by sending undulating motions across the greatly modified pectoral fins, and manta rays flap them like wings. The flying fish, *Exocoetus,* uses greatly enlarged pectoral fins to glide for up to 400 m (1312 ft) across the ocean surface after propelling itself with the enlarged ventral lobe of its caudal fin into the air to escape dolphins and other predators. Other fish have pectoral fins modified into fingerlike structures used to walk on the ocean floor.

Many fish in no great hurry propel themselves by undulation of the dorsal fin. Examples are triggerfish, sea horses, ocean sunfish, and trunkfish. The sunfish also uses its anal fin and stubby caudal fin to aid in the process, still without impressive results.

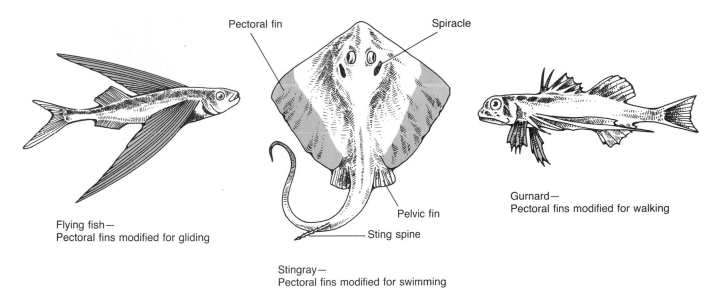

FIGURE 14–9
Pectoral Fin Modifications.

Modifications in Swimming Behavior

Some fish spend most of their time waiting patiently for prey and exert themselves only in short bursts as they lunge at the prey; others cruise relentlessly through the water seeking out prey. There is a marked difference in the nature of the musculature of fishes with these different styles of seeking out food. A grouper, representative of the **lungers,** has a truncate caudal fin (aspect ratio 3), and almost all its muscle tissue is white. Tuna, on the

other hand, are **cruisers,** and less than half their muscle tissue is white; it is mostly of a variety referred to as red (figure 14–10).

Red muscle fibers are much smaller in diameter (25–50 μm, or 0.01–0.02 in.) than **white muscle fibers** (135 μm, or 0.05 in.) and contain higher concentrations of **myoglobin,** a red pigment with an affinity for oxygen (O_2). These factors allow the red fibers to obtain a much greater oxygen supply than is possible for white fibers and therefore achieve a metabolic rate six times that of

A.

B.

FIGURE 14–10
Feeding Styles—Lunger and Cruiser. *A,* lungers like this tiger grouper sit patiently on the bottom and capture prey with quick, short lunges. Their muscle tissue is predominantly white. (Photo © Fred Bavendam/Peter Arnold Inc.) *B,* cruisers like these yellowfin tuna swim constantly in search of prey and capture it with short periods of high-speed swimming. Their muscle tissue is predominantly red. (Photo courtesy of National Marine Fisheries Service.)

white fibers. The red muscle tissue is abundant in cruisers that constantly swim. The lungers can get along quite well with little red tissue because they need not constantly move. White tissue, which fatigues much more rapidly than red tissue, is used by the tuna for short periods of acceleration while on the attack. It is also quite adequate for propelling the grouper and other lungers during their quick passes at prey.

The speed of fish is thought to be closely related to body length. For tuna, well-adapted for sustained cruising and short bursts of high-speed swimming, cruising speed averages about 3 body lengths/second; maximum speed of about 10 body lengths/second can be maintained for only one second. A yellowfin tuna, *Thunnus albacares,* has been clocked at 74.6 km/h (46 mi/h), or more than 20 body lengths/second, over a period of 0.19 second. Theoretically, a 4-m (13-ft) bluefin tuna, *Thunnus thynnus,* could reach speeds of up to 143.9 km/h (89.2 mi/h). Many of the toothed whales are known to be capable of high rates of speed. The porpoise *Stenella* has been clocked at 40 km/h (25 mi/h), and it is believed that the top speed of killer whales may be in excess of 55 km/h (34 mi/h).

As is true with many chemical reactions, metabolic processes can proceed at greater rates at higher temperatures. Although fish are generally considered to be **poikilothermic** (cold-blooded), having body temperatures that conform to the temperature of their environment, there are some fast swimmers with body temperatures above that of the surrounding water. Some are only slightly above the temperature of the water: the mackerel (*Scomber*), yellowtail (*Seriola*), and bonito (*Sarda*) have body temperature elevations of 1.3°, 1.4°, and 1.8°C (2.3°, 2.5°, and 3.2°F), respectively. Other fast swimmers that have much higher elevations are members of the mack-

erel shark genera, *Lamna* and *Isurus,* as well as the tuna, *Thunnus.* **Homeothermic** bluefin tuna have been observed to maintain a body temperature of 30°–32°C (86°–90°F), regardless of the temperature of the water in which they are swimming. Although these tuna are more commonly found in warmer water where the temperature difference is no more than 5°C (9°F), body temperatures of 30°C (86°F) have been measured in bluefin tuna swimming in 7°C (45°F) water.

Why do these fish exert so much energy to maintain their body temperatures at high levels when other fishes do quite well with ambient body temperatures? Their mode of behavior is that of a cruiser, and any adaptation (high temperature and metabolic rate) that can increase the power output of their muscle tissue helps them search out and capture prey.

The mackerel sharks and tuna are aided in conserving the heat energy needed to maintain their high body heats by a modification of the circulatory system (figure 14–11). While most fish have a **dorsal aorta,** just beneath the vertebral column, that provides blood to the swimming muscles, the mackerel sharks and tuna have additional **cutaneous arteries** just beneath the skin on either side of the body. As cool blood flows into red muscle tissue, its temperature is increased by heat generated by muscle metabolism (muscle contractions). A fine network of tiny blood vessels within the muscle tissue is designed to minimize heat loss. The vessels that return the blood to the **cutaneous vein,** parallel to the cutaneous artery along the side of the fish, are all paired with small vessels carrying blood into the muscle tissue. In this way, the warm blood leaving the tissue helps to heat the cooler blood entering from the cutaneous artery.

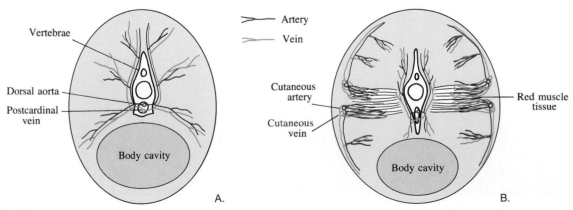

FIGURE 14–11
Circulatory Modifications in Warm-Blooded Fishes. *A,* most cold-blooded (poikilothermic) fishes have major blood vessels arranged in the pattern represented on the left. All blood flows to muscle tissue from the dorsal aorta and returns to the postcardinal vein beneath the vertebral column. *B,* warm-blooded (hoemothermic) fishes like the tuna have cutaneous arteries and veins that help maintain high blood temperature by using heat energy generated by contracting muscle tissue. Its blood vessel pattern is represented on the right.

A. Manatees

B. Elephant seals

C. Harbor seals

D. Sea otter

FIGURE 14–12
Marine Mammals—Manatee, Pinnipeds, and Sea Otter. (Photo *A* © Fred Bavendam/Peter Arnold, Inc.; *B,* © Tim Ragen; *C,* © *C.* Allan Morgan/Peter Arnold, Inc.; *D,* © Jeff Foott/Bruce Coleman, Inc.)

MARINE MAMMALS

Ocean mammals include the coast-dwelling herbivores, the **sirenans** (sea cows—dugongs and manatees), and a variety of carnivorous forms (figure 14–12). Spending all or most of their lives in coastal waters and coming ashore to breed are the **sea otter** and **pinnipeds** (sea lions, seals, and walruses). The truly oceanic mammals are the **cetaceans**—whales, porpoises, and dolphins (figure 14–13). Since mammals evolved from reptiles on land

some 200 million years ago and no marine forms are known earlier than 60 million years ago, it is believed that all marine forms evolved from some ancient land-dwellers.

Cetaceans

Well-known as the mammals best adapted to life in the oceans, the 75-plus species of cetaceans are of two basic types. The toothed whales, the **odontoceti,** include

305

FIGURE 14–13
Baleen and Toothed Whales. No creature in the sea captures the human imagination like the whales, huge mammals that returned to the sea about 60 million years ago. Two suborders are present in today's oceans, the toothed whales (Odontoceti) and baleen whales (Mysticeti). Represented here are members of each group drawn to scale. The toothed whales—bottle-nosed dolphin, narwhal, killer whale, and sperm whale—are active predators and may track down their prey by use of echolocation. The baleen whales—humpback whale, right whale, and blue whale—fill their mouths with water and

ATLANTIC RIGHT WHALE (Eubalaena glacialis)
This cold-water whale was the first recorded target of whaling by Basque seamen during the middle ages. Similar species are found near Japan and in high southern latitudes. Length is up to 18 m (60 ft)

SPERM WHALE (Physeter catodon)
Found mostly in tropical waters, this deep diver has a huge snout that contains a large amount of oil. Length of male is to 19 m (63 ft). Length of female is to 10.5 m (35 ft)

force it out between long baleen slats that hang down from the upper jaw. Small fish, krill, and other plankton are trapped behind the baleen. The Pacific gray whale has short baleen slats and feeds by sucking in sediment from the shallow bottom of its North Pacific feeding grounds. Thus it strains out benthic amphipods and other invertebrates. Although many great whale populations were pushed to near extinction by nineteenth- and twentieth-century whaling efforts, it appears all will survive this threat. The major threat to their survival may now be the disruption of their feeding and breeding grounds.

BOTTLE-NOSED DOLPHIN (*Tursiops truncatus*)
Found in the North Atlantic, this is one of the many species of beaked dolphins. Length is to 3 m (10 ft).

KILLER WHALE (*Grampus orca*)
Cosmopolitan in distribution, this whale is unique in that is not only feeds on fish but also on seals, birds, and other whales. The male grows to twice the female's size. Length of male is to 9 m (31 ft).

NARWHAL (*Monodon monoceros*)
The scientific name, which means one tooth, one horn, is accurate for the male of this Arctic whale. Length is to 6 m (20 ft).

PACIFIC GRAY WHALE (*Rhachianectes glaucus*)
This primitive baleen whale with a small head and very reduced dorsal fin stays close to shore. It even enters the surf zone. Length is to 14 m (45 ft).

HUMPBACK WHALE (*Megaptera nodosa*)
The great communicators of the Cetacea have a cosmopolitan distribution. The genus name is derived from their extraordinary winglike flippers. Length is to 16 m (52 ft).

BLUE WHALE (*Sibbaldus musculus*)
The largest animal to inhabit the earth will consume over five tons of krill per day. Length is up to 30 m (100 ft).

sperm whales, porpoises, and dolphins. The baleen whales, the **mysticeti,** probably evolved from the toothed whales some 30 million years ago. In place of teeth, they have baleen, plates of horny material, that hang down from the upper jaw to serve as a sieve. The body of cetaceans is more or less cigar-shaped, nearly hairless, and insulated with a thick layer of blubber. The forelimbs are modified into flippers that may be moved only at the "shoulder" joint. The hind limbs are vestigial, not attached to the rest of the skeleton, and not externally visible. The skull is highly modified, with one nasal opening (toothed) or two openings (baleen) near the top. They propel themselves by vertical movements of a horizontal fluke (tail fin).

The baleen whales include three families. The **gray whale** has short, coarse baleen, no dorsal fin, and only two to five ventral grooves beneath the lower jaw. The **rorqual whales** have short baleen and many ventral grooves and are divided into two subfamilies. The balaenopterids, with long, slender bodies, small sickle-shaped dorsal fins, and flukes with smooth edges, include the minke, Baird's, Bryde's, sei, fin, and blue whales. The megapterids, or humpback whales, have a more robust body, long flippers, flukes with uneven trailing edges, tiny dorsal fins, and tubercles on the head. The third family includes the **right whales,** with long, fine baleen, broad triangular flukes, no dorsal fin, and no ventral grooves. The "northern" right whale is the baleen whale that is most threatened by extinction. The southern right whale of the Southern Hemisphere and the bowhead whale that remains near the Arctic pack-ice edge are the other members of this family. The smallest and least known of the baleen whales is the pigmy right whale. It is rare and found only in the cooler temperate waters of the Southern Hemisphere.

The toothed whales are predators that feed mostly on smaller fish and squid, although the killer whale is known to feed on a variety of larger animals, including other whales. The baleen whales feed primarily by filtering crustaceans at depths ranging from the surface down to and including the sediment of shallow ocean basins.

Modifications to Increase Swimming Speed. Cetaceans' muscles are not vastly more powerful than those of other mammals, so it is believed that their ability to swim at high speed must result from modifications that reduce frictional drag. To illustrate the importance of streamlining in reducing the energy requirements of swimmers, a small dolphin would require muscles five times more powerful than it has to swim at 40 km/h (25 mi/h) in turbulent flow than in streamlined laminar flow. In addition to having a streamlined body, cetaceans are believed to achieve near-laminar flow of water across their bodies with the aid of a specialized skin structure. The skin is composed of two layers: a soft outer layer that is 80 per-

cent water and has narrow canals filled with spongy material and a stiffer inner layer composed mostly of tough connective tissue. The soft layer tends to reduce the pressure differences at the skin-water interface by compressing under regions of higher pressure and expanding in regions of low pressure.

Modifications to Allow Deep Diving. Where humans can free-dive to a maximum recorded depth of about 100 m (328 ft) and hold their breaths in rare instances for 6 minutes, the sperm whale, *Physeter,* is known to dive deeper than 2200 m (7216 ft), and the North Atlantic bottle-nosed dolphin, *Hyperoodon ampullatus,* can stay submerged for up to two hours.

One reason whales can make deep dives and stay submerged for long periods is their ability to alternate between periods of eupnea (normal breathing) and apnea (cessation of breathing). The periods of apnea occur while the animal is submerged. To understand how some cetaceans are able to go for long periods without breathing requires a general knowledge of the lungs and their associated structures. Air that is inhaled passes through the trachea, into the bronchi, and through a series of smaller bronchioles and alveolar ducts, to tiny terminal chambers, the alveoli (figure 14–14). The alveoli are lined by a thin alveolar membrane that is in contact with a dense bed of capillaries; the exchange of gases between the inspired air and the blood occurs across the alveolar membrane. Some cetaceans have an exceptionally large concentration of capillaries surrounding the alveoli, which have muscles that move air against the membrane by repeatedly contracting and expanding. Cetaceans take from 1 to 3 breaths per minute while resting, compared to about 15 in humans. Because of the longer time inhaled breath is held, the large capillary mass in contact with the alveolar membrane, and the circulation of the air by muscular action, many cetaceans can extract almost 90 percent of the O_2 in each breath, compared to the 4 to 20 percent extracted by terrestrial mammals.

To use this large amount of O_2, which can be taken into the blood efficiently during long periods under water, cetaceans may apply two strategies. They (1) store large amounts of O_2 and (2) can reduce oxygen use. The storage of so much O_2 is possible because prolonged divers have a much greater blood volume per unit of body mass than those that dive for only short periods. Compared to terrestrial animals, some cetaceans have twice as many red blood cells per unit of blood volume and up to 9 times as much myoglobin in the muscle tissue. Thus large supplies of O_2 can be chemically stored in the **hemoglobin** of the red blood cells and the myoglobin.

Additionally, the muscles that start a dive with a significant O_2 supply can continue to function through an-

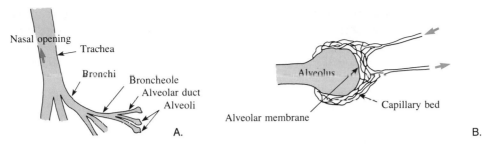

FIGURE 14–14
Cetacean Modifications to Allow Prolonged Submergence. *A,* basic lung design. Air enters the lung through the trachea, and oxygen is absorbed into the blood through the walls of the alveoli. *B,* oxygen exchange in the alveolus. A dense mat of capillaries receives oxygen through the alveolar membrane. Because air is normally held in the lungs of cetaceans for up to 1 minute before exhaling, as much as 90 percent of the oxygen can be extracted. Whales have up to twice as many red blood cells and 9 times as much myoglobin in their muscle tissue as terrestrial mammals in which to store this oxygen.

aerobic respiration when the O_2 is used up. The muscle tissue is relatively insensitive to high levels of carbon dioxide, a product of aerobic respiration, and lactic acid, a product of anaerobic respiration.

Because the swimming muscles can function without oxygen during a dive, they and other organs, such as the digestive tract and kidneys, may be sealed off from the circulatory system by constriction of key arteries. The circulatory system services primarily the heart and brain. Because of the decreased circulatory requirements, the heart rate can be reduced by 20 to 50 percent of its normal rate. However, recent research has shown that no such reduction in heart rate occurs when the common dolphin (*Delphinus delphis*), the white whale (*Delphinapterus leucas*), and the bottle-nosed dolphin (*Tursiops truncatus*) dive.

Another problem with deep and prolonged dives is the absorption of compressed gases into the blood. When humans make dives using compressed air, the prolonged breathing of compressed air, particularly N_2, can result in nitrogen narcosis, or decompression sickness (the bends). Nitrogen narcosis is an effect similar to drunkenness, and it can occur when a diver either goes too deep or stays too long at depths greater than 30 m (98 ft). If a diver surfaces too rapidly, the lungs cannot remove the excess gases fast enough, and the reduced pressure may cause small bubbles to form in the blood and tissue. The bubbles interfere with blood circulation, and the resulting decompression sickness can cause excruciating pain, severe physical debilitation, or even death.

Cetaceans and other marine mammals do not suffer from these difficulties. Their main defense against absorbing too much N_2 seems to be a more flexible rib cage. By the time a cetacean has reached a depth of 70 m (230 ft), the rib cage has collapsed under the 8 atmospheres of pressure. The lungs within the rib cage also collapse, removing all air from the alveoli. Since most absorption of gases by the blood occurs across the alveo-

lar membrane, the blood cannot absorb the compressed gases, and the problem of nitrogen narcosis is avoided.

It is, however, possible that the collapsible rib cage is not the main defense against the bends. A research effort designed to increase the nitrogen concentration in the lungs of a trained dolphin and give it the bends put enough nitrogen into the tissue of the animal to give a human a severe case of the bends. The dolphin suffered no ill effects. This whale, as well as other species, may have simply evolved an insensitivity to this gas not present in other mammals.

Use of Sound. It has long been known that cetaceans, none of which possess vocal cords, make a variety of sounds. Speculations that the sounds may represent aspects of behavior from echolocation to a highly developed language have been presented. To the present, what we do know about the use of sound by cetaceans is limited.

The baleen whales produce sounds at frequencies generally below 5,000 Hz. They are believed to be produced by air passing through the trachea. Gray whales produce pulses that could conceivably be used for echolocation and moans that may serve to maintain contact with other gray whales.

The rorqual whales produce moans that last from one to many seconds. These sounds are in the 10 to 20 Hz range and are probably used to communicate over distances of up to 50 km (31 mi). There is some evidence that the moans of fin whales may be related to reproductive behavior, but much more observation is required before much can be said about the message content of the full range of sounds. The repertoire of sounds produced by humpback whales ranges from chirps to moans. Repeated sequences of sounds termed "songs" are produced primarily by adult males on breeding grounds and possibly communicate the males' readiness to mate and challenge other males.

Right whales and bowhead whales, the "chatterboxes" of the whale population, generate complex suites of moans during feeding and migration and on breeding grounds. Some sounds have been correlated with specific social behavior patterns. For example, the bowhead males produce "songs" that may be similar in function to those of the humpback males.

High-pitched and more-varied sounds are produced by toothed whales. The sounds fall into two categories: clicks and whistles. Both are used for communication, and some clicks are used for echolocation in which the animal's brain involuntarily determines the distance to an object by multiplying the velocity with which a sound signal travels and returns from the object by the time it takes to reach it and dividing this product by 2 (figure 14–15).

$$D = \frac{V \times T}{2}$$

Although all marine mammals have good vision, conditions within the marine environment often limit its effectiveness. In coastal waters, where suspended sediment and dense plankton blooms make the water turbid, and in the deeper waters, where light is limited or absent, echolocation would surely be beneficial in pursuit of prey or the location of objects in the water. Using lower-frequency clicks at great distance and higher frequency at closer range, the bottle-nosed dolphin, *Tursiops truncatus,* can detect a school of fish at distances greater than 100 m (328 ft) and an individual 13.5-cm (5.3 in.) fish at a distance of 9 m (30 ft).

Sperm whales produce intense clicks in series that are called codas, which may serve as individual signatures. The "signature whistles" of some dolphins may serve a similar purpose. Very high intensity clicks are produced by some dolphins and sperm whales and may be used to stun prey. There is, however, no direct evidence of such behavior. Although there is no doubt that many species of toothed whales communicate using sound, there is no evidence of language in the sense of an organized structure and sequences of sound corresponding to the human use of sentence structure. The clicks of the bottle-nosed dolphin are of a frequency range that is partly audible to humans; however, some are of ultra-

sonic frequencies that can be repeated up to 800 times per second. Large toothed whales emit lower-frequency clicks at lower repetition rates. The killer whale, *Orcinus orca,* produces from 6 to 18 clicks per second, and the sperm whale, *Physeter catodon,* may produce from less than one to more than 40 clicks per second. Each click of the sperm whale contains 9 pulses, or clickettes. It has been estimated that the sperm whales can detect their main prey, squid, from a distance of up to 400 m (1312 ft) by use of their low-frequency scanning clicks.

How the pulses of clicks are produced and how the returning sound is received are not fully understood. However, there are some theories that attempt to answer both questions for certain species of toothed whales. Two of these theories of sound production are discussed in figure 14–16.

Whereas the human ear is sensitive to frequencies from 16 Hz to 20 kHz, some toothed whales respond to frequencies as high as 150 kHz. In most mammals, the bony housing of the inner ear structure is fused to the skull, so that, when submerged, sounds transmitted through the water are first picked up by the skull and travel to the hearing structure from many directions. This structure makes it impossible for such mammals to locate accurately the source of the sounds. Obviously such an arrangement for the hearing structure would not suit an animal that depended on echolocation to accurately determine the positions of objects in the water. All cetaceans have evolved structural features that insulate the inner ear housing, the tympanic bulla, from the rest of the skull. In the toothed whales, the tympanic bullae are separated from the rest of the skull by an appreciable distance and surrounded by an extensive system of air sinuses. The sinuses are filled with an insulating emulsion of oil, mucus, and air and are surrounded by fibrous connective tissue and venous networks.

There is some disagreement on how sound reaches the inner ear to be processed. Some investigators believe it passes through the external auditory canal as it does in other mammals, yet this canal is usually blocked by a plug of ear wax in baleen whales. In toothed whales it may be nearly or completely covered by skin. A thin flaring of the lower jaw, less than 0.1 mm (0.004 in.) thick, is directed toward the tympanic bulla on either side of the head and connected with it by a fat- or oil-containing body. It is

FIGURE 14–15
Echolocation. Clicking sound signals are generated by cetaceans and bounced off objects in the ocean to determine their size, shape, and distance.

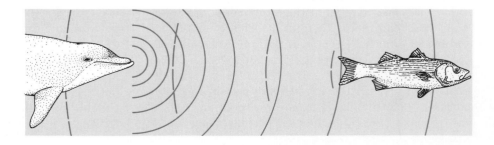

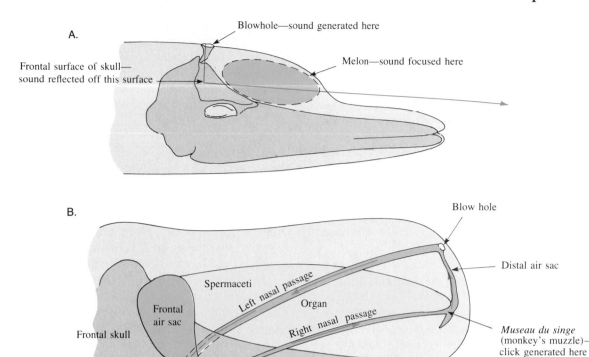

FIGURE 14–16
Generation of Echolocation Clicks in Small-Toothed and Sperm Whales. *A,* in small-toothed whales, the clicks may be generated within the blowhole mechanism, reflected off the frontal surface of the skull, and focused by the fatty melon into a forward-directed beam. *B,* structures that may be related to the generation of clicks by sperm whales. Air may pass from the trachea through the right nasal passage across the *museau du singe,* where clicks are generated. It passes along the distal air sac past the closed blow hole and returns to the lungs along the left nasal passage. The initial click is followed by 8 progressively weaker clicks that may result from sound energy bouncing back and forth between the frontal air sac and the distal air sac being transmitted through the oil-filled spermaceti organ.

believed that sound is picked up by the thin, flaring jawbone and passed to the inner ear via the connecting fat body. Tests of acoustical sensitivity have indicated the lower jaw is up to 6 times as sensitive as the external auditory canal.

GROUP BEHAVIOR

Depending on the size, feeding behavior, nature of the reproductive process, and other requirements of maintaining the species, many animals inhabiting the open ocean have developed patterns of group behavior that allow them to most efficiently exploit their environment. Two of these patterns are **schooling** and **migration.**

Schooling

Obtaining food occupies most of the time of many inhabitants of the open ocean. It is achieved by the fast and agile animals through active predation; other animals move more leisurely while filtering small food particles from the water. Examples of predators and filter feeders can be found in populations of pelagic animals from the tiny zooplankton to the massive whales. Patches of zooplankton may occur simply because the nutrients that support the phytoplankton they feed on may be in high concentrations in certain coastal waters. We usually do not refer to these patches as schools. The term school is usually reserved for well-defined social organizations of fish, squid, and crustaceans.

The number of individuals in a school can vary from a few larger predaceous fish to hundreds of thousands of small filter feeders. Within the school, individuals of the same size move in the same direction with equal spacing between each. This spacing is probably maintained through visual contact, and in the case of fish, by use of the **lateral line system** that detects vibrations of swimming neighbors. The school can turn abruptly or reverse direction as individuals at the flank or the rear of the school assume leadership positions (figure 14–17).

The advantage of schooling seems obvious from the reproductive point of view. During spawning, it assures that there will be males to release sperm to fertilize the eggs shed into the water or deposited on the bottom by females. However, most investigators believe the most important function of schooling in small fish is *protection from predators.* On first consideration, it seems illogical that the formation of schools would serve such a purpose. Aren't the smaller fish making it easier for the predator by forming a large mass? This is a difficult question; in fact, no experimental evidence provides the basis for an unequivocal answer. However, the consensus, based primarily on conjecture, is *no.*

The belief that small fish are safer from predators if they form schools is based in part on the following reasoning: Over two thousand species of fish are known to form schools. This fact alone indicates that this behavior has evolved and is so pervasive today among fish populations because, for fish with no other means of defense, it somehow provides a better chance of survival than swimming alone. How schooling is protective may grow out of the following considerations: (1) If members of a species form schools, they reduce the percentage of ocean volume in which a cruising predator might find one of their kind. (2) Should a predator encounter a large school, it is less likely to consume the entire unit than if it encounters a small school or an individual. (3) The school may appear as a single large and dangerous opponent to the potential predator and prevent some attacks. (4) Predators may find the continually changing position and direction of movement of fish within the school confusing, making attack particularly difficult for predators that can attack only one fish at a time. It is surely possible that there are other more subtle reasons for schooling. Whatever they might be, this gregarious behavior must enhance species survival because it is widely observed among pelagic animals.

Migration

Many oceanic animals undertake migrations of varying magnitudes. This behavior is observed among sea turtles, fish, and mammals. Some animals, for instance the sea turtles, return to dry beach deposits to lay their eggs. Other populations, including many of the baleen whales, migrate because the physical environment and nature of available food suitable for adults do not meet the needs of young members of their species.

Migratory routes of commercially important baleen whales have been well known since the mid-nineteenth century. The paths of these and other air-breathing mammals are easy to observe. However, the migratory paths of many less visible fish, even though they may also have high commercial value, have been more difficult to identify. Tagging studies have helped us understand the patterns of movement of these fish. Sampling the distribution patterns of eggs, larvae, young, and adult populations and radio-tracking individuals have also helped to identify migratory routes.

One aspect of migratory behavior that is of interest to investigators is orientation: How do migratory species orient themselves in time and space? To put the problem another way, How do they know where they are in relationship to where they want to go? How do they know when to leave their present location to get where they are going at the proper time?

Although these answers are still being sought, it is believed that all of the migratory species have an innate sense of time referred to as a biological clock. External manifestations of the biological clock of a species, called **circadian rhythms,** are physiological variations that occur independent of changes in the external environment. Circadian rhythms include cyclic changes in respiration rate and body temperature. However, some changes in the external environment can alter the circadian rhythms. It is possible that changes in food availability, water temperature, and length of daylight periods may serve to trigger seasonal migrations.

It seems that orientation in space must be a more complex problem than orientation to time. Animals such as mammals and turtles that migrate at the surface could well use their sight as a means of orientation. Gray whales that migrate close to land over a large part of their migratory route could possibly identify landmarks along the shore. When out of sight of land, mammals and turtles could use the relative positions of the sun, moon, and

FIGURE 14–17
Schooling. Soldier fish and butterfly fish on a reef, Maldives. (Photo © Thomas Ives/The Stock Market.)

stars to guide them on their way. Fish that migrate beneath the surface are thought to use smell, relative movement within ocean currents, and small induced currents (generated as the ions dissolved in ocean water are carried through Earth's magnetic field by ocean currents) to help them orient themselves in ocean space.

Many of the important food fish of the high northern latitudes deposit pelagic egg masses that are transported by currents. Such fish will actively migrate upcurrent during spring or summer and spawn in a location that will assure the current will carry the eggs back to a nursery ground where the food supply will be appropriate for newly hatched fry. Migrations of this type are usually over distances of a few tens or hundreds of kilometers. This behavior is observed in the population of Atlantic cod, *Gadus morhua,* that spawns along the south shore of Iceland. Adults that occupy feeding grounds along the north and east coasts of Iceland and around the southern tip of Greenland migrate to the spawning grounds in the late winter and early spring. This migration involves swimming against the East Greenland and Irminger currents and includes only mature adults at least 8 years of age. These spawning migrations are undertaken annually until the adults die at an age of 18 to 20 years (figure 14–18A).

The spawning grounds are bathed in the relatively warm water of the Irminger Current, a northward-flowing branch of the Gulf Stream. Each female releases up to 15 million eggs, which float in the surface current. The drifting eggs, carried by the East Greenland and Irminger currents, hatch in about two weeks. The larvae feed at midwater depths while the currents are carrying them to the adult feeding grounds. They will remain there, undertaking only short onshore-offshore migrations, until they mature and make their first spawning migration.

Longer migrations are undertaken by two Atlantic bluefin tuna populations that spawn in tropical waters near the Azores in the eastern Atlantic and the Bahamas in the western Atlantic. Little is known about where the newly hatched tuna feed, but the adults are known to move north along the seaward edge of the Gulf Stream in May and June and feed on the herring and mackerel populations off the coasts of Newfoundland and Nova Scotia. They may move even farther north, but the route used by them to return to the spawning grounds is unknown.

Anguilla, the Atlantic eel, undertakes what appears to be a single round trip migration during its lifetime. Its behavior is **catadromous:** After being spawned in the ocean, it enters freshwater streams for its adult life and returns near the end of its life to the ocean spawning site.

FIGURE 14–18
Migration Routes. *A,* migration routes of the Icelandic cod, bluefin tuna, and Atlantic eel (*Anguilla*). *B,* migration routes of the Chinook salmon and the California gray whale.

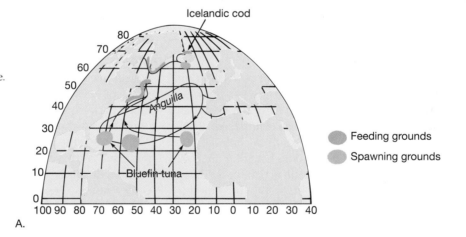

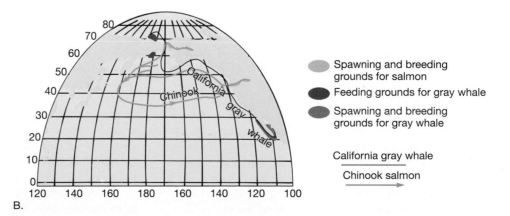

Spawning is thought to occur in the Sargasso Sea southeast of Bermuda. This spawning has not been observed, and the earliest stage of *Anguilla* that has been observed is the leaf-shaped, transparent leptocephalus larva. The 5-mm (0.2-in.)-long larvae are carried into the Gulf Stream and north along the North American coast. After one year in the ocean, some of the leptocephali metamorphose into young eels, elvers, and move into the streams of North America. Others spend another year migrating to the European coast before metamorphosing and moving into European streams. A third population spends another year moving into the Mediterranean Sea before undergoing metamorphosis. The American and European populations are considered by some to be different species—*A. rostrata* and *A. anguilla,* respectively—based primarily on the fact that the American eel has 8 to 10 fewer vertebrae than the European eel.

After spending up to ten years in the freshwater environment, the mature eels undergo yet another change. They develop a silvery color pattern and enlarged eyes that are typical of fish inhabiting the mesopelagic zone. They swim downriver to the ocean and presumably back to the spawning ground, where they are thought to die.

An opposite behavior pattern is characteristic of salmon of the north Atlantic and Pacific oceans. This **anadromous** behavior includes spawning in freshwater streams and spending most of their adult lives in the ocean. Pacific species die after spawning; Atlantic salmon return to the ocean.

Six species of Pacific salmon, *Oncorhynchus,* follow similar paths in their migrations (figure 14–19). Spawning occurs during the late summer and early fall. Eggs are deposited and fertilized in gravel beds far upstream. The chinook salmon spawns from as far south as central California to Alaska. The young chinook hatching in the spring may head downstream as a silvery, filter-feeding smolt during its first or second year. By the time it reaches the ocean, the young chinook has become a predator, feeding on smaller fish. After spending about four years in the North Pacific, mature salmon of more than 10 kg (22 lb) in weight and 1 m (3.28 ft) in length return to their home streams to spawn (figure 14–18B). Although it is not known how salmon achieve this homing, most investigators think the most likely sources of information processed by the salmon during their return trip to spawning grounds are odors and currents.

Some of the longest migrations known to occur in the open ocean are the seasonal migrations of baleen whales. Many baleen species feed in the colder high-latitude waters and breed and calve in warm tropical waters. Feeding occurs during summer while the long hours of sunlight illuminating the nutrient-rich waters produce a vast population of crustaceans to be fed upon. Only the vast supplies of food available in these waters make it

FIGURE 14–19
Varieties of Salmon. Top, chum (*Oncorhynchus keta*); center, coho (*O. kisutch*); bottom, chinook (*O. tschawyscha*). (Courtesy U.S. Fish and Wildlife Service.)

possible for the whales to maintain themselves during the long migrations and mating and calving season when feeding is minimal.

Although it is widely believed that many baleen whales calve in warm waters because the young could not survive in the cold waters of the feeding grounds, this may not be so. Bowhead whales live year-round at the Arctic pack-ice edge, and the smaller narwhal lives much of the year within the ice field. The fact that these whales can successfully calve in Arctic waters makes it difficult to understand why the larger calves of the species that are born in tropical waters could not also survive in the colder waters. The long migrations baleen whales now undertake may have resulted from productive feeding grounds that were once near the tropical calving grounds progressively moving poleward because of climate change. If this is the case, the breeding and calving grounds must be very important locations of social aggregation. Otherwise, they would surely have been abandoned for locations closer to the present feeding grounds. An alternate explanation for leaving the colder waters to calve is to avoid the killer whales that are more numerous in colder waters. They are a major threat to young whales. Because of the energy demands faced by female baleen whales in producing large offspring (the gestation period is up to 1 year) and providing them with fat-rich milk for several months, it is not uncommon for them to mate only once every two or three years.

The migration route of the *California gray whale* demonstrates how the conditions discussed above are met by whale migration patterns. Gray whales are moderately large whales, reaching lengths of 15 m (49 ft) and weighing over 30 metric tons. Populations of gray whales feed during the summer months in the extreme north Pacific Ocean, Sea of Okhotsk, Bering Sea, and Chukchi Sea. They are unique among baleen whales in that they do not feed by straining pelagic crustaceans and small fish from the water; instead, they stir up bottom sediment with their snouts and feed on bottom-dwelling amphipods (figure 14–20).

The western populations winter along the coast of Korea, but the population known as the California gray whale migrates from the Chukchi and Bering seas to winter in lagoons of the Pacific coast of Baja California and the Mexican mainland coast near the southern end of the Gulf of California (figure 14–18B). The migration usually begins in September when pack ice begins to form over the continental shelf areas that are their feeding grounds. This migration, which is the longest known migration undertaken by any mammal, may involve a round trip

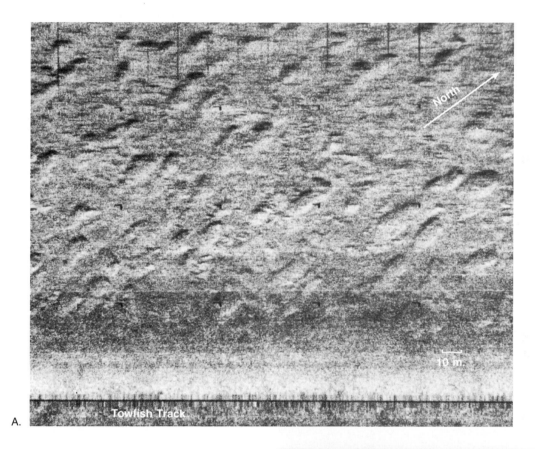

A.

FIGURE 14–20
California Gray Whale Feeding. *A,* side-scan sonograph showing pits on the floor of the northern Bering Sea created by California gray whales feeding on bottom-dwelling amphipods and other invertebrates. The smaller 2- to 3-m (6.5- to 10-ft)-long pits produced by the initial feeding do not show up well at this scale. All of the readily visible pits have been enlarged by currents that flow in a northerly direction. The pits show this north-south alignment. (Sonograph courtesy of William K. Sacco, Yale University.) *B,* amphipod mat. A mat of mucus-lined burrows is created by ampeliscid amphipods that are the preferred prey of California gray whales. (Photo courtesy of Kirk R. Johnson, Yale University.)

B.

distance of 22,000 km (13,662 mi). First to leave are the pregnant females. They are followed by a procession of mature females that are not pregnant, immature females, mature males, and immature males. After cutting through the Aleutian Islands by way of Unimak Pass, they follow the coast throughout their southern journey. Traveling at an average rate of about 200 km/day (124 mi/day), most of the whales reach the lagoons of Baja California by the end of January.

In these warm-water lagoons, the pregnant females give birth to 2-ton calves. The calves nurse and put on weight quickly during the next two months. While the calves are nursing, the mature males breed with the mature females that did not bear calves. Late in March, the return to the feeding grounds begins, with the procession order reversed. Most of the whales are back in the feeding grounds by the end of June, and they feed on prodigious amounts of amphipods to replenish their depleted store of fat and blubber before the next trip south.

Reproduction

As in the examples just described, reproduction of pelagic animals involves bringing the males and females of the species together for this purpose on some periodic basis. These gatherings usually occur during the warmer spring or summer months, when water temperatures are higher and primary productivity is at its peak.

Most of the invertebrates and fish that inhabit the open ocean are **oviparous:** They lay eggs that hatch in the open water or on the bottom into larval forms that become meroplankters. Animals that reproduce in this way usually produce enormous numbers of eggs—15 million eggs per season in the migrating Icelandic cod—because most will be lost to predators before they hatch or the larvae will be eaten by other members of the plankton community. However, some oviparous fish may produce only one or a few well-protected eggs per season; the cartilaginous sharks, skates, and rays include many such species (figure 14–21).

The females of other fish species keep their fertilized eggs in their reproductive tract until they hatch. Although the young come into the world live, the process by which the embryos develop is the same as for those that develop from eggs that are laid. This behavior is called **ovoviviparous,** describing a method of reproduction in which the eggs are incubated internally. The sea horse and pipefish exemplify a special twist to internal incubation where the female deposits the eggs in a ventral pouch of the male (figure 14–22). The male carries the eggs until they hatch, and the young fish enter the ocean directly from this pouch. The additional protection against loss of eggs through pelagic predation results in the reduction of the number of eggs that must be produced; for example, the dogfish shark, *Squalus,* usually produces fewer than a dozen eggs at a time.

Viviparous animals give birth to live young, but the behavior is considered by most to require also that the young be given more than space in the mother's reproductive tract to incubate. Part of the behavior includes providing embryonic nutrient in addition to the yolk of the egg. Some sharks and rays have projections from the wall of the uterus called villi that secrete a rich milk. With this additional nutrition, the stingray, *Pteroplatea,* produces young that enter the ocean with a mass up to 50 times that of the mass of the yolk of the egg from which they came. In the case of the white-tip shark, *Carcharhinus,* this additional nutrition passes to the embryo through a yolk sac attached to the wall of the uterus. Mammals exhibit the highest level of viviparous behavior in that the embryo is encased in the placental sac. The mother's blood flows through the placenta, and the embryo receives nutrition from the mother through an umbilical cord by which it is attached to the placenta; essentially all nutrition is provided by the mother. This

FIGURE 14–21
Egg Case. Swell shark with yolk sac. (Photo © Alex Kerstitch/Bruce Coleman, Inc.)

FIGURE 14–22
Ovoviviparous Behavior in Sea Horses. The female deposits the eggs in the ventral pouch of the male, where they are protected during incubation. (Photo by Larry Ford.)

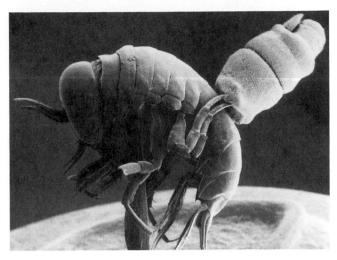

FIGURE 14–23
Captive Pteropod Protects Pelagic Amphipod from Predators. Scanning electron micrograph of the amphipod, *Hyperiella dilatata*, carrying its captive pteropod, *Clione limacina*, which provides chemical protection against predatory fish. (Photo courtesy of James B. McClintock with permission from *Nature*, Vol. 346, 462–464, 1990. Photo by Phil Oshal.)

behavior provides a high degree of protection to the developing young but puts a high energy demand on the mother. Thus mammalian births typically involve only one or a few young who will remain dependent on their mothers for protection and food for some time after their birth.

Abduction: A Unique Form of Protection from Predators

We will close out this chapter on the adaptations of animals that live in the pelagic environment with an unusual example of how a small marine organism protects itself from predators.

In 1990, pelagic Antarctic amphipods were observed to capture pteropods (sea butterflies) and carry them around on their backs beneath the Antarctic ice. Some investigation revealed that the 1.25 cm (0.5 in.)

amphipod, *Hyperiella dilatata*, uses the 0.65 cm (0.25 in.) pteropod, *Clione limacina*, as a chemical defense against being eaten by predator fish. The pteropod contains a chemical that tastes awful, and while the amphipod has it captured, fish will not eat it (figure 14–23).

The fish that prey on the amphipods depend on sight to find their food. Therefore, it is not surprising that up to 75 percent of the amphipods observed at depths of less than 9 m (28 ft) were carrying pteropods on their backs. In the darker waters at 50 m (160 ft), only 6 percent of the amphipods observed were protected by their foul-tasting captives.

This pattern of capturing another organism for the purpose of using its chemical defense against predators does not fit into any of the categories of symbiotic relationships previously observed between marine species. It is believed that the pteropods are eventually released unharmed although they may experience only a short reprieve until they are once again held hostage.

SUMMARY

Frictional resistance to sinking—a major factor in helping the tiny plankton stay near the surface—is not a major factor in keeping the larger nekton from sinking; they depend primarily on buoyancy or swimming. Rigid gas containers found in some cephalopods and the expandable swim bladders of many bony fishes are adaptations that help increase buoyancy. Many invertebrates such as the jellyfish, tunicates, and arrowworms have soft, gelatinous bodies that reduce their density. The Portu-

guese man-of-war also has a gas-filled float that supports this pelagic colony. Many of these invertebrate forms are weak swimmers and depend primarily on buoyancy to maintain their positions near the surface.

Strong swimmers, the nekton—squid, fish, and mammals—depend on this expenditure of energy to obtain food in the water column. Squid swim by trapping water in their mantle cavities and forcing it out through a siphon. Most fish swim by creating a wave of body curvature that passes from the front to the back of the body and provides a forward thrust. The caudal fin is the most important in providing thrust, while the paired pelvic and pectoral fins are used for maneuvering. The vertically oriented dorsal and anal fins serve primarily as stabilizers. The efficiency with which a caudal fin produces thrust is indicated by its aspect ratio, equal to the square of the fin height divided by the fin area. The rounded caudal fin, found on the sculpin, is flexible, has a low aspect ratio of about 1, and can be used for maneuvering at slow speeds. The lunate fin is rigid, has a high aspect ratio of about 10, and is of little use in maneuvering. It is very efficient in producing thrust for fast swimmers such as the tuna.

Fish such as groupers are lungers that sit motionless and make short, quick passes at passing prey. They have mostly white muscle tissue. Tuna and other cruisers, constantly swimming in search of prey, possess mostly red muscle tissue. Red fibers have a great affinity for oxygen and tire less rapidly than white fibers, enabling tuna to maintain a cruising speed of about 3 body lengths/second. Using their white muscles to help in short periods of rapid swimming, they can reach speeds of up to 10 body lengths/second. Although most fish are poikilothermic (cold-blooded), the tuna, *Thunnus*, is homeothermic; that is, it maintains a body temperature well above the temperature of the water.

The best-adapted mammals for life in the open ocean are the Cetacea. The toothed whales (Odontoceti) and baleen whales (Mysteceti) propel their highly streamlined bodies through the water with vertical movement of their horizontal flukes.

Whales are able to dive deep and stay submerged for unusually long periods of time because of modifications that allow them to absorb 90 percent of the O_2 in the air they inhale, store large quantities of oxygen, possibly reduce oxygen use, and collapse their lungs below depths of 100 m (328 ft).

Most whales and many other marine mammals are thought to use echolocation in finding their way through the ocean and locating prey. The clicking sounds emitted by whales are bounced off objects, and the animal can determine the size, shape, and distance of the objects by the nature of the returning signals and the time elapsed.

Schooling of active swimmers such as fish, squid, and crustaceans is not fully understood, but it has obvious advantages as a reproductive strategy. Schooling more likely serves a protective function, although the ways in which it may meet this end are speculative.

Migrations are observed among sea turtles, fish, and mammals and are related to the needs of reproduction and finding food. It is believed that orientation is maintained during migrations through use of visible landmarks, smell and Earth's magnetic field. Baleen whales may migrate from their cold-water summer feeding grounds to warm low-latitude lagoons in winter so their young can be born into warm water; they might not be able to survive the cold water of higher latitudes at birth. Fishes such as the Atlantic cod swim upcurrent to deposit their eggs so that when they hatch the cod fry will be in water where suitable food is available.

The North Atlantic eel, *Aguilla*, is catadromous: It spawns in the deep waters of the Sargasso Sea and spends its adult life in freshwater streams of North America and Europe. The Atlantic and Pacific salmon are anadromous: They spawn in the freshwater streams of North America, Europe, and Asia and spend their adult life in the open ocean.

Most fish are oviparous, depositing their eggs in the ocean. Some sharks and rays maintain their eggs in a body cavity until they hatch; this ovoviviparous behavior provides a greater protection for the eggs. Where oviparous fish may produce millions of eggs, ovoviviparous fish may produce fewer than a dozen. The stingray, white-tip shark, and mammals are viviparous, not only providing space in their bodies for the eggs to develop but also providing nutrition in addition to the egg yolk. In mammals, the young will still be dependent on their mothers for nutrition for some time after birth.

A unique method of defense against predators is practiced by an Antarctic amphipod that captures a foul-tasting pteropod and carries it around on its back.

KEY TERMS

Anadromous (p. 314)	Lateral line system (p. 311)	Poikilothermic (p. 304)
Arrowworm (p. 300)	Lunger (p. 303)	Polyp (p. 297)
Aspect ratio (p. 301)	Medusa (p. 297)	Red muscle fiber (p. 303)
Catadromous (p. 313)	Migration (p. 311)	Right whale (p. 308)
Cetacea (p. 305)	Myoglobin (p. 303)	Rorqual whale (p. 308)
Circadian rhythms (p. 312)	Myomere (p. 301)	Schooling (p. 311)
Coelenterates (p. 297)	Mysticeti (p. 308)	Scyphozoan (p. 298)
Cruiser (p. 303)	Nematocyst (p. 297)	Sea otter (p. 305)
Ctenophore (p. 298)	Ocelli (p. 298)	Siphonophore (p. 297)
Cutaneous artery (p. 304)	Odonticeti (p. 305)	Sirenia (p. 305)
Cutaneous vein (p. 304)	Otocyst (p. 298)	Swim bladder (p. 297)
Dorsal aorta (p. 304)	Oviparous (p. 316)	Tunicate (p. 298)
Gray whale (p. 308)	Ovoviviparous (p. 316)	Viviparous (p. 316)
Hemoglobin (p. 308)	Pinniped (p. 305)	White muscle fiber (p. 303)
Homeothermic (p. 304)	Pneumatic duct (p. 297)	

QUESTIONS AND EXERCISES

1. Discuss how the rigid gas chambers in cephalopods may be more effective in limiting the depth to which they can descend than the flexible swim bladders of bony fish.
2. Describe the body form and lifestyle of the following plankters: *Physalia,* jellyfish, tunicates, ctenophorans, and arrowworms.
3. What are the major structural and physiological differences between the fast-swimming cruisers and the lungers that patiently lie in wait for their prey?
4. List the modifications that are thought to allow some cetaceans to (1) dive to great depths without suffering the bends and (2) stay submerged for long periods of time.
5. Describe the process by which the sperm whale may produce echolocation clicks.
6. Although there is disagreement about how it is achieved, discuss what most investigators believe to be the method by which sound reaches the inner ear of toothed whales.
7. Summarize the reasons some investigators believe schooling increases the safety of fishes from predators.
8. What are the methods believed to be used by migrating animals to maintain their orientation?
9. Why do fishes such as Icelandic cod swim upcurrent to spawn?
10. How are the migrations of the North Atlantic eels and Pacific salmon fundamentally different?
11. Discuss possible reasons for the California gray whales leaving the cold-water feeding grounds during the winter season.
12. Compare the reproductive behavior of oviparous, ovoviviparous, and viviparous animals.

REFERENCES

Carey, F. G. 1973. Fishes with warm bodies. *Scientific American* 228:2, 36–44.

Denton, E. J., and Shaw, T. I. 1962. The buoyancy of gelatinous marine animals. *Journal of Physiology* 161:14P–15P.

George, D., and George, J. 1979. *Marine life: An illustrated encyclopedia of invertebrates in the sea.* New York: Wiley–Interscience.

Kanwisher, J. W., and Ridgway, S. H. 1983. The physiological ecology of whales and porpoises. *Scientific American* 248:6, 110–21.

MacGinitie, G. E., and MacGinitie, N. 1968. *Natural history of marine animals.* 2nd ed. New York: Mcgraw-Hill.

McClintock, J. B., and Janssen, J. 1990. Pteropod abduction as a chemical defence in a pelagic Antarctic amphipod. *Nature* 346:6283, 462–64.

Mrosovsky, N.; Hopkins-Murphy, S. R.; and Richardson, J. I. 1984. Sex ratio of sea turtles: Seasonal changes. *Science* 225:4663, 739–40.

Norris, K. S., and Harvey, G. W. 1972. A theory for the function of the spermaceti organ of the sperm whale (*Physeter catodon* L). *Animal orientation and navigation,* 397–417. Washington, D.C.: National Aeronautics and Space Administration.

Pike, G. C. 1962. Migration and feeding of the gray whale (*Eschrichtius gibbosus*). *Journal of the Fisheries Research Board of Canada* 19:815–38.

Royce, W.; Smith, L. S.; and Hartt, A. C. 1968. Models of oceanic migrations of Pacific salmon and comments on guidance mechanisms. *Fishery Bulletin* 66:441–62.

Thorson, G. 1971. *Life in the sea.* New York: Mcgraw-Hill.

Würsig, B. 1989. Cetaceans. *Science* 244:4912, 1550–57.

SUGGESTED READING

Sea Frontiers

Bachand, R. G. 1985. Vision in marine animals. 31:2, 68–74. An overview of the types of eyes possessed by marine animals.

Bleecker, S. E. 1975. Fishes with electric know-how. 21:3, 142–48. A survey of fishes that use electrical fields to navigate, to capture prey, and to defend themselves.

Bushnell, P. G., and Holland, K. N. 1989. Tunas: Athletes in a can. 35:1, 42–48. The physiology of various tuna species, specifically that related to high-speed swimming, is discussed.

Hersh, S. L. 1988. Death of the dolphins: Investigating the east coast die-off. 34:4, 200–207. The 1987–88 east coast die-off of dolphins is discussed by a marine mammalogist.

Kleen, S. 1989. The diving seal: A medical marvel. 35:6, 370–74. The physiology of seals that allows them to dive deep and stay submerged for long periods of time is covered.

Klimley, A. P. 1976. The white shark: A matter of size. 22:1, 2–8. Describes procedure used by Dr. John E. Randall for determining the size of sharks from the perimeter of the upper jaw and height of teeth.

Lineaweaver, T. H., III. 1971. The hotbloods. 17:2, 66–71. Discusses the physiology of the bluefin tuna and sharks that have high body temperatures.

Maranto, G. 1988. The Pacific walrus. 34:3, 152–59. A summary of the natural history of walruses of the Bering Sea and the role of humans as their predators.

Netboy, A. 1976. The mysterious eels. 22:3, 172–82. An informative discussion describing what is known of the migrations of the catadromous eels and their importance as a fishery.

O'Feldman, R. 1980. The dolphin project. 26:2, 114–18. A description of the response of Atlantic spotted dolphin to musical sounds.

Reeve, M. 1971. The deadly arrowworm. 17:3, 175–83. This important group of plankton is described in terms of body structure and life style.

Volger G., and Volger, S. 1988. Northern elephant seals. 34:6, 342–47. This article deals primarily with the history of the elephant seal populations off the coast of California and Baja California.

Scientific American

Denton, E. 1960. The buoyancy of marine animals. 303:11, 118–28. How various marine animals use buoyancy to maintain their position in the water column.

Donaldson, L. R., and Joyner, T. 1983. The salmonid fishes as a natural livestock. 249:1, 50–69. The genetic adaptability of the salmonid fishes may help them adapt to "ranching" operations.

Gosline, J. M., and DeMont, M. E. 1985. Jet propelled swimming in squids. 252:1, 96–103. By using jet propulsion resulting from expelling water through their siphons, squids can move as fast as the speediest fishes.

Gray, J. 1957. How fishes swim. 197:2, 48–54. The roles of musculature and fins in the swimming of fishes.

Horn, M. H., and Gibson, R. N. 1988. Intertidal fishes. 258:1, 64–71. A discussion of a wide variety of fishes that inhabit tide pools.

Kooyman, G. L. 1969. The Weddell seal. 221:2, 100–107. The life style and problems related to this mammal's living in water permanently covered with ice.

Leggett, W. C. 1973. Migration of shad. 228:3, 92–100. The routes of the anadromous shad in the rivers and Atlantic waters are described, along with factors that may control the migrations.

Rudd, J. T. 1956. The blue whale. 195:6, 46–65. A description of the ecology of the largest animal that ever lived.

Sanderson, S. L., and Wassersug, R. 1990. Suspension-feeding vertebrates. 263:3, 96–102. The ecology of filter-feeding fishes and cetaceans is covered.

Shaw, E. 1962. The schooling of fishes. 206:6, 128–36. A consideration of theories seeking to explain the schooling behavior of fishes.

Whitehead, H. 1985. Why whales leap. 252:3, 84–93. Whales seem to communicate with their spectacular lunges above the ocean's surface.

Würsig, B. 1988. The behavior of baleen whales. 258:4, 102–107. A summary of the facts concerning the behavior of baleen whales.

Zapol, W. M. 1987. Diving adaptations of the Weddell seal. 256:6, 100–107. The physiological adaptations that enable the Weddell seal to make deep, long dives are discussed.

Chapter 15

Animals of the Benthic Environment

More than 98 percent of the 200,000 species of animals found in the ocean live in the two-dimensional world of the ocean floor. Ranging from the rocky, sandy, and muddy environments of the intertidal zone to the muddy deposits of deep ocean trenches more than 11 km (6.8 mi) deep, the ocean floor provides a varied environment that is home for a diverse benthic community.

Living at or near the interface of the ocean floor and the ocean water, the success of an organism is closely tied to its ability to cope with the physical conditions of the water, the ocean floor, and the other members of the biological community. The vast majority of benthic species live on the shallow bottom of the continental shelf, and approximately 400 of the known 196,000 benthic species are found in the hadal zone of the deep-ocean trenches.

One of the most prominent variables affecting species diversification is temperature. We have previously discussed its effect on species diversity on a latitudinal basis; however, even at the same latitude, a significant difference in the number of benthic species is found on

Dusky anemone fish among the tentacles of their sea anemone host, *Amphieron melanopus,* in the Coral Sea, Australia. (Photo © Fred Bavendam/Peter Arnold, Inc.)

opposite sides of an ocean basin because of the effect of ocean currents on the coastal water temperature. Along the European coast where the Gulf Stream warms the water from the northern tip of Norway to the Spanish coast, over three times the number of species of benthos exist than are found along the similar latitudinal range of the Atlantic coast of North America, where the Labrador Current cools the water as far south as Cape Cod.

Figure 15–1 shows that the distribution of benthic biomass is patterned after the distribution of photosynthetic productivity in the surface waters shown in figure 13–8. This tells us that life on the ocean floor is very much dependent upon the primary photosynthetic productivity of the ocean-surface waters. We will nonetheless discuss some special benthic communities that are an exception to this general condition.

ROCKY SHORES

Most of the rocky shore is confined to the supralittoral and littoral zones. The supralittoral corresponds with the backshore, or *spray zone,* above the spring high tide line and is covered by water only during storms. The littoral zone, also known as the foreshore or the **intertidal zone,** lies between the high and low tidal extremes. Along most shores, the intertidal zone can be divided into the high tide zone, mostly dry—covered by the highest high tide but not by the lowest high tides; the middle tide zone, exposed and covered equally—covered by all high tides and exposed during all low tides; and the low tide zone, mostly wet—covered during the highest low tides and exposed during the lowest low tides (figure 15–2).

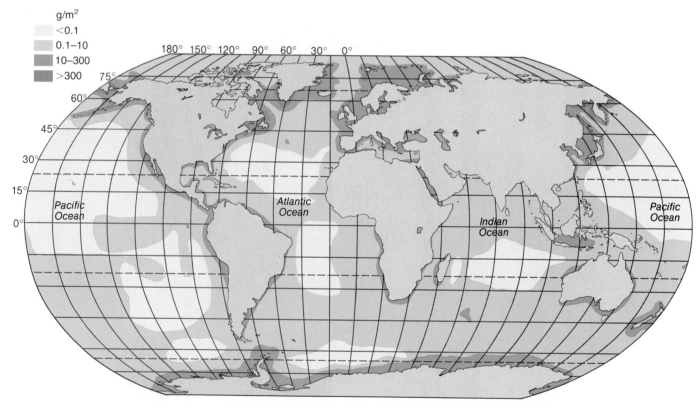

FIGURE 15–1
Distribution of Benthic Biomass of the World Ocean (g/m², wet weight). Although the distribution pattern shown here is defensible in concept, it is based on too few samples to be regarded as accurate. The pattern is similar to that of photosynthetic pigment distribution shown in figure 3–8. This suggests that most of the benthic community is directly dependent on the photosynthetic production in the surface waters of the oceans for its nutritional support. High concentrations of biomass associated with hydrothermal vents and cold seeps are not reflected in this distribution. The concentrations beneath the centers of the subtropical gyre waters are 20,000 times less than those of the continental-shelf benthos. Up to 60 percent of the organic matter reaching the deep-ocean floor may enter the food chain through bacteria, and deep-sea bacteria may metabolize organic matter at a rate at least 10 times slower than bacteria on the continental shelf. This could reduce the rates of biological productivity on the deep-ocean floor to levels of 1/200,000 that occurring on the continental shelves. (After Zenkevitch et al., 1971.)

Along rocky shores these divisions of the intertidal zone are often obvious due to the sharp boundaries between populations of attached organisms. Because each centimeter of the littoral rocky shore has a significantly different character from the centimeter above and below it, evolution has been able to produce organisms with the abilities to withstand very specific degrees of exposure to the atmosphere. This condition results in the most finely defined biozones known in the marine environment.

Supralittoral (Spray) Zone

Throughout the world the most obvious inhabitant of the rocky supralittoral zone is the periwinkle snail. The supralittoral zone can easily be identified if one considers the middle of the periwinkle belt to be the boundary between the littoral and the supralittoral. The genus *Littorina* includes species able to breathe air, like land snails. They need the spray of breaking waves as their only tie to the sea, since they are viviparous (give birth to live young) rather than depositing their eggs into the water as do more marine periwinkles and other snails (figure 15–2).

The periwinkles feed by rasping algae and lichens off the rocks of the spray zone with their filelike radula, a calcium carbonate mouthpiece common to many snails.

Hiding among the cobbles and boulders covering the floors of sea caves well above the high tide line are isopods belonging to the genus *Ligia;* their common names are rock lice and sea roaches. Neither name is particularly flattering to these little scavengers that reach lengths of 3 cm (1.2 in.) and scurry about at night feeding on organic debris.

Another indicator of the spray zone is a distant relative of the periwinkle snails, a limpet belonging to the genus *Acmaea*. Having a flattened conical shell, the limpets feed in a manner similar to that of the periwinkles.

High Tide Zone

Unlike the periwinkles, some of which give birth to live young, the buckshot barnacles of the **high tide zone** cannot abandon the sea for dry land. Their limit is the high tide shoreline. These hermaphroditic animals fertilize their neighbor's eggs by inserting a penis into a neighboring barnacle. The fertilized eggs hatch within the housing of the barnacle and are released into the water as nauplii larvae; these eventually transform into cypris larvae and abandon the planktonic existence to attach to the rocky shore. In addition to needing water for the planktonic existence of their larval forms, barnacles also are confined to their marine existence through their method of filter feeding.

The most conspicuous plants in the high tide zone are members of the genus *Fucus* in colder latitudes and *Pelvetia* in warmer latitudes (figure 15–2). Both have thick cell walls to reduce water loss during periods of low tide. These plants continue to flourish in the middle tide zone, where the variety of life forms is much greater than in the upper tide zone. Not only does the variety increase, but the total biomass is also much greater; there is, therefore, a greater competition for rock space by sessile forms.

Middle Tide Zone

On a clean rocky shore, the *Fucus* or *Pelvetia* establishes itself before the sessile animal forms; however, it seems doomed once barnacles or mussels move in. Although the smaller acorn barnacles usually found in the upper tide zone may occasionally be found in the upper reaches of the **middle tide zone,** larger species of the genus *Balanus* (figure 15–2) are the more common acorn barnacles of the middle tide zone. The barnacle most characteristic of this zone is the goose barnacle, genus *Pollicipes* (figure 15–2), with two valves composed of numerous plates that house the animal, which attaches itself to the rock surface by a long muscular neck. The total length of this structure may exceed 10 cm (4 in.). Even more successful than barnacles in the competition for space in the middle tide zone are the various species of mussels belonging to the genera *Mytilus* and *Modiolus;* they will attach to bare rock, algae, or barnacles. Settling on these surfaces as larval forms, they attach themselves by tough proteinous byssal threads. Given time, mussels would eventually overgrow all other sessile forms.

But life in the intertidal zone is not simple. Mussels are fed upon by predators such as sea stars and carnivorous snails. Two common genera of sea stars are the *Pisaster* and *Asterias*.

Sea stars feed by attaching tube feet that radiate from the mouth at the center of the bottom side of the animal. There are five double rows located in an oral groove beneath each ray or arm. To get to the mussel tissue protected by a calcium carbonate bivalve covering, sea stars exert a continuous pull on the valves by alternating the use of the tube feet such that some are always pulling while others are resting. The mussel is eventually fatigued and can no longer hold the valves closed. The valves open ever so slightly, and the sea star everts its stomach, slips it through the crack, and digests the mussel without having to take it into its mouth (figure 15–3).

Boring carnivorous snails feed on both barnacles and mussels. To feed on the mussels, the snail must use its radula to drill a hole through the shell, through which it can rasp away the flesh. The more widely distributed carnivorous snails are the dog-whelk, *Nucella*, and the similar Pacific coast species *Thais lamellosa*.

FIGURE 15–2

The Rocky Shore. *A,* a typical rocky intertidal zone and some organisms typically found in its subzones (not to scale). *B,* periwinkles *(Littorina)* nestled in a depression near the upper limit of the high-tide zone. This behavior helps reduce exposure to direct sunlight. *C,* rock louse *(Ligia).* *D,* rough keyhole limpet *(Diodora aspera)* with encrusting red algae, *Lithothamnion.* (Photo © Norbert Wu/Peter Arnold, Inc.) *(Acmaea).* *E,* buckshot barnacles *(Chthamalus).* *F, Fucus filiformes.* (Photo © Breck P. Kent.) Sharing the middle tide zone with these rock weeds are *G,* acorn barnacles *(Balanus)* and *I,* goose barnacles *(Pollicipes)* and mussels. *H, Pelvetia fastigata.* (Photo © Eda Rogers.)

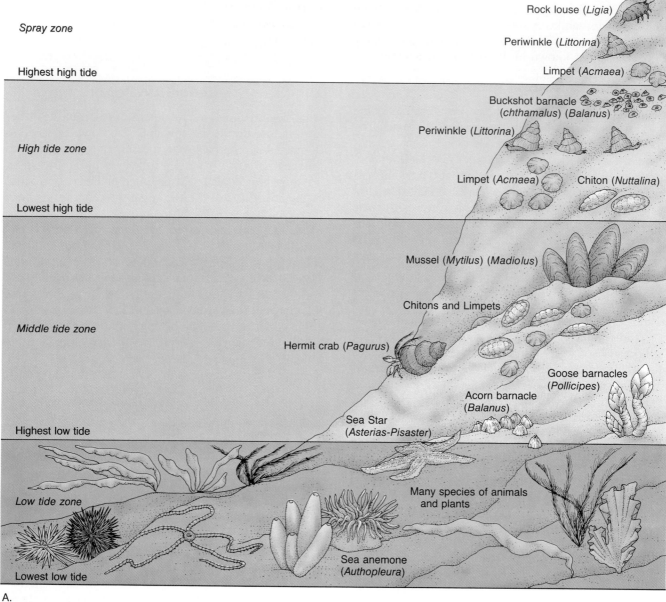

Spray zone

Highest high tide

High tide zone

Lowest high tide

Middle tide zone

Highest low tide

Low tide zone

Lowest low tide

Rock louse (*Ligia*)

Periwinkle (*Littorina*)

Limpet (*Acmaea*)

Buckshot barnacle (*chthamalus*) (*Balanus*)

Periwinkle (*Littorina*)

Limpet (*Acmaea*)   Chiton (*Nuttalina*)

Mussel (*Mytilus*) (*Madiolus*)

Chitons and Limpets

Hermit crab (*Pagurus*)

Goose barnacles (*Pollicipes*)

Acorn barnacle (*Balanus*)

Sea Star (*Asterias-Pisaster*)

Many species of animals and plants

Sea anemone (*Authopleura*)

A.

B.

C.

D.

E.

F.

G.

H.

I.

FIGURE 15-3
Sea Star Feeding on a Mussel.
(Photo © Joy Sparr/Bruce Coleman, Inc.)

The dominant feature of the middle tidal zone along most rocky coasts is a mussel bed that thickens toward the bottom until it reaches an abrupt bottom limit. This may be so pronounced that it appears as though an invisible horizontal plane has prevented the mussels from growing below this depth. Protruding from the mussel bed will be numerous goose barnacles, and concentrated in the lower levels of the bed will be the sea stars browsing on the mussels. Less conspicuous forms common to the mussel beds are varieties of algae, hydroids, worms, clams, and crustaceans.

Where the rock surface flattens out within the middle tidal zone, tide pools trapping water as the tide ebbs serve as interesting microecosystems containing a wide variety of organisms. The largest member of this community will often be the sedentary relative of the jellyfish, the sea anemone. Shaped like a sack, anemones have a flat pedal disc that provides a suction attachment to the rock surface. Directed upward, the open end of the sack has the only opening to the gut cavity, the mouth, surrounded by rows of tentacles. The tentacles are covered with *cnidoblast* cells which contain a stinging threadlike *nematocyst* that automatically is released to penetrate any organism that brushes against the tentacles. Unlike their pelagic relatives, the tide pool coelenterates do not have nematocysts long enough to penetrate human skin and therefore are not dangerous (figure 15-4).

Swimming in the tide pools are a variety of small fish. The opaleye reaches lengths of 5 cm (2 in.) in the pools and grows larger in deeper water offshore. It is easily identified by the tiny white spots on its back on either side of the dorsal fin. The woolly sculpin is a per-

manent resident of tide pools and grows longer than 15 cm (6 in.). It is covered with hairlike cirri and usually is resting on the bottom or walking on its large pectoral fins. Readily identified by its blunt head and continuous dorsal fin is the 15-cm (6-in.) rockpool blenny (figure 15-5).

The most interesting inhabitant of the tide pools is the hermit crab, *Pagurus*. With a well-armored pair of claws and cephalothorax, hermit crabs have a soft abdomen that is protected by some hard protective container, usually a snail shell. The abdomen has even developed a curl to the right to make it fit properly into snail shells. Once in the snail shell, the crab can close off the opening with its large claws. However, one primary enemy is the octopus, which is not bothered by the shell that the hermit crab has adopted. The octopus can cover the crab with the web surrounding its mouth and release a poisonous secretion that paralyzes the crab so that it can be pulled from the shell. Male hermit crabs may be seen fighting over females or carrying them around by grasping the edge of the shell in which the female lives. The male is waiting for the female to molt so he may deposit his sperm on her abdomen. She will later use the sperm to fertilize her eggs (figure 15-6).

In tide pools near the lower limit of the middle tide zone, sea urchins (figure 15-6), with their five-toothed structure centered on the bottom side of a hard spherical covering that supports many spines, may be found feeding on algae. The hard protective covering is called a test and is composed of fused calcium carbonate plates that are perforated to allow tube feet and dermal gills to pass through.

FIGURE 15–4
Sea Anemone. *A,* sea anemone. (Photo by Shane Anderson.) *B,* basic body plan of an anemone and detail of the stinging mechanism.

A.

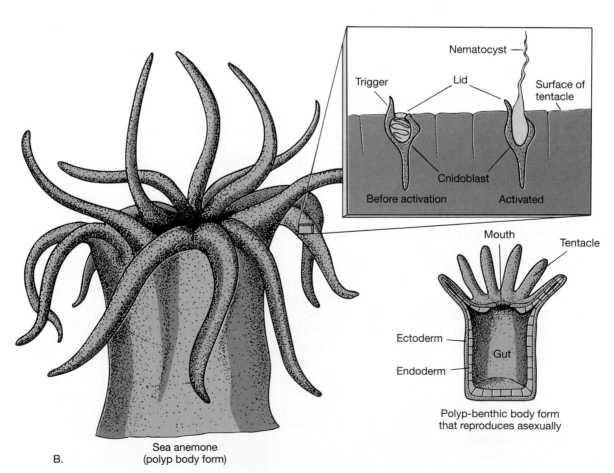

Nematocyst

Trigger

Lid

Surface of tentacle

Cnidoblast

Before activation

Activated

Mouth

Tentacle

Ectoderm

Gut

Endoderm

Polyp-benthic body form
that reproduces asexually

Sea anemone
(polyp body form)

B.

FIGURE 15-5
Rockpool Blenny. (Photo © Larry Lipsky/Bruce Coleman, Inc.)

Low Tide Zone

Unlike the upper and middle tide zones, the **low tide zone** is dominated by plants rather than animals. A diverse community of animals exists, but they are less obvious because they are hidden by the great variety of seaweeds and surf grass, *Phylospadix* (figure 15–7). The encrusting red algae, *Lithothamnion,* which is also seen in middle zone tide pools, becomes very abundant in the lower tide pools. In temperate latitudes, moderate-sized red and brown algae provide a drooping canopy beneath which much of the animal life is found.

Scampering from crevice to crevice and in and out of tide pools across the full range of the intertidal zone are various species of shore crabs (figure 15–8). Scavengers that help keep the shore clean, these creatures can spend long periods of time out of the sea. Shore crabs spend most of the daylight hours hiding in cracks or beneath overhangs. At night they engage in most of their eating activity, shoveling in algae as rapidly as they can tear it from the rock surface with their chelae, large front claws. Shore crabs need to return to ocean water only periodically to wet their gills, and females also must deposit the eggs they carry under their abdomens when it is time for the eggs to hatch. A characteristic they share with many crustaceans is that of **autotomy,** the ability to shed appendages involuntarily along a predetermined breaking plane as necessary for survival. It is therefore not unusual to see them with a claw missing or much smaller than its paired counterpart. With each molt the new limb grows until it ultimately becomes full-sized.

SEDIMENT-COVERED SHORE

The Sediment

The sediment-covered shore includes what are commonly called *beaches, salt marshes,* or *mud flats* that represent lower energy environments. As the energy level decreases, the particle size and sediment slope decrease, and sediment stability increases; the energy level refers to the strength of the longshore current. Another characteristic of sediments closely related to particle size is **permeability,** the ease with which fluid can move through the deposit. Because of the strength of cohesive attraction between flat clay-sized particles and the greater angularity of small silt- and fine sand-sized particles compared with the more rounded shape of coarser sand-sized particles, the permeability of sediments increases with increased particle size. Therefore, breaking waves can more readily percolate down through coarse sands and replace the oxygen used up by animals buried in the sediment. This readily available oxygen also enhances the process of bacterial decomposition of dead tissue.

Life in the Sediment

Fragments of kelp plants broken from their rocky substrate by heavy wave action are commonly piled high on beaches during storms. Thus, some of the suspension feeders that filter plankton from the clear water of the rocky shore are replaced by deposit feeders that eat detritus on many sandy beaches and ingest sediment in mud flats.

A.

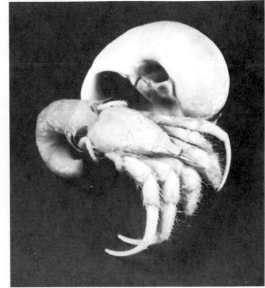

B.

C.

D.

FIGURE 15–6
Hermit Crab and Sea Urchin. *A,* hermit crab *(Pagurus)* in *Maxwellia gemma* shell. (Photo by James McCullagh.) *B,* hermit crab out of its shell; note the soft curved abdomen. *C,* sea urchins burrowed into the bottom of a lower middle tide zone tide pool. *D,* the bottom side of this dead sea urchin shows the mouth at the center and the hard calcium carbonate test beneath the spiny exterior.

Life on and in the sediment requires very different adaptations than on the rocky coast. The sandy beach supports fewer species than the rocky shore, and mud flats fewer still; however, the number of individuals may be as high. In the low tide zone of some beaches and on mud flats, as many as 5000–8000 burrowing clams have been counted in 1 m² (10.8 ft²). Burrowing is the most successful adaptation for life in the sediment-covered shore, so life is less visible. By burrowing a few centimeters beneath the surface, organisms can find a stable environment where they are not bothered by fluctuations of temperature and salinity and the threat of desiccation.

Burrowing in the sediment does not prevent animals from suspension feeding or straining plankton from

FIGURE 15–7
Algal Canopy of the Low Tide Zone Exposed during an Extremely Low Tide.

A.

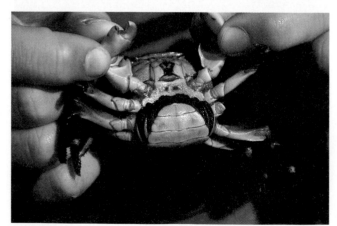

B.

FIGURE 15–8
Shore Crabs. *A,* coral crab, or queen crab, Bonaire Island, Netherlands Antilles. (Photo © Fred Bavendam/Peter Arnold, Inc.) *B,* Shore crab, *Pachygrapsis crassipes,* female with eggs. (Photo © Eda Rogers.)

the clear water above the sediment as do many of the animals of the rocky shore. By the use of various techniques, the water above the sediment surface can be stripped of its plankton content by buried animals. Deposit feeders collect debris from the sediment surface and eat their way through the sediment, digesting its organic content and excreting inorganic particles. Burrowing carnivores chemically sense the presence of their prey in the darkness of the infaunal habitat (figure 15–9).

The Sandy Beach

Bivalve Mollusks.   Bivalve mollusks, or pelecypods, are well adapted for life in the sediment. The greatest variety of clams are found buried beneath the low tide region of sandy beaches; their numbers decrease as the sands become muddier. The bivalve mollusks possess a soft body, a portion of which is called the mantle; this part secretes the calcium carbonate lateral valves that hinge together on the dorsal region of the body. The foot extends anteriorly to dig into the sediment and pull the valves and posterior siphons down after it. The tip of the foot is then pumped full of blood to make it swell and anchor the mollusk, while retractor muscles running the length of the foot contract and pull the animal down. How deep the bivalve can bury itself depends on the length of its siphons; they must reach above the sediment surface to pull in water from which plankton will be filtered. Oxygen is also extracted in the gill chamber before the water is expelled through the excurrent siphon. Undigestible particulate matter is forced back out the incurrent siphon periodically by quick muscular contractions (figure 15–10).

Annelid Worms.   A variety of annelids, or segmented worms, are also well adapted for life in the sediment.

FIGURE 15–9
Modes of Feeding along the Sediment-Covered Shore. *A,* suspension feeding. This method is used by clams that bury themselves in sediment and extend siphons through the surface to pump in overlying water. They feed by filtering plankton and other organic matter that is suspended in the water. *B,* deposit feeding. Some deposit feeders such as the segmented worm *Arenicola* feed by ingesting the sediment and extracting organic matter contained within it. Others, like the amphipod *Orchestoidea,* feed on more concentrated deposits of organic matter, detritus, found on the sediment surface. *C,* carnivorous feeding. The sand star, *Astropecten,* cannot climb rocks like its sea star relatives, but it can burrow rapidly into the sand, where it feeds voraciously on crustaceans, mollusks, worms, and other echinoderms.

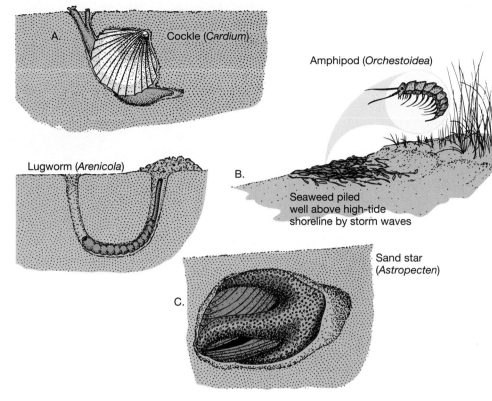

FIGURE 15–10
How a Clam Burrows. Clams exposed at the sediment surface will quickly burrow into the sediment by *(A)* extending the point-shaped foot into the sediment and *(B)* forcing the foot deeper into the sediment and using this increasing leverage to bring the exposed, shell-clad body toward vertical. When the foot has penetrated deep enough, a bulbous anchor forms at the tip *(C)*, and a quick muscular contraction pulls the entire animal into the sediment *(D)*. The siphons are then pushed up above the sediment to pump in water from which the clam will extract food and oxygen *(E)*.

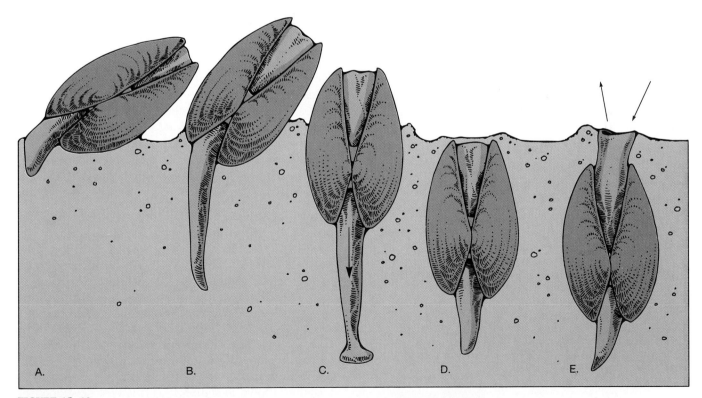

Most common of the sand worms is *Arenicola* sp., or lugworm. It lives in a U-shaped burrow, the walls of which are strengthened by mucus. With its finely hooked parapods gripping the sides of the tube, it moves forward and backward. The worm moves forward to feed and extends its proboscis up into the head shaft of the burrow to loosen sand with quick pulsing movements. A cone-shaped depression forms at the surface over the head end of the burrow as sand continually slides into the burrow and is ingested by the worm. As the sand passes through the digestive tract, the organic content is digested, and the processed sand is deposited as castings at the surface surrounding the opening or openings to the tail shaft (figure 15–9).

Crustaceans. Staying high on the beach and feeding on kelp cast up by storm or high tide waves, numerous amphipod crustaceans called *beach hoppers* are found. They are known to jump distances more than 2 m (6.6 ft). A common genus is *Orchestoidea*, which usually range from 2 to 3 cm (0.8–1.2 in.) in length. Laterally flattened, they usually spend the day buried in the sand or hidden in the kelp on which they feed. They become active at night and may form large clouds above the masses of seaweed on which they are feeding (figure 15–9).

A larger crustacean that may be harder to find on sandy beaches is a member of the variety of sand crabs belonging to the genera *Blepharipoda, Emerita,* and *Lepidopa.* Ranging in length from 2.5 to 8 cm (1 to 3 in.), they move up and down the beach near the shoreline. Burying their bodies into the sand and leaving their long, curved, V-shaped antennae pointing up the beach slope, these little crabs filter food particles out of the water (figure 15–11).

Echinoderms. Echinoderms are represented in the beach deposits by the sand star, *Astropecten,* and heart urchins. Members of the genus *Astropecten* are called sand stars because they are well adapted to prey on invertebrates buried in the low tide region of sandy beaches.

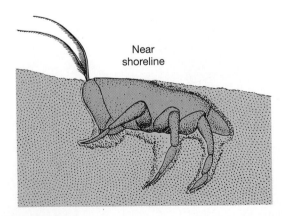

FIGURE 15–11
Sand Crab, *Emerita.*

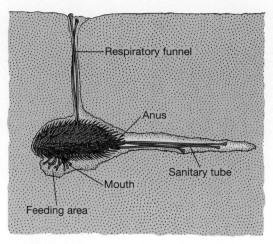

FIGURE 15–12
Heart Urchin, *Echinocardium.*

With five rays and a smooth dorsal surface, sand stars have sturdy spines along the margins of the rays and tube feet tapered to a point to aid them in moving through the sediment (figure 15–9).

The most widely distributed heart urchins belong to the genus *Echinocardium.* More flattened and elongated than the sea urchins of the rocky shore, the heart urchins live buried in the sand near the low tide line. They gather sand grains into the mouth, where the coating of organic matter is scraped off and ingested (figure 15–12).

Meiofauna. Living in the spaces between sediment particles are tiny organisms ranging in size from 0.1 to 2 mm (0.004 to 0.08 in.) in length (figure 15–13). They feed primarily on bacteria removed from the surface of sediment particles. The **meiofauna** population, composed primarily of polychaetes, mollusks, arthropods, and nematodes, are found in sediment from the intertidal zone to the deep-ocean trenches.

Intertidal Zonation. Although it requires more investigation to observe it, there is a characteristic faunal distribution across the intertidal range of sediment-covered shores similar to that observed on rocky shores. Although the species of animals found in each of the corresponding intertidal zones are appropriately different, the distribution of the life forms across the intertidal rocky and sediment-covered shores have two characteristics in common: The maximum number of species and the greatest biomass are found near the low tide shoreline, and both decrease toward the high tide shoreline. This zonation, shown in figure 15–14, is best developed on steeply sloping, coarse sand beaches and is less readily identified on the gentler sloping, fine sand beaches. The tiny clay-sized particles of the mud flat produce a deposit with essentially no slope, which eliminates the possibility of zonation in this protected, low-energy environment.

 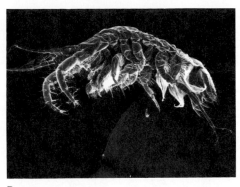 

A.   B.   C.

FIGURE 15–13
Meiofauna. Scanning electron micrographs of *A,* nematode head (×804). Image width is 0.06 mm. The projections and pit on right side are possibly sensory structures. *B,* amphipod (×20). This 3-mm long organism builds a burrow of cemented sand grains. *C,* a 1-mm long polychaete (×55) with its proboscis extended. (Courtesy of Howard J. Spero, University of South Carolina.)

The Mud Flat

Two widely distributed plants associated with the mud flat are eelgrass, *Zostera,* and turtle-grass, *Thalassia,* which occupy the low tide zone and adjacent shallow sublittoral regions bordering the flats. Numerous openings at the surface of mud flats attest to a large population of bivalve mollusks and other invertebrates.

Quite visible and interesting inhabitants of the mud flats are the fiddler crabs, *Uca,* living in burrows that may be more than 1 m (3 ft) deep. Relatives of the shore crabs, they usually measure no more than 2 cm (0.8 in.) across the carapace. Fiddler crabs get their name from one outsized claw, up to 4 cm (1.6 in.) long, on the male. This large claw is waved around in such a manner the crab seems to be fiddling. The females have two normal-sized claws. The large claw of the male is used to court females and fight off other competing males. They feed by extracting organic matter from the mud and show a degree of dislike for the sea by jumping into their burrows and closing the mud door when the tide rises to cover the mud flat (figure 15–15).

THE SHALLOW OFFSHORE OCEAN FLOOR

Extending from the spring low-tide shoreline to the seaward edge of the continental shelf is an environment that is mainly sediment-covered, although bare rock exposures may be found locally near shore. This region conforms roughly with what has been described as the sublittoral zone.

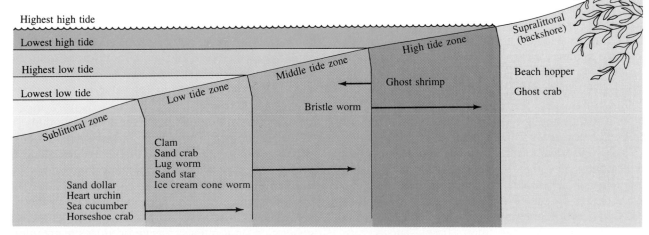

FIGURE 15–14
Intertidal Zonation on the Sediment-Covered Shore. Zonation is best displayed on coarse sand beaches with steep slopes. As the sediment becomes finer and the beach slope decreases, zonation becomes less distinct. It disappears entirely on mud flats.

FIGURE 15–15
Life of the Mud Flat. *Zostera,* eelgrass, and *Uca,* fiddler crab.

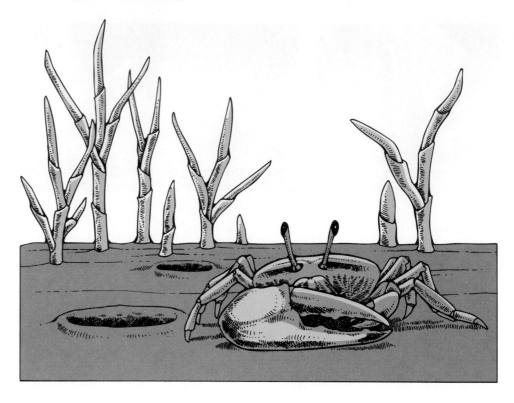

The Rocky Bottom (Sublittoral)

A rocky bottom within the shallow inner sublittoral region will usually be covered with algae. Along the Pacific coast of North America, the giant bladder kelp, *Macrocystis,* is known to attach to rocks at depths as great as 30 m (100 ft) if the water is clear enough to allow sunlight to support plant growth at this depth. The giant bladder kelp and another fast-growing kelp, *Nereocystis,* often form bands of kelp forest along the Pacific coast. Smaller tufts of red and brown algae are found on the bottom and living epiphytically (one plant living nonparasitically on another plant) on the kelp fronds. Epifauna commonly found growing along with the algal tufts on the fronds are hydroids and bryozoan colonies. All these smaller life forms serve as food for many of the animals found living within the kelp forest community. Nudibranchs are important as predators on the hydroids and bryozoan colonies. Surprisingly, very few animals feed directly on the living kelp plant. Among those that do are the large sea hare, *Aplysia,* and sea urchins (figure 15–16).

Large crustaceans called *lobsters* are common to rocky bottoms. They are a somewhat varied group with robust external skeletons. The chelaless, or spiny lobsters, are named for their spiny carapace and are characterized by two very large and spiny antennae that have noise-making devices near their base. The genus *Palinurus* is found at depths greater than 20 m (65 ft) along the coast of Europe, reaches lengths up to 50 cm (20 in.), and is a delicacy. The species *Panulirus argus* is found in the Caribbean and is sometimes observed to migrate single file for unknown reasons over distances of several kilometers. *Panulirus interruptus* is the spiny lobster of the American west coast. Other related species are found in the Juan Fernandez Islands and along the coasts of New Zealand, Australia, and South Africa. All the spiny lobsters are taken for food, but none are as highly regarded as the so-called true lobsters belonging to the genus *Homarus.* Although they are scavengers like their spiny relatives, the true lobsters—which include the American lobster, *Homarus americanus*—also feed on live animals, including mollusks, crustaceans, and members of their own species. Migrating into deep water during winter and returning to nearshore waters in summer, the American lobster is nocturnal. Related species are found along the European coast from Norway to the Mediterranean and the coast of South Africa, and the American lobster is found from Labrador to Cape Hatteras (figure 15–17).

Oysters are sessile bivalve mollusks found in estuarine environments. They prefer locations where there is a steady flow of clean water to provide plankton and oxygen. Having great commercial importance throughout the world, they have been closely studied. Oyster beds consist of the empty shells of generation after generation cemented to rock bottom or one another, with the living generation on top. Each female produces many millions of eggs each year, which become planktonic larvae when fertilized. After a few weeks as plankton, the larvae settle to attach themselves to the bottom. The larvae prefer live

FIGURE 15–16
Life of the Sublittoral Zone. *A,*
giant bladder kelp, *Macrocystis,* and
B, sea hares, *Aplysia californica,* and
sea urchins in a kelp forest. *C,* purple
sea urchin, *Strongylocentrotus purpuratus.* (Photos *A* and *C* by Mia Tegner; *B* by James McCullagh.)

A.

B.

C.

A.

B.

FIGURE 15—17
Spiny and American Lobsters. *A,* spiny lobster. (Photo by B. Kiwala.) *B,* American lobster. (Photo by Harold W. Pratt/Biological Photo Service.)

oyster shells, dead oyster shells, and rock, in that order, as a substrate for attachment. Oysters serve as prey for a variety of sea stars, fishes, crabs, and boring snails that drill through the shell and rasp away the soft tissue of the oyster (figure 15—18).

Coral Reefs

Although corals are found throughout the ocean, coral accumulations that might be classified as reefs are restricted to the warmer-water regions where the average monthly temperature exceeds 18°C (64°F) throughout the year (figure 15—19). Such temperature conditions are found primarily between the tropics, although reefs grow to latitudes approaching 35°N and S on the western mar-

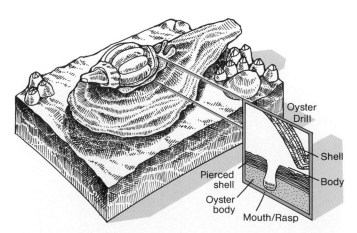

FIGURE 15—18
An Oyster Drill Feeding on an Oyster.

gins of ocean basins where warm-water masses move into the high-latitude areas and raise average temperatures. Also shown in figure 15—19 is the greater diversity of reef-building corals on the western side of ocean basins. More than 50 genera of corals are found in a broad area of the western Pacific Ocean and a narrow belt of the western Indian Ocean. Fewer than 30 genera are found in the Atlantic Ocean with the greatest diversity occurring in the Caribbean Sea. The explanation for this pattern of diversity may be related to past ocean current patterns, but such an explanation is controversial.

Because of changes in wave energy, salinity, water depth, temperature, and other less obvious factors, there is a well-developed vertical and horizontal zonation of the reef slope. These zones are readily identified by the assemblages of plant and animal life found in and near them (figure 15—20).

Not only coral, but algae, mollusks, and foraminifers make important contributions to the reef structure. Reef-building corals are **hermatypic.** They have a mutualistic relationship with the algae **zooxanthellae** that live within the tissue of the coral polyp. Not only reef-building corals but other reef animals have a similar symbiotic relationship with algae. These heterotrophs that derive part of their nutrition from algal symbionts are called **mixotrophs.** They include corals, foraminifers, sponges, and mollusks (figure 15—21). Because light is essential for algal photosynthesis, reef-building corals are restricted to clear, shallow bottom waters. The algae contribute not only to nourishing the coral, but could contribute to the calcification capability of corals by extracting carbon dioxide from the animals' body fluids. This would increase the concentration of the carbonate ion

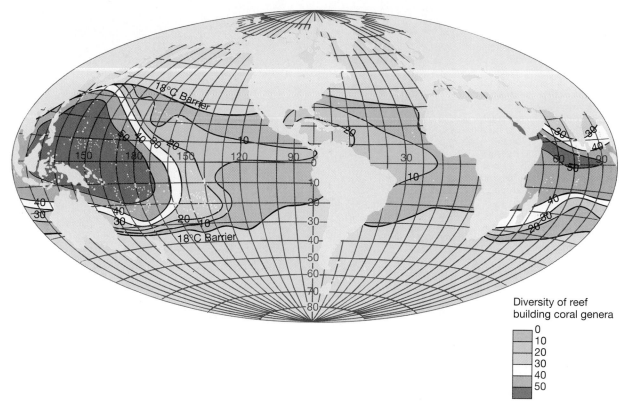

FIGURE 15–19
Coral Reef Distribution and Diversity. Coral reef development is restricted to the low-latitude area between the two 18°C (64°F) temperature lines shown on the map. Minimum water temperatures of 18°C in surface waters of the Northern and Southern hemispheres occur in February and August, respectively. In each ocean basin, the coral reef belt is wider and the diversity of coral genera is greater on the western side of the ocean basins. (After Stehli and Wells, 1971.)

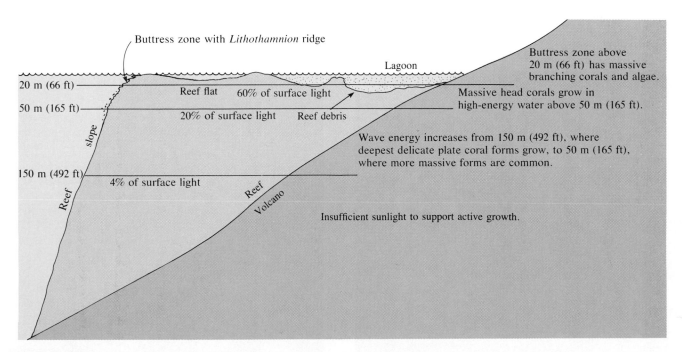

FIGURE 15–20
Coral Reef Zonation.

FIGURE 15–21
Coral Reef Inhabitants That Depend on Algal Symbionts. *A,* polyps of a reef coral extend their tentacles to capture tiny planktonic organisms from the surrounding water. However, for most reef-building corals, most of their nourishment is provided by symbiotic algae, zooxanthellae, that live in the coral tissue. *B,* the blue-gray sponge on the left is *Niphates digitalis,* a totally heterotrophic sponge. On the right is *Angelas* sp., which contains some cyanobacterial symbionts. It is, like most reef-building corals, a mixotroph. Photo was taken in 20 m (60 ft) of water at Carrie Bow Cay, Belize. *C,* a giant clam, *Tridacna gigas.* These suspension feeders also depend on symbiotic algae living in the mantle tissue. (Photos *A* by Christopher Newbert, *B* by C. R. Wilkerson, Australian Institute of Marine Science, *C* by Ken Lucas/Biological Photo Service.)

A.

B.

C.

needed for the precipitation of calcium carbonate. The corals contribute to the relationship by providing a supply of nutrients to the zooxanthellae. This exchange between algal symbionts and corals, as well as other mixotrophs, supports high levels of biological productivity within the reef community even though the level of nutrient concentration in the surrounding water is low.

Reef growth also requires that the water have a relatively normal salinity and that it be free from particulate matter. Therefore, we see very little coral reef growth near the mouths of rivers that lower the salinity and carry large quantities of suspended material that would choke the reef colony.

Coral reefs actually contain up to three times as much plant as animal biomass. The zooxanthellae account for less than 5 percent of the reef's plant mass; most of the rest is filamentous green algae. However, zooxanthellae account for up to 75 percent of the biomass of reef-building corals and provide the coral with up to 90 percent of its nutrition.

The greatest depth to which active coral growth extends is 150 m (492 ft), below which there is not enough sunlight to support hermatypic mutualism. Water motions are not great at these depths, so relatively delicate plate corals can live in the outer slope of the reef from 150 m (492 ft) up to about 50 m (164 ft), where light intensity exceeds 4 percent of the surface intensity (figure 15–20). From 50 m to about 20 m (66 ft), the strength of water motion from breaking waves increases on the side of the reef facing into the prevailing current flow. Correspondingly, the mass of coral growth and the strength of the coral structure supporting it increase toward the top of this zone, where light intensity exceeds 20 percent of surface value. Head corals are dominant. The buttress zone extends from 20 m to the low-tide line; within it more-massive branching corals, *Acropora,* and encrusting algae such as the red alga, *Lithothamnion,* withstand the crashing waves. Light intensity exceeds 60 percent of surface intensity in the buttress zone. The waves cut surge channels across the algal ridge, which is inhabited by only a few animals—such as snails, limpets, and the slate pencil urchin, *Heterocentrotus*—that can withstand the constant beating of the surf. The surge channels extend down the reef slope as debris channels that carry the products of wave erosion. Small reef fish find protection from larger predators such as sharks, barracuda, and jacks in the debris grooves extending between the buttress ridges. The reef flat extends across the reef beyond the lagoons of atolls and barrier reefs. Here the reef may be under a few centimeters to a few meters of water at low tide. A variety of beautiful reef fish inhabit this shallow water. The sand of reef debris and foraminifer tests fills in the deeper holes and provides a home for sea cucumbers, worms, and a variety of mollusks. In the protected water behind the *Lithothamnion* ridge lagoonal reefs may form, where species of *Porites* and *Acropora* grow into beautiful large colonies. Gorgonian coral, anemones, crustaceans, mollusks, and echinoderms of great variety also will be found in the lagoon reef (figure 15–22).

An obvious example of commensalism is the behavior of the shrimpfish, which swims head down among the long slender spines of the reef sea urchins. The urchin's spines are a significant deterrent to any predator, and the sea urchin is neither hindered nor aided by the presence of the little fish. The clown fish receives a similar protection by swimming among the tentacles of two species of sea anemones. This relationship is believed to be mutual since the anemones benefit by the clown fish serving as bait to draw other fish within reach of anemone tentacles and actually carrying food to them (figure 15–23).

Coral Reefs and Nutrient Levels. When human populations increase in lands adjacent to coral reefs, the reefs deteriorate. There are many aspects of human behavior that would obviously damage the reef, but one of the more subtle is the inevitable increase in the nutrient levels of the reef waters. As nutrient levels increase in reef waters, the dominant benthic community changes. At low nutrient levels the hermatypic corals and other reef animals that contain algal symbionts thrive. As nutrient levels in the water increase, conditions favor development of fleshy benthic plants at moderate nutrient levels and suspension feeders at high nutrient levels. At high nutrient levels, phytoplankton mass exceeds the benthic plant mass, and benthic populations tied to the phytoplankton food web dominate. The clarity of water is reduced by increased phytoplankton biomass, and the fast-growing members of the phytoplankton-based ecosystem destroy the reef structure by overgrowing the slow-growing coral and by **bioerosion.** Bioerosion by sea urchins and sponges is particularly effective in damaging the reef.

The Crown-of-Thorns Phenomenon. Since 1962 the crown-of-thorns sea star, *Acanthaster planci,* has caused destruction of living coral on many reefs throughout the western Pacific Ocean (figure 15–24). Some investigators believe this is a modern phenomenon brought about by the activities of humans. However, there is little evidence to point to such a cause. A 1989 study of the Great Barrier Reef has indicated that during the past 8,000 years *A. planci* has been even more abundant on the reefs studied than it is today. If this is true, the sea star may be an integral part of the reef ecology in this region rather than a destructive upstart taking advantage of human actions that have modified the reef in some way favorable to its proliferation.

Bleaching of Coral Reef Communities. The bleaching or loss of color by coral reef organisms has occurred on a local basis numerous times in the past. However, a

A.

FIGURE 15–22
Nonreefbuilding Inhabitants. *A,* some of the more unusual of the great variety of colorful and strange fishes associated with coral reef and other shallow-water, tropical environments are the puffers. When a potential predator comes too close, they gulp water to fill a ventral stomach pouch, expanding their loose, scaleless skin to produce a spherical shape. Actually, puffers are seldom bothered by predators as they have viscera and skin that often contain deadly neurotoxin. Though they are feeble swimmers, puffers like this swelled-up member of the genus *Arothron* are virtually free from attack. *B,* many members of the coral family do not secrete the calcium carbonate of the reef builders. An example is a soft gorgonian coral, shown with feeding polyps extending from its branches. (Photos by Christopher Newbert.)

B.

FIGURE 15–23
Mutualism. The clown fish, which lives unharmed among the stinging tentacles of the sea anemone, may pay for this protection by bringing food to the anemone. This symbiotic relationship that benefits both participants is called mutualism. (Photo by Christopher Newbert.)

mass mortality of at least 70 percent of the corals along the Pacific Central American coast occurred as a result of a bleaching episode associated with the severe El Niño of 1982–83. Although the cause is not known for certain, this eastern Pacific bleaching may have been caused by the increased water temperatures associated with the El Niño event. Whatever the cause, it will take many years for these reefs to recover. Two species of Panamanian coral became extinct during this event.

One thing is clear, the direct cause of the bleaching is the expulsion of symbiont algae, zooxanthellae. Essentially all reef-building corals and some other reef mixotrophs are nourished by these algae that live within their tissue. The loss of this source of nourishment can be life threatening to the reef dwellers.

Figure 15–25 shows the bleaching of a round starlet coral, *Siderastrea siderea,* on Enrique Reef, Puerto Rico. The photograph was taken in November 1987. Normally a rusty brown color, this coral has been bleached on its lower left half.

The broad scope of human-induced changes throughout Earth; increased concentrations of green-

house gases in the lower atmosphere; depletion of ozone in the upper atmosphere; pollution of the oceans with petroleum, plastics, sewage, and so on; and the bleaching of coral reefs may all be symptoms of a worldwide pathology that will require major modifications in human behavior before it can be cured.

DEEP-OCEAN FLOOR

The Physical Environment

The deep-ocean floor includes the bathyal, abyssal, and hadal zones described in chapter 12. Light is present in only the lowest concentrations above 1000 m (3280 ft) and absent below this depth. Everywhere the temperature is low, rarely exceeding 3°C (37°F) and falling as low as −1.8°C (28.8°F) in the high latitudes. Pressure exceeds 200 atmospheres on the oceanic ridges, ranges between 300 and 500 atmospheres on the deep-ocean abyssal plains, and climbs to over 1000 atmospheres in the deepest trenches. (The pressure is one atmosphere at the ocean surface and increases by one atmosphere for each

FIGURE 15–24
Crown-of-Thorns Sea Star. This crown-of-thorns sea star is being attacked by one of its few predators, the Pacific triton. (Photo © Christopher Newbert.)

FIGURE 15–25
Bleached Coral. This round starlet coral, *Siderastrea siderea,* on Enrique Reef, Puerto Rico, has been bleached across the lower left side of the image. (Photo by Dr. Lucy Bunkley Williams.)

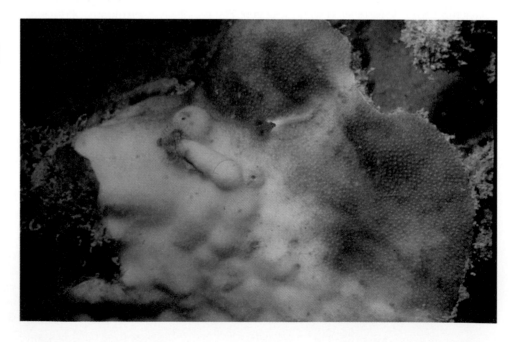

FIGURE 15–26
Alvin Approaches a Hydrothermal Vent Community Typical of That Found on the East Pacific Rise. In the lower left corner are a sea anemone and three octacorals, which are close relatives of the anemone. They may feed primarily on sulfur-oxidizing bacteria suspended in the water near the vents. Grenadier, or rattail fish, are common in the deep ocean and are usually the first to arrive at bait placed on the deep-ocean floor. Vestimentiferan tube worms *(Riftia)* and giant clams *(Calypotogena)* do not possess guts: They are nourished by chemosynthetic bacteria that live in their tissues. White brachyuran crabs swarm over the lava pillows, and a "black smoker" spews hot (350°C) sulfide-rich water from its metallic sulfide chimney.

10 m, or 33 ft, of depth.) Much of the deep-ocean floor is covered by at least a thin layer of sediment ranging from the mudlike abyssal clay deposits of the abyssal plains and deep trenches, through the oozes of the oceanic ridges and rises, to some coarse sediment deposited as turbidites on the continental rise. Near the crests of the oceanic ridges and rises and down the slopes of seamounts and oceanic islands, sediment is absent and basaltic ocean crust is the substrate.

Deep-Sea Hydrothermal Vent Biocommunities

In 1977 the first active hydrothermal vent field was discovered at a depth below 2500 m (8200 ft) in the Galápagos Rift, near the equator in the eastern Pacific Ocean (figures 15–26 and 15–27). Water temperatures in the immediate area of the vents ranged between 8° and 12°C (46° and 54°F), while the normal bottom-water tempera-

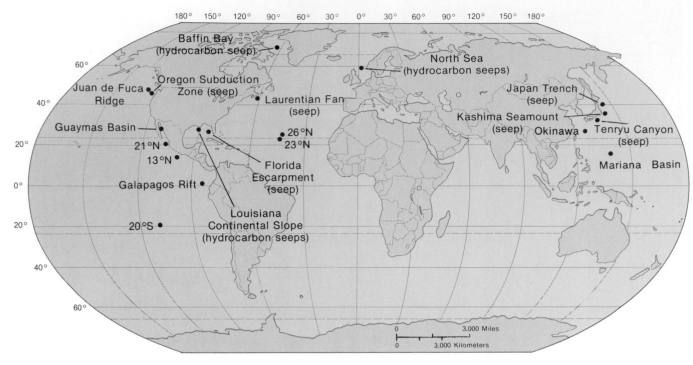

FIGURE 15–27
Hydrothermal Vents, Cold Seeps, and Hydrocarbon Seeps Known to Support Biocommunities.

ture at these depths is about 2°C (36°F). These vents were found to support the first known **hydrothermal vent biocommunities,** consisting of unusually large organisms for these depths. The more prominent members of this community are tube worms with tubes over 1 m long, giant clams up to 25 cm (10 in.) in length, large mussels, and two varieties of white crabs.

At 21°N latitude on the East Pacific Rise, south of the tip of Baja California, tall chimneys were found to belch out 350°C (662°F) vent water so rich in metal sulfides that it was black. These chimney vents, first observed in 1979, were composed primarily of sulfides of copper, zinc, and silver and came to be called black smokers. A new species of vent fish associated with some vent fields was first observed here.

The most important members of these biocommunities are the **sulfur-oxidizing bacteria** that synthesize, through chemosynthesis, the organic molecules that feed the community. The bacteria use the chemical energy released by sulfur oxidation much as marine plants use the sun's energy to carry on photosynthesis. Although some animals feed on the bacteria and larger prey, many of them depend primarily on a symbiotic relationship with the bacteria. For instance, the tube worms and giant clams depend entirely on sulfur-oxidizing bacteria that live symbiotically in their tissues.

In 1981 the Juan de Fuca Ridge biocommunity was observed. Although vent fauna at this site is less robust

than at the Galápagos Rift and on the East Pacific Rise, the metallic sulfide deposits associated with the vents aroused much interest because they are so near the U.S. coast (figure 15–28).

The Guaymas Basin, in the Gulf of California, provided the first observation, in 1982, of hydrothermal vents beneath a thick layer of sediment. Of particular interest is the fact that sediment and sulfide samples recovered were saturated with hydrocarbons, which may have entered the food chain through bacterial uptake. The abundance and diversity of life may exceed that of the rocky bottom vents. A phenomenon similar to that of the Guaymas Basin was observed in 1987 at the Mariana Basin, a small spreading-center system beneath a sediment-filled back-arc basin (figure 15–27).

Subsequent exploration has revealed numerous hydrothermal vent biocommunities in the Pacific Ocean, including one in the Southern Hemisphere on the East Pacific Rise at 20°S. The first active hydrothermal vents with associated biocommunities in the Atlantic Ocean were discovered in 1985 at depths below 3600 m (11,800 ft) near the axis of the Mid-Atlantic Ridge, which divides the Atlantic Ocean in half. The predominant fauna of these vents consists of shrimps that have no eye lens but can detect levels of light emitted by the black smoker chimneys that is not visible to the human eye (figure 15–29). After these Atlantic discoveries, at 23°N and 26°N latitude, it now seems likely that similar biocommunities

A.

B.

C.

FIGURE 15–28
Hydrothermal Vent Biocommunities of the Juan de Fuca Ridge, Guaymas Basin, and Mariana Back-Arc Basin. *A,* vestimentiferan tube worms, *Ridgeia phaeophiale,* found at Juan de Fuca vents. (Photo by Robert W. Embly, courtesy of William Chadwick, Jr., Oregon State University, Hatfield Marine Science Center.) *B,* at the Guaymas Basin in the Gulf of California, the tube worms, *Riftia pachyptila,* intergrown with a mat of bacteria. The mats are composed of very large bacteria about 150 μm in diameter. (Photo by Robert Hessler, courtesy of Scripps Institution of Oceanography, University of California, San Diego.) *C,* this view of the community of the Mariana Back-Arc Basin includes a new genus and species of sea anemone, *Marianactis bythios,* a new family, genus, and species of gastropod, *Alviniconcha hessleri* (the first known snail to contain chemosynthetic bacterial symbionts), and the galatheid crab, *Munidopsis marianica.* (Photo courtesy of Robert Hessler.)

will be found in significant numbers in all the world's oceans.

A relict vent area found on the Galápagos spreading center indicates vent communities may have a relatively short life span. This inactive vent was identified by an accumulation of dead clams. It appears that when the vent becomes inactive and the hydrogen sulfide that serves as the source of energy for the community is no longer available, the community dies.

Low-Temperature Seep Biocommunities

Three additional submarine spring environments have also been found to have chemosynthetically supported biocommunities. During the summer of 1984, an investigation of an ambient-temperature, **hypersaline seep** (46.2‰) at the base of the Florida Escarpment in the Gulf of Mexico (figure 15–30) revealed a biocommunity similar in many respects to the hydrothermal vent communities. The seeping water appears to flow from joints at the base of the limestone escarpment and move out across the clay deposits of the abyssal plain at a depth of about 3200 m (10,496 ft).

The hydrogen sulfide–rich waters support a number of white bacterial mats that carry on chemosynthesis in a fashion similar to that of the bacteria of the hydrothermal vents. These and other chemosynthetic bacteria may provide most of the support for a diverse community of animals. The community includes holothurians, sea stars, shrimp, snails, limpets, brittle stars, anemones, vestimentiferan worms, galatheid crabs, clams, mussels, and zoarcid fish.

Drilling into the limestone rocks of the platform east of the escarpment revealed fluids with temperatures

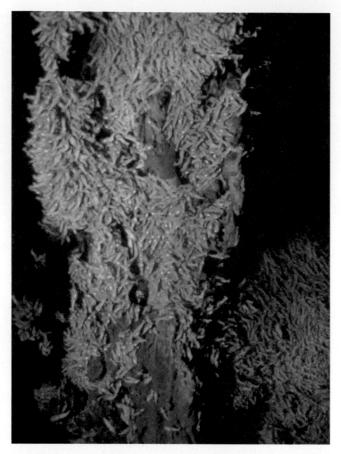

FIGURE 15–29
Atlantic Ocean Hydrothermal Vent Organisms. Swarm of particulate-feeding shrimp, the predominant animals observed at hydrothermal vents near 23°N and 26°N on the Mid-Atlantic Ridge. This swarm was photographed at 26°N. (Courtesy of Peter A. Rona, NOAA.)

of up to 115°C (239°F) and salinities of 250‰. This dense water apparently seeps deep into the jointed rocks of the platform and is trapped by the impermeable clay deposited on the ocean floor at the base of the escarpment. It then flows out onto the surface clays, mixes with seawa-

ter, and produces the spring water with salinity intermediate between that of the platform fluids and ocean water.

Also observed in 1984 were dense biological communities associated with oil and gas seeps on the Gulf of Mexico continental slope (figure 15–31). Trawls at depths of between 600 and 700 m (1968 and 2296 ft) recovered epifauna and infauna similar to those observed at the hydrothermal vents and the hypersaline seep at the base of the Florida Escarpment. Subsequent investigations have identified seeps with associated communities to depths of 2200 m (7620 ft) on the continental slope. Carbon isotope analysis indicates the **hydrocarbon-seep fauna** is based on a chemosynthetic productivity that derives its energy from hydrogen sulfide and/or methane. Bacterial oxidation of methane results in the production of $CaCO_3$ slabs found here and at other hydrocarbon seeps shown in figure 15–27.

Finally, a third environment of **subduction-zone seep** communities was observed from *Alvin* during the summer of 1984. It is located at a subduction zone of the Juan de Fuca Plate at the base of the continental slope off the coast of Oregon (figure 15–32). Here, the trench is filled with sediments. At the seaward edge of the slope, clastic sediments are folded into a ridge. At the crest of the ridge, pore water escapes from the two-million-year-old folded sedimentary rocks into a thin overlying layer of soft sediment. Occurring at a depth of 2036 m (6678 ft) the seeps produce water that is only slightly warmer (about 0.3°C, or 0.5°F) than ambient conditions. The vent water contains CH_4 (methane) that is probably produced by decomposition of organic material in the sedimentary rocks. The methane serves as the source of energy for bacteria that oxidize it and chemosynthetically produce food for themselves and the rest of the community, which contains many of the same genera found at other vent and seep sites. During 1985, similar communities were located in subduction zones of the Japan Trench and Peru-Chile Trench. All the seeps are located on the landward side of the trenches at depths from 1300 to 5640 m (4264 to 18,499 ft).

FIGURE 15–30
Hypersaline Seep Biocommunity at Base of Florida Escarpment. *A,* rectangle is the location of the seep and biocommunity. *B,* seismic reflection profile of limestone Florida Escarpment and abyssal bedded sediments at its base. Arrow marks location of seep. *C,* Florida Escarpment seep biocommunity of dense mussel beds covers much of the image. White dots are small gastropods on mussel shells. At lower right are tube worms covered with hydrozoans and galatheid crabs. Fractures in the escarpment limestone are visible along top. (Courtesy of C. K. Paull, Scripps Institution of Oceanography, University of California, San Diego.)

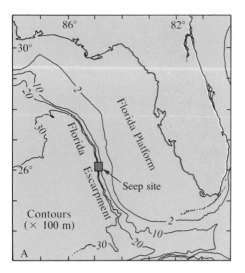

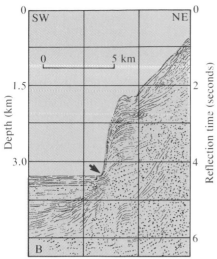

FIGURE 15–31
Hydrocarbon Seeps on the Continental Slope of the Gulf of Mexico. *A,* locations of known hydrocarbon seeps with associated biocommunities. *B,* chemosynthetic mussels and tube worms from the Bush Hill seep. The mussel has demonstrated that it can grow with methane as its sole carbon and energy source. The tube worm is dependent on hydrogen sulfide. (Courtesy of Charles R. Fisher, Penn State University.) *C,* Alaminos Canyon site (Neptune's Garden) discovered in 1990 is a significant oil seep with a new species of mussel that probably depends on methane and two species of vestimentiferan tube worms that depend on hydrogen sulfide. One of the tube worms may be a new species. (Courtesy of Charles R. Fisher, Penn State University.)

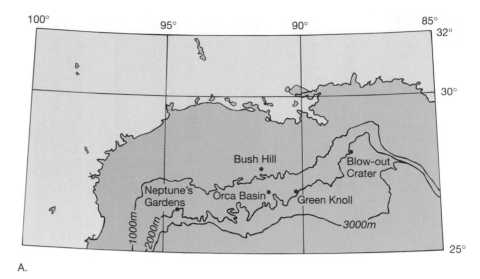

A.

B.

C.

FIGURE 15–32
Locations of Vent Communities off the Coast of Oregon. The communities are associated with the subduction zone of the Juan de Fuca Plate. Sediment filling the trench is folded into a ridge with vents at its crest. (Courtesy of L. D. Klum, Oregon State University.)

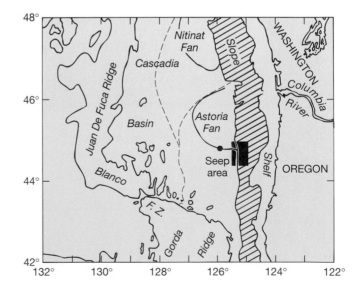

SUMMARY

Over 98 percent of the 200,000 species of marine animals live on the ocean floor. Most of these live on the continental shelf, with only 400-odd species found in deep-ocean trenches. The importance of temperature on the development of species diversity is reflected in the fact that there are three times as many species of benthos along the European coast warmed by the Gulf Stream than along a similar length of North American coast cooled by the Labrador Current.

Because of tidal motions, the littoral (intertidal) zone can be divided into the high tide zone (mostly dry), middle tide zone (equally wet and dry), and the low tide zone (mostly wet). The littoral is bounded by the supralittoral (covered only by storm waves) and the sublittoral, which extends below the low tide shoreline.

The supralittoral zone along rocky coasts is characterized by the presence of the periwinkle snail, the rock louse, and the limpet.

The barnacles most characteristic of the high tide zone are the tiny buckshot barnacles. The most conspicuous plants of the high tide zone are *Fucus* along colder shores and *Pelvetia* in warmer latitudes. These plants become more abundant in the middle tide zone, and the diversity and abundance of the flora and fauna in general increase toward the lower intertidal zone.

Larger species of acorn barnacles are found in the middle tide zone. A common assemblage of rocky middle tide zones includes the goose barnacle, the mussels, and sea stars. Joining the sea stars as predators on the barnacles and mussels are carnivorous snails.

Tide pools within the middle tide zone commonly house sea anemones, fishes such as the opaleye, woolly sculpin, and rockpool blenny, and hermit crabs. At the lower limit of the middle tide zone, sea urchins become numerous.

The low tide zone in temperate latitudes is characterized by a variety of moderately sized red and brown algae providing a drooping canopy for the animal life.

Scampering across the entire intertidal zone are the scavenging shore crabs.

The only sediments usually found along rocky shores are a few cobbles and boulders. Moving into more protected segments of the shore, lower levels of wave energy allow deposition of sand and mud. Due to the higher permeability of sand, sand deposits are usually well oxygenated, and mud deposits are anaerobic below a thin surface layer.

Compared to the rocky shore, the diversity of species is reduced on and in beach deposits and is quite restricted in mud deposits. This does not mean that the abundance of life is reduced. Although life is less visible, up to 8000 burrowing clams have been recovered from 1 m² (10.8 ft²) of mud flat. Suspension feeders (filter feeders) characteristic of the rocky shore are still found in the sediment, but there is a great increase in the relative abundance of deposit feeders that ingest sediment and detritus.

Bivalve mollusks are well suited for sediment-covered shores. The lugworm is a deposit feeder that ingests sand and extracts whatever organic content is available.

Amphipods called beach hoppers can be found feeding on the kelp deposited high on the beach by storm waves. Sand crabs filter food from the water with their long, curved antennae while buried in the sand near the shoreline.

Echinoderms are represented in the sandy beach by the sand star, which feeds on buried invertebrates, and the heart urchin, which scrapes organic matter from sand grains.

As is true for the rocky shore, the diversity of species and abundance of life on the sediment-covered shore increases toward the low tide shoreline. Zonation is best developed on steep, coarse sand beaches and is probably missing on mud flats with little or no slope.

Attached to the rocky sublittoral bottom just beyond the shoreline is a band of algae including large kelp plants. Growing on the large fronds of the kelp are small varieties of algae, hydroids, and bryozoans. Nudibranchs feed on the small hydroids and bryozoans, and the sea hare and sea urchin feed on the kelp and other algae. Spiny lobsters are common to rocky bottoms in the Caribbean and along the West Coast, and the American lobster is found from Labrador to Cape Hatteras.

Oyster beds found in estuarine environments consist of individuals that attach themselves to the bottom or to the empty shells of previous generations.

Above a depth of 150 m (492 ft) in clear, nutrient-poor tropical waters, living coral reef can be found off the shores of islands and continents. Reef-building corals and other mixotrophs are hermatypic, containing algal symbionts, zooxanthellae, in their tissues. Delicate varieties are found at 150 m, and they become more massive near the surface, where wave energy is higher. The top 20 m (66 ft), the buttress zone, is reinforced with calcium carbonate deposited by algae such as *Lithothamnion,* producing an algal ridge. Waves cut surge channels across the ridge, and debris channels extend down the reef front below the surge channels. Many varieties of commensalism and mutualism are found within the coral reef biological community. There is some evidence that the crown-of-thorns may be a natural phenomenon and the potentially lethal "bleaching" of coral reefs may result from the expulsion of symbiont algae under stress of elevated temperature.

Although little is known of the deep-ocean benthos, it is clear that it is much more varied than previously thought.

An important discovery of the hydrothermal vent communities made in 1977 on the Galápagos Rift has shown that, at least locally, chemosynthesis is an important means of primary productivity at these vents and at cold-water and hydrocarbon seeps.

KEY TERMS

Autotomy (p. 328)
Bioerosion (p. 339)
High tide zone (p. 323)
Hydrocarbon seep biocommunity
    (p. 346)
Hydrothermal vent biocommunity
    (p. 344)

Hypersaline seep (p. 345)
Intertidal zone (p. 322)
Low tide zone (p. 328)
Meiofauna (p. 332)
Middle tide zone (p. 323)
Mixotroph (p. 336)

Permeability (p. 328)
Subduction zone seep biocommunity
    (p. 346)
Sulfur-oxidizing bacteria (p. 344)
Zooxanthellae (p. 336)

QUESTIONS AND EXERCISES

1. Discuss the general distribution of life in the ocean. Include species diversity (variety of types of organisms) between the pelagic and benthic environments and within the benthic environment.

2. Diagram the intertidal zones of the rocky shore and list characteristic organisms of each zone.

3. Describe the dominant feature of the middle tide zone along rocky coasts, the mussel bed. Include a discussion of other organisms found in association with the mussels.

4. List and describe crabs found within the rocky shore intertidal zone.

5. Discuss how sediment stability and permeability of sandy and muddy shores differ.

6. Other than predation, discuss the two types of feeding styles that are characteristic of the rocky, sandy, and muddy shores. One of these feeding styles is rather well represented in all these environments; name it and give an example of an organism that uses it in each environment.

7. How does the diversity of species on sediment-covered shores compare with that of the rocky shore? Can you think of any reasons why this should be so? If you can, discuss them.

8. In which intertidal zone of a steeply sloping, coarse sand beach would you find the following: clams, beach hoppers, ghost shrimp, sand crabs, and heart urchins?

9. What relationships exist among intertidal zonation, sediment particle size, and beach slope?

10. Discuss the dominant species of kelp, their epifauna, and animals that feed on kelp in the Pacific coast kelp forest.

11. Where in the world are spiny lobsters and American lobsters (and their related species) found?

12. Discuss the preferred environment, reproduction, and threats to survival of larval and adult forms of oysters.

13. Describe the environment suited to development of coral reefs.

14. Describe the zones of the reef slope, the characteristic coral types, and the physical factors related to zonation.

15. As one moves from the shoreline to the deep-ocean floor, what changes in the physical environment can be expected?

16. Discuss the most important barriers to the migration of shallow-water benthos into the deep ocean and to the reverse migration.

17. What are the major differences between the conditions and biocommunities of the hydrothermal vents and the cold seeps? How are they similar?

REFERENCES

Chadwick, W. W.; Embley, R. W.; and Fox, C. G. 1991. Evidence for volcanic eruption on the southern Juan de Fuca ridge between 1981 and 1987. *Nature* 350, 416–18.

Childress, J. J.; Fisher, C. R.; Brooks, J. M.; Kennicutt, M. C., II; Bidigare, R.; and Anderson, A. E. 1986. A methanotrophic marine molluscan (Bivalvia, Mytilidae) symbiosis: Mussels fueled by gas. *Science* 233:4770, 1306–8.

Fisher, C. R. 1990. Chemoautotrophic and methanotrophic symbioses in marine invertebrates. *Reviews in Aquatic Sciences* 2:3 and 4, 399–436.

George, D., and George, J. 1979. *Marine life: An illustrated encyclopedia of invertebrates in the sea.* New York: Wiley-Interscience.

Hallock, P., and Schlager, W. 1986. Nutrient excess and the demise of coral reefs and carbonate platforms. *Palaios* 1:389–98.

Hessler, R. R.; Ingram, C. L.; Yayanos, A. A.; and Burnett, B. R. 1978. Scavenging amphipods from the floor of the Philippine Trench. *Deep-Sea Research* 25:1029–47.

Hessler, R. R.; and Lonsdale, P. F. 1991. Biogeography of Mariana Trough hydrothermal vent communities. *Deep-Sea Research* 38:2, 185–99.

Kennicutt, M. C., II; Brooks, J. M.; Bidigare, R. R.; Fay, R. R.; Wade, T. L.; and McDonald, T. J. 1985. Vent-type taxa in a hydrocarbon seep region on the Louisiana slope. *Nature* 317:6035, 351–53.

Klum, L. D.; Suess, E.; Moore, J. C.; Carson, B.; Lewis, B. T.; Ritger, S. D.; Kadko, D. C.; Thornburg, T. M.; Embley, R. W.; Rugh, W. D.; Massoth, G. J.; Langseth, M. G.; Cochrane, G. R.; and Scamman, R. L. 1986. Oregon subduction zone: Venting, fauna and carbonates. *Science* 231:4738, 561–66.

MacGinitie, G. E., and MacGinitie, N. 1968. *Natural history of marine animals.* 2nd ed. New York: McGraw-Hill.

Ricketts, E. F.; Calvin, J.; and Hedgpeth, J. 1968. *Between Pacific tides.* Stanford, Calif.: Stanford University Press.

Rona, P. A.; Klinkhammer, G.; Nelson, T. A.; Trefry, J. H.; and Elderfield, H. 1986. Black smokers, massive sulphides and vent biota at the Mid-Atlantic Ridge. *Nature* 321:6065, 33–37.

Walbran, P. D.; Henderson, R. A.; Jull, A. J. T.; and Head, M. J. 1989. Evidence from sediments of long-term *Acanthaster planci* predation on corals of the Great Barrier Reef. *Science* 245:4920, 847–50.

Williams, E. H., Jr.; Goenaga, C.; and Vicente, V. 1987. Mass bleaching on Atlantic coral reefs. *Science* 238:4830, 877–78.

Yonge, C. M. 1963. *The sea shore.* New York: Atheneum.

Zenkevitch, L. A.; Filatove, A.; Belyaev, G. M.; Lukyanove, T. S.; and Suetove, I. A. 1971. Quantitative distribution of zoobenthos in the world ocean. *Bulletin der Moskauer Gen der Naturforscher, Abt. Biol.* 76, 27–33.

ZoBell, C. E. 1968. Bacterial life in the deep sea. In Proceedings of the U.S.-Japan Seminar on Marine Microbiology, August 1966, Tokyo. *Bulletin Misaki Marine Biology, Kyoto Inst. Univ.* 12:77–96.

SUGGESTED READING

Sea Frontiers

Alper, J. 1990. The methane eaters. 36:6, 22–29. An account of the exploration of hydrocarbon seep biocommunities in the Gulf of Mexico.

Coleman, N. 1974. Shell-less molluscs. 20:6, 338–42. A description of nudibranchs, gastropods without shells.

George, J. D. 1970. The curious bristle-worms. 16:5, 291–300. The variety of worms belonging to the class *Polychaeta* of phylum Annelida is described. The discussion includes locomotion, feeding, and reproductive habits of the various members of the class.

Gibson, M. E. 1981. The plight of *Allopora.* 27:4, 211–18. The reason the author believes the unusual California hydrocoral is headed for the endangered species list is the topic of this article.

Humann, P. 1991. Loving the reef to death. 37:2, 14–21. The many ways divers in particular and humans in general degrade the coral reef environment are discussed.

Ruggiero, G. 1985. The giant clam: Friend or foe? 31:1, 4–9. The ecology and behavior of the giant clam *(Tridacna gigas)* is discussed in reference to whether it is a danger to divers.

Shinn, E. A. 1981. Time capsules in the sea. 27:6, 364–74. The method by which geologists determine past climatic and environmental conditions by studying coral reefs is discussed.

Viola, F. J. 1989. Looking for exotic marine life? Don't leave the dock. 35:6, 336–341. A description of marine life living on and near pilings for a boat dock are described along with good color photos.

Winston, J. E. 1990. Intertidal space wars. 36:1, 47–51. Florida intertidal life is discussed.

Scientific American

Caldwell, R. L., and Dingle, H. 1976. Stomatopods. 234:1, 80–89. Presents the ecology of these interesting crustaceans that have appendages specialized for spearing and smashing prey.

Feder, H. A. 1972. Escape responses in marine invertebrates. 227:1, 92–100. Discusses the surprisingly rapid movements and other responses made by invertebrates to the presence of predators. Some interesting photographs accompany the text, which describes the escape responses of limpets, snails, clams, scallops, sea urchins, and sea anemones.

Wicksten, M. K. 1980. Decorator crabs. 242:2, 146–57. Describes how species of spider crabs use materials from their environment to camouflage themselves.

Yonge, C. M. 1975. Giant clams. 232:4, 96–105. The distribution and general ecology of the tridacnid clams, some of which grow to lengths well over 1 m (3.3 ft), are investigated.

Appendix I

Scientific Notation

To simplify writing very large and very small numbers, scientists indicate the number of zeros by scientific notation, in which one integer is placed to the left of the decimal and a multiplication times a power of 10 tells which direction and how far the decimal would be moved to write the number out in its long form. For example:

$$2.13 \times 10^5 = 213,000$$

or

$$2.13 \times 10^{-5} = 0.0000213$$

Further examples showing numbers that are powers of 10 are:

$$
\begin{aligned}
1,000,000,000 &= 1.0 \times 10^9 \quad \text{or } 10^9 \\
1,000,000 &= 1.0 \times 10^6 \quad \text{or } 10^6 \\
1,000 &= 1.0 \times 10^3 \quad \text{or } 10^3 \\
100 &= 1.0 \times 10^2 \quad \text{or } 10^2 \\
10 &= 1.0 \times 10^1 \quad \text{or } 10^1 \\
1 &= 1.0 \times 10^0 \quad \text{or } 10^0 \\
0.1 &= 1.0 \times 10^{-1} \text{ or } 10^{-1} \\
0.01 &= 1.0 \times 10^{-2} \text{ or } 10^{-2} \\
0.001 &= 1.0 \times 10^{-3} \text{ or } 10^{-3} \\
0.000001 &= 1.0 \times 10^{-6} \text{ or } 10^{-6} \\
0.000000001 &= 1.0 \times 10^{-9} \text{ or } 10^{-9}
\end{aligned}
$$

To add or subtract numbers written as powers of 10, they must be converted to the same power:

Addition

$$
\begin{array}{r}
2.1 \times 10^3 \\
+1.0 \times 10^5
\end{array}
=
\begin{array}{r}
0.021 \times 10^5 \\
+1.000 \times 10^5 \\
\hline
1.021 \times 10^5
\end{array}
$$

Subtraction

$$
\begin{array}{r}
3.4 \times 10^4 \\
-2.0 \times 10^3
\end{array}
=
\begin{array}{r}
3.4 \times 10^4 \\
-0.2 \times 10^4 \\
\hline
3.2 \times 10^4
\end{array}
$$

To multiply or divide numbers written as powers of 10, the exponents are added or subtracted:

Multiplication

$$
\begin{array}{r}
6.04 \times 10^2 \\
\times \; 2.1 \times 10^4 \\
\hline
12.684 \times 10^6
\end{array}
$$

Division

$$
\begin{array}{r}
3.0 \times 10^3 \\
\div 1.5 \times 10^2 \\
\hline
2.0 \times 10^1
\end{array}
$$

Appendix II

The Metric System and Conversion Factors

Units of Length
1 micrometer (μm) = 10^{-6} m = 0.0000394 in.
1 millimeter (mm) = 10^{-3} m = 0.0394 in. = 10^3 μm
1 centimeter (cm) = 10^{-2} m = 0.394 in. = 10^4 μm
1 meter (m) = 10^2 cm = 39.4 in. = 3.28 ft = 1.09 yd = 0.547 fath
1 kilometer (km) = 10^3 m = 0.621 statute mi = 0.540 nautical mi

Units of Volume
1 liter (l) = 10^3 cm^3 = 1.0567 liquid qt = 0.264 U.S. gal
1 cubic meter (m^3) = 10^6 cm^3 = 10^3 l = 35.32 ft^3 = 264 U.S. gal
1 cubic kilometer (km^3) = 10^9 m^3 = 10^{15} cm^3 = 0.24 statute mi^3 = 0.157 nautical mi^3

Units of Area
1 square centimeter (cm^2) = 0.151 in.^{2}
1 square meter (m^2) = 10.76 ft^2
1 square kilometer (km^2) = 0.292 nautical mi^2 = 0.386 statute mi^2

Units of Time
1 hour = 60 min = 3600 s
1 day = 8.64 $\times$ 10^4 s (mean solar day) = 24 h
1 year (y) = 8765.8 h = 3.156 $\times$ 10^7 s = 365.25 solar days

Units of Mass
1 gram (g) = 0.035 oz
1 kilogram (kg) = 10^3 g = 2.205 lb
1 metric ton (MT) or tonne (t) = 10^6 g = 2205 lb = 1.10 U.S. short tons

Units of Speed
1 centimeter per second (cm/s) = 0.0328 ft/s
1 meter per second (m/s) = 2.24 statute mi/h = 1.94 kt = 3.281 ft/s = 3.60 km/h
1 kilometer per hour (km/h) = 27.8 cm/s = 0.55 kt = 0.909 ft/s
1 knot (1 nautical mile per hour, kt) = 1.15 statute mi/h = 0.51 m/s

Units of Temperature

Celsius (°C)	Fahrenheit (°F)	Kelvin (K)	
−237.2	−459.7	0	Absolute zero (lowest possible temperature)
0	32	273.2	Freezing point of water
100	212	373.2	Boiling point of water

Conversions: °F = (1.8 $\times$ °C) + 32   °C = (°F − 32)/1.8   K = °C + 273.2

Appendix III

The Geologic Time Table

Era	Period	Epoch	Age (Ma)	Development of Life	Orogenic Events
Cenozoic	Quaternary	Recent			Pacific Coast Ranges form
		Pleistocene	1.6 —		
	Tertiary	Pliocene	5.3 —		
		Miocene	23.7 —		
		Oligocene	36.6 —		
		Eocene	57.8 —		
		Paleocene	66.4 —	First primates	Rocky Mts. form
Mesozoic	Cretaceous		144 —		
	Jurassic		208 —	First birds	Sierra Nevada form
	Triassic		245 —	First mammals First flowering plants	
Paleozoic	Permian		286 —	First reptiles	
	Pennsylvanian		320 —	First insects	Appalachian Mts. form
	Mississippian		360 —		
	Devonian		408 —	First amphibians	
	Silurian		438 —	First land plants	
	Ordovician		505 —	First vertebrates (fish)	
	Cambrian		570 —	First abundant animal fossils	
Precambrian			4500 —	First algae, bacteria	

Appendix IV Periodic Table of the Elements

Legend

```
      1         Atomic number
      H         Symbol of element
   1.0080       Atomic weight
  Hydrogen      Name of element
```

- ■ Inert gas
- ▓ Gas
- □ Liquid
- Solid — all others

Light Metals · Transitional Elements · Heavy Metals · Nonmetals

Period	I A	II A	III B	IV B	V B	VI B	VII B	VIII B			I B	II B	III A	IV A	V A	VI A	VII A	VIII A
1	1 **H** 1.0080 Hydrogen																	2 **He** 4.003 Helium
2	3 **Li** 6.939 Lithium	4 **Be** 9.012 Beryllium											5 **B** 10.81 Boron	6 **C** 12.011 Carbon	7 **N** 14.007 Nitrogen	8 **O** 15.9994 Oxygen	9 **F** 18.998 Fluorine	10 **Ne** 20.183 Neon
3	11 **Na** 22.990 Sodium	12 **Mg** 24.31 Magnesium											13 **Al** 26.98 Aluminum	14 **Si** 28.09 Silicon	15 **P** 30.974 Phosphorus	16 **S** 32.064 Sulfur	17 **Cl** 35.453 Chlorine	18 **Ar** 39.948 Argon
4	19 **K** 39.102 Potassium	20 **Ca** 40.08 Calcium	21 **Sc** 44.96 Scandium	22 **Ti** 47.90 Titanium	23 **V** 50.94 Vanadium	24 **Cr** 52.00 Chromium	25 **Mn** 54.94 Manganese	26 **Fe** 55.85 Iron	27 **Co** 58.93 Cobalt	28 **Ni** 58.71 Nickel	29 **Cu** 63.54 Copper	30 **Zn** 65.37 Zinc	31 **Ga** 69.72 Gallium	32 **Ge** 72.59 Germanium	33 **As** 74.92 Arsenic	34 **Se** 78.96 Selenium	35 **Br** 79.909 Bromine	36 **Kr** 83.80 Krypton
5	37 **Rb** 85.47 Rubidium	38 **Sr** 87.62 Strontium	39 **Y** 88.91 Yttrium	40 **Zr** 91.22 Zirconium	41 **Nb** 92.91 Niobium	42 **Mo** 95.94 Molybdenum	43 **Tc** (99) Technetium	44 **Ru** 101.1 Ruthenium	45 **Rh** 102.90 Rhodium	46 **Pd** 106.4 Palladium	47 **Ag** 107.870 Silver	48 **Cd** 112.40 Cadmium	49 **In** 114.82 Indium	50 **Sn** 118.69 Tin	51 **Sb** 121.75 Antimony	52 **Te** 127.60 Tellurium	53 **I** 126.90 Iodine	54 **Xe** 131.30 Xenon
6	55 **Cs** 132.91 Cesium	56 **Ba** 137.34 Barium	57 to 71	72 **Hf** 178.49 Hafnium	73 **Ta** 180.95 Tantalum	74 **W** 183.85 Tungsten	75 **Re** 186.2 Rhenium	76 **Os** 190.2 Osmium	77 **Ir** 192.2 Iridium	78 **Pt** 195.09 Platinum	79 **Au** 197.0 Gold	80 **Hg** 200.59 Mercury	81 **Tl** 204.37 Thallium	82 **Pb** 207.19 Lead	83 **Bi** 208.98 Bismuth	84 **Po** (210) Polonium	85 **At** (210) Astatine	86 **Rn** (222) Radon
7	87 **Fr** (223) Francium	88 **Ra** 226.05 Radium	89 to 103															

Lanthanide series

57 **La** 138.91 Lanthanum	58 **Ce** 140.12 Cerium	59 **Pr** 140.91 Praseodymium	60 **Nd** 144.24 Neodymium	61 **Pm** (147) Promethium	62 **Sm** 150.35 Samarium	63 **Eu** 151.96 Europium	64 **Gd** 157.25 Gadolinium	65 **Tb** 158.92 Terbium	66 **Dy** 162.50 Dysprosium	67 **Ho** 164.93 Holmium	68 **Er** 167.26 Erbium	69 **Tm** 168.93 Thulium	70 **Yb** 173.04 Ytterbium	71 **Lu** 174.97 Lutetium

Actinide series

89 **Ac** (227) Actinium	90 **Th** 232.04 Thorium	91 **Pa** (231) Protactinium	92 **U** 238.03 Uranium	93 **Np** (237) Neptunium	94 **Pu** (242) Plutonium	95 **Am** (243) Americium	96 **Cm** (247) Curium	97 **Bk** (249) Berkelium	98 **Cf** (251) Californium	99 **Es** (254) Einsteinium	100 **Fm** (253) Fermium	101 **Md** (256) Mendelevium	102 **No** (254) Nobelium	103 **Lw** (257) Lawrencium

Appendix V The Phylogenetic Tree

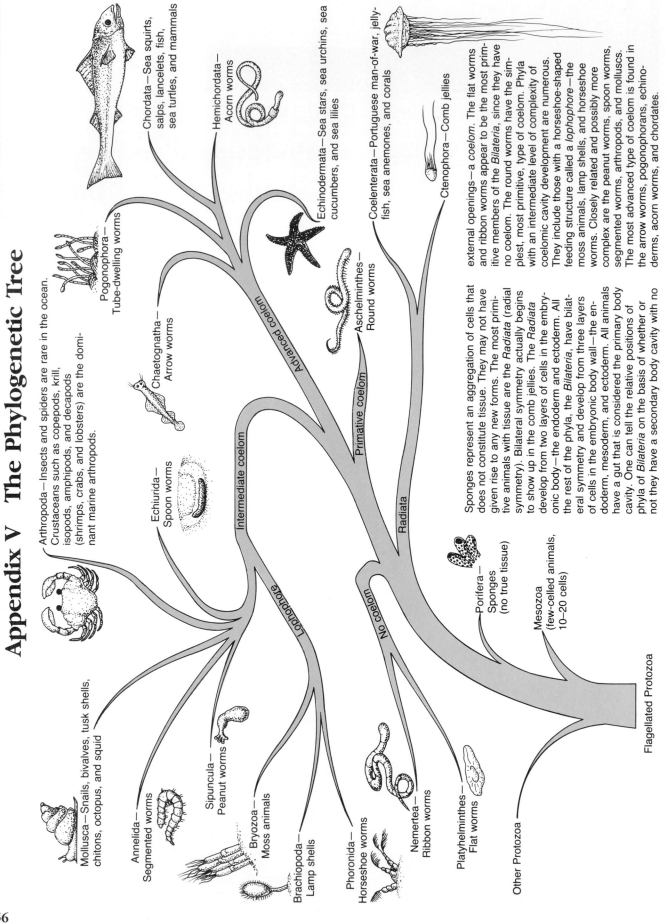

Arthropoda—Insects and spiders are rare in the ocean. Crustaceans such as copepods, krill, isopods, amphipods, and decapods (shrimps, crabs, and lobsters) are the dominant marine arthropods.

Pogonophora—Tube-dwelling worms

Chaetognatha—Arrow worms

Echiurida—Spoon worms

Intermediate coelom

Lophophore

Mollusca—Snails, bivalves, tusk shells, chitons, octopus, and squid

Annelida—Segmented worms

Sipuncula—Peanut worms

Bryozoa—Moss animals

Brachiopoda—Lamp shells

Phoronida—Horseshoe worms

Nemertea—Ribbon worms

Platyhelminthes—Flat worms

Other Protozoa

Chordata—Sea squirts, salps, lancelets, fish, sea turtles, and mammals

Hemichordata—Acorn worms

Echinodermata—Sea stars, sea urchins, sea cucumbers, and sea lilies

Coelenterata—Portuguese man-of-war, jelly-fish, sea anemones, and corals

Ctenophora—Comb jellies

Advanced coelom

Primitive coelom

Radiata

Aschelminthes—Round worms

No coelom

Porifera—Sponges (no true tissue)

Mesozoa (few-celled animals, 10–20 cells)

Flagellated Protozoa

Sponges represent an aggregation of cells that does not constitute tissue. They may not have given rise to any new forms. The most primitive animals with tissue are the *Radiata* (radial symmetry). Bilateral symmetry actually begins to show up in the comb jellies. The *Radiata* develop from two layers of cells in the embryonic body—the endoderm and ectoderm. All the rest of the phyla, the *Bilateria*, have bilateral symmetry and develop from three layers of cells in the embryonic body wall—the endoderm, mesoderm, and ectoderm. All animals have a gut that is considered the primary body cavity. One can tell the relative positions of phyla of *Bilateria* on the basis of whether or not they have a secondary body cavity with no external openings—a *coelom*. The flat worms and ribbon worms appear to be the most primitive members of the *Bilateria*, since they have no coelom. The round worms have the simplest, most primitive, type of coelom. Phyla with an intermediate level of complexity of coelomic cavity development are numerous. They include those with a horseshoe-shaped feeding structure called a *lophophore*—the moss animals, lamp shells, and horseshoe worms. Closely related and possibly more complex are the peanut worms, spoon worms, segmented worms, arthropods, and molluscs. The most advanced type of coelom is found in the arrow worms, pogonophorans, echinoderms, acorn worms, and chordates.

356

Appendix VI

Taxonomic Classification of Common Marine Organisms

Kingdom Monera

Organisms without nuclear membranes; nuclear material is spread throughout cell; predominantly unicellular.

Phylum Schizophyta Smallest known cells; bacteria (1500 species).

Phylum Cyanophyta Blue-green algae; chlorophyll *a*, carotene and phycobilin pigments (200 species).

Kingdom Protista

Organisms with nuclear material confined to nucleus by a membrane.

Phylum Chrysophyta Golden-brown algae; includes diatoms, coccolithophores, and silicoflagellates; chlorophyll *a* and *c*, xanthophyll, and carotene pigments (6000⁺ species).

Phylum Pyrrophyta Dinoflagellate algae; chlorophyll *a* and *c*, xanthophyll, and carotene pigments (1100 species).

Phylum Chlorophyta Green algae; chlorophyll *a* and *b* and carotene pigments (7000 species).

Phylum Phaeophyta Brown algae; chlorophyll *a* and *c*, xanthophyll, and carotene pigments (1500 species).

Phylum Rhodophyta Red algae; chlorophyll *a*, carotene, and phycobilin pigments (4000 species).

Phylum Protozoa Nonphotosynthetic, heterotrophic protists (27,400 species).
Class Mastigophora Flagelleted; dinoflagellates (5200 species).
Class Sarcodina Ameboid; foraminiferans and radiolarians (11,500 species).
Class Ciliophora Ciliated (6000 species).

Kingdom Fungi

Phylum Mycophyta Fungi, lichens; most fungi are decomposers found on the ocean floor, whereas lichens inhabit the upper intertidal zones (3000 species of fungi, 160 species of lichen).

Kingdom Metaphyta

Multicellular, complex plants.

Phylum Tracheophyta Vascular plants with roots, stems, and leaves that are serviced by special cells that carry food and fluids (287,200 species).
Class Angiospermae Flowering plants with seeds contained in a closed vessel (275,000 species).

Kingdom Metazoa

Multicellular animals.

Phylum Porifera Sponges; spicules are the only hard parts in these sessile animals that do not possess tissue (10,000 species).
Class Calcarea Calcium carbonate spicules (50 species).
Class Desmospongiae Skeleton may be composed of siliceous spicules or spongin fibers or be nonexistent (9500 species).
Class Sclerospongiae Coralline sponges; massive skeleton composed of calcium carbonate, siliceous spicules, and organic fibers (7 species).
Class Hexactinellida Glass sponges; six-rayed with siliceous spicules (450 species).

Phylum Cnidaria (Coelenterata) Radially symmetrical, two-cell, layered body wall with one opening to gut cavity; polyp (asexual, sexual, benthic) and medusa (sexual, pelagic) body forms (10,000 species).
Class Hydrozoa Polypoid colonies such as pelagic Portuguese man-of-war and benthic Obelia common; me-

dusa present in reproductive cycle but reduced in size (3000 species).

Class Scyphozoa Jellyfish; medusa up to 1 m in diameter is dominant form; polyp is small if present (250 species).

Class Anthozoa Corals and anemones possessing only polypoid body form and reproducing asexually and sexually (6500 species).

Phylum Ctenophora Predominantly planktonic comb jellies; basic eight-sided radial symmetry modified by secondary bilateral symmetry (80 species).

Phylum Platyhelminthes Flatworms; bilateral symmetry; hermaphroditic (25,000 species).

Phylum Nemertea Ribbon worms; as long as 30 m; benthic and pelagic (800 species).

Phylum Nematoda Roundworms; marine forms are primarily free-living and benthic; most 1 to 3 mm in length (5000 marine species).

Phylum Rotifera Ciliated, unsegmented forms less than 2 mm in length (1500 species, only a few marine).

Phylum Bryozoa (Ectoprocta) Moss animals; benthic, branching or encrusting colonies; lophophore feeding structure (4500 species).

Phylum Branchiopoda Lamp shells; lophophorate benthic bivalves (300 species).

Phylum Phoronida Horseshoe worms; 24-cm-long lophorate tube worms that live in sediment of shallow and temperate shallow waters (15 species).

Phylum Sipuncula Peanut worms, benthic (325 species).

Phylum Echiura Spoon worms; sausage shaped with spoon-shaped proboscis; burrow in sediment or live under rocks (130 species).

Phylum Pogonophora Tube-dwelling, gutless worms 5 to 80 cm long; absorb organic matter through body wall (100 species).

Phylum Tardigrada Marine meiofauna that have the ability to survive long periods in a cryptobiotic state (diversity is poorly known).

Phylum Mollusca Soft bodies possessing a muscular foot and mantle that usually secretes calcium carbonate shell (75,000 species).

Class Monoplacophora Rare trench-dwelling forms with segmented bodies and limpetlike shells (10 species).

Class Polyplacophora Chitons; oval, flattened body covered by eight overlapping plates (600 species).

Class Gastropoda Large, diverse group of snails and their relatives; shell spiral if present (64,500 species).

Class Bivalvia Bivalves; includes mostly filter-feeding clams, mussels, oysters, and scallops (7500 species).

Class Aplacophora Tusk shells; sand-burrowing organisms that feed on small animals living in sand deposits (350 species).

Class Cephalopoda Octopus, squid, and cuttlefish which possess no external shell except in the genus *Nautilus* (600 species).

Phylum Annelida Segmented worms in which musculature, circulatory, nervous, excretory, and reproductive systems may be repeated in many segments; mostly benthic (10,000 marine species).

Phylum Arthropoda Jointed-legged animals with segmented body covered by an exoskeleton (30,000 marine species).

Subphylum Crustacea Calcareous exoskeleton, two pairs of antennae; cephlon, thorax, and abdomen body parts; includes copepods, ostracods, barnacles, shrimp, lobsters, and crabs (26,000 species).

Subphylum Chelicerata Horseshoe crabs (4 species).
Class Merostomata
Class Pycnogonide Sea spiders.

Subphylum Uniramia Insects; genus *Halobites* is the only truly marine insect.

Phylum Chaetognatha Arrowworms; mostly planktonic, transparent and slender; up to 10 cm long (50 species).

Phylum Echinodermata Spiny-skinned animals; benthic animals with secondary radial symmetry and water vascular system (6000 species).

Class Asteroidea Starfishes; free-living, flattened body with five or more rays with tube feet used for locomotion; mouth down (1600 species).

Class Ophiuroidea Brittle stars and basket stars; prominent central disc with slender rays; tube feet used for feeding; mouth down (200 species).

Class Echinoidea Sea urchins, sand dollars, heart urchins; free-living forms without rays; calcium carbonate test; mouth down or forward (860 species).

Class Holothuroidea Sea cucumbers; soft bodies with radial symmetry obscured; mouth forward (900 species).
Class Crinoidea Sea lillies, feather stars; cup-shaped body attached to bottom by a jointed stalk or appendages; mouth up (630 species).

Phylum Hemichordata Acorn worms and pterobranchs; primitive nerve chord; gill slits; benthic (90 species).

Phylum Chordata Notochord; dorsal nerve chord and gills or gill slits (55,000 species).

Subphylum Urochordata Tunicates; chordate characteristics in larval stage only; benthic sea squirts and planktonic thaliaceans and larvaceans (1375 species).

Subphylum Cephalochordata Amphioxus or lancelets; live in coarse temperate and tropical sediment (25 species).

Subphylum Vertebrata Internal skeleton; spinal column of vertebrae; brain (52,000 species).

Class Agnatha Lampreys and hagfishes; most primitive vertebrates with cartilaginous skeleton, no jaws, and no scales (50 species).
Class Chondrichthyes Sharks, skates, and rays; cartilaginous skeleton; 5 to 7 gill openings; placoid scales (625 species).
Class Osteichthyes Bony fishes; cycloid scales; covered gill opening; swim bladder common (30,000 species).
Class Amphibia Frogs, toads, and salamanders; Asian mud flat frogs are the only amphibians that tolerate marine water (2600 species).
Class Reptilia Snakes, turtles, lizards, and alligators; orders Squamata (snakes) and Chelonia (turtles) are major marine groups (6500 species).
Class Aves Birds; many live on and in the ocean but all must return to land to breed (8600 species).
Class Mammalia Warm-blooded; hair; mammary glands; bear live young; marine representatives found in the orders Sirenia (sea cows, dugong), Cetacea (whales), and Carnivora (sea otter, pinnipeds) (4100 species).

Appendix VII

Latitude and Longitude Determination

FIGURE 1

Determining Latitude. The method of determining latitude used by Pytheas was to measure the angle between the horizon and the North Star, a star that is directly above the North Pole. Latitude north of the equator is the angle between the two sightings. The present North Star is Polaris. Similar determinations may be made in the Southern Hemisphere by using the Southern Cross as Polaris is used in the Northern Hemisphere.

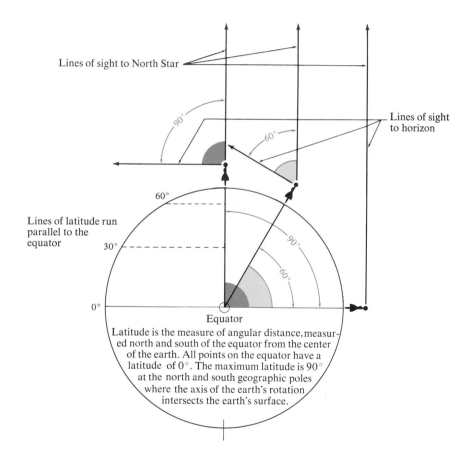

Lines of sight to North Star

Lines of sight to horizon

Lines of latitude run parallel to the equator

60°

30°

0°

90°

60°

90°

60°

Equator

Latitude is the measure of angular distance, measured north and south of the equator from the center of the earth. All points on the equator have a latitude of 0°. The maximum latitude is 90° at the north and south geographic poles where the axis of the earth's rotation intersects the earth's surface.

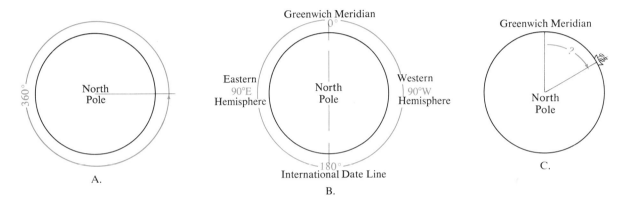

FIGURE 2
Determining Longitude. View of Earth from outer space, looking down on North Pole. *A*, as Earth turns on its rotational axis, it moves through 360° of angle per 24 h. *B*, a prime meridian that runs through Greenwich, England, was selected as the reference meridian, and Earth was divided into a Western and Eastern hemisphere. After John Harrison's chronometer was developed, many ships carried it showing the time on the Greenwich Meridian—Greenwich time. *C*, since Earth rotates through 15° of angle, longitude, per h (360° ÷ 24 h = 15°/h), a ship's captain could easily determine his longitude each day at noon. For example, a ship sets sail west across the Atlantic Ocean, checking its longitude each day at noon (when the sun crosses the meridian running directly overhead). One day when the sun is at the noon position, the captain checks the chronometer. It reads 16:18 h (0:00 is midnight and 12:00 is noon). What is the ship's longitude? Longitude solution:

16:18 h = 4:18 p.m.
Earth rotates through ¼° (15′) of angle per minute of time.
(One degree of arc is divided into 60 min)
4 h × 15°/h = 60° of longitude
18 minutes of time × 15 minutes of angle/minute of time = 270 minutes of arc
270 min ÷ 60 min/degree (°) = 4.5° of longitude
60° + 4.5° = 64.5° W longitude

Appendix VIII

Roots, Prefixes, and Suffixes*

a- not, without
ab-, abs- off, away, from
abysso- deep
acanth- spine
acro- the top
aero- air, atmosphere
aigial- beach, shore
albi- white
alga- seaweed
alti- high, tall
alve- cavity, pit
amoebe- change
an- without, not
annel- ring
annu- year
anomal- irregular, uneven
antho- flower
apex- tip
aplysio- sponge, filthiness
aqua- water
arachno- spider
arena- sand
arthor- joint
asthen- weak, feeble
astro- star
auto- self
avi- bird
bacterio- bacteria
balaeno- whale
balano- acorn
barnaco- goose
batho- deep
bentho- the deep sea
bio- living
blast- a germ
botryo- bunch of grapes
brachio- arm
branchio- gill
broncho- windpipe
bryo- moss
bysso- a fine thread

calci- limestone ($CaCO_3$)
calori- heat
capill- hair
cara- head
carno- flesh
cartilagi- gristle
caryo- nucleus
cat- down, downward
cen-, ceno- recent
cephalo- head
ceta- whale
chaeto- bristle
chiton- tunic
chlor- green
choano- funnel, collar
chondri- cartilage
cilio- small hair
circa- about
cirri- hair
clino- slope
cnido- nettle
cocco- berry
coelo- hollow
cope- oar
crusta- rind
cteno- comb
cyano- dark blue
cypri- Venus, lovely
-cyst bladder, bag
-cyte cell
deca- ten
delphi- dolphin
di- two, double
diem- day
dino- whirling
diplo- double, two
dolio- barrel
dors- back
-dram- run, a race
echino- spiny
eco- house, abode

ecto- outside, outer
edrio- a seat
en- in, into
endo- innter, within
entero- gut
epi- upon, above
estuar- the sea
eu- good, well
exo- out, without
fauna- animal
fec- dregs
fecund- fruitful
flacci- flake
flagell- whip
flora- flower
fluvi- river
fossili- dug up
fuco- red
geno- birth, race
geo- Earth
giga- very large
globo- ball, globe
gnatho- jaw
guano- dung
gymno- naked, bare
halo- salt
haplo- single
helio- the sun
helminth- worm
hemi- half
herbi- plant
herpeto- creeping
hetero- different
hexa- six
holo- whole
homo-, homeo- alike
hydro- water
hygro- wet
hyper- over, above, excess
hypo- under, beneath
ichthyo- fish

*Source: D. J. Borror, 1960. *Dictionary of Word Roots and Combining Forms*. Palo Alto, Calif.: National Press Books.

-idae members of the animal family of
infra- below, beneath
insecti- cut into
insula- an island
inter- between
involute- intricate
iso- equal
-ite rock
juven- young
juxta- near to
kera- horn
kilo- one thousand
lacto- milk
lamino- layer
larvi- ghost
latent hidden
latero- side
lati- broad, wide
lemni- water plant
limno- marshy lake
lipo- fat
litho- stone
litorial near the seashore
lopho- tuft
lorica- armor
luci- light
luna moon
lux light
macro- large
mala- jaw, cheek
mamilla- teat
mandibulo- jaw
mantle- cloak
mari- the sea
masti- chewing
mastigo- whip
masto- breast, nipple
maxillo- jaw
madi- middle
medus- a jellyfish
mega- great, large
meio- less
meridio- noon
meros- part
meso- middle
meta- after
meteor- in the sky
-meter measure
-metry science of measuring
mid- middle
milli- thousandth

mio- less
moll- soft
mono- one, single
-morph form
myo- muscle
myst- mustache
nano- dwarf
necto- swimming
nemato- thread
neo- new
neph- cloud
nerito- sea nymph
noct- night
-nomy the science of
nucleo- nucleus
nutri- nourishing
o-, oo- egg
ob- reversed
ocellus- little eye
octa- eight
oculo- eye
odonto- teeth
oiko- house, dwelling
oligo- few, scant
-ology science of
omni- all
ophi- a serpent
opto- the eye, vision
orni- a bird
-osis condition
oto- hear
ovo- egg
pan- all
para- beside, near
pari- equal
pecti- comb
pedi- foot
penta- five
peri- all around
phaeo- dusky
pholado- lurking in a hole
-phore carrier of
photo- light
phyto- plant
pinnati- feather
pisci- fish
plani- flat, level
plankto- wandering
pleisto- most
pleuro- side
plio- more

pluri- several
pneuma- air, breath
pod- foot
poikilo- variegated
poly- many
polyp many footed
poro- channel
post- behind, after
-pous foot
pre- before
pro- before, forward
procto- anus
proto- first
pseudo- false
ptero- wing
pulmo- lung
pycno- dense
quadra- four
quasi- almost
radi- radial
rhizo- root
rhodo- rose-colored
sali- salt
schizo- split, division
scyphi- cup
semi- half
septi- partition
sessil- sedentary
siphono- tube
-sis process
spiro- spiral, coil
stoma- mouth
strati- layer
sub- below
supra- above
symbio- living together
taxo- arrangement
tecto- covering
terra- earth
terti- third
thalasso- the sea
trocho- wheel
tropho- nourishment
tunic- cloak, covering
turbi- disturbed
un- not
vel- veil
ventro- underside
xantho- yellow
xipho- sword
zoo- animal

Glossary

Abiotic environment The nonliving components of an ecosystem.

Absolute dating The use of radioisotope half-lives to determine the age of rock units in years within 2 or 3 percent.

Abyssal clay Deep-ocean (oceanic) deposits containing less than 30 percent biogenous sediment.

Abyssal hill Volcanic peaks rising less than 1 km (0.621 mi) above the ocean floor.

Abyssal hill province Deep-ocean regions, particularly in the Pacific Ocean, where oceanic sedimentation rates are so low that abyssal plains do not form and the ocean floor is covered with abyssal hills.

Abyssal plain A flat depositional surface extending seaward from the continental rise or oceanic trenches.

Abyssal zone The benthic environment between 4000 and 6000 m (13,120 and 19,680 ft).

Abyssopelagic Open-ocean (oceanic) environment below 4000 m depth.

Adiabatic Pertaining to a change in the temperature of a mass resulting from compression or expansion. It requires no addition of heat to or loss of heat from the substance.

Agulhas Current A warm current that carries Indian Ocean water around the southern tip of Africa and into the Atlantic Ocean.

Algae One-celled or many-celled plants that have no root, stem, or leaf systems. Simple plants.

Amino acid One of more than 20 naturally occurring compounds that contain NH_2 and COOH groups. They combine to form proteins.

Amphidromic point A nodal or no-tide point in the ocean or sea around which the crest of the tide wave rotates during one tidal period.

Amphineura Class of mollusks with eight dorsal calcareous plates. Chitons.

Amphipoda Crustacean order containing laterally compressed members such as the "sand hoppers."

Anadromous Pertaining to a species of fish that spawns in fresh water, then migrates into the ocean to grow to maturity.

Anaerobic respiration Respiration carried on in the absence of free oxygen (O_2). Some bacteria and protozoans carry on respiration this way.

Annelida Phylum of elongated segmented worms.

Anomalistic month The time required for the moon to go from perigee to perigee, 27.5 days.

Anoxic Without oxygen.

Antarctic bottom water A water mass that forms in the Weddell Sea, sinks to the ocean floor, and spreads across the bottom of all oceans.

Antarctic Circle The latitude of 66.5° south.

Antarctic Circumpolar Current The eastward-flowing current that encircles Antarctica and extends from the surface to the deep-ocean floor. The largest volume current in the oceans.

Antarctic Convergence The zone of convergence along the northern boundary of the Antarctic Circumpolar Current where the southward-flowing boundary currents of the subtropical gyres converge on the cold Antarctic waters.

Antarctic Divergence The zone of divergence separating the westward-flowing East Wind Drift and the easterly-flowing West Wind Drift.

Antarctic Intermediate Water Antarctic zone surface water that sinks at the Antarctic convergence and flows north at a depth of about 900 m beneath the warmer upper-water mass of the South Atlantic subtropical gyre.

Antilles Current This warm current flows north seaward of the Lesser Antilles from the north equatorial current of the Atlantic Ocean to join the Florida Current.

Antinode Zone of maximum vertical particle movement in standing waves where crest and trough formation alternate.

Aphelion The point in the orbit of a planet or comet where it is farthest from the sun.

Aphotic zone Without light. The ocean is generally in this state below 1000 meters (3280 ft).

Apogee The point in the orbit of the moon or an artificial satellite that is farthest from Earth.

Aragonite A form of $CaCO_3$ that is less common and less stable than calcite. Pteropod shells are usually composed of aragonite.

Archipelagic apron A gently sloping sedimentary feature surrounding an oceanic island or seamount.

Arctic Circle The latitude of 66.5° north.

Arrowworm A member of the phylum Chaetognatha. They average about 1 cm in length and are an important member of the plankton.

Aschelminthes Phylum of wormlike pseudocoelomates.

Aspect ratio The index of propulsive efficiency obtained by dividing the square of fin height by fin area.

Assimilative capacity The level of concentration to which the ocean can accumulate foreign substances without harming marine life or humans that use the ocean or its resources.

Asthenosphere A plastic layer in the upper mantle 80–200 km deep which may allow lateral movement of lithospheric plates and isostatic adjustments.

Atlantic-type margin The passive trailing edge of a continent that is subsiding due to lithospheric cooling and increasing sediment load.

Atoll A ring-shaped coral reef growing upward from a submerged volcanic peak. It may have low-lying islands composed of coral debris.

Atom The smallest particle of an element that can combine with similar particles of other elements to produce compounds.

Autolytic decomposition Decomposition of organic matter that is achieved by enzymes present in the tissue. The enzymes are triggered to begin their work at death of the tissue.

Autotomy Pertaining to the ability of some organisms to slough off certain body parts as a defensive mechanism.

Autotroph Plants and bacteria that can synthesize organic compounds from inorganic nutrients.

Autumnal equinox The passage of the sun across the equator as it moves from the Northern Hemisphere into the Southern Hemisphere, September 23.

Backshore The inner portion of the shore, lying landward of the mean spring tide high water line. Acted upon by the ocean only during exceptionally high tides and storms.

Bacterioplankton Bacteria that live as plankton.

Bacteriovore Organisms that feed on bacteria.

Bar-built estuary A shallow estuary (lagoon) separated from the open ocean by a bar deposit such as a barrier island. The water in these estuaries usually exhibits vertical mixing.

Barnacle *See* Cirripedia.

Barrier flat Lying between the salt marsh and dunes of a barrier island, it is usually covered with grasses and even forests if protected from overwash for sufficient time.

Barrier island A long, narrow, wave-built island separated from the mainland by a lagoon.

Barrier reef A coral reef separated from the nearby landmass by open water.

Basalt A dark-colored volcanic rock characteristic of the ocean crust. Contains minerals with relatively high iron and magnesium content.

Bathyal zone The benthic environment between the depths of 200 and 4000 m (656 and 13,120 ft). It includes mainly the continental slope and the oceanic ridges and rises.

Bathymetry The study of ocean depth.

Bathypelagic zone The pelagic environment between the depths of 1000 and 4000 m (3,280 and 13,120 ft).

Bay barrier A marine deposit attached to the mainland at both ends and extending entirely across the mouth of a bay, separating the bay from the open water.

Beach Sediment seaward of the coastline through the surf zone that is in transport along the shore and within the surf zone.

Benguela Current The cold eastern boundary current of the South Atlantic subtropical gyre.

Benthic Pertaining to the ocean bottom.

Benthos The forms of marine life that live on the ocean bottom.

Benthic nepheloid layer (BNL) A layer of turbid water created by the resuspension of sediment from the ocean floor by bottom currents.

Bioerosion Erosion of reef or other solid bottom material by the activities of organisms.

Biogenous sediment Sediment containing material produced by plants or animals, e.g., coral reefs, shell fragments, and housing of diatoms, radiolarians, foraminifera, and coccolithophores.

Biogeochemical cycles The natural cycling of compounds among the living and nonliving components of an ecosystem.

Bioluminescence Light produced by chemical reaction. Found in bacteria, phytoplankton, and metazoans.

Biomass The total mass of a defined organism or group of organisms in a particular community or the ocean as a whole.

Biotic community The living organisms that inhabit an ecosystem.

Body wave A longitudinal or transverse wave which transmits energy through a body of matter.

Boiling point The temperature at which a substance changes state from a liquid to a gas at a given pressure.

Bore A steep-fronted tide crest that moves up a river in association with high tide.

Bosphorus A narrow strait between the Black Sea and the Sea of Marmara through which Mediterranean and Black Sea water may mix.

Boundary current The northward- or southward-flowing currents that form the western and eastern boundaries, respectively, of the subtropical circulation gyres.

Brazil Current The warm western boundary current of the South Atlantic subtropical gyre.

Breaker zone Region where waves break at the seaward margin of the surf zone.

Breakwater Any artificial structure constructed to protect a coastal region from the force of ocean waves.

Bryozoa Phylum of colonial animals that often share one coelomic cavity. Encrusting and branching forms secrete a protective housing (zooecium) of calcium carbonate or chitinous material. Possess lophophore feeding structure.

Buoyancy The ability or tendency to float or rise in a liquid.

Calcareous Containing calcium carbonate.

Calcite The most common form of $CaCO_3$.

Calcium carbonate, $CaCO_3$ A chalk-like substance secreted by many organisms in the form of coverings or skeletal structures.

Calorie Unit of heat defined as the amount of heat required to raise the temperature of 1 g of water 1°C.

Canary Current The cold eastern boundary current of the North Atlantic subtropical gyre.

Capillarity The action by which a fluid, such as water, is drawn up in small tubes as a result of surface tension.

Capillary wave Ocean waves whose wavelength is less than 1.74 cm. The dominant restoring force for such waves is surface tension.

Carapace Chitinous or calcareous shield that covers the cephalothorax of some crustaceans. Dorsal portion of a turtle shell.

Carbohydrate An organic compound containing the elements carbon, hydrogen, and oxygen with the general formula $(CH_2O)n$.

Carbonate compensation depth The depth at which carbonate particles falling from above are dissolved.

Caribbean Current The warm current that carries equatorial water across the Caribbean Sea into the Gulf of Mexico.

Carnivore An animal that depends on other animals solely or chiefly for its food supply.

Carotin A red to yellow pigment found in plants.

Catadromous Pertaining to a species of fish which spawns in the sea and then migrates into fresh water to grow to maturity.

Celsius temperature scale $0°C = 273.16$ K; $0°C = $ freezing point of water; $100°C = $ boiling point of water.

Centrifugal force A fictional force that seems to make an object move away from the center of a curved path it is following. It results from the application of a centripetal force that acts against the inertia of the object.

Centripetal force A center-seeking force that tends to make rotating bodies move toward the center of rotation.

Cephalopoda A class of the phylum Mollusca with a well-developed pair of eyes and a ring of tentacles surrounding the mouth. The shell is absent or internal on most members. The class includes the squid, octopus, and nautilus.

Cephalothorax The head and midbody region of crustaceans.

Cetacea An order of marine mammals that includes the whales.

Chaetognatha A phylum of elongate transparent wormlike pelagic animals commonly called *arrowworms*.

Chela Arthropod appendage modified to form a pincer.

Chemical energy A form of potential energy stored in the chemical bonds of compounds.

Chemosynthesis A process by which bacteria synthesize organic molecules from inorganic nutrients using chemical energy released from the bonds of some chemical compound by oxidation.

Chesapeake Bay The largest coastal plain estuary in the United States; created by the flooding of the river valleys of the Susquehanna River and its tributaries. It opens into the Atlantic Ocean at Norfolk, Va.

Chiton Common name for any member of the class of mollusks, Amphineura, with eight dorsal plates.

Chloride ion, Cl⁻ A chlorine atom that has become negatively charged by gaining one electron.

Chlorinity The amount of chloride ion and ions of other halogens in ocean water expressed in parts per thousand by weight, ‰.

Chlorophyll A group of green pigments that make it possible for plants to carry on photosynthesis.

Chlorophyta Green algae. Characterized by the presence of chlorophyll and other pigments.

Chrysophyta An important phylum of planktonic algae, including the diatoms. The presence of chlorophyll is masked by the pigment carotin that gives the plants a golden color.

Cilium Short hairlike structures common on lower animals. Beating in unison, they may create water currents that carry food toward the mouth of an animal or may be used for locomotion.

Circadian rhythm Behavioral and physiological rhythms of organisms related to the 24-hour day. Sleeping and waking patterns are an example.

Circumpolar Current Eastward-flowing current that extends from the surface to the ocean floor and encircles Antarctica.

Cirripedia An order of crustaceans with up to six pairs of thoracic appendages that strain food from the water. They are the barnacles that attach themselves to the substrate and secrete an external calcareous housing.

Clastic A rock or sediment composed of broken fragments of preexisting rocks. Two common examples are beach deposits and sandstone.

Clay A term relating to particle size between silt and colloid. Clay minerals are hydrous aluminum silicates with plastic, expansive, and cation exchange properties.

Cnidoblast Stinging cell of phylum Coelenterata that contains the stinging mechanism (nematocyst) used in defense and capturing prey.

Coast A strip of land that extends inland from the coastline as far as marine influence is evidenced in the landforms.

Coastal plain estuary An estuary formed by rising sea level flooding a coastal river valley.

Coastal upwelling The movement of deeper nutrient-rich water into the surface water mass as a result of windblown surface water moving offshore.

Coastline Landward limit of the effect of the highest storm waves on the shore.

Cobalt-rich manganese crust Hydrogenous deposits found on the flanks of volcanic islands and seamounts.

Coccolith Tiny calcareous discs averaging about 3 μm in diameter that form the cell wall of coccolithophores.

Coccolithophore A microscopic planktonic form of algae, encased by a covering composed of calcareous discs (coccoliths).

Coelenterata Phylum of radially symmetrical animals that includes two basic body forms, the medusa and the polyp. Includes jellyfish (medusoid) and sea anemones (polypoid).

Coelom Secondary body cavity (gut cavity is primary cavity). Forms within the mesoderm in higher animals, is lined with peritoneum, and contains vital organs.

Cold front A weather front in which a cold air mass moves into and under a warm air mass. It creates a narrow band of intense precipitation.

Colonial animals Animals that live in groups of attached or separate individuals. Groups of individuals may serve special functions.

Columbia River estuary An estuary at the border between the states of Washington and Oregon that has been most adversely affected by the construction of hydroelectric dams.

Comb jelly Common name for members of the phylum Ctenophora. (*See* Ctenophora.)

Commensalism A symbiotic relationship in which one party benefits and the other is unaffected.

Compensation depth, CaCO$_3$ The depth at which the amount of CaCO$_3$ produced by the organisms in the overlying water column is equal to the amount of CaCO$_3$ the water column can dissolve. There will be no CaCO$_3$ deposition below this depth.

Compensation depth, O$_2$ The depth where the oxygen produced by photosynthesis is equal to the oxygen requirements of plant respiration. A plant population cannot be sustained below this depth, which will be greater in the open ocean than near the shore due to the relatively deeper light penetration in the open ocean.

Compound A substance containing two or more elements combined in fixed proportions.

Condensation The conversion of water from the vapor to the liquid state. When it occurs, the energy required to vaporize the water is released into the atmosphere. This is about 585 cal/g of water at 20°C.

Conduction The transmission of heat by the passage of energy from particle to particle.

Conjunction The apparent closeness of two heavenly bodies. During the new moon phase, Earth and the moon are in conjunction on the same side of the sun.

Conservative property A property of ocean water that the water attains at the surface and is changed only by mixing and diffusion after the water sinks below the surface.

Constancy of composition, Rule of The major constituents of ocean-water salinity are found in the same relative concentrations throughout the ocean-water volume.

Constructive interference A form of wave interference in which two waves come together in phase, e.g., crest to crest, to produce a greater displacement from the still-water line than that produced by either of the waves alone.

Consumers The animal populations within an ecosystem that consume the organic mass produced by the producers.

Continent About one-third of Earth's surface that rises above the deep-ocean floor to be exposed above sea level. Continents are composed primarily of granite, an igneous rock of lower density than the basaltic oceanic crust.

Continental borderland A highly irregular portion of the continental margin that is submerged beneath the ocean and is characterized by depths greater than those characteristic of the continental shelf.

Continental drift A term applied to early theories supporting the possibility the continents are in motion over Earth's surface.

Continental rise A gently sloping depositional surface at the base of the continental slope.

Continental shelf A gently sloping depositional surface extending from the low water line to the depth of a marked increase in slope around the margin of a continent or island.

Continental slope A relatively steeply sloping surface lying seaward of the continental shelf.

Convection In a fluid being heated unevenly, the warmer part of the mass will rise and the cooler portions will sink. If the heat source is stationary, cells may develop as the rising warm water cools and sinks in regions on either side of the axis of rising.

Convergence There are polar, tropical, and subtropical regions of the oceans where water masses with different characteristics come together. Along these lines of convergence, the denser mass will sink beneath the others.

Convergent plate boundary A lithospheric plate boundary where adjacent plates converge, producing ocean trench-island arc systems, ocean trench-continental volcanic arcs, or folded mountain ranges.

Copepoda An order of microscopic to nearly microscopic crustaceans that are important members of the zooplankton in temperate and subpolar waters.

Coral A group of benthic anthozoans that exist as individuals or in colonies and secrete CaCO$_3$ external skeletons. Under the proper conditions corals may produce reefs composed of their external skeletons and the CaCO$_3$ material secreted by varieties of algae associated with the reefs.

Core The core of Earth is composed primarily of iron and nickel. It has a liquid outer portion 2270 km thick and a solid inner core with a radius of 1216 km.

Coriolis effect A small force resulting from Earth's rotation causes particles in motion to be deflected to the right in the Northern Hemisphere and to the left in the Southern Hemisphere.

Cosmogenous sediment All sediment derived from outer space.

Cotidal lines Lines connecting points where high tide occurs simultaneously.

Crest (wave) The portion of an ocean wave that is displaced above the still-water line.

Cromwell Current A ribbon-like eastward-flowing current embedded in the South Equatorial Current that flows from Samoa to the Galápagos Islands.

Cruiser Fish such as the bluefin tuna that constantly cruise the pelagic waters in search of food.

Crust Unit of Earth's structure that is composed of basaltic ocean crust and granitic continental crust. The total thickness of the crustal units may range from 5 km beneath the ocean to 50 km beneath the continents.

Crustacea A class of phylum Arthropoda that includes barnacles, copepods, lobsters, crabs, and shrimp.

Crystalline rock Igneous or metamorphic rocks. These rocks are made up of crystalline particles with orderly molecular structures.

Ctenophora A phylum of gelatinous organisms that are more or less spheroidal with biradial symmetry. These exclusively marine animals have eight rows of ciliated combs for locomotion and most have two tentacles for capturing prey.

Current A horizontal movement of water.

Cutaneous artery The artery that runs down both sides of some cruiser-type fish to help maintain a constant elevated temperature in the myomere musculature used for swimming.

Cutaneous vein The vein that runs down both sides of some cruiser-type fish to help maintain a constant elevated temperature in the myomere musculature used for swimming.

Cypris Advanced free-swimming larval stage of barnacles. After attaching to substrate, it metamorphoses into an adult.

Davidson Current A northward-flowing current along the Washington-Oregon coast that is driven by geostrophic effects on a large freshwater runoff.

DDT An insecticide that caused damage to marine bird populations in the 1950s and 60s. Its use is now banned throughout most of the world.

Decapoda 1. An order of crustaceans with five pairs of thoracic "walking legs." Includes crabs, shrimp, and lobsters. 2. Suborder of cephalopod mollusks with 10 arms that includes squids and cuttlefish.

Declination The angular distance of the sun or moon above or below the plane of Earth's equator.

Decomposers Primarily bacteria that break down nonliving organic material, extract some of the products of decomposition for their own needs, and make available the compounds needed for plant production.

Deep boundary current Relatively strong deep currents flowing across the continental rise along the western margin of ocean basins.

Deep scattering layer A layer of marine organisms in the open ocean that scatter signals from an echo sounder. It migrates daily from depths of slightly over 100 m at night to more than 800 m during the day.

Deep-sea fan A large fan-shaped deposit commonly found on the continental rise seaward of such sediment-laden rivers as the Amazon, Indus, or Ganges-Brahmaputra.

Deep-sea system Includes all benthic environments beneath the littoral (sublittoral, bathyal, abyssal, and hadal).

Deep water The water beneath the permanent thermocline (pycnocline) that has a uniformly low temperature.

Deep-water wave Ocean wave traveling in water that has a depth greater than one-half the average wave length. Its speed is independent of water depth.

Delta A low-lying deposit at the mouth of a river.

Denitrifying bacteria Bacteria that reduce oxides of nitrogen to produce free nitrogen (N_2).

Density Mass per unit volume of a substance. Usually expressed as *grams per cubic centimeter*. For ocean water with a salinity of 35‰ at 0°C, the density is 1.028 g/cm^3.

Density, In situ (σ or sigma) Density of ocean water *in place*.

Density, Potential ($\sigma\Theta$ or sigma theta) Density of ocean water with the adiabatic effect removed. It is always less than *in situ* density except at the surface where the adiabatic effect is zero.

Desalination The removal of salt ions from ocean water to produce pure water.

Destructive interference A form of wave interference in which two waves come together out of phase, e.g., crest to trough, and produce a wave with less displacement than the larger of the two waves would have produced alone.

Detritus Any loose material produced directly from rock disintegration. (Organic: material resulting from the disintegration of dead organic remains.)

Diatom Member of the class Bacillariophyceae of algae that possesses a wall of overlapping silica valves.

Diatomite A deposit composed primarily of the frustules of diatoms.

Dipolar Having two poles. The water molecule possesses a polarity of electrical charge with one pole being more positive and the other more negative in electrical charge.

Diffraction Any bending of a wave around an obstacle that cannot be interpreted as refraction or reflection.

Diffusion A process by which fluids move through other fluids from areas of high concentration to areas in which they are in lower concentrations by random molecular movement.

Dinoflagellates Single-celled microscopic organisms that may possess chlorophyll and belong to the plant phylum Pyrrophyta (autotrophic) or may ingest food and belong to the class Mastigophora of the animal phylum Protozoa (heterotrophic).

Discontinuity An abrupt change in a property such as temperature or salinity at a line or surface.

Disphotic zone The dimly lit zone, corresponding approximately to the mesopelagic, in which there is not enough light to carry on photosynthesis; sometimes called the twilight zone.

Dissolved oxygen Oxygen that is dissolved in ocean water.

Distillation A method of purifying liquids by heating them to their boiling point and condensing the vapor.

Distributary A small stream flowing away from a main stream. Such streams are characteristic of deltas.

Diurnal inequality The difference in the heights of two successive high waters or two successive low waters during a lunar (tidal) day.

Diurnal tide A tide with one high water and one low water during a tidal day.

Divergence A horizontal flow of water from a central region, as occurs in upwelling.

Divergent plate boundary A lithospheric plate boundary where adjacent plates diverge, producing an oceanic ridge or rise (spreading center).

Doldrums A belt of light variable winds within 10°–15° of the equator, resulting from the vertical flow of low density air within this equatorial belt.

Dolphin 1. A brilliantly colored fish of the genus *Coryphaena*. 2. The name applied to the small, beaked members of the cetacean family, Delphinidae.

Dorsal Pertaining to the back or upper surface of most animals.

Dorsal aorta For most fish, this is the only major artery that runs the length of the fish through openings in the vertebrae and supplies blood. Some pelagic cruisers also have a cutaneous artery.

Downwelling In the open or coastal ocean where Ekman transport causes surface waters to converge or impinge on the coast, surface water will be carried down beneath the surface.

Drifts Thick sediment deposits on the continental rise produced where the deep boundary current slows and loses sediment when it changes direction to follow the base of the continental slope.

Drowned beach An ancient beach now beneath the coastal ocean because of rising sea level.

Dunes Coastal deposits of sand lying landward of the beach and deriving their sand from onshore winds that transport beach sand inland.

Dynamical tide theory The theory of tidal behavior that takes into account friction between the ocean water and the ocean floor, the effects of changing depth of the ocean floor, and the interference of the continents on the passage of tidal waves.

Dynamic topography A surface configuration resulting from the geopotential difference between a given surface and a reference surface of no motion. A contour map of this surface is useful in estimating the nature of geostrophic currents.

Earthquake A sudden motion or trembling in the earth caused by the sudden release of slowly accumulated strain by faulting (movement along a fracture in Earth's crust) or volcanic activity.

East Pacific Rise A fast-spreading divergent plate boundary extending southward from the Gulf of California through the eastern South Pacific Ocean.

East Wind Drift The coastal current driven in a westerly direction by the polar easterly winds blowing off of Antarctica.

Eastern boundary current Equatorward-flowing cold drifts of water on the eastern side of all subtropical gyres.

Ebb current During a decrease in the height of the tide, the ebb current flows seaward.

Echinodermata Phylum of animals that have bilateral symmetry in larval forms and usually a five-sided radial symmetry as adults. Benthic and possessing rigid or articulating exoskeletons of calcium carbonate with spines, this phylum includes sea stars, brittle stars, sea urchins, sand dollars, sea cucumbers, and sea lilies.

Echo sounding Determining the depth of water by measuring the time required for a sonic or ultrasonic signal to travel to the bottom and back to the ship that emitted the signal.

Ecliptic The plane of the center of the Earth-moon system as it orbits around the sun.

Ecological efficiency Efficiency with which energy is transferred from one trophic level to the next. The ratio of the amount of protoplasm added to a trophic level to the amount of food required to produce it.

Ecosystem All the organisms in a biotic community and the abiotic environmental factors with which they interact.

Ectoderm Outermost layer of cells in an animal embryo. In vertebrates it gives rise to skin, nervous system, sense organs, etc.

Eddy A current of any fluid forming on the side of a main current. It usually moves in a circular path and develops where currents encounter obstacles or flow past one another.

Ekman spiral A theoretical consideration of the effect of a steady wind blowing over an ocean of unlimited depth and breadth and of uniform viscosity. The result is a surface flow at 45° to the right of the wind in the Northern Hemisphere. Water at increasing depth will drift in directions increasingly to the right until at about 100 m depth it is moving in a direction opposite to that of the wind. The net water transport is 90° to the wind, and speed decreases with depth.

Ekman transport The net transport of surface water set in motion by wind. Due to the Ekman spiral phenomenon, it is theoretically in a direction 90° to the right and 90° to the left of the wind direction in the Northern Hemisphere and Southern Hemisphere, respectively.

El Niño A southerly flowing warm current that generally develops off the coast of Ecuador shortly after Christmas. Occasionally it will move farther south into Peruvian coastal waters and cause the widespread death of plankton and fish.

El Niño-Southern Oscillation The correlation of El Niño events with an oscillatory pattern of pressure change in a persistent high pressure cell in the southeastern Pacific Ocean and a persistent low pressure cell over the East Indies.

Electrolysis A separation process by which salt ions are removed from salt water through water impermeable membranes toward oppositely charged electrodes.

Electromagnetic energy Energy that travels as waves or particles with the speed of light. Different kinds possess different properties based on wavelength. The longest wavelengths belong to radio waves, up to 100 km in length. At the other end of the spectrum are cosmic rays with greater penetrating power and wavelengths of less than 0.000001 μm.

Electromagnetic spectrum The spectrum of radiant energy emitted from stars and ranging between cosmic rays with wavelengths less than 10–11 cm and very long waves with wavelengths in excess of 100 km (62 mi).

Element One of a number of substances, each of which is composed entirely of like particles, atoms, that cannot be broken into smaller particles by chemical means.

Emergent shoreline A shoreline resulting from the emergence of the ocean floor relative to the ocean surface. It is usually rather straight and characterized by marine features usually found at some depth.

Endoderm Innermost cell layer of an embryo. Develops into digestive and excretory system and forms lining for respiratory system, etc., in vertebrates.

Endothermic reaction A chemical reaction that absorbs energy. For example, energy is stored in the organic products of the chemical reaction photosynthesis.

Entropy A thermodynamic quantity that reflects the degree of randomness in a system. It increases in all natural systems.

Environment The sum of all physical, chemical, and biological factors to which an organism or community is subjected.

Epicenter The point on Earth's surface that is directly above the focus of an earthquake.

Epifauna Animals that live on the ocean bottom, either attached or moving freely over it.

Epipelagic zone A subdivision of the oceanic province that extends from the surface to a depth of 200 m (656 ft).

Epitheca The top valve of a diatom frustule.

Equatorial countercurrent Eastward-flowing currents found between the north and south equatorial currents in all oceans, but particularly well developed in the Pacific Ocean.

Exclusive Economic Zone (EEZ) A coastal zone 200 nautical miles wide over which the coastal nation has jurisdiction over mineral resources, fishing, and pollution. If the continental shelf extends beyond 200 miles, the EEZ may be up to 350 miles in width.

Exothermic reaction A chemical reaction that liberates energy. For example, the energy stored in the products of photosynthesis is released by the chemical reaction respiration.

Extrusive rocks Igneous rocks that flow out onto Earth's surface before cooling and solidfying (lavas).

Fahrenheit temperature scale (°F) Freezing point of water is 32°; boiling point of water is 212°.

Falkland Current A northward-flowing cold current found off the southeastern coast of South America.

Fan A gently sloping, fan-shaped feature normally located near the lower end of a canyon.

Fast ice Sea ice that is attached to the shore and therefore remains stationary.

Fat An organic compound formed from alcohol glycerol and one or more fatty acids; a lipid, it is a solid at atmospheric temperatures.

Fathom A unit of depth in the ocean, commonly used in countries using the English system of units. It is equal to 1.83 m, or 6 ft.

Fault A fracture or fracture zone in Earth's crust along which displacement has occurred.

Fault block A crustal block bounded on at least two sides by faults. Usually elongate; if it is down-dropped, it produces a graben; if uplifted, it is a horst.

Fauna The animal life of any particular area or of any particular time.

Fecal pellet Exrement of planktonic crustaceans that assist in speeding up the descent rate of sedimentary particles by combining them into larger packages.

Ferromagnesian Minerals rich in iron and magnesium.

Fetch 1. Pertaining to the area of the open ocean over which the wind blows with constant speed and direction, thereby creating a wave system. 2. The distance across the fetch (wave-generating area) measured in a direction parallel to the direction of the wind.

Fishery assessment The conduct of research on the ecological and economic factors related to a fishery and the application of the knowledge gained to its regulation.

Fjord A long, narrow, deep, U-shaped inlet that usually represents the seaward end of a glacial valley that has become partially submerged after the melting of the glacier.

Flagellum A whiplike living process used by some cells for locomotion.

Floe A piece of floating ice other than fast ice or icebergs. May range in dimension from about 20 cm across to more than a kilometer.

Flood current A tidal current associated with increasing height of the tide, generally moving toward the shore.

Flora The plant life of any particular area or of any particular time.

Florida Current A warm current flowing north along the coast of Florida. It becomes the Gulf Stream.

Folded mountain range Mountain ranges formed as a result of the convergence of lithospheric plates. They are characterized by masses of folded sedimentary rocks that formed from sediments deposited in the ocean basin that was destroyed by the convergence.

Food chain The passage of energy materials from producers through a sequence of a herbivore and a number of carnivores.

Food web A group of interrelated food chains.

Foraminifera An order of planktonic and benthic protozoans that possess protective coverings, usually composed of calcium carbonate.

Forced wave A wave that is generated and maintained by a continuous force such as the gravitational attraction of the moon.

Foreshore The portion of the shore lying between the normal high and low water marks—the intertidal zone.

Fortnight Half a synodic month (29.5 days), or about 14.75 days. Normally used in reference to a period of time equal to two weeks. The time that elapses between new moon and full moon.

Fossil Any remains, print, or trace of an organism that has been preserved in Earth's crust.

Fracture zone An extensive linear zone of unusually irregular topography of the ocean floor, characterized by large seamounts, steep-sided or asymmetrical ridges, troughs, or long, steep slopes. Usually represents ancient, inactive transform fault zones.

Free wave A wave created by a sudden rather than a continuous impulse that continues to exist after the generating force is gone.

Freezing The process by which a liquid is converted to a solid at its freezing point.

Freezing point The temperature at which a liquid becomes a solid under any given set of conditions. The freezing point of water is 0°C under atmospheric pressure.

Fringing reef A reef that is directly attached to the shore of an island or continent. It may extend more than 1 km from shore. The outer margin is submerged and often consists of algal limestone, coral rock, and living coral.

Frustule The siliceous covering of a diatom, consisting of two halves (epitheca and hypotheca).

Fucoxanthin The reddish brown pigment that gives brown algae its characteristic color.

Full moon The phase of the moon that occurs when the sun and moon are in opposition, on opposite sides of Earth.

Fully developed sea The maximum average size of waves that can be developed for a given wind speed when it has blown in the same direction for a minimum duration over a minimum fetch.

Galápagos Rift A divergent plate boundary extending eastward from the Galápagos Islands toward South America. The first deep-sea hydrothermal vent biocommunity was discovered here in 1977.

Galaxy One of the billions of large systems of stars that make up the universe.

Gaseous state A state of matter in which molecules move by translation and only interact through chance collisions.

Gastropoda A class of mollusks, most of which possess an asymmetrical, spiral one-piece shell and a well-developed flattened foot. A well-developed head will usually have two eyes and one or two pairs of tentacles. Includes snails, limpets, abalone, cowries, sea hares, and sea slugs.

Geostrophic current A current that grows out of Earth's rotation and is the result of a near balance between gravitational force and the Coriolis effect.

Gill A thin-walled projection from some part of the external body or the digestive tract used for respiration in a water environment.

Glacial epoch The Pleistocene epoch, the earlier of two divisions of the Quaternary period of geologic time. During this time high latitude continental areas now free of ice were covered by continental glaciers.

Glacier A large mass of ice formed on land by the recrystallization of old compacted snow. It flows from an area of accumulation to an area of wasting where ice is removed from the glacier by melting.

Glauconite A group of green hydrogenous minerals consisting of hydrous silicates of potassium and iron.

Global plate tectonics The process by which lithospheric plates are moved across Earth's surface to collide, slide by one another, or diverge to produce the topographic configuration of Earth.

Gondwanaland A hypothetical protocontinent of the Southern Hemisphere named for the Gondwana region of India. It included the present continental masses Africa, Antarctica, Australia, India, and South America.

Graded bedding Stratification in which each layer displays a decrease in grain size from bottom to top.

Gradient The rate of increase or decrease of one quantity or characteristic relative to a unit change in another. For example, the slope of the ocean floor is a change in elevation (a vertical linear measurement) per unit of horizontal distance covered. Commonly measured in m/km.

Gran method A system involving the observation of the change in dissolved oxygen within paired transparent and opaque bottles containing phytoplankton and suspended in the ocean surface water to determine the base of the euphotic zone.

Granite A light-colored igneous rock characteristic of the continental crust. Rich in nonferromagnesian minerals such as feldspar and quartz.

Gravitational force The force of attraction that exists between any two bodies in the universe that is proportional to the product of their masses and inversely proportional to the distance between the centers of their masses.

Gravity wave A wave for which the dominant restoring force is gravity. Such waves have a wavelength of more than 1.74 cm, and their speed of propagation is controlled mainly by gravity.

Gray whale Pacific baleen whales of the species *Rhachianectes glaucus* that feed in the Chukchi Sea and Bering Sea and breed and calve in the warm waters of lagoons in Baja California and Japan.

Greenhouse effect The heating of Earth's atmosphere that results from the absorption by components of the atmosphere such as water vapor and carbon dioxide of infrared radiation from Earth's surface.

Groin A low artificial structure projecting into the ocean from the shore to interfere with longshore transportation of sediment. It usually has the purpose of trapping sand to cause the buildup of a beach.

Gross ecological efficiency The amount of energy passed on from a trophic level to the one above it divided by the amount it received from the one below it.

Gross primary production The total carbon fixed into organic molecules through photosynthesis or chemosynthesis by a discrete autotrophic community.

Gulf Stream The high-intensity western boundary current of the North Atlantic Ocean subtropical gyre that flows north off the east coast of the United States.

Guyot A tablemount, a conical volcanic feature on the ocean floor that has had the top truncated to a relatively flat surface.

Gyre A circular spiral form. Used mainly in reference to the circular motion of water in each of the major ocean basins centered in subtropical high pressure regions.

Habitat A place where a particular plant or animal lives. Generally refers to a smaller area than environment.

Hadal Pertaining to the deepest ocean environment, specifically that of ocean trenches deeper than 6 km.

Hadal zone Pertaining to the deepest ocean benthic environment, specifically that of ocean trenches deeper than 6 km (3.7 mi).

Half-life The time required for half the atoms of a sample of a radioactive isotope to decay to an atom of another element.

Halocline A layer of water in which a high rate of change in salinity in the vertical dimension is present.

Headland A steep-faced irregularity of the coast that extends out into the ocean.

Heat Energy moving from a high temperature system to a lower temperature system. The heat gained by the one system may be used to raise its temperature or to do work.

Heat budget The equilibrium that exists on the average between the amount of heat absorbed by Earth and its atmosphere in one year and the amount of heat radiated back into space in one year.

Heat capacity Usually defined as the amount of heat required to raise the temperature of 1 g of a substance 1°C.

Heat energy Energy of molecular motion. The conversion of higher forms of energy such as radiant or mechanical energy to heat energy within a system increases the heat energy within the system and the temperature of the system.

Heat flow (flux) The quantity of heat flow to Earth's surface per unit of time.

Hemoglobin A red pigment found in red blood corpuscles that carries oxygen from the lungs to tissue and carbon dioxide from tissue to lungs.

Herbivore An animal that relies chiefly or solely on plants for its food.

Hermaphroditic Pertaining to the possession of functional male and female reproductive organs by an animal. It is rare for both systems to function at the same time.

Hermatypic coral Reef-building corals that have symbiotic algae in their ectodermal tissue. They cannot produce a reef structure below the euphotic zone.

Heterotroph Animals and bacteria that depend on the organic compounds produced by other animals and plants as food. Organisms not capable of producing their own food by photosynthesis.

High tide zone That portion of the littoral zone that lies between the lowest high tides and highest high tides that occur in an area. It is, on the average, exposed to desiccation for longer periods each day than it is covered by water.

High water (HW) The highest level reached by the rising tide before it begins to recede.

Higher high water (HHW) The higher of two high waters occurring during a tidal day where tides are mixed.

Higher low water (HLW) The higher of two low waters occurring during a tidal day where tides are mixed.

Highly stratified estuary A relatively deep estuary in which a significant volume of marine water enters as a subsurface flow. A large volume of freshwater stream input produces a widespread low surface-salinity condition which produces a well developed halocline throughout most of the estuary.

Holoplankton Organisms that spend their entire life as members of the plankton.

Homeotherm An animal that maintains a precisely controlled internal body temperature using its own internal heating and cooling mechanisms.

Homologous A basic similarity of structures in different organisms resulting from a similar embryonic origin and development, e.g., the foreflippers of a seal and the arms and hands of a person are homologous.

Hook A spit or narrow cape of sand or gravel with an end that bends landward to form a "hook."

Horse latitudes The latitude belts between 30° and 35° north and south where winds are light and variable, since the principal movement of air masses at these latitudes is one of vertical descent. The climate is hot and dry.

Hot spot The relatively stationary surface expression of a persistent jet of molten mantle material rising to the surface.

Hurricane A tropical cyclone in which winds reach speeds in excess of 120 km/h (73 mi/h). Generally applied to such storms in the North Atlantic Ocean, eastern North Pacific Ocean, Caribbean Sea, and Gulf of Mexico. Such storms in the western Pacific Ocean are called *typhoons*.

Hydrated To be chemically combined with water or surrounded by water.

Hydrocarbon Any organic compound consisting only of hydrogen and carbon. Crude oil is a mixture of hydrocarbons.

Hydrocarbon seep biocommunity Deep bottom-dwelling community associated with a hydrocarbon seep from the ocean floor. The community depends on methane and sulfur-oxidizing bacteria as producers. The bacteria may live free in the water, on the bottom, or symbiotically in the tissues of some of the animals.

Hydrogen bond An intermolecular bond that forms within water because of the dipolar nature of water molecules.

Hydrogenous sediment Sediment that forms from precipitation from ocean water or ion exchange between existing sediment and ocean water. Examples are manganese nodules, phosphorite, glauconite, phillipsite, and montmorillonite.

Hydrologic cycle The cycle of water exchange among the atmosphere, land, and ocean through the processes of evaporation, precipitation, runoff, and subsurface percolation.

Hydrothermal spring Vents of hot water found primarily along the spreading axes of oceanic ridges and rises.

Hydrothermal vent Ocean water that percolates down through fractures in recently formed ocean floor is heated by underlying magma and surfaces again through these vents. They are usually located near the axis of spreading on oceanic ridges and rises.

Hydrothermal vent biocommunity Deep bottom-dwelling community associated with a hydrothermal vent. The hot water vent is usually associated with the axis of a spreading center, and the community is dependent on sulphur-oxidizing bacteria that may live free in the water, on the bottom, or symbiotically in the tissue of some of the animals of the community.

Hydrozoa A class of coelenterates that characteristically exhibits alternation of generations with a sessile polypoid colony giving rise to a pelagic medusoid form by asexual budding.

Hypersaline lagoon Shallow lagoons such as Laguna Madre, which may become hypersaline due to little tidal flushing and seasonal variability in freshwater input. High evaporation rates and low freshwater input can result in very high salinities.

Hypersaline seep At the base of the Florida Escarpment there is a biocommunity that depends on methane and sulfur-oxidizing bacteria as producers. The bacteria may live free in the water, on the bottom, or symbiotically in the tissue of some of the animals within the biocommunity.

Hypertonic Pertaining to the property of an aqueous solution having a higher osmotic pressure (salinity) than another aqueous solution from which it is separated by a semipermeable membrane that will allow osmosis to occur. The hypertonic fluid will gain water molecules through the membrane from the other fluid.

Hypotheca The lower valve of a diatom frustule.

Hypotonic Pertaining to the property of an aqueous solution having a lower osmotic pressure (salinity) than another aqueous solution from which it is separated by a semipermeable membrane that will allow osmosis to occur. The hypotonic fluid will lose water molecules through the membrane to the other fluid.

Ice floe *See* floe.

Ice shelf A thick layer of ice with a relatively flat surface which is attached to and nourished by a continental glacier from one side. The shelf, which is for the most part afloat, may extend above water level by more than 50 m along its seaward cliff formed by the break-off of large tabular chunks of ice that become icebergs.

Iceberg A massive piece of glacier ice that has broken from the front of the glacier (calved) into a body of water. It floats with its tip at least 5 m above the water's surface and at least $\frac{4}{5}$ of its mass submerged.

Igneous rock One of the three main classes into which all rocks are divided, i.e., igneous, metamorphic, and sedimentary. Rock that forms from the solidification of molten or partly molten material (magma).

In situ In place, i.e., in situ density of a sample of water is its density at its original depth.

Inertia Newton's first law of motion. It states that a body at rest will stay at rest and a body in motion will remain in a uniform motion in a straight line unless acted on by some external force.

Infauna Animals that live buried in the soft substrate (sand or mud).

Infrared radiation Electromagnetic radiation lying between the wavelengths of 0.8 μm and about 1000 μm. It is bounded on the shorter wavelength side by the visible spectrum and on the long side by microwave radiation.

Inner sublittoral Pertaining to the inner continental shelf, above the intersection with the euphotic zone, where attached plants grow.

Insolation The rate at which solar radiation is received per unit of surface area at any point at or above Earth's surface.

Interface A surface separating two substances of different properties, i.e., density, salinity, or temperature. In oceanography, it usually refers to a separation of two layers of water with different densities caused by significant differences in temperature and/or salinity.

Interface wave An orbital wave that moves along an interface between fluids of different density. An example is ocean surface waves moving along the interface between the atmosphere and the ocean which is one thousand times more dense.

Intermolecular bond A relatively weak bond that forms between molecules of a given substance. The hydrogen bond and the van der Waals bonds are intermolecular bonds.

Internal wave A wave that develops below the surface of a fluid, the density of which changes with increased depth. This change may be gradual or occur abruptly at an interface.

Intertidal zone Littoral zone, the foreshore. The ocean floor covered by the highest normal tides and exposed by the lowest normal tides and the water environment of the tide pools within this region.

Intertropical Convergence Zone Zone where northeast trade winds and southeast trade winds converge. Averages about 5°N in the Pacific and Atlantic oceans and 7°S in the Indian Ocean.

Intrusive rocks Igneous rocks such as granite that cool slowly beneath Earth's surface.

Invertebrate Animal without a backbone.

Ion An atom that becomes electrically charged by gaining or losing one or more electrons. The loss of electrons produces a positively charged cation, and the gain of electrons produces a negatively charged anion.

Ionic bond A chemical bond formed as a result of the electrical attraction between ions of unlike charge.

Irminger Current A warm current that branches off from the Gulf Stream and moves up along the west coast of Iceland.

Island arc system A linear arrangement of islands, many of which are volcanic, usually curved so the concave side faces a sea separating the islands from a continent. The convex side faces the open ocean and is bounded by a deep-ocean trench.

Island mass effect As surface current flows past an island, surface water is carried away from the island on the down-current side. This water is replaced in part by upwelling of water on the down-current side of the island.

Isohaline Of the same salinity.

Isopoda An order of dorso-ventrally flattened crustaceans that are mostly scavengers or parasites on other crustaceans or fish.

Isostasy A condition of equilibrium, comparable to buoyancy, by which Earth's brittle crust floats on the plastic mantle.

Isotherm Line connecting points of equal temperature.

Isothermal Of the same temperature.

Isotonic Pertaining to the property of having equal osmotic pressure. If two such fluids were separated by a semipermeable membrane that will allow osmosis to occur, there would be no net transfer of water molecules across the membrane.

Isotope One of several atoms of an element that has a different number of neutrons, and therefore a different atomic mass, than the other atoms, or isotopes, of the element.

Jellyfish 1. Free-swimming, umbrella-shaped medusoid members of the coelenterate class, Scyphozoa. 2. Also frequently applied to the medusoid forms of other coelenterates.

Jet stream An easterly-moving air mass at an elevation of about 10 km (6.2 mi). Moving at speeds that can exceed 300 km/h (186 mi/h), the jet stream follows a wavy path in the midlatitudes and influences how far polar air masses may extend into the lower latitudes.

Jetty A structure built from the shore into a body of water to protect a harbor or a navigable passage from being shoaled by deposition of longshore (littoral) drift material.

Juan de Fuca Ridge A divergent plate boundary off the Oregon-Washington coast.

Kelp Large varieties of Phaeophyta (brown algae).

Kelvin temperature scale (K) 0 K = −273.16°C. One degree on the Kelvin scale equals the same temperature range as one degree on the Celsius scale. 0 K is the lowest temperature possible.

Key A low, flat island composed of sand or coral debris that accumulates on a reef flat.

Kinetic energy Energy of motion. It increases as the mass or speed of the object in motion increases.

Knot (kt) Unit of speed equal to 1 nautical mile per hour, approximately 51 cm/s.

Krill A common name frequently applied to members of crustacean order, Euphausiacea (euphausids).

La Niña An event where the surface temperature in the waters of the eastern South Pacific fall below average values. It usually occurs at the end of an El Niño-Southern Oscillation event.

Labrador Current A cold current flowing south along the coast of Labrador in the northeastern Atlantic Ocean.

Lagoon A shallow stretch of seawater partly or completely separated from the open ocean by an elongate narrow strip of land such as a reef or barrier island.

Laguna Madre A hypersaline lagoon behind Padre Island along the south Texas coast.

Laminar flow Flow in which water, or any fluid, flows in parallel layers or sheets. The direction of flow at any point does not change with time. Nonturbulent flow.

Langmuir circulation A cellular circulation set up by winds that blow consistently in one direction with speeds in excess of 12 km/h. Helical spirals running parallel to the wind direction are alternately clockwise and counterclockwise.

Larva An embryo that is on its own before it assumes the characteristics of the adults of the species.

Latent heat The quantity of heat gained or lost per unit of mass as a substance undergoes a change of state (liquid to solid, etc.) at a given temperature and pressure.

Latent heat of evaporation The heat energy that must be added to one gram of a liquid substance to convert it to a vapor at a given temperature below its boiling point. For water, it is 585 cal at 20°C.

Latent heat of melting The heat energy that must be added to one gram of a substance at its melting point to convert it to a liquid. For water, it is 80 cal.

Latent heat of vaporization The heat energy that must be added to one gram of a substance at its boiling point to convert it to a vapor. For water, it is 540 cal.

Lateral line system A sensory system running down both sides of fishes to sense subsonic pressure waves transmitted through ocean water.

Latitude Location on Earth's surface based on angular distance north or south of the equator. Equator, 0°; North Pole, 90° N; South Pole, 90° S.

Laurasia A hypothetical protocontinent of the Northern Hemisphere. The name is derived from Laurentia, pertaining to the Canadian Shield of North America, and Eurasia of which it was composed.

Lava Fluid magma coming from an opening in Earth's surface, or the same material after it solidifies.

Law of gravitation *See* gravitational force.

Leeuwin Current A warm current that flows south out of the East Indies along the western coast of Australia.

Leeward Direction toward which the wind is blowing or waves are moving.

Levee 1. Natural levees are low ridges on either side of river channels that result from deposition during flooding. 2. Artificial levees are built by human beings.

Light-year The distance traveled by light during one year at a speed of 300,000 km/s (186,000 mi/s). It equals 9.8 trillion kilometers or 6.2 trillion miles.

Limestone A class of sedimentary rocks composed of at least 80 percent carbonates of calcium or magnesium. Limestones may be either biogenous or hydrogenous.

Limpet A mollusk of the class Gastropoda that possesses a low conical shell that exhibits no spiraling in the adult form.

Lipid Fats which, along with proteins and carbohydrates, are the principal structural components of living cells.

Liquid state A state of matter in which a substance has a fixed volume but no fixed shape.

Lithogenous sediment Sediment composed of mineral grains derived from the rocks of continents and islands and transported to the ocean by wind and running water.

Lithosphere The outer layer of Earth's structure, including the crust and the upper mantle to a depth of about 200 km. It is this layer that breaks into the plates that are the major elements of the theory of plate tectonics.

Lithothamnion ridge A feature common to the windward edge of a reef structure, characterized by the presence of the red algae, *Lithothamnion*.

Littoral zone The benthic zone between the highest and lowest spring-tide shorelines; the intertidal zone.

Lobster Large marine crustacean used as food. *Homarus americanus* (American lobster) possesses two large chelae (pincers) and is found off the New England coast. *Panulirus* sp. (spiny lobsters or rock lobsters) have no chelae but possess long spiny antennae effective in warding off predators. *P. argus* is found off the coast of Florida and in the West Indies, while *P. interruptus* is common along the coast of southern California.

Longitude Location on Earth's surface based on angular distance east or west of Greenwich Meridian (0° long.). 180° longitude is the International Date Line.

Longitudinal wave A wave phenomenon where particle vibration is parallel to the direction of energy propagation.

Longshore current A current located in the surf zone and running parallel to the shore as a result of waves breaking at an angle on the shore.

Longshore drift The load of sediment transported along the beach from the breaker zone to the top of the swash line in association with the longshore current.

Lophophore Horseshoe-shaped feeding structure bearing ciliated tentacles characteristic of the phyla Bryozoa, Brachiopoda, and Phoronidea.

Low tide zone That portion of the intertidal zone that lies between the lowest low-tide shoreline and the highest low-tide shoreline.

Low water (LW) The lowest level reached by the water surface at low tide before the rise toward high tide begins.

Lower high water (LHW) The lower of two high waters occurring during a tidal day where tides are mixed.

Lower low water (LLW) The lower of two low waters occurring during a tidal day where tides are mixed.

Lunar day The time interval between two successive transits of the moon over a meridian, approximately 24 h and 50 min of solar time.

Lunar hour One twenty-fourth of a lunar day. About 62.1 min.

Lunar tide The part of the tide caused solely by the tide-producing force of the moon.

Lunger Fish such as groupers that sit motionless on the ocean floor waiting for prey to appear. A quick burst of speed over a short distance suffices to capture the prey.

Macroplankton Plankton larger than 2 cm (0.8 in.) in their smallest dimension.

Magma Fluid rock material from which igneous rock is derived through solidification.

Magnetic anomaly Distortion of the regular pattern of Earth's magnetic field, resulting from the various magnetic properties of local concentrations of ferromagnetic minerals in Earth's crust.

Magnetic dip The dip of magnetite particles in rock units of Earth's crust relative to sea level. It is approximately equivalent to latitude.

Manganese nodules Concretionary lumps containing oxides of iron, manganese, copper, and nickel found scattered over the ocean floor.

Mangrove swamp A marshlike environment that is dominated by mangrove trees. They are restricted to latitudes below 30°.

Mantle The relatively plastic zone rich in ferromagnesian minerals between the core and crust of Earth.

Marginal sea A semienclosed body of water adjacent to a continent and floored by submerged continental crust.

Mariculture The application of the principles of agriculture to the production of marine organisms.

Marine terrace Wave-cut benches that have been exposed above sea level by a drop in sea level.

Marsh An area of soft, wet land. Flat land periodically flooded by salt water, common in portions of lagoons.

Mean high water (MHW) The average height of all the high waters occurring over a 19-year period.

Mean low water (MLW) The average height of the low waters occurring over a 19-year period.

Mean sea level (MSL) The mean surface water level determined by averaging all stages of the tide over a 19-year period, usually determined from hourly height observations along an open coast.

Mean tidal range The difference between mean high water and mean low water.

Meander A sinuous curve, bend, or turn in the course of a current.

Mechanical energy Energy manifested as work being done; the movement of a mass some distance.

Mediterranean circulation Circulation characteristic of bodies of water with restricted circulation with the ocean that results from an excess of evaporation as compared to precipitation and runoff. Surface flow is into the restricted body of water with a subsurface counterflow as exists between the Mediterranean Sea and the Atlantic Ocean.

Medusa Free-swimming, bell-shaped coelenterate body form with a mouth at the end of a central projection and tentacles around the periphery. Reproduces sexually.

Meiofauna Small species of animals that live in the spaces among particles in a marine sediment.

Meridian of longitude Great circles running through the north and south poles.

Meroplankton Planktonic larval forms of organisms that are members of the benthos or nekton as adults.

Mesoderm Primitive cell layer of embryo that develops between endoderm and ectoderm. In vertebrates, it gives rise to the skeleton, muscles, circulatory, and excretory systems, and most of the reproductive system.

Mesopelagic That portion of the oceanic province 200–1000 m deep. Corresponds approximately with the disphotic (twilight) zone.

Metamorphic rock Rock that has undergone recrystallization while in the solid state in response to changes of temperature, pressure, and chemical environment.

Metaphyta Kingdom of many-celled plants.

Metazoa Kingdom of many-celled animals.

Microplankton Net plankton. Plankton not easily seen by the unaided eye, but easily recovered from the ocean with the aid of a silk-mesh plankton net.

Mid-Atlantic Ridge A slow-spreading divergent plate boundary running north-south and bisecting the Atlantic Ocean.

Middle tide zone That portion of the intertidal zone that lies between the highest low-tide shoreline and the lowest high-tide shoreline.

Migration Long journeys undertaken by many marine species for the purpose of successful feeding and reproduction. The stimuli that cause the animals to initiate their migrations are for the most part still unknown.

Minamata Bay, Japan The site of the 1953 occurrence of human poisoning by mercury contained in marine food victims consumed.

Mineral An inorganic substance occurring naturally on Earth and having distinctive physical properties and a chemical composition that can be expressed by a chemical formula. The term is also sometimes applied to organic substances such as coal and petroleum.

Mixed interference A pattern of wave interference in which there is a combination of constructive and destructive interference.

Mixed layer The surface layer of the ocean water mixed by wave and tide motions to produce relatively isothermal and isohaline conditions.

Mixed tide A tide having two high waters and two low waters per tidal day with a marked diurnal inequality. Such a tide may also show alternating periods of diurnal and semidiurnal components.

Mixotroph An organism that depends on a combination of autotrophic and heterotrophic behavior to meet its energy requirements. Many coral reef species exhibit such behavior.

Mohorovicic discontinuity A sharp compositional discontinuity between the crust and mantle of Earth. It may be as shallow as 5 km below the ocean floor or as deep as 60 km beneath some continental mountain ranges.

Molecular motion Molecules move in three ways: vibration, rotation, and translation.

Molecule The smallest particle of an element or compound that, in the free state, retains the characteristics of the substance.

Mollusca Phylum of soft unsegmented animals usually protected by a calcareous shell and having a muscular foot for locomotion. Includes snails, clams, chitons, and octopuses.

Molt Periodic shedding of exoskeleton by arthropods to permit growth.

Monera Kingdom of organisms that do not have nuclear material confined within a sheath but spread throughout the cell. Bacteria and blue-green algae.

Mononodal Pertaining to a standing wave with only one nodal point or nodal line.

Monsoons A name for seasonal winds derived from the Arabic word for season, *mausim*. The term was originally applied to winds over the Arabian Sea that blow from the southwest during summer and the northeast during winter.

Moraine A deposit of unsorted material deposited at the margins of glaciers. Many such deposits have become important economically as fishing banks after being submerged by the rising level of the ocean.

Mud Sediment consisting of silt and clay-sized particles smaller than 0.06 mm. Actually, small amounts of larger particles will also be present.

Mutualism A symbiotic relationship in which both participants benefit.

Mycota The kingdom of fungi. In the marine environment they are found living symbiotically with algae as lichen in the intertidal zone and as decomposers of dead organic matter in the open sea.

Myoglobin A red, oxygen-storing pigment found in the muscle tissue.

Myomere A muscle fiber.

Mysticeti The baleen whales.

Nadir The point on the celestial sphere directly opposite the zenith and directly beneath the observer.

Nanoplankton Plankton less than 50 μm in length that cannot be captured in a plankton net and must be removed from the water by centrifuge or special microfilters.

Nansen bottle A device used by oceanographers to obtain samples of ocean water from beneath the surface.

Nauplius A microscopic free-swimming larval stage of crustaceans such as copepods, ostracoids, and decapods. Typically has three pairs of appendages.

Neap tide Tides of minimal range occurring when the moon is in quadrature, first and third quarters.

Nearshore That zone from the shoreline seaward to the line of breakers.

Nektobenthos Those members of the benthos that can actively swim and spend much time off the bottom.

Nekton Pelagic animals such as adult squids, fish, and mammals that are active swimmers to the extent they can determine their position in the ocean by swimming.

Nematocyst The stinging mechanism found within the cnidoblast of members of the phylum Cnidaria (Coelenterata).

Neritic province That portion of the pelagic environment from the shoreline to where the depth reaches 200 m.

Neritic sediment That sediment composed primarily of lithogenous particles and deposited relatively rapidly on the continental shelf, continental slope, and continental rise.

Net primary production The primary production of plants after they have removed what is needed for their metabolism.

New moon The phase of the moon that occurs when the sun and the moon are in conjunction, on the same side of Earth.

New production Photosynthetic production supported by nutrients supplied from outside the immediate ecosystem by upwelling or other physical transport.

Niche The ecological role of an organism and its position in the ecosystem.

Niigata, Japan In the 1960s the site of mercury poisoning of humans by mercury-contaminated seafood.

Nitrogen fixation Conversion by bacteria of atmospheric nitrogen (N_2) to oxides of nitrogen (NO_2, NO_3) usable by plants in primary production.

Node The point on a standing wave where vertical motion is lacking or minimal. If this condition extends across the surface of an oscillating body of water, the line of no vertical motion is a nodal line.

Nonconservative property A property of ocean water attained at the surface and changed by processes other than mixing and diffusion after the water sinks below the surface. For example, dissolved oxygen content will be altered by biological activity.

Nonferromagnesian A group of common igneous rock-forming minerals that do not contain iron and magnesium.

North Atlantic Deep Water A deep-water mass that forms primarily at the surface of the Norwegian Sea and moves south along the floor of the North Atlantic Ocean.

Northeast Monsoon A northeast wind that blows off the Asian mainland onto the Indian Ocean during the winter season.

Norwegian Current A warm current that branches off from the Gulf Stream and flows into the Norwegian Sea between Iceland and the British Isles.

Nudibranch Sea slug. A member of the mollusk class Gastropoda that has no protective covering as an adult. Respiration is carried on by gills or other projections on the dorsal surface.

Nutrients Any number of organic or inorganic compounds used by plants in primary production. Nitrogen and phosphorus compounds are important examples.

Ocean Acoustical Tomography A method by which changes in water temperature may be determined by changes in the speed of transmission of sound. It has the potential to help map ocean circulation patterns over large ocean areas.

Ocean beach Beach on the open-ocean side of a barrier island.

Ocean Drilling Program In 1983, this program replaced the Deep Sea Drilling Project. It focuses more on drilling the continental margins using the drill vessel *JOIDES Resolution*.

Oceanic crust A mass of rock with basaltic composition that is about 5 km thick and may or may not extend beneath the continents.

Oceanic province That division of the pelagic environment where the water depth is greater than 200 m.

Oceanic ridge A linear, seismic mountain range that extends through all the major oceans, rising 1–3 km above the deep-ocean basins. Averaging 1500 km in width, rift valleys are common along the central axis. Source of new oceanic crustal material.

Oceanic sediment The inorganic abyssal clays and the organic oozes that accumulate on the deep-ocean floor slowly, particle by particle.

Oceanic spreading center The axes of oceanic ridges and rises that are the locations at which new lithosphere is added to lithospheric plates. The plates move away from these axes in the process of sea-floor spreading.

Ocelli Light-sensitive organ around the base of many medusoid bells

Odonticeti Toothed whales.

Offshore The comparatively flat submerged zone of variable width extending from the breaker line to the edge of the continental shelf.

Oligotrophic Areas such as the mid-subtropical gyres where there are low levels of biological production.

Omnivore An animal that feeds on both plants and animals.

Oolite A deposit formed of small spheres from 0.25 to 2 mm in diameter. They are usually composed of concentric layers of calcite.

Ooze A pelagic sediment containing at least 30 percent skeletal remains of pelagic organisms, the balance being clay minerals. Oozes are further defined by the chemical composition of the organic remains (siliceous or calcareous) and by their characteristic organisms (diatom ooze, foraminifera ooze, radiolarian ooze, pteropod ooze).

Opal An amorphous form of silica ($SiO_2 \cdot nH_2O$) that usually contains from 3 to 9 percent water. It forms the shells of radiolarians and diatoms.

Opposition The separation of two heavenly bodies by 180° relative to Earth. The sun and moon are in opposition during the full moon phase.

Orbital wave A wave phenomenon in which energy is moved along the interface between fluids of different density. The wave form is propagated by the movement of fluid particles in orbital paths.

Orthogonal lines Lines drawn perpendicular to wave fronts and spaced uniformly so equal amounts of energy are contained by the segments of the wave front lying between any two orthogonal lines in a series. The areas where energy is concentrated as the waves break on the shore can be identified by the convergence of the orthogonal lines.

Orthophosphate Phosphoric oxide (P_2O_5) can combine with water to produce orthophosphates ($3H_2O \cdot P_2O_5$ or H_3PO_4) that may be used by plants as nutrients.

Osmosis Passage of water molecules through a semipermeable membrane separating two aqueous solutions of different solute concentration. The water molecules pass from the solution of lower solute concentration into the other.

Osmotic pressure A measure of the tendency for osmosis to occur. It is the pressure that must be applied to the more concentrated solution to prevent the passage of water molecules into it from the less concentrated solution.

Osmotic regulation Physical and biological processes used by organisms to counteract the osmotic effects of differences in osmotic pressures of their body fluids and the water in which they live.

Ostracoda An order of crustaceans that are minute and compressed within a bivalve shell.

Otocyst Gravity sensitive organs around the bell of a medusa.

Outer sublittoral The continental shelf below the intersection with the euphotic zone where no plants grow attached to the bottom.

Oviparous Referring to embryological development in an egg that is deposited outside the mother's body where nutrition for development is provided by the yolk of the egg.

Ovoviviparous Referring to embryological development in the female reproductive tract where nutrition for development is provided by the yolk of the egg.

Oxygen compensation depth The depth in the ocean at which marine plants receive just enough solar radiation to meet their basic metabolic needs. It marks the base of the euphotic zone.

Oxygen utilization rate (OUR) The rate at which oxygen is being used by the demands of respiration and bacterial decomposition of dead organic material.

Pacific-type margin Leading edge of a continent that undergoes tetonic uplift as a result of lithospheric plate convergence.

Pack ice Any area of sea ice other than fast ice. Less than 3 m thick, it covers the ocean sufficiently that navigation is possible only by icebreakers.

Paleoceanography The study of the physical and biological changes of the oceans brought about by the changing shapes and positions of the continents.

Paleogeography The study of the historical changes of shapes and positions of the continents and oceans.

Pancake ice Circular pieces of newly formed sea ice from 30 cm to 3 m in diameter that form in early fall in polar regions.

Pangaea A hypothetical supercontinent of the geologic past that contained all the continental crust of Earth.

Panthalassa A hypothetical protoocean surrounding Pangaea.

Parapodia Flat protuberances on each side of most segments of polychaete worms. Most possess cirri and setae (bristlelike projections); may be modified for special functions such as feeding, locomotion, and respiration.

Parasitism A symbiotic relationship between two organisms in which one benefits at the expense of the other.

PCBs A group of industrial chemicals used in a variety of products; responsible for several episodes of ecological damage in coastal waters.

Pelagic environment The open ocean environment which is divided into the neritic province (water depth 0–200 m) and the oceanic province (water depth greater than 200 m).

Pelecypoda A class of mollusks characterized by two more or less symmetrical lateral valves with a dorsal hinge. These filter feeders pump water through the filter system and over gills through posterior siphons. Many possess a hatchet-shaped foot used for locomotion and burrowing. Includes clams, oysters, mussels, and scallops.

Perigee The point on the orbit of an Earth satellite (moon) that is nearest Earth.

Perihelion That point on the orbit of a planet or comet around the sun that is closest to the sun.

Permeability Capacity of a porous rock or sediment for transmitting fluid.

Petroleum A naturally occurring liquid hydrocarbon.

Phaeophyta Brown algae characterized by the carotenoid pigment fucoxanthin. Contains the largest members of the marine plant community.

Phosphorite A sedimentary rock composed primarily of phosphate minerals.

Photic zone The upper ocean in which the presence of solar radiation is detectable. It includes the euphotic and disphotic zones.

Photophore One of several types of light-producing organs found primarily on fishes and squids inhabiting the mesopelagic and upper bathypelagic zones.

Photosynthesis The process by which plants produce carbohydrate food from carbon dioxide and water in the presence of chlorophyll, using light energy and releasing oxygen.

Phycoerythrin A red pigment characteristic of the Rhodophyta (red algae).

Phytoplankton Plant plankton. The most important community of primary producers in the ocean.

Picoplankton Small plankton within the size-range of 0.2 to 2.0 μm in size. Composed primarily of bacteria.

Pinniped A suborder of marine mammals that includes the sea lions, seals, and walruses.

Plankton Passively drifting or weakly swimming organisms that are not independent of currents. Includes mostly microscopic algae, protozoa, and larval forms of higher animals.

Plankton bloom A very high concentration of phytoplankton, resulting from a rapid rate of reproduction as conditions become optimum during the spring in high latitude areas. Less obvious causes produce blooms that may be destructive in other areas.

Plankton net Plankton extracting device that is cone-shaped and typically of a silk material. It is towed through the water or lifted vertically to extract plankton down to a size of 50 μm.

Plume Rising jets of molten mantle material that create hot spots when they penetrate the crust of Earth.

Plunging breaker Impressive curling breakers that form on moderately sloping beaches.

Pneumatic duct An opening into the swim bladder of some fishes that allows rapid release of air into the esophagus.

Poikilotherm An organism whose body temperature varies with and is largely controlled by its environment.

Polar Pertaining to the polar regions.

Polar easterly winds Cold air masses that move away from the polar regions toward lower latitudes.

Polar emergence The emergence of low- and midlatitude temperature-sensitive deep-ocean benthos onto the shallow shelves of the polar regions where temperatures similar to that of their deep-ocean habitat exist.

Pollution (marine) The introduction of substances or energy into the marine environment that results in harm to the living resources of the ocean or humans that use these resources.

Polychaeta Class of annelid worms that includes most of the marine segmented worms.

Polynya A nonlinear opening in sea ice.

Polyp A single individual of a colony or a solitary attached coelenterate.

Population A group of individuals of one species living in an area.

Porifera Phylum of sponges. Supporting structure composed of $CaCO_3$ or SiO_2 spicules or fibrous spongin. Water currents created by flagella-waving choanocytes enter tiny pores, pass through canals, and exit through a larger osculum.

Potential (world) fishery The mass of fish that are supported annually by nutrients that originate outside of an ecosystem (world fishery ecosystems) and flow into it through upwelling or other processes. This amount of fish can be removed on an annual basis without degrading the ecosystem.

Precession Regarding the moon's orbit around Earth, the axis of this orbit slowly changes its direction and describes a complete cone every 18.6 years. This is accompanied by a clockwise rotation of the plane of the moon's orbit that is completed in the same time interval.

Precipitation In a meteorological sense, the discharge of water in the form of rain, snow, hail, or sleet from the atmosphere onto Earth's surface.

Primary productivity The amount of organic matter synthesized by organisms from inorganic substances within a given volume of water or habitat in a unit of time.

Prime meridian The meridian of longitude 0° used as a reference for measuring longitude. The Greenwich Meridian.

Producer The autotrophic component of an ecosystem that produces the food that supports the biocommunity.

Progressive wave A wave in which the waveform progressively moves.

Propagation The transmission of energy through a medium.

Protein A very complex organic compound made up of large numbers of amino acids. Proteins make up a large percentage of the dry weight of all living organisms.

Protista A kingdom of organisms that includes all one-celled forms with nuclear material confined to a nuclear sheath. Includes the animal phylum Protozoa and the phyla of algal plants.

Protoplasm The complicated self-perpetuating living material making up all organisms. The elements carbon, hydrogen, and oxygen constitute more than 95 percent; water and dissolved salts make up 50–97 percent of most plants and animals with carbohydrates, lipids (fats), and proteins constituting the remainder.

Protozoa Phylum of one-celled animals with nuclear material confined within a nuclear sheath.

Pseudopodia An extension of protoplasm in a broad, flat, or long needle-like projection used for locomotion or feeding. Typical of amoeboid forms such as foraminifera and radiolaria.

Pteropoda An order of pelagic gastropods in which the foot is modified for swimming and the shell may be present or absent.

Purse seine A curtainlike net that can be used to encircle a school of fish. The bottom is then pulled tight much the way a purse string is used to close a baglike purse.

Pycnocline A layer of water in which a high rate of change in density in the vertical dimension is present.

Pycnogonid A spiderlike arthropod found on the ocean bottom at all depths. The more commonly observed nearshore varieties are usually less than 1 cm across, while deeper water varieties may reach spreads of over 1 m.

Pyrrophyta A phylum of microscopic algae that possesses flagella for locomotion—the dinoflagellates.

Quadrature The first and third quarter moon phases occur when the sun and moon are in quadrature, at right angles to one another relative to Earth.

Quarter moon First and third quarter moon phases which occur when the sun and moon are in quadrature one week after the new moon and full moon phases, respectively.

Radiata A grouping of phyla with primary radial symmetry—phyla Coelenterata and Ctenophora.

Radioactive dating *See* absolute dating.

Radioactivity The spontaneous breakdown of the nucleus of an atom resulting in the emission of radiant energy in the form of particles or waves.

Radiolaria An order of planktonic and benthic protozoans that possess protective coverings usually made of silica.

Ray A cartilaginous fish in which the body is dorso-ventrally flattened, eyes and spiracles are on the upper surface, and gill slits are on the bottom. The tail is reduced to a whiplike appendage. Includes electric rays, manta rays, and stingrays.

Recruitment (fishery) The year-class (number of fish or mass of fish) of young adults added to a fishery following each spawning season.

Red muscle fiber Fine muscle fibers rich in myoglobin that are abundant in cruiser-type fishes.

Red tide A reddish brown discoloration of surface water, usually in coastal areas, caused by high concentrations of microscopic organisms, usually dinoflagellates. It probably results from increased availability of certain nutrients for various reasons. Toxins produced by the dinoflagellates may kill fish directly; or large populations of animal forms that spring up to feed on the plants, along with decaying plant and animal remains, may use up the oxygen in the surface water to cause asphyxiation of many animals.

Reef A consolidated rock (a hazard to navigation) with a depth of 20 m or less.

Reef flat A platform of coral fragments and sand on the lagoonal side of a reef that is relatively exposed at low tide.

Reef front The upper seaward face of a reef from the reef edge (seaward margin of reef flat) to the depth at which living coral and coralline algae become rare, 16–30 m.

Reflection The process in which a wave has part of its energy returned seaward by a reflecting surface.

Refraction The process by which the part of a wave in shallow water is slowed down to cause the wave to bend and tend to align itself with the underwater contours.

Regenerated production The portion of gross primary production that is supported by nutrients recycled within an ecosystem.

Relative dating The determination of whether certain rock units are older or younger than others by the use of fossil assemblages. It was not possible to tell the actual age of rocks by use of fossils until radiometric dating was developed.

Relict beach A beach deposit laid down and submerged by a rise in sea level. It is still identifiable on the continental shelf, indicating no deposition is presently taking place at that location on the shelf.

Relict sediment A sediment deposited under a set of environmental conditions that still remains unchanged although the environment has changed and it remains unburied by later sediment. An example is a beach deposited

near the edge of the continental shelf when sea level was lower.

Residence time The average length of time a particle of any substance spends in the ocean. It is calculated by dividing the total amount of the substance in the ocean by the rate of its introduction into the ocean or the rate at which it leaves the ocean.

Respiration The process by which organisms utilize organic materials (food) as a source of energy. As the energy is released, oxygen is used and carbon dioxide and water are produced.

Restoring force A force such as surface tension or gravity that tends to restore the ocean surface displaced by a wave to that of a still-water level.

Reverse osmosis A method of desalinating ocean water that involves forcing water molecules through a water-permeable membrane under pressure.

Reversing current The tide current as it occurs at the margins of landmasses. The water flows in and out for approximately equal periods of time separated by slack water where the water is still at high and low tidal extremes.

Rhodophyta Phylum of algae composed primarily of small encrusting, branching, or filamentous plants that receive their characteristic red color from the presence of the pigment phycoerythrin. With a worldwide distribution, they are found at greater depths than other algae.

Right whales Whales of the genus *Eubalaena* that were the favorite target of early whalers.

Rip current A strong narrow surface or near-surface current of short duration (up to 2 h) and high speed (up to 4 km/h) flowing seaward through the breaker zone at nearly right angles to the shore. It represents the return to the ocean of water that has been piled up on the shore by incoming waves.

Rise A long, broad elevation that rises gently and rather smoothly from the deep-ocean floor.

Rorqual whales Include the minke, Baird's, Bryde's, sei, fin, blue, and humpback whales. They are baleen whales with many ventral grooves.

Rotary current Tidal current as observed in the open ocean. The tidal crest makes one complete rotation during a tidal period.

Sabellid A member of the annelid family Sabellidae that lives in a tube composed of shell fragments, sand, and a glutinous material. Featherlike gills and feeding structures filter food from the water above the tube opening.

Salinity A measure of the quantity of dissolved solids in ocean water. Formally, it is the total amount of dissolved solids in ocean water in parts per thousand by weight after all carbonate has been converted to oxide, the bromide and iodide to chloride, and all the organic matter oxidized. It is normally computed from conductivity, refractive index, or chlorinity.

Salinometer An instrument that is used to determine the salinity of seawater by measuring its electrical conductivity.

Salpa Genus of pelagic tunicates that are cylindrical, transparent, and found in all oceans.

Salt Any substance that yields ions other than hydrogen or hydroxyl. Salts are produced from acids by replacing the hydrogen with a metal.

Salt marsh A relatively flat area of the shore where fine sediment is deposited and salt-tolerant grasses grow. One of the most biologically productive regions of Earth.

Salt wedge estuary A very deep river mouth with a very large volume of freshwater flow beneath which a wedge of salt water from the ocean invades. The Mississippi River is an example.

San Andreas Fault A transform fault that cuts across the state of California from the northern end of the Gulf of California to Pt. Arena north of San Francisco.

Sand Particle size of 0.0625–2 mm. It pertains to particles that lie between silt and granules on the Wentworth scale of grain size.

Sargasso Sea A region of convergence in the North Atlantic lying south and east of Bermuda where the water is very clear, deep blue in color, and contains large quantities of floating *Sargassum*.

Sargassum A brown alga characterized by a bushy form, substantial holdfast when attached, and a yellow brown, green yellow, or orange color. Two species, *S. fluitans* and *S. natans*, make up most of the macroscopic vegetation in the Sargasso Sea.

Scaphopoda A class of mollusks commonly called *tusk shells*. The shell is an elongate cone open at both ends. The conical foot surrounded by threadlike tentacles extends from the larger end to aid the animal in burrowing in fine sand or mud.

Scarp A linear steep slope on the ocean floor separating gently sloping or flat surfaces.

Scavenger An animal that feeds on dead organisms.

Schooling Well-defined social organizations of fish, squid, and crustaceans that aid them in survival. Often the precise benefit to the schooling species eludes the human observer.

Scyphozoa A class of coelenterates that includes the true jellyfish in which the medusoid body form predominates and the polyp is reduced or absent.

Sea 1. A subdivision of an ocean. Two types of seas are identifiable and defined. They are the *mediterranean seas,* where a number of seas are grouped together collectively as one sea, and *adjacent seas* that are connected individually to the ocean. 2. A portion of the ocean where waves are being generated by wind.

Sea anemone A member of the class Anthozoa whose bright color, tentacles, and general appearance make it resemble flowers.

Sea arch An opening through a headland caused by wave erosion. Usually develops as sea caves are extended from one or both sides of the headland.

Sea cave A cavity at the base of a sea cliff formed by wave erosion.

Sea cow An aquatic, herbivorous mammal of the order Sirenia that includes the dugong and manatee.

Sea cucumber A common name given to members of the echinoderm class Holotheuroidea.

Sea-floor spreading A process producing the lithosphere when convective upwelling of magma along the oceanic ridges moves the ocean floor away from the ridge axes at rates of from 1 to 10 cm (0.4 to 4 in.) per year.

Sea ice Any form of ice originating from the freezing of ocean water.

Sea otter A seagoing otter that has recovered from near extinction along the North Pacific coasts. It feeds primarily on abalone and sea urchins.

Sea snake A reptile belonging to the family Hydrophiidae with venom similar to that of cobras. They are found primarily in the coastal waters of the Indian Ocean and the western Pacific Ocean.

Sea state A description of the ocean surface that includes the average height of the highest one-third of the waves observed in a wave train, referred to by a numerical code.

Sea turtle Any of the reptilian order Testudinata found widely in warm water.

Sea urchin An echinoderm belonging to the class Echinoidea possessing a fused test (external covering) and well-developed spines.

Seamount An individual peak extending over 1000 m above the ocean floor.

Seasonal thermocline A thermocline that develops due to surface heating of the oceans in mid- to high latitudes. The base of the seasonal thermocline is usually above 200 m (656 ft).

Seawall A wall built parallel to the shore to protect coastal property from the waves.

Sediment Particles of organic or inorganic origin that accumulate in loose form.

Sediment maturity A condition in which the roundness and degree of sorting increase and clay content decreases within a sedimentary deposit.

Sedimentary rock A rock resulting from the consolidation of loose sediment, or a rock resulting from chemical precipitation, i.e., sandstone and limestone.

Seiche A standing wave of an enclosed or semienclosed body of water that may have a period ranging from a few minutes to a few hours, depending on the dimensions of the basin. The wave motion continues after the initiating force has ceased.

Seismic Pertaining to an earthquake or earth vibration, including those that are artificially induced.

Seismic sea wave *See* tsunami.

Seismic surveying The use of sound-generating techniques to identify features on or beneath the ocean floor.

Semidiurnal tide Tide having two high and two low waters per tidal day with small inequalities between successive highs and successive lows.

Tidal period is about 12 h and 25 min solar time. Semidaily tide.

Serpulid A polychaete worm belonging to the family Serpulidae that builds a calcareous or leathery tube on a submerged surface.

Sessile Permanently attached to the substrate and not free to move about.

Seta Hairlike or needlelike projections on the exoskeleton of arthropods. Similar structures exist on some annelids.

Shallow water wave A wave on the surface of the water whose wavelength is at least 20 times water depth. The bottom affects the orbit of water particles, and speed is determined by water depth.
$$S(m/s) = 3.1\sqrt{\text{water depth (m)}}.$$

Shelf break The depth at which the gentle slope of the continental shelf steepens appreciably. It marks the boundary between the continental shelf and continental rise.

Shelf ice Thick shelves of glacial ice that push out into Antarctic seas from Antarctica. Large tabular icebergs calve at the edge of these vast shelves.

Shoal Shallow.

Shore Seaward of the coast, extends from highest level of wave action during storms to the low water line.

Shoreline The line marking the intersection of water surface with the shore. Migrates up and down as the tide rises and falls.

Shoreline of emergence Shorelines that indicate a lowering of sea level by the presence of stranded beach deposits and marine terraces above it.

Shoreline of submergence Shorelines that indicate a rise in sea level by the presence of drowned beaches or submerged dune topography.

Side-scan sonar A method of mapping the topography of the ocean floor along a strip up to 60 km wide using computers and sonar signals that are directed away from both sides of the survey ship.

Silica Silicon dioxide (SiO_2).

Siliceous A condition of containing abundant silica (SiO_2).

Sill A submarine ridge partially separating bodies of water such as fjords and seas from one another or from the open ocean.

Silt A particle size of $1/128$–$1/16$ mm. It is intermediate between sand and clay.

Siphonophora An order of hydrozoan coelenterates that forms pelagic colonies containing both polyps and medusae. Examples are *Physalia* and *Velella*.

Sirenia An order of vegetarian marine mammals that includes the dugong and manatee.

Slack water Occurs when a reversing tidal current changes direction at high or low water. Current speed is zero.

Slick A smooth patch on an otherwise rippled surface caused by a monomolecular film of organic material that reduces surface tension.

Slightly stratified estuary An estuary of moderate depth in which marine water invades beneath the freshwater runoff. The two water masses mix so that the bottom water is slightly saltier than the surface water at most places in the estuary.

Sofar Channel Sound fixing and ranging channel. This is a low velocity sound travel zone that coincides with the permanent thermocline in low and midlatitudes.

Solar humidification A process by which ocean water can be desalinated by evaporation and the condensation of the vapor on the cover of a container. The condensate then runs into a separate container and is collected as fresh water.

Solar system The sun and the celestial bodies, asteroids, planets, and comets, that orbit around it.

Solar tide The partial tide caused by the tide-producing forces of the sun.

Solid state A state of matter in which the substance has a fixed volume and shape. A crystalline state of matter.

Solstice The time during which the sun is directly over one of the tropics. In the Northern Hemisphere the summer solstice occurs on June 21 or 22 as the sun is over the Tropic of Cancer, and the winter solstice occurs on December 21 or 22 when the sun is over the Tropic of Capricorn.

Solute A substance dissolved in a solution. Salts are the solute in salt water.

Solution A state in which a solute is homogeneously mixed with a liquid solvent. Water is the solvent for the solution that is ocean water.

Solvent A liquid that has one or more solutes dissolved in it.

Somali Current This current flows north along the Somali coast of Africa during the southwest monsoon season.

Sonar An acronym for *sound navigation and ranging*. A method by which objects may be located in the ocean.

Sounding Measuring the depth of water beneath a ship.

Southwest Monsoon A southwest wind that develops during the summer season. It blows off the Indian Ocean onto the Asian mainland.

Southwest Monsoon Current During the southwest monsoon season, this eastward-flowing current replaces the west-flowing North Equatorial Current in the Indian Ocean.

Species diversity The number or variety of species found in a subdivision of the marine environment.

Specific gravity The ratio of density of a given substance to that of pure water at 4°C and at atmospheric pressure.

Specific heat The quantity of heat required to raise the temperature of 1 g of a given substance 1°C. For water it is 1 cal.

Spermatophyta Seed-bearing plants.

Spicule A minute needlelike calcareous or siliceous form found in sponges, radiolarians, chitons, and echinoderms that acts to support the tissue or provide a protective covering.

Spilling breaker A type of breaking wave that forms on a gently sloping beach which gradually extracts the energy from the wave to produce a turbulent mass of air and water that runs down the front slope of the wave.

Spit A small point, low tongue, or narrow embankment of land commonly consisting of sand deposited by longshore currents and having one end attached to the mainland and the other terminating in open water.

Sponge *See* Porifera.

Spray zone The shore zone lying between the high-tide shoreline and the coastline. It is covered by water only during storms.

Spreading center A divergent plate boundary.

Spreading rate The rate of divergence of plates at a spreading center.

Spring tide Tide of maximum range occurring every fortnight when the moon is new or full.

Stack An isolated mass of rock projecting from the ocean off the end of a headland from which it has been detached by wave erosion.

Standard laboratory bioassay Tests conducted in a laboratory setting to determine the percentage of dying caused within a population by a given concentration of foreign material.

Standard seawater Ampules of ocean water for which the chlorinity has been determined by the Institute of Oceanographic Services in Wormly, England. The ampules are sent to laboratories all over the world so equipment and reagents used to determine the salinity of ocean water samples can be calibrated by adjustment until they give the same chlorinity as is shown on the ampule label.

Standing stock (crop) The biomass of a population present at any given time.

Standing wave A wave, the form of which oscillates vertically without progressive movement. The region of maximum vertical motion is an *antinode*. On either side are *nodes* where there is no vertical motion but maximum horizontal motion.

Stenohaline Pertaining to organisms that can withstand only a small range of salinity change.

Stenothermal Pertaining to organisms that can withstand only a small range of temperature change.

Storm surge A rise above normal water level resulting from wind stress and reduced atmospheric pressure during storms. Consequences can be more severe if it occurs in association with high tide.

Strait of Gibraltar The narrow opening between Europe and Africa through which the waters of the Atlantic Ocean and Mediterranean Sea mix.

Stranded beach An ancient beach deposit found above present sea level because of a lowering of sea level.

Streamlining The shaping of an object so it produces the minimum of turbulence while moving through a fluid medium. The teardrop shape displays a high degree of streamlining.

Subduction The process by which one lithospheric plate descends beneath another as they converge.

Subduction zone seep biocommunity Animals that live in association with seeps of pore water squeezed out of deeper sediments. They depend on sulfur-oxidizing bacteria that act as producers for the ecosystem.

Sublimation The transformation of the solid state of a substance to a vapor without going through the liquid phase, and vice versa.

Sublittoral zone That portion of the benthic environment extending from low tide to a depth of 200 m (656 ft); considered by some to be the surface of the continental shelf.

Submarine canyon A steep V-shaped canyon cut into the continental shelf or slope.

Submerged dune topography Ancient coastal dune deposits found submerged beneath the present shoreline because of a rise in sea level.

Submergent shoreline Shoreline formed by the relative submergence of a landmass in which the shoreline is on landforms developed under subaerial processes. It is characterized by bays and promontories and is more irregular than a shoreline of emergence.

Subneritic province The benthic environment extending from the shoreline across the continental shelf to the shelf break. It underlies the neritic province of the pelagic environment.

Suboceanic province Benthic environments seaward of the continental shelf.

Subpolar Pertaining to the oceanic region that is covered by sea ice in winter. The ice melts away in summer.

Substrate The base on which an organism lives and grows.

Subsurface chlorophyll maximum (SCM) The concentration of chlorophyll in many places in the ocean is highest immediately below the oxygen compensation depth.

Subsurface current A current usually flowing below the pycnocline, generally at slower speed and in a different direction from the surface current.

Subsurface oxygen maximum (SOM) Generally observed within the lower euphotic zone, the SOM may represent supersaturations of oxygen of up to 120 percent as a result of photosynthetically produced oxygen.

Subtropical Pertaining to the oceanic region poleward of the tropics (about 30° latitude).

Subtropical Convergence The zone of convergence that occurs within all

subtropical gyres as a result of Ekman transport driving water toward the interior of the gyres.

Subtropical gyre The trade winds and westerly winds initiated in the subtropical regions of all ocean basins, with the influence of the Coriolis effect, set large regions of ocean water in motion. They rotate clockwise in the Northern Hemisphere and counterclockwise in the Southern Hemisphere, and they are centered in the subtropics.

Sulfur A yellow mineral composed of the element sulphur. It is commonly found in association with hydrocarbons and salt deposits.

Sulfur-oxidizing bacteria Bacteria that support many deep-sea hydrothermal vent and cold water seep biocommunities by using energy released by oxidation to synthesize organic matter chemosynthetically.

Summer solstice In the Northern Hemisphere, it is the instant when the sun moves north to the Tropic of Cancer before changing direction and moving southward toward the equator, June 21.

Supralittoral zone The splash or spray zone above the spring high-tide shoreline.

Surf zone The region between the shoreline and the line of breakers where most wave energy is released.

Surface tension The tendency for the surface of a liquid to contract owing to intermolecular bond attraction.

Swash A thin layer of water that washes up over exposed beach as waves break at the shore.

Swell A free ocean wave by which energy put into ocean waves by wind in the sea is transported with little energy loss across great stretches of ocean to the margins of continents where the energy is released in the surf zone.

Swim bladder A gas-containing, flexible, cigar-shaped organ that aids many fishes in attaining neutral buoyancy.

Symbiosis A relationship between two species in which one or both benefit or neither or one is harmed. Examples are commensalism, mutualism, and parasitism.

Tectonic estuary An estuary, the origin of which is related to tectonic deformation of the coastal region.

Tectonics Deformation of Earth's surface by forces generated by heat flow from Earth's interior.

Temperate Pertaining to the oceanic region where pronounced seasonal change occurs (about 40° to 60° latitude).

Temperature A direct measure of the average kinetic energy of the molecules of a substance.

Temperature of maximum density The temperature at which a substance reaches its highest density. For water, it is 4°C.

Territorial Sea A 12-nautical-mile-wide strip of ocean adjacent to land over which the coastal nation has control over the passage of ships.

Tethered-float breakwater Floating hollow spheres tethered to the bottom and acting as an upside-down pendulum to extract energy from waves near the shore without interfering with the longshore drift of sediment.

Tethys Sea An ancient body of water that separated Laurasia to the north and Gondwanaland to the south. Its location was approximately that of the present Alpine-Himalayan mountain system.

Thermocline A layer of water beneath the mixed layer in which a rapid change in temperature can be measured in the vertical dimension.

Thermohaline circulation The vertical movement of ocean water driven by density differences resulting from the combined effects of variations in temperature and salinity.

Tidal bore A steep-fronted wave that moves up some rivers when the tide rises in the coastal ocean.

Tidal range The difference between high tide and low tide water levels over any designated time interval, usually one lunar day.

Tide Periodic rise and fall of the surface of the ocean and connected bodies of water resulting from the gravitational attraction of the moon and sun acting unequally on different parts of Earth.

Tide-generating force The magnitude of the centripetal force required to keep all particles of Earth with identical mass moving in identical circular paths required by the movements of the Earth-moon system is identical. This force is provided by the gravitational attraction between the particles and the moon. This gravitational force is identical to the required centripetal force only at the center of Earth. For ocean tides, the horizontal component of the small force that results from the difference between the required and provided forces is the tide-generating force on that individual particle. These forces are such that they tend to push the ocean water into bulges toward the tide-generating body on one side of Earth and away from the tide-generating body on the opposite side of Earth.

Tide wave The long-period gravity wave generated by tide-generating forces described above and manifested in the rise and fall of the tide.

Tintinnid A ciliate protozoan of the family Tintinnidae with a tubular to vase-shaped outer shell.

Tissue An aggregate of cells and their products developed by organisms for the performance of a particular function.

Tombolo A sand or gravel bar that connects an island with another island or the mainland.

Topography The configuration of a surface. In oceanography it refers to the ocean bottom or the surface of a mass of water with given characteristics.

Total allowable catch The permissible annual catch of a given species of fish that will not degrade the fishery given the knowledge of the ecosystem from which the fishery is being removed.

Trade winds The air masses moving from subtropical high pressure belts toward the equator. They are northeasterly in the Northern Hemisphere and southeasterly in the Southern Hemisphere.

Transform fault A fault characteristic of oceanic ridges along which they are offset.

Transform plate boundary The shear boundary between two lithospheric plates formed by a transform fault.

Transitional crust Thinned section of continental crust at the trailing edge of a continent created by the breaking apart of an ancient continent over a newly formed spreading center.

Transitional wave A wave moving from deep water to shallow water that has a wavelength more than twice the water depth but less than 20 times the water depth. Particle orbits are beginning to be influenced by the bottom.

Transverse wave A wave in which particle motion is at right angles to energy propagation.

Trawl A sturdy bag or net that can be dragged along the bottom to catch fish or a variety that can be towed at various depths above the bottom for the same purpose.

Trench A long, narrow, and deep depression on the ocean floor, with relatively steep sides.

Trophic level A nourishment level in a food chain. Plant producers constitute the lowest level, followed by herbivores and a series of carnivores at the higher levels.

Tropic of Cancer The latitude of 23.5° north.

Tropic of Capricorn The latitude of 23.5° south.

Tropical Pertaining to the regions of the tropics (about 23.5° latitude).

Tropical tide A tide occurring twice monthly when the moon is at its maximum declination north and south of the equator. It is in the tropical regions where tides display their greatest diurnal inequalities.

Trough The part of an ocean wave that is displaced below the still-water line.

Tsunami Seismic sea wave. A long-period gravity wave generated by a submarine earthquake or volcanic event. Not noticeable on the open ocean but builds up to great heights in shallow water.

Tube worms *See* Sabellidae and Serpulidae.

Tunicates Members of the chordate subphylum Urochordata, which includes sacklike animals. Some are sessile (sea squirts) while others are pelagic (salps).

Turbidite A sediment or rock formed from sediment deposited by turbidity currents characterized by both horizontally and vertically graded bedding.

Turbidity A state of reduced clarity in a fluid caused by the presence of suspended matter.

Turbidity current A gravity current resulting from a density increase brought about by increased water turbidity. Possibly initiated by some sudden force such as an earthquake, the turbid mass continues under the force of gravity down a submarine slope.

Turbulent flow Flow in which the flow lines are confused heterogeneously due to random velocity fluctuations.

Typhoon A severe tropical storm in the western Pacific.

Ultraplankton Plankton for which the greatest dimension is less than 5 μm. Very difficult to separate from the water.

Ultrasonic Sound frequencies above those that can be heard by humans (above 20,000 cycles per second).

Ultraviolet radiation Electromagnetic radiation shorter than visible radiation and longer than X rays. The approximate range is 1–400 nanometers (nm).

Upper water Includes the mixed layer and the permanent thermocline. It is approximately the top 1000 m of the ocean.

Upwelling The process by which deep, cold, nutrient-laden water is brought to the surface, usually by diverging equatorial currents or coastal currents that pull water away from the coast.

Valence The combining capacity of an element measured by the number of hydrogen atoms with which it will combine.

van der Waals force Weak attractive force between molecules resulting from the interaction between the nuclear particles of one molecule and the electrons of another.

Vector A physical quantity that has magnitude and direction. Examples are force, acceleration, and velocity.

Ventral Pertaining to the lower or under surface.

Vernal equinox The passage of the sun across the equator as it moves from the Southern Hemisphere into the Northern Hemisphere, March 21.

Vertebrata Subphylum of chordates that includes those animals with a well-developed brain and a skeleton of bone or cartilage; includes fish, amphibians, reptiles, birds, and animals.

Vertically mixed estuary Very shallow estuaries such as lagoons in which fresh water and marine water are totally mixed from top to bottom so that the salinity at the surface and the bottom is the same at most places within the estuary.

Viviparous Referring to embryonic development inside the mother's reproductive tract with the provision of nutrition to the embryo by the mother.

Viscosity A property of a substance to offer resistance to flow. Internal friction.

Walker Circulation The pattern of atmospheric circulation that involves the rising of warm air over the East Indies low pressure cell and its descent over the high pressure cell in the southeastern Pacific Ocean off the coast of Chile. It is the weakening of this circulation that accompanies an El Niño event which has led to the development of the term El Niño-Southern Oscillation event.

Warm front A weather front in which a warm air mass moves into and over a cold air mass producing a broad band of gentle precipitation.

Water mass A body of water identifiable from its temperature, salinity, or chemical content.

Wave A disturbance that moves over the surface or through a medium with a speed determined by the properties of the medium.

Wave-cut bench A gently sloping surface produced by wave erosion and extending from the base of the wave-cut cliff out under the offshore region.

Wave-cut cliff A cliff produced by landward cutting by wave erosion.

Wave dispersion The separation of waves as they leave the sea area by wave size. Larger waves travel faster than smaller waves and thus leave the sea area first to be followed by progressively smaller waves.

Wave frequency The number of waves that pass a fixed point in a unit of time, usually one second.

Wave height Vertical distance between a crest and the preceding trough.

Wave period The elapsed time between the passage of two successive wave crests past a fixed point.

Wave steepness Ratio of wave height to wavelength.

Wave train A series of waves from the same direction.

Wavelength Horizontal distance between two corresponding points on successive waves, such as from crest to crest.

Weathering A process by which rocks are broken down by chemical and mechanical means.

West Australian Current This cold current forms the eastern boundary current of the Indian Ocean subtropical gyre. It is separated from the coast by the warm Leeuwin Current except during El Niño-Southern Oscillation events when the Leeuwin Current weakens.

West Wind Drift The surface portion of the Antarctic Circumpolar Current driven in an easterly direction around Antarctica by the strong westerly winds.

Westerly winds The air masses moving away from the subtropical high pressure belts toward higher latitudes. They are southwesterly in the Northern Hemisphere and northwesterly in the Southern Hemisphere.

Western boundary current Poleward-flowing warm currents on the western side of all subtropical gyres.

Western boundary undercurrent (WBUC) A bottom current that flows along the base of the continental slope eroding sediment from it and redepositing the sediment on the continental rise. It is confined to the western boundary of deep-ocean basins.

Westward intensification Pertaining to the intensification of the warm western limb of the subtropical gyre currents that is manifested in higher velocity and deeper flow compared to the cold eastern boundary currents that drift leisurely toward the equator.

Wetlands Biologically productive regions bordering estuaries and other protected coastal areas. They are usually salt marshes at latitudes greater than 30° and mangrove swamps at lower latitudes.

White muscle fiber Thick muscle fibers with relatively low concentrations of myoglobin that make up a large percentage of the muscle fiber in lunger-type fishes.

Wind-driven circulation Any movement of ocean water that is driven by winds. This includes most horizontal movements in the surface waters of the world's oceans.

Windrows Rows of floating debris aligned parallel to the direction of the wind that result from Langmuir circulation.

Windward The direction from which the wind is blowing.

Winter solstice The instant the southward-moving sun reaches the Tropic of Cancer before changing direction and moving north back toward the equator.

Zenith That point on the celestial sphere directly over the observer.

Zeolite A group of aluminosilicate minerals that easily lose or gain water of hydration and have great capacity for ion exchange.

Zooplankton Animal plankton.

Zooxanthellae A form of algae that lives as a symbiont in the tissue of corals and other coral reef animals and provides varying amounts of their required food supply.

Index